For further volumes:
http://www.springer.com/series/5112

Series Editor

Professor Rajkumar Roy,
Department of Enterprise Integration School of
Industrial and Manufacturing Science, Cranfield University,
Cranfield, Bedford, MK43 0AL, UK

Other titles published in this series

Cost Engineering in Practice
John McIlwraith

IPA – Concepts and Applications in Engineering
Jerzy Pokojski

Strategic Decision Making
Navneet Bhushan and Kanwal Rai

Product Lifecycle Management
John Stark

From Product Description to Cost: A Practical Approach
Volume 1: The Parametric Approach
Pierre Foussier

From Product Description to Cost: A Practical Approach
Volume 2: Building a Specific Model
Pierre Foussier

Decision-Making in Engineering Design
Yotaro Hatamura

Composite Systems Decisions
Mark Sh. Levin

Intelligent Decision-making Support Systems
Jatinder N.D. Gupta, Guisseppi A. Forgionne and Manuel Mora T.

Knowledge Acquisition in Practice
N.R. Milton

Global Product: Strategy, Product Lifecycle Management and the Billion Customer Question
John Stark

Enabling a Simulation Capability in the Organisation
Andrew Greasley

Network Models and Optimization
Mitsuo Gen, Runewei Cheng and Lin Lin

Management of Uncertainty
Gudela Grote

Introduction to Evolutionary Algorithms
Xinjie Yu and Mitsuo Gen

Data Mining
Y. Yin, I. Kaku, J. Tang and J. Zhu

Irene Ng · Glenn Parry · Peter J. Wild
Duncan McFarlane · Paul Tasker
Editors

Complex Engineering Service Systems

Concepts and Research

Editors
Prof. Irene Ng
University of Exeter Business School
Streatham Court Rennes Drive
Exeter EX4 4PU
UK
e-mail: irene.ng@exeter.ac.uk

Prof. Duncan McFarlane
Institute for Manufacturing
Alan Reece Building
University of Cambridge
Charles Babbage Road 17
Cambridge CB3 0FS
UK
e-mail: dcm@eng.cam.ac.uk

Dr. Glenn Parry
University of the West of England
Bristol Business School
Frenchay Campus Coldharbour Lane
Bristol BS16 1QY
UK
e-mail: Glenn.Parry@uwe.ac.uk

Prof. Paul Tasker
Institute for Manufacturing
Alan Reece Building
University of Cambridge
Charles Babbage Road 17
Cambridge CB3 0FS
UK
e-mail: pht25@cam.ac.uk

Peter J. Wild
PaCT Laboratory
Department of Psychology
University of Northumrbia
Newcastle upon Tyne
UK
e-mail: peter.wild@northumbria.ac.uk

ISSN 1619-5736
ISBN 978-0-85729-188-2 e-ISBN 978-0-85729-189-9
DOI 10.1007/978-0-85729-189-9
Springer London Dordrecht Heidelberg New York
British Library Cataloguing in Publication Data.
A catalogue record for this book is available from the British library.

Cover design: eStudio Calamar, Berlin/Figueres

Printed on acid-free paper

Springer is part of Springer Science+Business Media (www.springer.com)

Because things are the way they are, things will not stay the way they are

Bertolt Brecht

Foreword

Creating value in any service market depends on many key factors. These include the ability to work closely with customers, partners and suppliers; the ability to anticipate customers' future requirements; and the ability to integrate and optimise business processes, people, tools and information to create value-added service systems and solutions.

In today's business environment, increased pressure on budgets means that customers are increasingly looking for greater value for money, and long-term service contracts to support complex engineering products are becoming the norm. Customers for complex engineering products are not passive recipients of goods; they recognise the need for close integration of service systems with their own business systems and are taking an active role in working with their suppliers to ensure those services deliver the outputs they need in an affordable way.

BAE Systems is not only one of the world's largest manufacturers in the defence, security and aerospace sectors; it is also one of the largest service providers in this industry. We have some of the best engineers in the world, creating highly complex engineering solutions for our customers, and providing support and services for these products requires complex engineering service systems. As well as integrating industrial capabilities, complex service systems are often embedded within a customer's organisation, with multiple supply chains and an extensive network of subordinate service providers.

The field of complex engineering service systems is a developing area of interest for both ourselves, as industrial practitioners, and the academic researchers with whom we collaborated in the areas described in this book. As an organisation we have had the privilege, along with the UK Engineering and Physical Sciences Research Council, of supporting the research undertaken by the Support Service Solutions: Strategy and Transformation Project (S4T), upon which this book is based.

The research carried out in this field has helped us to explore and address the complexity of the challenging new environment in which our business operates to support our customers better. Working with the S4T researchers and academics has given us insights and different perspectives into what underpins value-added service offerings.

Complex engineering service systems do require new ways of thinking; changes in mindset and an evolution in business models, processes, organisation, tools and information management, to deliver continually improving performance over product lifecycles that span decades. Our commitment to research in this field is about investment in our future, to maximise our potential and to provide the highest levels of service and best value for money for our customers.

I am proud that BAE Systems has supported the work described in this book. It follows many years of association between people in the company and the University of Cambridge and its Institute for Manufacturing, and other institutions. The result is a very powerful mix of academic research, innovation, rigour and above all systems thinking, all being driven by clear business requirements as we break new ground in this field. I am delighted to see the results of these collaborations being published.

Peter Fielder
B Eng(Hons), MA Mgmt, FIET, C Eng, MAPM, Hon FAPM
Managing Director
Performance Excellence
BAE Systems

Acknowledgements

The editors would like to acknowledge the significant contribution made by the S4T—Support Service Solutions: Strategy and Transition project, funded by BAE Systems and the Engineering and Physical Sciences Research Council (EPSRC), who together have made this book possible. We are particularly indebted to Peter Fielder in BAE Systems whose initial vision and long-term support enabled the academic interest and research on which this book is based, and to Greg Bolan who continues to lead on the company's adoption of research outputs into good operations practice. We would also like to extend our thanks and appreciation to the many organisations and people who supported and contributed to the research in terms of both finances and their time.

We have relied upon the insights and experience of many practitioners in BAE Systems, MBDA and the Ministry of Defence. Particular thanks must go to those who have led the industrial contributions to the work themes and to those who have contributed to several themes; Louise Wallwork, Jenny Cridland, John Barrie, Paul Johnson, Simon Wilson, Paul Gregory, Jonathan Murphy, Ed Smith, Dave Hogan, Barrie Walters, Andrew Matters, Stewart Leinster-Evans, Ian Davies, Scot Colquhoun, Phil Wardle, Simon Davis-Poynter, Steve Debonnaire and Simon Hall.

The editors would like to thank the authors for their hard work in producing the chapters. Additional thanks also goes to the Editorial Review Board for their contributions in reviewing the manuscripts. Thanks also to Yin F Lim who has supported the production of the book, and to Chris Pearson for his contribution to the book and all his work supporting and guiding the S4T project.

We would like to thank all those colleagues from industry, government and academia who have helped to formulate our thinking with regard to Complex Engineering Service, in particular the insights and encouragement offered by Steve Vargo, Roger Maull and Andy Neely.

Finally, all the editors and the authors would like to thank our families. Organising, directing and researching for the S4T programme as well as putting a book together is unavoidably intrusive on family life. Producing a book with so many contributors is always a challenging experience and this would have not been possible without our families' ongoing support and understanding.

Contents

Chapter 1
Towards a Core Integrative Framework for Complex Engineering Service Systems

Irene Ng, Glenn Parry, Duncan McFarlane and Paul Tasker

Abstract Complex Engineering Service provision is a developing area for both practitioners and academics. Delivery requires an integrated offering, drawing upon company, customer and supplier resources to deliver value that is an integration of complex engineered assets, people and technology. For a business to present a sustainable value proposition, managers are required to develop a diverse skill set, working dynamically across previously separated business areas with established company boundaries. In this chapter we will present a framework for complex engineering service system that is value-centric and that conceptually integrates the chapters of this book. The framework proposes that the provision of service requires companies to be capable of working together with their clients to create value through three integrated transformations: people, information and materials & equipment. Successful provision of complex engineering service solutions therefore requires the integration and mastery of many different disciplines that bring about these transformations as well as understanding the

I. Ng (✉)
University of Exeter Business School, Exeter, Devon, UK
e-mail: Irene.Ng@exeter.ac.uk

G. Parry (✉)
Bristol Business School, University of the West of England, Bristol, UK
e-mail: glenn.parry@uwe.ac.uk

D. McFarlane
Institute for Manufacturing, University of Cambridge, Cambridge, UK
e-mail: dcm@cam.ac.uk

P. Tasker
The University of Cambridge, Cambridge, UK
e-mail: pht25@cam.ac.uk

P. Tasker
Cranfield University, Cranfield, UK

P. Tasker
University of Kent, Canterbury, Kent, UK

I. Ng et al. (eds.), *Complex Engineering Service Systems*, Decision Engineering,
DOI: 10.1007/978-0-85729-189-9_1,

interactions and links between both the transformations and the disciplines. The challenge laid out in this chapter and developed throughout this book explores this new environment, providing guidance and identifying areas requiring future work.

1.1 Introduction

Provision of goods has been a hallmark of manufacturing since the start of the industrial era. Indeed, as early as 1776, Adam Smith proposed that the wealth of nations was built upon a country's ability to produce an excess quantity of goods and then export this excess to generate wealth. This provided the foundation for the dominant view of goods as the staple for value creation.

The key category of manufactured goods is 'equipment': systems generated to provide a transforming function of their own. As equipment provision has become more complex and as competition heightened, firms have felt the pressure to add value, predominantly through the provision of services. Research has shown that manufacturers provide services in the form of training, integration with clients' capabilities, consultancy and other services related to the provision of equipment (Ren 2009). Indeed, for many manufacturers to remain viable, research has shown that they may need to diversify into the provision of services (Neely 2008). This provision has been commonly referred to as the servitization of manufacturing.

Servitization has been discussed widely, frequently through an examination of the move by manufacturers to generate greater returns by providing through-life support for their products (Vandermerwe and Rada 1988; Matthyssens and Vandembempt 1988; Anderson and Narus 1995). The hazards and enablers to the process of servitization have also been studied (Oliva and Kallenborg 2003; Mills et al. 2008). However, due to the established paradigm that production of goods is the basis of wealth creation, much of the discussion and analysis of engineering service has been through the lens of goods-based thinking:

> because manufacturing has been the dominant economic force of the last century, most managers have been educated through experience and/or formal education to think about strategic management in product-oriented terms. Unfortunately, a large part of this experience is irrelevant to the management of many service businesses (Thomas 1978)

This raises the challenge for academics to question the assumptions upon which conclusions are being drawn.

1.2 Theoretical Foundations

Servitization has resulted in combinations of offerings to generate value from both products and services in bundled packages. These combinations of products and services have been called Product Service Systems (PSS). Baines et al. (2007)

defines PSS such that they embody "*an integrated product and service offering that delivers value in use*", highlighting the importance of value. This introductory chapter provides a value-centric integrated framework for Complex Engineering Service (CES) systems which aims to deliver value to the customer through a system of people, processes, assets and technology and the interactions between them rather than the function of the individual components themselves. Such a value is emergent from the CES system and not from a linear chain of operations optimised individually. The understanding of CES systems requires individuals within organisations to develop new skill sets as traditional boundaries are challenged and this presents further challenges as the current component, business unit or functional operation of firms creates power bases that may provide resistance to change. The trans-disciplinary challenge is also true for academics wishing to understand and capture the nature of this new CES system as here too, reductionism is the dominant logic, with teaching split into subject disciplines without focus on the interaction between them.

The move from design and manufacture of equipment and its corresponding capabilities to a combination of activities and assets to achieve consistent and high value outcomes is crucial to a world of depleting resources and to the global sustainability movement.

The concept of value has long been discussed in academic literature. Organisations have been called upon to deliver superior customer value as a major source of competitive advantage (Payne and Holt 2001; Eggert et al. 2006; Liu et al. 2005; Ulaga and Eggert 2006). Similarly, value and customer orientation is echoed amongst academics in different fields (Cannon and Homburg 2001; Chase 1978; Amit and Zott 2001; Ramirez 1999; Kim and Mauborgne 1999). Indeed, Ravald and Gronroos (1996) claimed that a firm's ability to provide superior value is regarded as one of the most successful competitive strategies in the nineties. Within business-to-business (B2B) literature, delivering superior customer value assists firms in developing and maintaining strategic buyer–seller relationships (Liu et al. 2005), resulting in loyalty (Bolton and Drew 1991) and the potential to grow margins and profits (Butz and Goodstein 1996). From the practitioner's domain, Drucker (1993) proposed that what value means to the customer is one of the most important questions a business should ask. Thus, practitioners and academics alike have stressed the importance of delivering customer value as the key to success.

The traditional notion of value is that of exchange value which underpins the traditional customer–producer relationships, where each party exchanges one kind of value for another (Bagozzi 1975), with something in exchange for something else. However, contemporary literature has moved the discussion of value away from this understanding to the concept of value-in-use (see Vargo and Lusch 2004, 2008; Schneider and Bowen 1995), which is evaluated by the customer rather than the currency for the transfer of ownership of a particular "good". Value-in-use, described by Marx, is "value only in use, and is realised only in the process of consumption" [Marx 1867 (2001), p. 88]. In this regard, value and quality are therefore significantly harder to conceptualise due to the requirement for the customer to contribute to the service creation (Parasuraman et al. 1985), the notion of *value co-creation*, as we will see from the book's chapters. Thus, as proposed by Ballantyne and Varey (2006), the exchange value implicitly includes an estimate of the value-in-use of any "good" and activity that has been contractually exchanged or promised for consumption. Sellers of services must focus upon use through a relationship, ensuring the clients remain satisfied after the point of exchange through constructive engagement, resolving their complaints and meeting their future needs. Consequently, whether benefits to customers are attained through tangible goods or through the activities of firms, a customer-focused orientation would focus on value-in-use, delivered by the outcomes rendered by a combination of equipment and activities. Where activities are often tacit and heuristically driven and the equipment is highly complex in engineering terms, the capability to design and deliver value becomes a challenge. This is thus the book's foundational premise.

A goods-centric legacy from the industrial era has embedded processes, systems and knowledge for the production of tangible products, which has been effective in delivering high quality equipment, to the level of 'six-sigma' and the like (Nonthaleerak and Henry 2008). As manufacturers add 'service' to the body of goods-centric knowledge, the tendency is to treat services as an extension of that body of knowledge. This is theoretically problematic for three reasons.

First, value in use for service activities would immediately imply an inseparability of production and consumption (de Brentani 1991; Ng 2009). By its logical extension, the delivery of value-in-use in service could happen at any and all encounters with the customer, as the customer 'uses' the service. A goods-centric legacy of linear production processes towards some tangible end may not hold for value that is amorphously delivered through a multitude of touchpoints with the customer.

Second, the *use* of tangible goods to achieve benefits is often conducted by the customer away from the firm that manufactured it. Thus, for the manufacturing of goods the responsibility of the firm ends at production or when the ownership of the product has been transferred. In the delivery of service activities, however, the firm's responsibilities often *include* the customer where the customer's capability to use the service becomes the firm's responsibility as well, so that beneficial outcomes could be attained. A goods-centric mindset with boundaries of where 'production' ends may imply that the firm is only responsible for the delivery of 'service activities' which they undertake when faced with their customer. Such a mindset results in a lack of motivation to truly understand how customers co-create value with the firm, resulting in poor service outcomes.

Third, achievement of excellent service outcomes, as opposed to excellent product outcomes, is through the contribution of resources by both the firm and the customer. Traditional manufacturing systems, processes and knowledge frequently exclude the customer resources in delivering a manufactured good. It may even be proposed as a necessity to achieve consistent, high quality tangible goods. This approach may need to be changed and the access to and integration with a customer's systems, processes and knowledge are proposed as a necessity for the delivery of high quality service.

To achieve a better understanding of this approach to service without the baggage of goods-logic, some academic researchers have turned to service management research. Service management research, like research into production, has a long history (Huang et al. 2009). Early work focussed upon goods as tangible objects and services as intangible and a form of 'performance' (Say 1803; Senior 1863) and the concept that production and consumption are separate for goods, but may be instantaneous for service (Hicks 1942). The intangibility and interaction between producer and consumer formed two of the key building blocks for the IHIP service definition [intangible, heterogeneous, inseparable and perishable] which has been a touchstone for many service researchers (Kotler 2003). However, much of this research is focused on service contexts that are inherently more intangible in nature, such as hospitality, tourism, banking or telecommunication. Even where tangible goods are involved, the goods are a component of the service. In the way manufacturing researchers treat activities (services) as an extension of manufactured goods, service management researchers often treat tangible goods as merely a part of service activities, indicating that a service provision would range from greater tangibility e.g., cosmetics, to highly intangible e.g., teaching and education, on a spectrum (Shostack 1977). Such a point of view is also theoretically problematic.

While it is always possible to strategically distinguish between equipment and people-based service offerings (Thomas 1978), it may not be meaningful. This is because the operation of equipment may require different levels of skills. Thus, equipment design and manufacture has to consider what is appropriate for the market need and the cost of provision, whilst at the same time considering the skills of the customer. For example most vending machines or carwashes are fully automated and require very little skill to operate; dry cleaning equipment and the projector in a cinema require relatively unskilled operators; civil aircraft require highly skilled operators. The point is that from a value perspective, equipment within a service environment is not unchangeable, particularly for engineering. They are designed for ease of use and appropriate operability. Thus when customers use equipment or goods that include service activity provision it is essential that the combined offering delivered is integrated effectively to best serve the customer's requirement as well as being efficiently constructed for the firm. This means that the equipment could be engineered and designed to facilitate service activities provision just as processes and activities executed by service personnel could be redesigned for better outcomes. For example, designing vehicles for ease of service and repair became part of the focus of the thinking of German Porsche automobile company, reducing the cost of service and maintenance at a later time (Womack and Jones 1996, pp. 194). Particularly where service activities are tied to equipment provision such as in the case of complex engineering equipment in healthcare (MRI) or defence (fastjets), the understanding of use and outcomes required by the customer over time could allow the organisation to change business models, charging for use or outcomes (e.g., power by the hour® by Rolls Royce) instead of equipment ownership. By not separating equipment from activities and instead, focusing on benefits and the value offering in totality, firms could also innovate for better outcomes and achieve efficiency gains from redesigned and re-engineered equipment that enable better service activities. There is still great scope in both manufacturing and service fronts for exploring interactions between equipment and services in creating service as an experience (Pine and Gilmore 1998). More advanced knowledge on how equipment and service activities integrate and interact to deliver the outcomes is required.

Much of the research behind services and systems integration resides within the information, communications and technology (ICT) literature where systems integration has traditionally been about linking together different computing systems and software applications physically or functionally. Service related research in the field of ICT has been growing significantly in recent years. Zhao et al. (2007b) has provided comprehensive definitions on service computing, which refers to an emerging area of computer sciences and engineering that includes a collection of techniques, such as web services, service-oriented architectures, and the associated computational techniques. Various other similar terms have been used, such as services computing, service engineering, software as a service (SaaS), service-oriented computing and service-centric computing (Zhao et al. 2007a). Moreover, the issue of service-oriented technologies such as web services (Curbera et al. 2003, Brown et al. 2006; Demirkan and Goul 2006) and service-oriented architectures (Spohrer et al. 2007) has also been the attention of many information systems (IS) researchers who have endeavoured to design an integrated service offering.

There are two theoretical criticisms of ICT research, both of which have come from within the ICT domain. First is its focus on modularity and mechanistic designs. ICT research often endeavours to modularise the service offering, where modules are designed such that component interfaces are standardised, and interdependencies amongst components are decoupled (e.g., Ulrich 1995; Sanchez and Mahoney 1996) so as to enable the outsourcing of design and production of components and subsystems, within a predefined system architecture. Thus, ICT research has often taken on the mechanistic, modular approach towards technology, processes or people into some sort of component-based architecture, so that the mixing and matching of modules could provide a best-of-breed system (Lyons 2008). In a complex system offering of goods, people, activities and technology, research has shown that components do not normally 'click and play'. Indeed, a full specification of an entire system often does not plan for interactions that could result in unpredictable emergent properties (Prencipe et al. 2003; Ng et al. 2009).

Second, ICT research often over-emphasises the importance of technology in value offering. To achieve value-in-use, the integrated offering of goods, activities, people and technologies need to be effective in an *integrated* fashion, grounded on how the customer could *best engage* with the service to achieve benefits, *as well as* what is easiest or efficient for the firm where, in certain cases, less technology could result in better outcomes.

In summary, the idea of 'adding services' to manufactured goods is based on a flawed assumption that service activities are mere extensions of knowledge acquired in manufacturing goods. Similarly, service provision should not look upon equipment manufacture as exogenous to the offering, even if it is less flexible to redesign. A firm may have bundled resources that may be catalogued, but the services those resources deliver are only realised during activity. Finally, technology may play a key role, but only as a way to enable value co-creation. From all these perspectives, the need for a revolutionary rather than an evolutionary approach is necessary.

1.3 Complex Engineering Service Systems: Core Transformations

This book aims to reconcile the various streams of research through a value-centric approach. Woodruff (1997) presented the following definition of customer value:

> Customer value is a customer's perceived preference for and evaluation of those product attributes, attribute performances, and consequences arising from use that facilitate (or block) achieving the customer's goals and purposes in use situations.

Taking this value-centric approach, what operations need to be undertaken to deliver value? A value-centric approach must therefore put value-in-use at the centre of what the firm needs to deliver, in partnership with the customer. Consequently, to achieve value in use, the firm has to ask how value is created and understand the role of the customer within that space (Lengnick-Hall 1996). Operations Management literature has proposed that firms deliver three generic transformations. These three generic types of operations are often used to distinguish between organisation types. They are categorised on the basis of their transformation process such as 'material-processing operations', 'information processing operations', and 'people-processing operations'—and academics have discussed the various managerial challenges which differ across the three archetypes (Morris and Johnston 1987; Ponsignon et al. 2007).

The three transformations are explained below:

- Transform materials and equipment (i.e., manufacturing and production, store, move, repair, install, discard materials and equipment through supply chain, repairs, obsolescence management, predictive maintenance)
- Transform information (i.e., design, store, move, analyse, change information through knowledge management, information, communication and technological strategies, data strategies in equipment management)
- Transform people (i.e., train use, change use, build trust through education, influence, build relationships, change mindsets)

Operations literature has usually considered one type of transformation to be dominant (e.g., Slack et al. 2004). Hence, hotels, schools and hospitals are about transforming people, manufacturing and production are about transforming

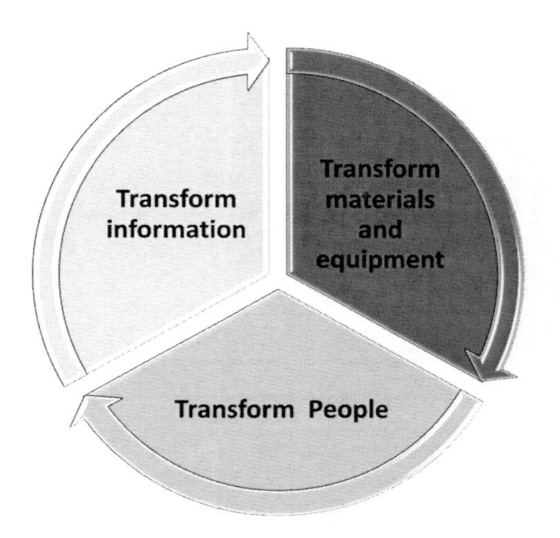

Fig. 1.1 Simple core integrative framework (CIF-lite) for complex engineering service systems

materials and equipment and media and information services such as Reuters and CNN are about transforming information.

While the three transformations are useful to depict a particular industry or sector, this model has an extended applicability to engineering services. In the service and support of complex engineering equipment or systems, such as airports or city transportation, what the customer considers as value, or experiences as value in use, may no longer be delivered through only one form of transformation but *simultaneously through all three*, particularly if the contracts are outcome based (Ng et al. 2009). For example, consider the maintenance repair and overhaul service of complex equipment such as ships, fastjets, submarines or tanks. Materials and equipment transformation would be concerned with repair, maintenance, supply chain and logistics, information transformation would be concerned with technology, information systems and communication with the customer, and people transformation would be concerned with how customer and the firm employees learn, react, use the equipment and interact with one another. Even in early research (e.g., Lengnick-Hall 1996; McDaniel and Morris 1978), customers have been proposed as a key outcome of transformation activities. To meet the full value in use of the firm's offering three integrated simultaneous transformations are required. This provides us with a Core Integrative Framework (CIF) and is shown in Fig. 1.1. The nomenclature reflects the development of a more detailed CIF, bringing in key learning from the chapters, which is presented at the end of the book.

When tasked to deliver the three transformations as the firm's value proposition to the customer, its respective delivery processes now interact with one another and there is no guarantee that the processes and knowledge for each transformation will complement each other if the delivery of all three was not designed into the system. In a CES system the three core transformations form part of the integrated value proposition, delivered through a value constellation (Normann and Ramirez 1993) and co-created with the customer.

The knowledge required to understand the various issues in complex engineering service systems is, therefore, twofold: first, component knowledge of each type of transformation; second, architectural or system knowledge which provides understanding of integration and how the value proposition will enable value co-creation with the customer. Both types of knowledge are essential to inform research into complex engineering service systems. However, research into component knowledge needs to be mindful of the whole, whilst research into system knowledge has to understand that the reduction of the system does not mean reduction of the system into components. Yet, there is a temptation to reduce the CES system into its components for ease of analysis and understanding. Indeed the standard scientific approach surrounds the 3R's of reduction, repeatability and refutation (Popper 1972). This has arisen essentially because many problems are complex and it becomes much easier for scientists to select some aspects of a problem for further detailed investigation. Science follows Descartes' advice to analyse problems piecemeal, that is, breaking down a phenomenon into its elemental parts. Accordingly, scientific thinking is very closely associated with analytical (divided into its constituent elements) thinking.

However, the reductionist approach is based on a number of assumptions that we should consider before applying it to the problems of engineering solutions that include equipment, human and technology interactions. The first and most crucial assumption is that when dividing the complex problem into separate parts, we assume that the elements of the whole are the same when examined independently as when they are examined as a whole. This needs careful consideration. If the elements are loosely connected then we can take them apart, analyse them, improve or change them and then put them back together and the whole will be improved. Whilst this may be true for the problems of simple mechanical systems, does this assumption hold for complex wholes? For example can we take out a part of the body e.g., the heart, modify it replace it back within the body and not expect effects elsewhere?

Lipsey and Lancaster (1956) and Goldratt (1994) have identified implications for the performance of parts where there is a close relationship. Goldratt pointed out the implications of optimising one part of a whole process that was not the limiting step. In his theory of constraints he points out that optimising the performance of a process step upstream of the bottleneck will only increase work in progress and working harder downstream is limited by the output of the bottleneck (Goldratt 1994). Sprague (2007) sums this up neatly proposing that "Optimizing the supply chain means convincing elements within that system to accept local sub optimums for the good of the whole" (Sprague 2007). We argue that this holds for any system rather than just supply chains. Thus, if we want to understand the performance of the whole service system and if we have begun the understanding by reducing to components, we would in essence be making three highly questionable assumptions; first, the connections between the parts must be very weak; second, the relationship between the parts must be linear so that the parts can be summed together to make the whole; and third, optimising each part will optimise the whole. Our understanding of the three concurrent transformations would reject these assumptions as, with value co-creation and customer variability as core factors, complexity becomes an inherent attribute of a through-life engineering service system.

Complex systems may be characterised by the interdependence that exists between the parts which make up a whole (Anderson 1999). Managers may create the value proposition but must remain flexible and adapt to embrace change as outcomes are emergent (Kao 1997; Snowden and Boone 2007; Santos 1998). Organisational effectiveness is increased by ‘fit’ between structures and organising mechanisms and the context in which they operate (Drago 1998; Brodbeck 2002). But single organisations rarely own or control all the capabilities necessary to deliver complex product systems. Bundling service in the business model increases the complexity, which can be compounded through social and political complexity between internal and external parties (Gann and Salter 2000). Gatekeepers stand at an organisation’s boundaries and translate information between the internal and external world of the enterprise (Lissack 1997). Standard processes of interaction would provide greater clarity of communication, but the application of rigid procedures would destroy the adaptable nature of the system. The gate keepers’ interaction with other business units, suppliers and customers requires greater understanding and study.

Competitive advantage may be gained through creating the capability to continuously adapt and co-evolve within the complex environments created, embedding a system capable of undergoing continuous metamorphosis in order to respond to a dynamic business landscape (Brodbeck 2002; Lewin et al. 1999). However, the rewards for the suppliers may not correlate with their capability as it is suggested that it is the customer’s perception of complexity, not that of the supplier which determines its contractual behaviour (Wikström et al. 2009). The dynamics and complexity of the system are further influenced by two key variables, both driven by value co-creation with the customer. First, contracts delivering service and support of equipment and people may range from merely supplying parts to delivering the availability of the equipment or to delivering the full capability of the customer. Thus, the degree of value co-creation could be contractually bound. Complexity arises when the firm shares resources across multiple contracts with different degrees of partnerships with the customer.

Second, Woodruff (1997) and Ng (2007, 2008) observed that customer value concepts differ because of time and context. For example, the firm and the customer could be engaged in set tasks and activities together and yet the benefits are different because of the context. This is the case for customers that face high environmental variability such as in defence where the support of equipment and people has to be designed to cater for delivery in a diverse set of environment, from Afghanistan to training around the barracks. Thus, customer variability in realising value needs to be factored into the design of the service system.

These system dynamics need to be recognised and organisational competency developed to meet the evolving customer need through transformation. Transformation is an active process, implying a change of state from a current to a future condition. The notion of service transformation, taken from a manufacturing

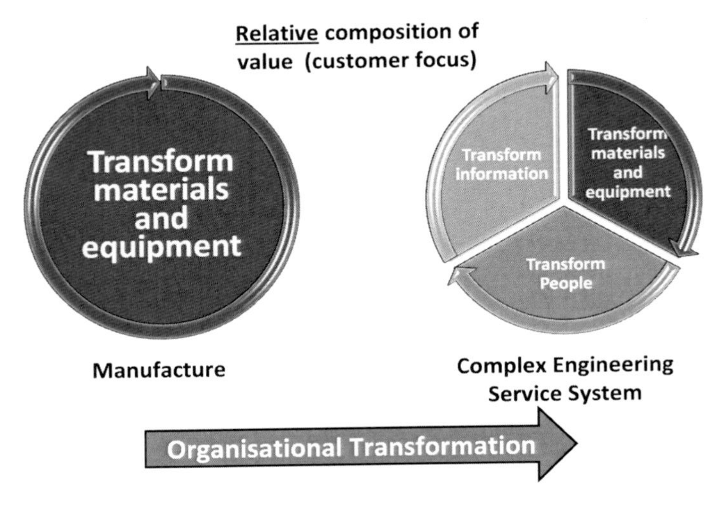

Fig. 1.2 Transformation challenge from manufacturer to partner in integrative service delivery enterprise

perspective, is embodied in Fig. 1.2, where the manufacturing organisation is engaged in a transformation from that which primarily delivers the value of equipment manufacturing (i.e., the transformation of materials and equipment set within the context of value in exchange) as the main value to the customer, to that which delivers the value of transforming materials/equipment, information *and* people (i.e., all three concurrently constituting value, working *with* the provider understanding value in use), and *in partnership with the customer,* through the realisation of value co-creation.

Thus, we define *CES Competency* as the ability of the firm to design, deliver and manage the entire CES system that is able to carry out the three core transformations above in a consistent, stable and profitable manner, co-creating value in partnership with the customer. The development of CES competency is core to companies engaged in the provision of complex engineering service. The competence to work in this dynamic environment will be highly valued by customers who realise greater value when working through partnerships.

1.4 Practice Implications

This book arose out of a research project entitled Service Support Solutions: Strategy and Transition (S4T), a £2m project co-funded by the UK Engineering and Physical Sciences Research Council and BAE Systems. Yet, the research on which this book is based draws on a wide base of academic experience and is directly applicable to other sectors struggling to contain and reduce the cost of those services which are dependent on the sustainment of complex, long life engineering assets.

In the case of defence, government and taxpayer interest is in the ability to deploy a defence service, essentially to provide the warfighter with the ability to have the required military effect in accordance with sovereign political will. This necessarily involves the cooperative working of a range of "customer" and other public sector agents with a wide range of industrial agents working at a national, if not a global scale. Each actor in the network is providing services to others all focused on the overall objective of providing military capability with a central theme of sustaining complex engineering assets, infrastructure and logistics from which the value in use is as dependent on the customer and on the training of operators, as it is on industrial efficiency. Operational and usage information is required to effectively configure service delivery. The driving question is: how is the overall service network best configured, how are its capabilities best integrated and how are the underpinning assets best sustained to assure availability of the desired overall service (military capability) at minimum cost?

It is suggested that there is little difference in principle between the defence perspective on CES systems and that of other industries where a user service is dependent on a complex and integrated delivery network reliant on complex, long-life engineering systems. Obvious examples are: utilities—electricity, water and gas—all of which have a complexity of service and/or "commodity" providers and distributors largely sharing a complex and ageing infrastructure; mass transportation—particularly rail with a similarly complex range of service providers with separate but common infrastructure provision; or medical services—particularly at regional and national level with many users, service agents and a wide range of assets and infrastructure.

Although the primary research base for this book has been defence, it is argued that the resulting insights will have a broad applicability. However, the book is developed from action-based research such that the underpinning case studies are mostly (but not invariably) based on defence and BAE Systems' programmes in particular. Much use is made of the Tornado ATTAC programme (Availability Transformation: Tornado Aircraft Contract) for case study material: this was initially a specific response by BAE Systems and the Ministry of Defence to the escalating cost of supporting the UK fleet of fast jets with early success recognised by the National Audit Office (NAO 2007). Whilst the ATTAC approach is becoming the norm for the provision of "sustainment services" in the defence air sector, it is arguably helping to point the way towards the development of more integrated defence services which will be a better value for the UK taxpayer and is identifying the issues associated with the co-creation of value between user, customer/owner and service providers which need to be addressed in such complex service systems. Clearly much more that can be done and the principle aim of this book is to identify potential further steps for both practitioners and researchers interested in developing the industry's capability to deliver increasingly effective CES systems.

The work reported in the book does draw on other case studies: all are introduced within the chapters which draw on particular insights or data so that the individual chapters are able to stand alone. Defence air is perhaps disproportionately represented because it is acknowledged, in many respects, to be leading the practice in the field. Other studies covered in defence are naval, land and missile systems, in the civil sector, developing experience in commercial air and some "business to consumer" services reliant on logistics and technology such as breakdown recovery. Tentative conclusions are drawn on a case-by-case basis. Whilst the integrating framework provides an overall architecture within which the chapter subjects, case studies and the resulting insights may fit, it is too early to draw any overall conclusions.

One thing is clear: there is a developing imperative for users, owners and operators of complex infrastructure and other engineering assets to improve "value in use", and this is dependent on the development of service delivery networks that are highly integrated, both behaviourally and organisationally, towards the co-creation of value and which have efficient (or lean) engineering management of the assets on which the service depends at their centre.

We consider the key dimensions of CES systems as: service performance and operational management, the service supply network configuration and capability and the engineering associated with efficiently sustaining the assets in service over very long lifetimes. Service performance and operational management is about ensuring that the customer and stakeholder needs are met and that users experience a value in use to at least match their expectations. These services are invariably provided as a result of integrating capabilities and individual outcomes across an extensive network of agents—within a complex customer organisation, with multiple supply channels and with a multitude of subordinate service providers. And the engineering is truly challenging—these systems are being asked to operate at the extremes of their performance availability over lifetimes spanning decades. Frequent updates are needed to sustain or improve performance, operating safety or environmental impact whilst increasingly efficient means need to be found to deliver system maintenance which usually represents a significant proportion of the overall service cost.

Although the case studies on which the research reported in this book is based cover only aspects of this overall picture, it is suggested the complexity of defence systems provides a good start towards understanding the imperatives in what is essentially a new field of study and practice. In describing the case studies in each chapter, authors have tried to reduce the "defence jargon" as much as possible in order to help understanding of potential broader application. Where insights and tentative conclusions are drawn these have, wherever possible, been expressed in generic language.

1.5 Three Core Transformations as a Structure

The Core Integrative Framework (CIF) is foundational to research into the provision of CES systems, providing a basis for communication, understanding and integration. It was with reference to the framework and transformations that this

Fig. 1.3 Transformation challenge from manufacturer to partner in integrative complex engineering service system

book and the contributions from authors were shaped. The outline of the book is mapped onto the CIF in Fig. 1.3.

Part I of the book will address Organisational Transformation. The work explores the meaning of Service Enterprise and how transformation may take place. The challenges involved when transforming a complex multi-organisational service enterprise are examined. The section further presents developmental tools for enterprise transformation. 'Enterprise Imaging' describes a methodology used to visualise multi-organisational entities that co-create value. The output of the tool is an image of the organisation that is to be transformed, including partners, contractors and sub-organisations. A developmental tool for complexity management is then presented, based upon a framework capturing the factors identified by ATTAC managers as making their enterprise complex. It is proposed that this framework is used for discussion of complexities, leading to identification and removal of unnecessarily complex structures and hence lowering cost and focussing effort on complexities that are necessary to value creation. The section concludes with an exploration of the service aspirations and fears of the ATTAC contract stakeholders. For long term value to be co-created it is proposed that the entirety of the required service, including but not limited to the service explicitly described in the contract, is articulated. This section includes all three forms of transformation captured in the CIF.

Part II provides some of the foundations necessary for service contract development and delivery. Initial work identifies the capabilities required to co-create value within a 'value-web'. Capabilities are mapped against core attributes of value co-creation. It is proposed that this matrix be used to define a firm's service capability. The following chapter maps the major classes of uncertainty within service contracts against service delivery cost drivers. The resultant uncertainty-based cost framework provides information which may be used for service contract development and negotiation. A study of how incentives in service contracts impact upon organisational behaviour is presented. Service contract incentives encourage flexibility and adaptability, but this may contradict cost reduction objectives. Performance is dependent upon the customer and, therefore, incentive mechanisms may include customer performance. The section concludes with an analysis of the dependence of community level relationships upon successful service delivery. The work presents the transition observed in two contracts, from inter-personal relationships to community level relationships as the contracts matured. The section touches upon all three transformations as work covers capability, uncertainty of costing, incentives in contracts and relational governance.

Part III is focussed upon the challenges of transforming information. The section begins by introducing a model for the identification of service information requirements. Combining the aspects of service supply networks and service lifecycles, this 12-box model utilises a traffic light approach to signal how well current solutions address information requirements. The section continues with the analysis of how computer-based simulations play an important role in understanding trade-offs between service affordability and performance when developing service contracts. An application of discrete event simulation to service systems for complex engineering products is described. Finally a service information blueprint is introduced as a means of defining service contracts, processes and relationships between the two.

Part IV is product transformation focused. The work presented shows how products may be better managed for service provision, beginning by introducing service support solutions used by the UK MoD to deliver contracts for both availability and capability. An overview of techniques to plan equipment availability and maintenance is then given to raise awareness of techniques that may strengthen support solutions. This is followed by a piece of work focused on the applicability and implementation of predictive maintenance, providing details of combining prognostic modelling with Condition Based Maintenance (CBM) and how this may improve the repair and maintenance service provided for complex systems. The section continues with work on operational requirements for component replacement decision analysis and explores approaches to estimate maximum availability whilst preventing component failures. Research is then presented into delay time modelling that may be used to optimise maintenance service intervals with respect to specific criteria, such as system cost, downtime and reliability. Work using simulation methods to develop service procedures follows. This shows how simulated service operations may inform support outcomes in terms of learning, cost and process. A framework for the development of a Maintenance Dashboard is then proposed which delivers status information to stakeholders whilst an asset is 'active', aiding consistent decision making. In the concluding chapter of this section modernisation of platforms through the insertion of new technology is examined. This requires interaction between three principal stakeholders: acquisition authority; product-service system provider; and end-user. A transformation mapping approach has been developed which brings these three groups together to set their joint vision and plan activities.

The final section of the book, Part V, returns to explore in more detail the question we have begun to address in this chapter: how do we produce integrating frameworks that link together the different disciplinary areas required to deliver service?

It is noted that the transformation of people element of the CIF does not have a specific section within this book. Many of the sections contribute to knowledge in this area, but there is scope for further and more detailed work to be done. As further service type contracts are put out to tender, further work may focus upon the transformation of people, as they transform from a product to a service mindset.

Whilst the book reflects the work of a large number of academic and industrial contributors, it does not claim to provide the full picture of complex engineering service systems. Research in this area is ongoing and this book aims to set conceptual foundations upon which others can build. We seek to provide an overview and sufficient detail to engage the reader in the subject, both through the research done to date and in identifying the gaps and continuing questions. We hope our book represents a starting point for future researchers and practitioners engaging in the challenge of designing, delivering and evaluating Complex Engineering Service Systems.

References

R. Amit, C. Zott, Value creation in e-business. Strateg. Manag. J. **22**(6/7), 493–520 (2001)

P. Anderson, Applications of complexity theory to organizational science. Org. Sci. **10**(3), 216–232 (1999)

J.C. Anderson, J.A. Narus, Capturing the value of supplementary services. Harv. Bus. Rev. (1995), pp. 75–83 (Jan–Feb)

R.P. Bagozzi, Marketing as exchange. J. Mark **39**, 32–39 (1975). Oct

T.S. Baines, H.W. Lightfoot, S. Evans, A. Neely, R. Greenough, J. Peppard, R. Roy, E. Shehab, A. Braganza, A. Tiwari, J.R. Alcock, J.P. Angus, M. Bastl, A. Cousens, P. Irving, M. Johnson, J. Kingston, H. Lockett, V. Martinez, P. Michele, D. Tranfield, I.M. Walton, H. Wilson, State-of-the-art in product service-systems. Proc. I Mech. E. Part B. J. Eng. Manuf. **221**(10), 1543–1552 (2007)

D. Ballantyne, R.J. Varey, Creating value-in-use through marketing interaction: The exchange logic of relating, communicating and knowing. Mark Theory **6**(3), 335–348 (2006)

R.N. Bolton, J.H. Drew, A longitudinal analysis of the impact of service changes on customer attitudes. J. Mark. **55**(1), 1–9 (1991)
P.W. Brodbeck, Complexity theory and organisation procedure design. Bus. Proc. Manag. J. **8**(4), 377–402 (2002)
G.W. Brown, H. Demirkan, M. Goul, M. Mitchell, Towards the service-oriented enterprise vision: Bridging industry and academics. Panel presentation, American conference on information systems (Acapulco, Mexico, 2006) Aug 4–6
H.E. Butz, L.D. Goodstein, Measuring customer value: Gaining the strategic advantage. Org. Dynam. **24**(3), 63–76 (1996)
J.P. Cannon, C. Homburg, Buyer–supplier relationships and customer firm costs. J. Mark. **65**, 29–43 (2001). January
R.B. Chase, Where does the customer fit in a service operation? Harv. Bus. Rev. **56**(6), 137–142 (1978)
F. Curbera, R. Khalaf, N. Mukhi, S. Tai, S. Weerawarana, The next step in web services. Commun. ACM **46**(10), 29–34 (2003)
U. de Brentani, Success factors in developing new business services. Eur. J. Mark. **25**(2), 33–59 (1991)
H. Demirkan, M. Goul, Towards the service-oriented enterprise vision: Bridging industry and academics. Comm. Assoc. Inf. Syst. **18**(26), 546–556 (2006)
W.A. Drago, Mintzberg's 'pentagon' and organisation positioning. Manag. Res. News **21**(4/5), 30–40 (1998)
P.F. Drucker, *Management tasks, responsibilities, practices* (Harper Collins, New York, 1993)
A. Eggert, W. Ulaga, F. Schultz, Value creation in the relationship life cycle: A quasi-longitudinal analysis. Ind. Mark. Manag. **35**(1), 20–27 (2006)
D.M. Gann, A.J. Salter, Innovation in project-based, service enhanced firms: The construction of complex products and systems. Res. Policy **29**, 955–972 (2000)
E.M. Goldratt, *Theory of constraints* (Gower Publications, Aldershot, 1994)
J.R. Hicks, *The social framework* (University Press, Oxford, 1942)
X. Huang, L.B. Newnes, G.C. Parry, A critique of product and service based systems. The 7th International Conference on Manufacturing Research (ICMR09) (University of Warwick, UK, 2009) Sept 8–10
J. Kao, *Jamming: The art and discipline of business creativity* (Harper Collins, New York, 1997)
C.W. Kim, R. Mauborgne, Strategy, value innovation, and the knowledge economy. MIT Sloan Manag. Rev. **40**(3), 41–54 (1999)
P. Kotler, *Marketing Management, 11th ed.,* (Prentice Hall, Upper Saddle River, 2003)
C.A. Lengnick-Hall, Customer contributions to quality: A different view of the customer oriented firm. Acad. Manag. Rev. **21**(3), 791–824 (1996)
A.Y. Lewin, C.P. Long, T.N. Carroll, The co-evolution of new organisational forms. Org. Sci. **10**(5), 535–550 (1999)
R.G. Lipsey, K. Lancaster, The general theory of second best. Rev. Econ. Stud. **24**(1), 11–32 (1956)
M.R. Lissack, Of chaos and complexity: Managerial insights from a new science. Manag. Decis. **35**(3), 205–219 (1997)
A.H. Liu, M.P. Leach, K.L. Bernhardt, Examining customer value perceptions of organizational buyers when sourcing from multiple vendors. J. Bus. Res. **58**(5), 559–568 (2005)
M.H. Lyons, Developing a science of services and systems. BT Tech. J. **26**(1), 112–122 (2008)
K. Marx, Capital: A critique of political economy. Penguin Books in association with New Left Review, Harmondsworth, Middlesex (1867 [2001])
R.R. McDaniel Jr., S.A. Morris, Management in human service systems. Working paper 78-47 (University of Texas at Austin, 1978)
J. Mills, E. Neaga, G. Parry, V. Crute, Toward a framework to assist servitization strategy implementation. *Proceedings of the POMS 19th annual conference*, May 9–12, La Jolla, (California, 2008)

B. Morris, R. Johnston, Dealing with inherent variability: The difference between manufacturing and service? Int. J. Oper. Prod. Manag. **7**(4), 13–22 (1987)

National Audit Office, (2007) Transforming logistics support for fast jets. Report by The Comptroller and Auditor General | HC 825 Session 2006-2007 | 17 July 2007 Available at: http://www.nao.org.uk/publications/nao_reports/06-07/0607825.pdf. Accessed Feb 2011

A.D. Neely, *Exploring the financial consequences of the servitization of manufacturing.* (Accepted for publication in Oper Manag Res, 2008)

I.C.L. Ng, Advanced demand and a critical analysis of revenue management. Serv. Ind. J. **27**(5), 525–548 (2007)

I.C.L. Ng, *The pricing and revenue management of services: A strategic approach* (Routledge, London, 2008)

I.C.L. Ng, A demand-based model for the advanced and spot pricing of services. J. Prod. Brand. Manag. **18**(7), 517–528 (2009)

I.C.L. Ng, R. Maull, N. Yip, Outcome-based contracts as a driver for systems thinking and service-dominant logic in service science: Evidence from the defence industry. Eur. Manage. J. **27**(6), 377–387 (2009)

I.C.L. Ng, J. Williams, A. Neely, Outcome-based contracting: Changing the boundaries of B2B customer relationships. Advanced Institute of Management (AIM) Research Executive Briefing Series. Available at: http://www.aimresearch.org/index.php?page=alias-3 (2009)

P. Nonthaleerak, L. Henry, Exploring the six sigma phenomenon using multiple case study evidence. Int. J. Oper. Prod. Manag. **28**(3), 279–303 (2008)

R. Normann, R. Ramirez, From value chain to value constellation: Designing interactive strategy. Harv. Bus. Rev. (July–August), pp. 65–77 (1993)

R. Oliva, R. Kallenborg, Managing the transition from products to services. Int. Jof. Serv. Ind. Manag. **14**(2), 160–172 (2003)

A. Parasuraman, V.A. Zeithaml, L.L. Berry, A conceptual-model of service quality and its implications for future-research. J. Mark. **49**(4), 41–50 (1985)

A. Payne, S. Holt, Diagnosing customer value: Integrating the value process and relationship marketing. British J. Manag. **12**(2001), 159–182 (2001)

B. Pine, J. Gilmore, Welcome to the experience economy. Harv. Bus. Rev. (July–August): (1998), pp. 97–105

F. Ponsignon, P.A. Smart, R.S. Maull, Service delivery systems: A business process perspective. Proceedings of the POMS 18th annual conference (London, 2007) July 12–13

K.R. Popper, *Objective knowledge: An evolutionary approach* (Clarendon Press, Oxford, 1972)

A. Prencipe, A. Davies, M. Hobday, *The business of systems integration* (Oxford University Press, Oxford, 2003)

R. Ramirez, Value co-production: Intellectual origins and implications for practice and research. Strateg. Manag. J. **20**(1), 49–65 (1999)

A. Ravald, C. Grönroos, The value concept and relationship marketing. Eur. J. Mark. **30**(2), 19–30 (1996)

G. Ren, *Service business development in manufacturing companies: Classification characteristics and implications.* Ph.D. Dissertation. (University of Cambridge, 2009)

R. Sanchez, J.T. Mahoney, Modularity, flexibility, and knowledge management in product and organization design. Strateg. Manag. J. **17**, 63–76 (1996). winter special issue

M. Santos, Simple, yet complex. CIO Enterprise Magazine, Enterprise Section 2. Available online at: www.cio.com/magazine (1998)

J.B. Say, A treatise on political economy. 1st America ed. Reprints of Economic Classics, 1964. Augustus M. Kelly, New York (1803)

B. Schneider, D.E. Bowen, *Winning the service game* (Harvard Business School Press, Boston, 1995)

N.W. Senior, *Political economy*, 5th edn. (Charles Griffin and Co, London, 1863)

G.L. Shostack, Breaking free from product marketing. J. Mark. **41**(2), 73–80 (1977). April

N. Slack, S. Chambers, R. Johnston, *Operations management*, 4th edn. (Prentice Hall Financial Times, Harlow, 2004)

D.J. Snowden, M.E. Boone, A leader's framework for decision making. Harv. Bus. Rev. (Nov), pp. 69–76 (2007)

J. Spohrer, P.P. Maglio, J. Bailey, D. Gruhl, Steps toward a science of service systems. IEEE Comput. **40**(1), 71–78 (2007)

L.G. Sprague, Evolution of the field of operations management. J. Oper. Manag. **25**(2), 219–238 (2007)

D.R.E. Thomas, Strategy is different in service businesses. Harv. Bus. Rev. **56**(3), 158–165 (1978)

W. Ulaga, A. Eggert, Value-based differentiation in business relationships: Gaining and sustaining key-supplier status. J. Mark. **70**(1), 119–136 (2006)

K. Ulrich, The role of product architecture in the manufacturing firm. Res. Policy **24**, 419–440 (1995)

S. Vandermerwe, J. Rada, Servitization of business: Adding value by adding services. Eur. Manag. J. **6**(4), 314–324 (1988)

S.L. Vargo, R.F. Lusch, Evolving to a new dominant logic for marketing. J. Mark. **68**, 1–17 (2004)

S.L. Vargo, R.F. Lusch, Service-dominant logic: Continuing the evolution. J. Acad. Mark. Sci. **36**(1), 1–10 (2008)

K. Wikström, M. Hellström, K. Artto, J. Kujala, S. Kujala, Services in project-based firms—four types of business logic. Int. J. Proj. Manag. **27**(2), 113–122 (2009)

J. Womack, D. Jones, *Lean thinking* (Free Press, London, 1996)

R.W. Woodruff, Customer value: The next source for competitive advantage. J. Acad. Mark. Sci. **25**(2), 139–153 (1997)

J.L. Zhao, M. Tanniru, L. Zhang, Service computing as the foundation for enterprise agility: Overview of recent advances and introduction to the special issue. Inf. Syst. Front. 9 (March), pp. 1–8 (2007a)

J.L. Zhao, C. Hsu, H.K. Jain, J.C. Spohrer, M. Tanniru, Bridging service computing and service management: How MIS contributes to service orientation. Proceedings of the 28th International Conference on Information Systems (ICIS), Montréal, Dec 9–12. Inf. Syst. Front. 9(1): 1–8 (2007b)

Part I
Organisation and Enterprise

Valerie Purchase, John Mills and Glenn Parry

Complex engineering service systems inevitably involve complex organisational solutions which go beyond the boundaries of single organisations. The challenges inherent in the delivery of complex, co-created services are considerable and re-quire a significant shift in perspective and practices among the partners.

Each chapter in this section describes work undertaken within the S4T programme work package tasked with examining the 'organisational transformation' necessary for the delivery of through-life support services. The chapters are based on an in depth case study of the ATTAC programme, an early example of the move to complex engineering availability service contract by a firm with a strong heritage of design and manufacture. It provided a clear opportunity to examine organisational transformation in action. The successful transformation of complex service systems is a difficult and challenging undertaking for all stakeholders as the literature review within the first chapter highlights. BAE Systems and their partners have met those challenges and are breaking new ground in their ways of co-creating value. The case study represents a stage in this service journey and was offered by BAE Systems and their partners to enable learning from some of the challenges identified and addressed in their transformation.

ATTAC is a long-term, whole-aircraft availability contract where BAE Systems takes prime responsibility to provide Tornado aircraft with a complex service of depth maintenance and upgrades, incentivised to achieve defined levels of available aircraft, spares and technical support at a target cost. However, within this research project it quickly became apparent that no single organisation delivered ATTAC. Instead the programme was delivered by a 'multi-organisational service enterprise' encompassing a complex network of co-dependent organisational entities that must collaborate to deliver the contract KPIs and to co-create further value via ongoing service improvement and cost reduction. Within this section we examine some of the challenges involved in the management and transformation of this complex service enterprise. The chapters share a research community comprising academic, provider and client contributors, and present a range of perspectives and insights from our case study work.

The chapters also share recognition of the need to adopt a holistic multi-organisational enterprise perspective in organising for the delivery of complex engineering services.

In the first chapter entitled Service Enterprise Transformation, the partnered organisation of ATTAC is briefly described, and the case study methodology introduced. The chapter then examines the challenges in transforming a complex multi-organisational service enterprise. The transformation is explored in three parts: understanding the drivers and challenges inherent in the move to service; examining the need for a 'holistic enterprise perspective' for service delivery in complex engineering systems; and finally capturing the challenges in undertaking such a complex transformation process.

One of the key challenges for the management of complex service enterprises is identifying the organisational participants and enterprise boundaries. Chapter 2 on Enterprise Imaging describes a methodology used to visualise the complex partnership of multi-organisational entities that participate in the co-creation of the ATTAC support service. This enterprise level visual management tool sheds light on the significant level of organisational complexity involved, yet makes explicit the sub organisations within both client and provider that must co-create the overall service. In this way it answers the question "What organisation needs to be transformed?"

Complex services involve complex organisational solutions but what factors create this complexity? The research theme of complexity was given broader consideration in the third chapter; Complexity Management identifies the set of factors that add to the complexity of the ATTAC availability contract. The complexity factors are described in context and a framework is presented which clusters them into six key areas. It is proposed that this framework may then be used as a tool for analysis and management of complexity.

Optimising organisational and enterprise solutions to deliver services, implies that the required services within such contracts are clearly known to all parties. Within the final chapter in this section, the need for recognition and exploration of enterprise stakeholder aspirations and fears for service provision is explored. A preliminary analysis of case study findings enables the translation of public sector client aspirations and fears into the services necessary for a complex, long-term service availability contract. Here we explore the operational services demanded by the contract, the support services required to deliver the operational service, which may or may not be explicitly described in the contract, and strategic services highly desired by the client but absent from the contract. This research begins to identify and focus on enterprise level services requiring co-creation by combinations of organisations from client and provider communities.

The research programme which underpins this section, like the ATTAC programme itself, has been undertaken by a multi-organisational team of academics, BAE Systems' on-base service management team, MoD, and organisational sponsors and contributors from the BAE Systems' back offices. This 'research enterprise' and agenda are continuing to explore issues relating to the management of complex service enterprises. This will involve making contributions to practice; theory building in the areas of complex co-creation of value in service enterprises; and supporting the generation of conceptual frameworks and 'enterprise level' tools to improve service enterprises leadership and performance.

Chapter 2
Service Enterprise Transformation

Valerie Purchase, Glenn Parry and John Mills

Abstract This chapter examines the challenges in transforming a complex multi-organisational service enterprise. It builds on a review of relevant literature and an empirical analysis of the early experience and lessons learned by industry and MoD partners in the ATTAC (ATTAC (Availability Transformation: Tornado Aircraft Contract) is a long-term, whole-aircraft availability contract where BAE Systems take prime responsibility to provide Tornado aircraft with depth support and upgrades, incentivised to achieve defined levels of available aircraft, spares and technical support at a target cost.) enterprise which delivers a through life support programme in the defence sector. The transformation will be explored within three sections. The first illustrates and further develops current understanding of the drivers and challenges inherent in the move to service. In the second section, the need for a 'holistic enterprise perspective' for service delivery in complex engineering systems is discussed and illustrated through the ATTAC case study. Finally, the challenges in undertaking such a complex transformation process are discussed. The frameworks created may support future service enterprise leaders in identifying and communicating to all stakeholders the key drivers for the transition to through-life support services and assessing the key barriers which may be faced in managing their own transformation.

2.1 Introduction

Many organisations are in the process of moving from a product- to a service-based business model. Such changes are driven by a variety of factors including the opportunity to create a more stable source of revenue; to establish a source of

V. Purchase (✉) · G. Parry · J. Mills
School of Communication, University of Ulster, Newtownabbey, Co. Antrim, UK
e-mail: vc.purchase@ulster.ac.uk

I. Ng et al. (eds.), *Complex Engineering Service Systems*, Decision Engineering,
DOI: 10.1007/978-0-85729-189-9_2,

competitive advantage and differentiation; and the opportunity through services to build or maintain customer relationships. However, for companies who have a long history as a product-focused business, such changes present a considerable challenge in transforming processes, culture and relationships with customers. Within this monograph, the focus is on complex engineering services which often require multiple partners involved in co-creating complex customer solutions beyond the capabilities of any individual organisation. In other words, the delivery of complex engineering services requires a correspondingly complex multi-organisation 'service enterprise'.

This chapter explores service enterprise transformation for the move to service. A defence sector case study of a complex service enterprise, involving multiple customer, prime and supplier partners, will be used to illustrate such efforts to transform for service co-creation. The successful transformation of complex service systems is a difficult and challenging undertaking for all stakeholders as the literature review within the first chapter highlights. BAE Systems and their partners have met those challenges and are breaking new ground in their ways of co-creating value. The case study represents a stage in this service journey and was offered by BAE Systems and their partners to enable learning from some of the challenges identified and addressed in their transformation.

Following this introductory section, the chapter will be structured as follows:

- Section 2.2 will provide a brief overview of the ATTAC case study describing the methodology used and identifying an outline of key parties involved in delivering the programme;
- Section 2.3 will examine the key drivers in the shift from product to services, and challenges to implementation at the individual, organisational and inter-organisational levels;
- Section 2.4 will outline and discuss factors driving the adoption of a multi-organisational 'service enterprise' for complex delivery of customer value;
- Finally in Sect. 2.5, the case study findings on the partners' experience of the overall transformation and change cycle for ATTAC will be discussed. Partner behaviours, characteristics of each phase and changing relationships will be illustrated and lessons will be identified for future implementation of service contracts.

2.2 The ATTAC Case Study: Overview and Methodology

An in-depth case study of the ATTAC programme enabled researchers to examine the transition involved in implementing the ATTAC contract for through-life support. Through case study methodology phenomena can be examined in situ and the meanings actors bring to such phenomena can be better understood. Case study research is also useful when the aim of the research is to answer 'how' and 'why' questions (Yin 2003). This matches the wider aims of this research to gain an understanding of how and why such complex service provision contracts actually materialise in practice, as perceived by involved (and uninvolved) actors in the Provider and the Client. The ATTAC case study was chosen for two main reasons—it was the first of its scale and complexity between the Provider and the Client. Though both parties intended to continue to let and bid for such contracts, this first attempt was an opportunity for both parties to learn and this enabled the researchers to interview widely—six client and 22 provider interviews were conducted. These informants were classified into three groups—those involved in the design and implementation of the contract; those who supported implementation; and those who viewed the contract from a distance. The interviews were conducted in 2008 and were semi-structured, face-to-face taking an average of 1.5 h each. Interviews enable researchers to uncover how informants perceive and interpret situations and events (Bryman 2008). Themes covered were the scope of the contract; their role in implementation and the obstacles and enablers they met; their perceptions of issues in other areas of the implementation. The case study analysis presented in the chapter has been rigorously validated through a series of presentations to key customer and provider contract and support functions. Written reports have also been made available for validation and feedback (Yin 2003). A brief overview of the ATTAC programme and contributing partners is provided below.

The UK Ministry of Defence (MoD) is increasingly opening the support of military systems to private companies, and working in partnership with multiple organisations to deliver support. One example is ATTAC (Availability Transformation: Tornado Aircraft Contracts), a long-term, whole-aircraft availability contract where BAE Systems take prime responsibility to provide Tornado aircraft with depth support and upgrades, incentivised to achieve defined levels of available aircraft, spares and technical support at a target cost.

The support contract is delivered through a complex 'multi-organisational service enterprise' comprising a variety of on-base organisations at RAF Marham, supported by off-base organisations acting in partnership (Mills et al. 2009). The drivers for the adoption of this partnered approach was the need for reductions in the cost of providing this service and the belief that the service could be more effectively delivered through closer working between public and private sector partners.

From the time an aircraft is recalled for servicing, to the time it becomes available again for further duties, a wide variety of organisations and sub-organisations have collaborated in providing this 'availability service'.

BAE Systems are the prime service provider and perform many key roles either directly or through managing RAF personnel to deliver their services. Managed by BAE Systems, a 'fleet management' organisation provides the planning activities that translate the RAF Squadron requirements for Tornados into the schedule of aircraft through maintenance hangers. BAE Systems then manage the hanger activity, staffed by both BAE Systems and the customer *RAF Air Command* personnel, where the operational services are delivered.

Engineering support is managed by BAE Systems based at both RAF Marham and their other sites. This activity resolves technical queries and safety issues and is similarly staffed by RAF and industry personnel.

The *Defence Equipment & Support (DE&S)*-managed Tornado IPT (Integrated Project Team) contains solely MOD staff covering administration, engineering, logistics, and commercial support of ATTAC on behalf of the *Ministry of Defence*. This organisation is responsible for airworthiness and procurement and monitoring of contract performance.

Following maintenance, the aircraft may need to be repainted. A *third party company* provides a painting service, one of the later in-line processes in the delivery of maintained aircraft and, therefore, a significant dependency.

None of the support services would be possible without the variety of sub-organisations within *RAF Air Command* who both provide and are responsible for the hangars themselves, and their electrical/hydraulic power and information technology infrastructure.

A number of the *supply chain organisations* are also a critical part of this multi-organisational service enterprise. Spare components and systems are provided by both the prime and sub-tier supply organisations, which may deliver to the prime, to the customer or directly to the RAF squadron for aircraft on duty. Finally, a further organisation, the *Defence Storage and Distribution Agency*, is the sole provider of transport and off-base storage of Tornado parts.

The ATTAC services are co-created by a complex and inter-dependent multi-organisational service enterprise, which must align and coordinate activities to support delivery of the service for the RAF. However, in order to fulfil this role, the contributing partners and in particular BAE Systems have to undergo a significant transformation in their ways of working and indeed in the business model or rationale underpinning their practices. In the sections to follow, this transformation is explored in terms of the focus for transformation—the drivers and challenges in the move to service; the need for a holistic multi-organisational enterprise response; and finally, an exploration of the nature of the transformation process experienced by the service enterprise partners.

2.3 The Move to Service

Within this section, key drivers and challenges in the move to service are identified within the literature and an analysis of how such factors are evident in the ATTAC programme has also been provided. Key messages from interviews with the ATTAC partners are presented in the text to illustrate participants' perspectives on the drivers for transformation and the challenges faced. The analysis examines the drivers for the move to service more holistically and likewise, considers obstacles or challenges to achieving this transformation from the point of view of all parties.

2.3.1 Key Drivers in the Move to Service

What are the key drivers in the move from product to service orientation? Several studies have highlighted *financial factors* as drivers for a transition from products to services, such as the additional profits and revenues that can be achieved by manufacturers by focusing on aftermarket activities (Wise and Baumgartner 1999; Mathe and Shapiro 1993). Oliva and Kallenberg (2003) cite the following drivers: the generation of revenue from installed base of products with long-life cycles; services have higher margins than products; and services are a more stable source of revenue, resistant to economic cycles that drive investment and equipment purchases. Ward and Graves (2007) likewise identify financial factors as drivers for the move to Through-life Management. However in their study, the emphasis was less on increasing profits and revenues and more on cost reduction and revenue protection in the long term. Cost reduction included direct benefits for supplying companies and the reduction of cost of ownership for both military and civil customers. They emphasised that the aerospace industry has already supplemented product revenues with aftermarket revenues and are now more focused on agreeing through-life contracts, which seek to capture life-cycle business and revenues at the earliest possible stage and aim to reduce cost of ownership for customers. Such drivers are clearly in evidence within the ATTAC programme where interviewees were clear that the context for their business activities had changed. A change in the geopolitical climate ending the 'cold wars' have resulted in shifts within the defence sector from manufacture to service.

Several interviewees described the principal value proposition for the lead supplier as 'an opportunity to access ongoing support budgets for the Tornado aircraft'. It was recognised that the traditional stream of business was diminishing in that there would be 'no new manned air defence platforms in the foreseeable future'. The move to service was, therefore, being made to protect revenues. Such long-term support contracts were also perceived as representing a considerable opportunity for stable profit over time.

Returning to the research literature, *customers* are also demanding more services. Oliva and Kallenberg (2003) identify the role of changes within customer firms in prompting calls for service from their suppliers. They include factors, such as pressure to downsize, more outsourcing, focus on core competencies, increasing technological complexity and the consequent need for higher specialisation in influencing requirements for greater levels of through-life service. The nature of value for the customer would also seem to be changing. This is evidenced through the customer demand for capability, availability, operability and affordability (Ward and Graves 2007).

> Clients want more value and this value is connected to the use and performance of systems; they want solutions more than just products or services; they want to take advantage of their supplier's know-how and not just their product; they want an integrated and global offering and are reluctant to do business with several suppliers; finally, they want customised relationships Mathieu (2001, p. 458).

MoD is also changing its definition of customer value—emphasising availability and capability for future contracts. There is a willingness from MoD to transfer support of defence assets to the industry (Ministry of Defence 2005).

Cost reductions were a major driver for the ATTAC customer. The ATTAC contract offered savings of at least £510 million over the first 10 years. Savings arose from reductions in RAF and civilian personnel and the improvements a commercial organisation was expected to bring to task (Mills et al. 2009).

Competitive factors have also been suggested as key drivers in the move from a product to a service focus since services are more difficult to imitate, and are therefore a sustainable source of competitive advantage (Oliva and Kallenberg 2003). Mathieu (2001) also identified a number of competitive advantages in the transition from product to service delivery. Services, Mathieu suggested, offered strong competitive advantage through differentiation opportunities; can help in building industry barriers to entry; and services which directly enhance the value of the tangible product and support its application are more effective as competitive weapons than services of a more general nature. Superior service increases both first-time and repeat sales, enhancing the market share. Additionally, the level and quality of the services offered is an effective way to maintain on-going relationships. Hence, services also provide *marketing opportunities* and the opportunity to maintain strong customer relationships. There was evidence that relationship-building opportunities were recognised within the ATTAC programme where the delivery of support services was acknowledged as a factor in maintaining or strengthening the relationship with the customer and helping to secure the loyalty.

Finally, Ward and Graves (2007) additionally identified the need for *risk reduction or risk transfer* as a driving factor in the development of TLM contracts. They found some evidence particularly from defence sector companies that the adoption of TLM involves the transfer of risk from the customer down through the supply chain that assumes much higher levels of operating risk, financial risk and technological risk. The issue of transferring risk would seem to represent both a key driver and challenge for the ATTAC programme. The transfer of risk from the customer to the supply chain is more complex in the defence sector where there is a need to retain national capability.

Although the drivers for suppliers and customers have here been presented separately, there was a strong tendency for both supply and customer interviewees to express mutually beneficial drivers for the programme. There was common agreement among interviewees that the programme was developed to deliver 'more for less' in such a way as to provide benefits for all parties and a belief that the service could be more effective through closer working between public and private sector partners.

However, there was also clear recognition that the programme was a testing ground for new forms of contracting and collaborative delivery and that many challenges would need to be overcome to achieve this move to a service model. An overview of the key drivers for the move to service is presented in Table 2.1.

Table 2.1 Overview of key drivers in the move to service

Key drivers	Factors identified within research literature[a]	Driver for ATTAC
Financial factors	Opportunity for additional profits and revenues; higher margins for services; more stable source of revenue; cost reduction and revenue protection in the long term; reduced cost of ownership	The principal value proposition of the lead supplier is an opportunity to access ongoing support budgets for the Tornado aircraft. No new manned air defence platforms in the foreseeable future Considerable opportunity for stable profit in a growing world market
Customer factors	Pressure on the customer to downsize, outsource and focus on core competence; changing definitions of customer value—availability, capability, operability, affordability	Cost reductions were a major driver for the ATTAC customer. The ATTAC contract offered savings of at least £510 million over the first 10 years. Savings arose from reductions in RAF and civilian personnel and the improvements a commercial organisation was expected to bring to the task MoD is also changing its definition of customer value—emphasising availability and capability for future contracts. There is a willingness from MoD to transfer support of defence assets to industry
Competitive factors	Services difficult to imitate; source of competitive advantage and differentiation; can help create barriers to entry	BAE Systems may have also recognised that providing through-life services in the Defence sector would be difficult to imitate and provide competitive advantage
Marketing factors	Superior service increases first time and repeat sales; enhances market share; and supports maintenance of customer relationships	The successful implementation of the ATTAC programme may lead to further contracts, enhancing market share; and maintaining customer relationships. The contract is an enabler to maintain capability for upgrades to existing manned air defence assets
Risk reduction or risk transfer factors	Transfer of risk from the customer to the supply chain	The transfer of risk from the customer to supply chain is more complex in the defence sector where there is a need to retain national capability

[a] Key Sources: Mathieu (2001), Oliva and Kallenberg (2003), Gebauer et al. (2005) and Ward and Graves (2007)

Table 2.2 Summary of the challenges faced in the move to service

Individual	Organisational	Inter-organisational
Implementation barriers identified within the research literature		
Management tendency to overemphasise obvious, tangible characteristics over less tangible services potentially leading to under-resourcing services	Organisational culture may be an obstacle requiring a change from a traditional industrial to a service culture	The need to work with other service providers adds greater organisational complexity
Management scepticism that services will generate positive business outcomes	Complex organisational structures can be a barrier to seamless delivery of services	Suppliers and customers may retain traditional adversarial perspectives on the partnership relationships
Risk aversion limits estimation of potential rewards for service operations.	Need to develop information systems to support services which require greater levels of information sharing	The service vision needs to be communicated throughout the inter-organisational network of service providers
Resistance to the changes needed to traditional incentive and performance measurement systems	Support functions such as HR, Purchasing and Finance need to be persuaded to adapt from a product to service orientation	
Resistance to structural changes which threaten organisational units		

2.3.2 *Challenges Faced in the Transition to Service*

The research literature examining the transition to through-life services draws attention to a broad range of challenges to achieving this aim. Such challenges exist at the individual level where managers struggle to adopt the necessary service mindsets and behaviours; at the organisational level where structures, systems, governance and culture may all conflict with service needs; and at the inter-organisational level where partnerships can be difficult to form, traditional adversarial relationships dominate and suppliers or customers may need to be educated in different values and practices. The individual, organisational and inter-organisational barriers to the move from a product to a service orientation are summarised in Table 2.2 and discussed in detail in the following paragraphs. Evidence of the existence of such barriers within the ATTAC case study is assessed and any further obstacles identified.

Beginning at the *individual level*, Gebauer et al. (2005) point out that strong managerial motivation is necessary to support the scale and complexity of change towards servitization and through-life support. However, managers from a traditional manufacturing background have considerable scepticism which must be overcome in order for the move from products to services to generate positive outcomes for their businesses. They argue that such managers have a number of cognitive biases which mean they favour tangible products over rather intangible services processes.

> The overemphasis on obvious and tangible environmental characteristics explains, for example, why managers do not place a high valance (reward) on extending the service business, thus limiting the investment of resources in the service area. Scepticism of the economic potential explains why managers underestimate the probability that their efforts will result in successful performance. Risk aversion limits managerial expectations of estimating accurately that (successful) performance will result in the rewards Gebauer et al. (2005, p. 17).

Further, they suggest that behaviour aligning with such assumptions can become a self-fulfilling prophecy. In other words, managers place low investments of time and effort into service, leading to poor performance. This can then create further scepticism of the economic benefits of the move to a service model and exaggerated perceived risks, thus resulting in further low investment (Gebauer et al. 2005). Within the ATTAC programme, there was also evidence of a degree of scepticism among some individuals within both supplier and customer partners concerning the degree to which the contract would deliver benefits and whether partners would be able to move beyond traditional adversarial relationships. This scepticism seemed, however, to lessen as the process of transition progressed; this will be discussed in the later section on the process of transformation.

In addition to overcoming manager's preferences for working with the tangible, leaders of the transition to through-life support also need to recognise that some of the necessary adaptations to create a service environment will be 'politically sensitive'. For example, moving away from traditional incentive and performance systems can be challenging and disturbing for workers and managers. Mathieu (2001) also recognised such political sensitivities where changes would be seen as a threat to some organisational units and an opportunity for others. MoD efforts to reduce costs through the implementation of the ATTAC contract would no doubt have resulted in the major changes for individuals and units within the client organisations. In such circumstances there would naturally be strong resistance to change among some individuals. For BAE Systems managers, this move to service may likewise have threatened or enhanced 'empires'.

Finally, the ATTAC programme faced a further challenge relating to the individual career plans of those involved in the programme. For BAE Systems the issue faced was recruitment and retention for services on base, within an organisation where the traditional career pathways were in production and linked to 'visibility' at the main headquarter sites.

At the wider *organisational level* there are also a number of potential obstacles to the transition to through-life services. An organisation's culture may also be an obstacle to the move to through-life support. Mathieu (2001) argues that service culture is specific and different from the traditional manufacturing culture and that service firms would also tend to be organised differently from manufacturing firms. She argues that a manufacturing company will benefit from the implementation of a service strategy, only if it is successful in making the transition from a traditional industrial culture to a service culture where service quality is paramount.

The organisation's traditional internal efficiency metrics applied in manufacturing may also be inappropriate for many service activities (Gronroos and Ojasalo 2004). Many respondents within the ATTAC case study highlighted cultural factors as a major challenge in achieving the successful move to service.

In particular, on-base service contract staff noted very different timescales which were adopted by front and back office staff and the difficulties of changing such practices.

Services are often delivered through complex organisational structures which can be a potential barrier to the seamless delivery of services and support to the customer.

> The part of the business providing services or support may be different to the part of the business acting as the interface to customers and suppliers. This results in complex interactions and often conflict and delays between different parts of the organisation, Ward and Graves (2007, p. 474).

Geographical dispersion and a broader sense of fragmentation presented a very real problem for the ATTAC programme. This issue is discussed in the later section on transformation.

Likewise, the development of information systems to support advanced services, and through-life contracts is a key issue facing aerospace companies (Ward and Graves 2007). The information needs are complex and require greater levels of information sharing between the customer and the service provider.

> The move towards through-life management and the development of advanced services, such as equipment health monitoring, requires significantly higher levels of data sharing between customers, manufacturers and service providers. Contracts based on availability targets need detailed reliability, failure and usage data. Support chain management and integrated logistics support also require access to customer inventory data... However, respondents across the participating companies admit that data and knowledge sharing, whether within their own organisations or with customers and suppliers, could be improved (Ward and Graves 2007, p. 474).

Within ATTAC there was recognition that there needed to be significantly greater information sharing and openness. There was also a recognition that reporting mechanisms, while changing, often still reflected more traditional, predominantly product orientation of some of the managers within the business.

As Mathieu (2001) also suggests, support function services like Purchasing, HR and Finance also need to be persuaded to adapt from the embedded mindset and systems of a traditional equipment supply business to understand and respond to service needs. This challenge was also in evidence within the ATTAC case study where changes were required in the culture and practices of support staff from functions based in the corporate main sites towards better supporting the needs of the service contracts. Improving the transparency and visibility of service requirements and back office support processes are thought to be key enablers for this mindset change.

The transition to through-life solution provision often cannot be made by an individual company operating alone (Ward and Graves 2007), which, therefore, brings further *inter-organisational* challenges to adopting a through-life service approach.

> Where partnering is required with other companies within a group, other manufacturers or independent MRO and support providers, some of whom may be competitor organisations, the picture becomes even more complex (Ward and Graves 2007, p. 474).

The need to work in unison with other service partners adds a further layer of inter-organisational complexity to the journey that service companies must make. Some of the Aerospace case examples, presented in their study, were hindered in their move to through-life services by either their suppliers or customers who retained a traditional view of aftermarket activities. Within the ATTAC programme dependencies between partners are a critical issue, where a wide number of stakeholders are involved in service delivery.

Foote et al. (2001) even suggest that changing the basis of relationships with traditional clients may be so difficult that it is new clients without that baggage that should be preferred. For the ATTAC partners there is not an option of changing clients or indeed suppliers until contract review. Both parties have had to deal with historical relationships and stereotypes and forge new relationships over time. There is a recognition, however, that such relationships are largely between individuals. Respondents suggested that there is a need to move from positive individual relationships to broader organisational relationships.

In Table 2.3 the individual, organisational and inter-organisational barriers experienced within ATTAC are summarised and provide a framework that can be used to analyse and communicate relevant barriers to key stakeholders in the transformation.

2.4 Taking an 'Enterprise' Perspective

Complex engineering service systems inevitably involve complex organisational solutions, which go beyond the boundary of single organisations. The brief overview of the ATTAC programme makes clear the complex network of organisational entities involved in value delivery to the MoD customer. The complex multi-organisational nature of the ATTAC enterprise is illustrated and discussed in greater detail in the next chapter on 'enterprise imaging'. Within this and subsequent chapters in this section of the monograph, we have adopted a holistic 'service enterprise' perspective when discussing the ATTAC programme. This perspective views the boundaries of the enterprise as incorporating all inter-dependant parties involved in value delivery rather than a single and separate 'organisational entity' perspective. It is our contention that to understand the delivery of value in complex engineering service systems, the service enterprise needs to be viewed and managed from this holistic enterprise perspective.

Table 2.3 Summary of the challenges identified by the ATTAC service enterprise respondents in the move to service co-creation

Individual	Organisational	Inter-organisational
Implementation barriers		
Getting the right resources: Appropriate skills are scarce and there has been increasing competition for them internally and externally	Organisational culture: A move to a service mentality required a displacement of some traditional mindsets	The need to work with other service providers adds greater organisational complexity
Attracting resources to work on base in a service environment has been difficult	There is a division in understanding and culture between those who have 'lived' the project and those who are in back office positions	Complex relationship between key parties including BAE Systems 'on base' to 'back office', RAF, MoD and across the supply chain
Bases are geographically distant from BAE Systems main offices, adding to the challenge of people moving Resource requirements vary considerably from bid through contract, mobilisation to stabilisation	Improvement is needed in the relationship between Base and BAE Systems 'back office' enabling greater responsiveness and shift in locus of decision making	Greater trust was needed between all parties Partnership needed to be embedded and recognition that the relationships are everyone's responsibility Need to move from positive individual relationships to broader inter-organisational relationships
Leadership and transition management: No obvious single person or entity leads the ATTAC enterprise Senior management had a presence 'on-base' during contracting but have been perceived to have migrated away The transition was poorly understood	Complex organisational structures: ATTAC is difficult to integrate due to the complex process chain involving lots of people and fragmented organisational structure Geographic fragmentation causes a management complexity challenge	Suppliers and customers may retain traditional adversarial perspectives on the partnership relationships There is a lack of holistic service enterprise perspective and shared service vision

(continued)

Table 2.3 (continued)

Individual	Organisational	Inter-organisational
Resistance to structural changes which threaten organisational units and power structures	Co-location is critical for some teams to enable the ATTAC process to function smoothly	The service vision needs to be communicated throughout the inter-organisational network of service
Resistance to learning from the new service approach	The clear separation of the on- and off-base groups creates a 'them' and 'us' mentality	
	Support functions such as HR, Purchasing and Finance need to be persuaded to adapt from a product to service orientation	
	Need to develop information systems to support services which require greater levels of information sharing	
	Traditional product-focused governance must be crafted to better support availability contracts	

2.4.1 What is an 'Enterprise'?

> It is significant when a leader in aerospace or any industry asserts that a given set of activities - regardless of scale - must be viewed as an interconnected whole. That interconnected whole is an *enterprise* Murman et al. (2002, p. 8).

From Murman et al.'s perspective, it would seem that, in terms of distinct boundaries, an enterprise is whatever an enterprise leader says it is. This would seem to be a rather arbitrary boundary definition. However, while basing enterprise boundary definitions on leadership assertions may seem to devalue the concept, it may perhaps also be the concept's greatest strength. Leaders establish a coherent enterprise perspective to encourage the achievement of common goals. As Murman et al. (2002) argue, 'the meaning of enterprise is not always clear. Leaders and others have to assert the interdependence of various stakeholders, and make clear that they are part of a common enterprise' (p. 13). Nightingale (2000) attempts to capture the complex web of inter-related processes and organisations involved in her definition of 'enterprise',

> Enterprises are complex, highly integrated systems comprised of processes, organizations, information and supporting technologies, with multifaceted interdependencies and interrelationships across their boundaries Nightingale (2000).

This definition describes some of the elements of an enterprise and emphasises the necessary dependencies. It does not however, define the essential characteristics of an enterprise.

In the absence of clear definitions, the following defining characteristics are offered for a 'service enterprise'. A 'service enterprise' is an 'organising perspective' which is: (a) proposed by a group of leaders (b) as a means of establishing a holistic approach (c) among a number of disparate enterprise stakeholders (functions, organisations), for the (d) achievement of common complex significant purpose/mission (value creation and delivery, performance improvement), and which (e) cannot be delivered by single entity thinking and practices (interdependence).

This definition also emphasises the role of leadership in promoting a holistic enterprise perspective as a means of establishing coherence among a number of interdependent parties engaged in service delivery. The definition also echoes that commonly given to 'organisations' by organisation theorists who claim that 'organisations arise from activities that individuals cannot perform by themselves or that cannot be performed as efficiently and effectively alone as they can be with the organised efforts of a group' (Hatch and Cunliffe 2006).

2.4.2 The Need for Multi-Organisational Inter-Dependent Enterprises

In the earlier section, the drivers for the move from a product- to a service-oriented approach were discussed. Such a move might be undertaken by an individual company operating largely independently and as indicated, the move would face many challenges. Within complex engineering services such as ATTAC however, value is delivered by many partners operating interdependently and consequently, the move to service is significantly more complex. So why would organisations choose to adopt a multi-organisational, inter-dependant approach to delivering value?

In this section we will examine some of the key factors in the adoption of a multi-organisational enterprise approach to service delivery. Several inter-related factors will be considered including: the trend for organisations to narrow the scope of their activities and outsource all non-core activities; the growing requirement of customers to have holistic solutions; the increasing involvement of customers in the co-creation of value; and finally, the need to work collectively with others in the value chain to both reduce costs and improve performance.

2.4.2.1 Focus on core competence and outsourcing of non-core activities

Individual organisations are increasingly adopting a strategy of identifying and focusing on their core competence. This leads to a narrowing of capability and increased specialisation. A core competence perspective states that companies should differentiate between their competencies as "core", those that are essential to compete in the market and the firm is extremely good at, and "non-core" those that are not essential to compete in their chosen market (Lonsdale and Cox 2000). Only those competencies that are non-core should be performed externally by a third party company.

This trend is now well established in many sectors where prime contractors are increasingly relying on 'full service' suppliers for whole subsystems (Gadde and Jellbo 2002:43), as well as being in charge of managing and designing their own supply chain (Doran 2003).

The objective of this approach is that firms should strengthen and leverage their core competencies (Ellram and Billington 2001) and outsource non-core competencies. This is particularly the case when "the total costs of owning [them] are demonstrably higher than sourcing externally, and the associated risks of market failure or market power are not excessive" (Lonsdale and Cox 2000). The ATTAC programme provides evidence of the application of this principle in the defence sector where industry is being expected to take on roles which are considered 'non-core' for the MoD. However, there are risks within this approach, for instance where the MoD has a need to retain skill and capability for reasons of national security. There are consequently some reservations among ATTAC partners regarding the nature of activities appropriate for outsourcing and the need to maintain skills and capabilities within the armed forces.

2.4.2.2 'Holistic' customer solutions which cannot be delivered by individual companies

While individual organisations have been engaged in narrowing their strategic focus onto certain technologies, services or processes, customers are moving in the opposite direction, increasingly seeking total solutions and services. As highlighted in an earlier section, the MoD is clearly moving towards seeking total solutions for maintenance and upgrade of their various platforms as evidenced in the rise of through-life support and availability contracts (Ministry of Defence 2005). To meet this customer need, organisations seek to offer total, systemic product or service solutions. However, since the trend has been to outsource non-core activities, fewer organisations are able to provide a one-stop solution utilising their own resources alone. This problem has driven organisations in all sectors toward greater inter-organisational collaboration. The strategy deployed is to offer the customer a complete solution and to achieve this integration of each of the elements via close collaboration with a network of specialist external providers (Möller and Halinen 1999). The proposition that collaborative multi-organisational enterprises, rather than single companies, now compete is well supported in the literature (Akkermans et al. 1999; Lawrence 1999; McAfee et al. 2002).

2.4.2.3 Increasing involvement of customers in the 'co-creation of value'

The need to adopt a strategy of operating within integrated and inter-dependent 'service enterprises', is also prompted by the increasing involvement of customers in 'value co-creation'. This shift in view, where customers are part of the value creating enterprise has been highlighted by many writers. Vargo and Lusch (2006, 2008) described the shift as moving from a traditional goods centred or 'product-dominant logic' to an emerging 'service dominant logic'. In the former way of thinking, the customer was seen as the passive recipient of goods. Recent thinking has recognised that the customer is a co-producer or co-creator of value. Prahalad and Ramaswamy (2000, 2003) argued that companies should encourage the customer to be proactively involved in co-creation of value. They describe customers as being 'co-opted' into the design and delivery of services. Indeed, they suggest that the co-creation of value has shifted our ways of thinking about products and services and the boundaries between the provider and customer.

The ATTAC programme represents a complex engineering service which is truly co-created with the customer. For example, the hangar maintenance activities are managed by BAE Systems and resourced by a combination of BAE Systems, RAF and others. Maintenance activities are only made possible through a number of RAF provided services including electrical/hydraulic power and information technology infrastructure for the hangar buildings. The engineering support again provided by both BAE Systems and MoD personnel is necessary to resolve technical queries and safety issues. Finally, the service enterprise includes the

Defence Equipment & Support managed Tornado IPT (Integrated Project Team) containing solely MoD staff covering administration, engineering, logistics, and commercial support of ATTAC on behalf of the *Ministry of Defence*. This organisation is responsible for airworthiness and procurement and monitoring of contract performance. Such complex co-creation of value requires a holistic enterprise perspective to function effectively.

2.4.2.4 Need for collaborative cost reduction and performance improvement

Finally, companies are finding it necessary to engage to adopt a multi-organisational service enterprise perspective in their drive to reduce overall service costs. Globalisation and customer demands for better products and services at lower costs have led companies to seek ways to reduce the overall cost of delivering customer value. This pressure for cost reduction is explicitly built into the ATTAC contract and the enterprise partners are actively seeking to eliminate waste and generate improvements.

While individual company costs might be reduced by a single company approach, the impact of such cost reductions may be detrimental to the whole service system. Therefore, a multi-organisational service enterprise perspective is needed to achieve significant and sustainable cost reduction. Inter-organisational cost management techniques, including the use of target costing, may deliver significant benefits if they are adopted in an integrated manner (Slagmulder 2002). In the same vein, performance improvement in the overall service to customers is similarly only possible from a service enterprise perspective, in order to avoid 'islands of excellence' in an otherwise dissatisfactory service.

The need for a more holistic enterprise perspective and the challenges involved in achieving such an approach were widely recognised by all parties within the ATTAC programme. An enterprise approach involving greater integration would, for example, facilitate faster and more informed decision making than could be achieved in 'old functional or organisational silos'.

Progress has been made within the ATTAC programme towards achieving an enterprise perspective, but many felt that inter-organisational boundaries needed to be removed. Nevertheless, while the need for a holistic service enterprise perspective has been explored, achieving the goal of designing and managing such multi-organisational enterprises remain a significant challenge for all organisations including ATTAC.

There is perhaps a need for more effective communication of a common service vision within the overall ATTAC enterprise. Certainly, Ward and Graves (2007) emphasise the need for systematic and proactive communication of the service vision throughout the internal and external network of service providers. They highlight widely differing experiences of success in achieving such coherent communication. One company described within the study has 'embraced the through-life concept with conviction' and is actively implementing a service strategy externally and internally. This is being achieved through a combination of service dominated website and advertising; communication of their service strategy throughout their internal and external network; a training programme introducing service theory and knowledge; and recruitment policies emphasising service skills. For other companies within their study, the adoption of a service strategy to support through-life management does not appear to be a consistent strategic message. In a context such as the ATTAC programme where the customer plays such a strong role in the co-creation of value, it is difficult to determine whose role it is to develop and communicate the service vision and to lead the necessary transformation to achieve service goals.

In the next chapter of this book on 'enterprise imaging', a methodology is described for visualising and delineating the organisational entities involved in delivering service including the service providers, the customer and the supply chain. Regardless of where boundaries are drawn, it is clear that the move to service will require a significant transformation for all partner organisations. This transformation process is explored in the remaining part of this chapter.

2.5 The Transformation Process

Within the change management literature there have long been attempts to distinguish the levels of magnitude of change with many writers contrasting levels of significance or extent of the desired change and the impacts of such changes. These changes have been variously described as 'realignment' versus 'transformation' (Balogun and Hope-Hailey 2008); 'incremental' versus 'radical' (Baden-Fuller and Stopford 1995); or 'incremental change' versus 'reinvention' (Goss et al. 1998).

Transformation has been described as a 'change which cannot be handled within the existing paradigm and organisational routines: it entails a change in the taken for granted assumptions and the "ways of doing things around here". It is a fundamental change within the organization' (Balogun and Hope-Hailey 2008, p. 20).

Incremental change is not always enough for some businesses that require a more fundamental shift in their capabilities. 'These companies do not need to improve themselves; they need to reinvent themselves. Reinvention is not changing what is, but creating what is not (Goss et al. 1998, p. 85). They go on to suggest that reinvention involves 'altering your context' and that in doing so, this makes it possible for organisations to alter their culture and performance, and to do so sustainably.

In relation to the move from a product to a service business model, Mathieu (2001) suggests that

> the most ambitious service strategies are the ones which provide the manufacturing companies with the greatest benefits … Nevertheless, the most ambitious service strategies

> are also the riskiest because they have to support multiple costs associated with their implementation and challenges deeply embedded assumptions within the business (Mathieu 2001, p. 471).

The ATTAC service enterprise incorporating each of the organisational entities involved in delivering and supporting the programme would appear to be undergoing a significant transformation which has yet to be fully completed. The interview analysis with both industry and MoD partners provided evidence from the participants' perspective on the magnitude of change and the lack of recognition of the scale and complexity of the changes involved. There was also recognition that the service enterprise transformation process was significantly more complex than transforming a single organisation.

A lack of full recognition by leaders of the scale and impact of transformations on the participants seems to be common across all sectors.

> Executives have frequently underestimated the wrenching shift— the internal conflict and soul-searching— that goes hand in hand with a break from the present way of thinking and operating (Goss et al. 1998, p. 86).

There is certainly considerable evidence within the ATTAC case study that participants have found the transformation to a co-created service extremely challenging and that they have been through a cycle of changing reactions to this transformation. The stages through which the ATTAC enterprise has proceeded are presented in an analysis which builds upon the classical emotional change cycle first identified by Kubler-Ross in relation to the grieving process (1973, 2005) and widely adapted for organisational change processes (Scott and Jaffe 1994).

As the now classic change cycle suggests, those involved in change typically go through four phases in their emotional reactions to change—*denial*, where the impact of changes are underestimated; *resistance*, as the consequences of change are recognised; *exploration*, as those involved begin to make sense of the changes and experiment with new ways of working; and finally *commitment*, where the changing patterns are accepted and become embedded in normal practice (Scott and Jaffe 1994).

The findings of the ATTAC case study analysis demonstrated that participants went through a period of *'denial'* in the early bid development stages of ATTAC where the parties involved were unaware of the scale and scope of the transformation to new ways of working and relating.

At this early stage, BAE Systems and the MoD would seem to have been working closely from what might be described as an integrated 'service enterprise' perspective, where they attempted to understand and co-design what might be involved in a partnered process of service delivery. This was demonstrated through such measures as 'co-location' and 'having a joint mantra and joint mission statement'.

However, this holistic perspective may have been at least partially abandoned once the programme was underway in what could be described as the *resistance or 'discovery'* phase when the contract was being implemented. Realities began to be faced, blame allocated and some resistance experienced. Respondents suggested that some individuals were allocated to the service contract whose background was in building aircraft and who did not understand how maintenance differed from their normal context. Such individuals were resistant to the change to service.

During this difficult initial implementation period, the various enterprise partners seem to have reverted to 'single entity' rather than 'service enterprise' thinking, as they began to face the very real challenges of delivering the programme. This sense of fragmentation was evident not only between but within organisations as well. Such divisions were felt between on-base staff who directly managed the delivery of service and supporting functions who seemed not to understand the changing needs and timescales under which the service providers operated.

The perception among ATTAC personnel that they were 'out on a limb' operating in a very isolated way, was further reinforced by a geographical split. On-base personnel were supported by functions based on one of the company's main sites approximately 200 miles away and whom it was felt, did not understand the new service ways of working. It seemed to some respondents that senior corporate management which had a strong presence 'on base' during the initial bidding phase seemed also to have migrated away during this implementation phase. During this challenging period, the need for enterprise cohesion was expressed by many through a desire to be co-located in order to have a better communication and greater visibility of service needs.

The *exploration* phase, also evidenced within the ATTAC interviews, represents a recognition that things need to be done differently and also a greater clarity on the differences between old and new ways of working. It was for example, recognised that organisational pace and the processes needed to support it in design, development, and manufacture 'where lead times are measured in years', was significantly different to maintenance and upgrade, 'where lead times are measured in weeks or hours'. There was also acknowledgment that while the old business models were outdated, there remained significant vestiges of such models in the industry 'hierarchies' and 'power bases'.

At the time of data collection, efforts were being made to make the necessary changes towards new ways of working. There was evidence of a number of improvement activities and efforts to reconfigure the organisational arrangements in order to provide better service. While such efforts to make improvements seem to have been largely single organisation-led, there was, however, evidence of a return to a joint and holistic service enterprise perspective. When the findings of the ATTAC case study were fed back to all the parties involved, there was common agreement among the interviewee respondents and others present that the programme had not yet fully reached the *commitment stage* in the change cycle but that there were significant signs that they were getting there. The commitment stage would involve all service enterprise partners working to deliver and improve the service offered from a holistic enterprise perspective rather than optimising for individual organisational benefit. The ATTAC enterprise partners recognised that joint thinking was required to work as an enterprise both across and within

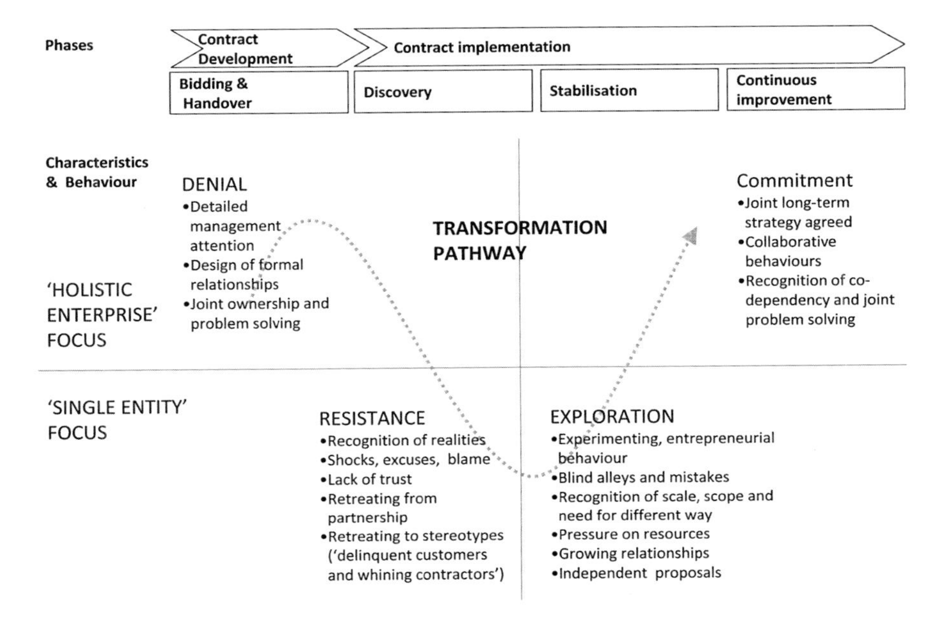

Fig. 2.1 Stages experienced in the ATTAC transformation (adapted from Scott and Jaffe 1994)

organisations, through the involvement of multi-disciplinary and multi-organisational teams rather than viewing issues as an 'industry—' or 'customer-problem'.

There was also clear evidence within the ATTAC enterprise of strong relationships and a common sense of purpose developing. Greater contact between service personnel, who had recent operational experience, and industry partners led to stronger 'connection' between the contract services and their customers' needs. This clear line of sight between service providers and users has helped to incentivise people to perform their best to deliver value to their customers. There was also a developing understanding that cohesion as a holistic service enterprise required time and increased openness to better understand partners motivations and ways of working.

As the ATTAC enterprise began to work more closely together, new questions were raised concerning how service enterprises should be managed, and who should lead such enterprises. Much has yet to be learnt in managing multi-organisational service enterprises.

Stages in the experience of transformation through contract development and implementation are summarised in Fig. 2.1, and characteristics and behaviours encountered at each stage are briefly outlined. As the enterprise partners have gone through the change cycle, they have moved from initial efforts to jointly manage from a holistic 'service enterprise perspective' to 'single entity thinking' while they struggled with the realities of implementation before seeing a return in later stages to the enterprise level management.

This model may be supportive to other complex service enterprises in understanding the cycle experienced in developing co-created services. Further research will clearly be necessary to determine if all enterprise stakeholders face a journey from initial enterprise cohesion in designing services; through a discovery phase where enterprise partnerships may be strained; to an eventual return to cohesion as partners come to understand their inter-dependence in co-creating value.

2.6 Conclusions and Future Work

Complex engineering service systems inevitably involve complex organisational solutions, which go beyond the boundary of single organisations. In this chapter, the need for complex multi-organisational service enterprises has been explored and questions have been raised regarding the challenges of managing such enterprises particularly where customers are heavily involved in value co-creation. Finally, the study has shed some light on stages in the transformation process and the experience of those involved. This study, and the related work presented in the chapters within this section, begins to uncover the significant challenges faced in 'organising' to deliver complex services. There is evidence of the need to clarify the service vision, including the key drivers for the move to service. Organisational leaders likewise must recognise and address the challenges faced in moving from a product-dominant focus to a service ethos. For complex services there are additional challenges in the need for leadership across a multi-organisational enterprise. Traditional relationships between the customer and suppliers must change as all parties recognise their inter-dependence in co-creating value. The transition model discussed highlights the evolving nature of organising to deliver service. Enterprise partners must experience a learning curve in discovering how to work together and deliver service in this new environment.

The challenges inherent in the delivery of complex, co-created services are considerable and require a significant shift in perspective and practices among the partners. This case study represents an early stage in this service journey for BAE Systems and their partners and has provided insights into some of the necessary transformations in ways of working and thinking. The research partnership underpinning the contents of the chapter is ongoing and continues to generate learning for the service science community. Many questions remain, concerning how best to manage the transformation of complex service enterprises. However, the frameworks created may support future service enterprise leaders in identifying and communicating to all stakeholders, the key drivers for the transition to through-life support services and assessing key barriers which may be faced during their own transformation.

2.7 Chapter Summary Questions

The chapter has highlighted the challenges faced by multi-organisational service enterprises in managing the transition from a product- to service-based business model for traditional manufacturing partners, their suppliers and customers.

- To what extent are the service enterprise partners clear about the driving forces and motivations for the move to service?
- What are the key individual, organisational and inter-organisational challenges faced in the transition to service?
- How can service enterprise partners be supported through the transition curve?

Acknowledgments The authors would like to acknowledge and thank their sponsors within BAE Systems, Louise Wallwork and Jenny Cridland, and John Barrie for their support and direction. We would also like to thank the ATTAC enterprise industry and MoD partners for their engagement with our work. This research was supported by BAE Systems and the UK Engineering and Physical Sciences Research Council via the S4T project.

References

H. Akkermans, P. Bogerd, B. Vos, Virtuous and vicious cycles on the road towards international supply chain management. Int. J. Oper. Prod. Manag. **19**(5/6), 565–581 (1999)

C. Baden-Fuller, J. Stopford, *Rejuvenating the mature business* (Routledge, London, 1995)

J. Balogun, V. Hope-Hailey, *Exploring strategic change*, 3rd edn. (Prentice Hall, London, 2008)

A. Bryman, *Social research methods*, 3rd edn. (Oxford University Press, Oxford, 2008)

D. Doran, Supply chain implications of modularization. Int. J. Oper. Prod. Manag. **23**(3), 316–326 (2003)

L. Ellram, C. Billington, Purchasing leverage considerations in the outsourcing decision. Eur. J. Purch. Supply. Manag. **7**, 15–27 (2001)

N.W. Foote, J. Galbraith, Q. Hope, D. Miller, Making solutions the answer. McKinsey Q. **3**, 84–93 (2001)

L. Gadde, O. Jellbo, System sourcing—Opportunities and problems. Eur. J. Purch. Supply. Manag. **8**, 43–51 (2002)

H. Gebauer, E. Fleisch, T. Friedli, Overcoming the service paradox in manufacturing companies. Eur. Manag. J. **23**(1), 14–26 (2005)

T. Goss, R. Pascale, A. Athos, The reinvention roller coaster: Risking the present for a powerful future. Harv. Bus. Rev. Change (Harvard Business School Press, Boston, 1998)

C. Gronroos, K. Ojasalo, Service productivity: Towards a conceptualization of the transformation of inputs into economic results in services. J. Bus. Res. **57**, 414–423 (2004)

M. Hatch, J. Cunliffe, *Organization theory: Modern, symbolic, and post-modern perspectives* (Oxford University Press, Oxford, 2006)

E. Kubler-Ross, *On death and dying* (Routledge, New York, 1973)

E. Kubler-Ross, *On grief and grieving: Finding the meaning of grief through the five stages of loss* (Simon & Schuster, New York, 2005)

F.B. Lawrence, Closing the logistics loop: A tutorial. Prod. Invent. Manag. J. **40**(1), 43–51 (1999)

C. Lonsdale, A. Cox, The historical development of outsourcing: The latest fad? Ind. Manag. Data Syst. **100**(9), 444–450 (2000)

H. Mathe, R.D. Shapiro, *Integrating service strategy in the manufacturing company* (Chapman & Hall, London, 1993)

R.B. McAfee, M. Glassman, E.D. Honeycutt Jr, The effects of culture and human resource management policies on supply chain management strategy. J. Bus. Logist. **23**(1), 1–18 (2002)

J. Mills, V. Crute, G. Parry, Enterprise imaging: Visualising the scope and complexity of large scale services. QUIS 11 – The Service Conference, Wolfsburg, Germany 11-14 June 2009

Ministry of Defence (MoD), Defence industrial strategy: Defence white paper, command 6697. Presented to Parliament by the Secretary of State for Defence. UK, December 2005

V. Mathieu, Service strategies within the manufacturing sector: Benefits, costs and partnership. Int J of Serv Ind Manag **12**(5), 451–475 (2001)

K. Moller, A. Halinen, Business relationships and networks: Managerial challenge of the network era. Ind. Mark. Manag. **28**(5), 413–427 (1999)

E. Murman, T. Allen, K. Bozdogan, J. Cutcher-Gershenfeld, H. McManus, D. Nightingale, E. Rebentisch, T. Shields, F. Stahl, M. Walton, J. Warmkessel, S. Weiss, S. Widnall, *Lean enterprise value: Insights from MIT's lean aerospace initiative* (Palgrave, London, 2002)

D.J. Nightingale, Lean enterprises—A systems perspective. MIT engineering systems division internal symposium, pp. 341–358 (Cambridge, MA, USA, 2000)

R. Oliva, R. Kallenberg, Managing the transition from products to services. Int. J. Serv. Ind. Manag. **14**(2), 160–172 (2003)

C.K. Prahalad, V. Ramaswamy, Co-opting customer competence. Harv. Bus. Rev. Jan–Feb (2000), pp. 79–87

C.K. Prahalad, V. Ramaswamy, The new frontier of experience innovation. MIT Sloan Manag. Rev. **44**(4), 12–18 (2003)

C. Scott, D.T. Jaffe, *Managing organisational change* (Kogan, London, 1994)

R. Slagmulder, Managing costs across the supply chain, in *Cost management in supply chains*, ed. by S. Seuring, M. Goldbach (Physica-Verlag, New York, 2002)

S.L. Vargo, R.F. Lusch, Service-dominant logic: What it is, what it is not, what it might be, in *The service-dominant logic of marketing: Dialog, debate and directions*, ed. by R.F. Lusch, S.L. Vargo (ME Sharpe, Armonk, 2006), pp. 43–56

S.L. Vargo, R.F.L. Lusch, Service-dominant logic: Continuing the evolution. J. Acad. Mark. Sci. **36**(1), 1–10 (2008)

Y. Ward, A. Graves, Through-life management: The provision of total customer solutions in the aerospace industry. Int. J. Serv. Technol. Manag. **8**(6), 455–477 (2007)

R. Wise, P. Baumgartner, Go downstream: The new profit imperative in manufacturing. Harv. Bus. Rev. Sept–Oct: (1999), pp. 133–141

R.K. Yin, *Case study research: Design and methods*, 3rd edn. (Thousand Oaks, Sage, 2003)

Chapter 3
Enterprise Imaging: Visualising the Scope and Dependencies of Complex Service Enterprises

John Mills, Glenn Parry and Valerie Purchase

Abstract This chapter develops a two dimensional "Enterprise Image" capable of assisting independent provider and client stakeholders to take a holistic perspective of their roles in complex support enterprises. Though the case study describes a complex, through-life, availability contract in the Defence sector (Empirical data are taken from ATTAC (Availability Transformation: Tornado Aircraft Contracts), a long term, whole-aircraft availability contract where BAE Systems take prime responsibility to provide Tornado aircraft with depth support and upgrades, delivering defined levels of available aircraft, spares and technical support at a target cost.), the image is believed to be applicable in other sectors. This is particularly the case in public sector to private sector contracts where both primary providers and clients have multiple aims and further independent organisations provide key inputs necessary for successful outcomes. In this environment the prime providers may manage combinations of their own and client staff at the client's premises using facilities provided by the client. Thus the provider may be directly dependent on actions by the client to fulfil the contract. This research contributes to provide a novel and structured mapping that illustrates the interfaces and dependencies that can emerge from complex, multi-organisational, contracts, provide a shared basis for co-operative discussion between stakeholders and thus raise opportunities for new resource configurations and integration mechanisms that create further value for clients and providers.

3.1 Introduction

In this chapter we take a lead from earlier chapters of this book in describing initial methods for supporting a common and shared holistic understanding of complex, support contracts, particularly applicable to public sector clients. While the need

J. Mills (✉) · G. Parry · V. Purchase
Institute for Manufacturing, University of Cambridge, Cambridge, UK
e-mail: jfm@eng.cam.ac.uk

I. Ng et al. (eds.), *Complex Engineering Service Systems*, Decision Engineering,
DOI: 10.1007/978-0-85729-189-9_3,

for management methods that take a holistic, multi-organisational perspective has been well rehearsed in notions like the "extended enterprise" (Dyer 2000)—sets of firms that collaborate to produce a product, or "virtual organisations" (Ahuya and Corley 1999)—and more recently in the previous chapter of this monograph, there are major obstacles:

> the mindset of many managers favours individual unit thinking over cross- functional and cross-firm thinking and performance measures which emphasize individual business success rather than supply chain success Spekman and Davies (2004)

This mindset has been common amongst managers within large, single organisations so the challenge for sets of independent organisations to co-ordinate towards "enterprise goals" is likely to be extremely challenging notwithstanding researchers' advocacy of a holistic, integrated view of supply chains regardless of organisational boundaries. Indeed, there are few suggestions as to how a reconfigured/realigned enterprise working together to co-produce and co-create value might be coordinated; how their decision processes might work; where the locus of control might lie; or how strategy might be deployed. Also unclear is what form such *reconfigured relationships* or *realigned management activities* might take; how they would come about; how they would function; and the dynamics of their operation. As a result, the transformation path from company-centric to collaborative enterprise functioning is also yet to be described or defined theoretically.

Within the Core Integrative Framework (CIF) discussed in the Introduction chapter, the need for management processes at a holistic, enterprise level that encourages co-creation of improved value between providers and clients is clear. This research takes that enterprise perspective, acknowledging both client and provider roles in specifying and designing management systems, organisations and attitudes that enable value co-creation. The clients for this research are inevitably senior managers in client and provider organisations because managing complex through-life service at an enterprise level will require methods and behaviours that will lie outside the rules, methods and assumptions of normal single organisation business practice. Developing approaches and solutions to these new requirements and business models are tasks only they can authorise.

A logical first step towards an enterprise perspective is to establish a shared understanding of the boundaries of the enterprise and to identify the requirements for enterprise level management processes between the major provider and client constituencies. Such a shared understanding would form the basis for identifying the interests and value propositions of the enterprise as a whole and its constituent organisations. This chapter sets out to develop an empirically derived visualisation of a complex through-life support enterprise. The research is a first step in developing a generic visualisation capable of providing an improved and shared understanding by clearly communicating the interfaces, leadership and organisation challenges that arise in multi-organisational service enterprises.

The requirement for such a visualisation emerged from a case study of the ATTAC contract focused on "organisational transformation" where the researchers, despite many interviews within the client and provider communities, failed to find a comprehensive representation of the ATTAC service delivery organisation. Further, a convincing description of how the contract was managed at an Enterprise level, above the major stakeholders managing their own activities, could not be found.

Simply speaking, the researchers could not answer the questions:

- What organisation needed transformation?
- Who was managing or will manage the Enterprise level of that transformation?

Local optima were thus likely to be developed that were not guaranteed to be supportive of the ultimate client requirement—a requirement that the researchers imagined to be concerned with the defence of the nation rather than the lower level yet necessary objectives of the contract—lower costs and availability of spares and aircraft for the client, alongside a reasonable profit for providers.

The chapter is organised into five sections presenting:

1. A review of research on the use of pictures to support the development of multi-organisational enterprise level understanding of the interactions between providers and clients, leading to a structured pictorial framework on which to place the organisations involved.
2. Case research methodology.
3. Case study analysis of the ATTAC support contract leading to the identification of six organisation types involved in that enterprise and their placement on the framework. This included sub organisations within both the prime service provider and the client as well as key third party organisations.
4. A critique of the visualisation, the questions it has raised and the benefits that accrued.
5. A discussion of opportunities for future research, particularly the development of a structured tool to enable joint stakeholder construction of their Enterprise Image.

3.2 Literature Review

This review addresses two areas of existing research:

- The use of pictures to aid the understanding of organisational phenomena
- Representations of multiple organisations in general and specifically representations of the interactions between service providers and their clients

3.2.1 Pictorial Representations of Organisational Phenomena

Meyer (1991) advocated visual approaches for collecting and representing the "fuzzy multi-dimensional constructs met when analyzing organisations" and the notion of multi-organisational enterprise is certainly one such construct. Research on the use of "boundary objects[1]" (Star and Griesemer 1989) to share knowledge across functional and divisional boundaries was extended by Carlile (2002) to show how, through their use, actors from various areas of knowledge and expertise can create a shared understanding of phenomena. In our case a pictorial representation of a complex multi-organisational enterprise may form a boundary object for actors involved in the organisations concerned. Further, from a practical perspective, there are other benefits to using a picture:

> Structured, pictorial approaches can enable data gathering and representation to be combined; can assist data analysis and therefore may enable a representation to be built and analysed in the time managers will make available Mills et al. (1998).

It is important to understand that the aim here is neither to produce a method that creates agreement on a picture/boundary object nor to produce a picture showing the flow of decisions, services, parts or repairs. The stakeholders have particular individual interests and roles that bias their perspective. The most that can be expected is a shared understanding of how each stakeholder *sees their world* and understands how other stakeholders *see theirs*. It is from the acknowledgement of these differences of perspective and discussion around them that a shared understanding may be formed.

Bresciani et al. (2008) have developed a "collaborative dimensions framework" which can be used to assess a visual artefact's effectiveness or otherwise in promoting "collaboration in circumstances that involve distributed knowledge". They identify seven dimensions based on the work of Green (1989) and Hundhausen (2005):

1. Visual impact	The extent to which the diagram is attractive and is facilitating attention and recall
2. Clarity	The ability of the diagram to be self-explanatory and easily understandable with reduced cognitive effort
3. Perceived finishedness	The extent to which the visualisation resembles a final, polished product
4. Directed focus	The extent to which the diagram draws attention to one or more items
5. Inference support	The extent to which new insights are generated as a result of the constraints of the visualisation form
6. Modifiability	The degree to which the items in the visualisation can be altered in response to the dynamics of discussion
7. Discourse management	Control over the discussion and work flow

[1] Artefacts, documents and perhaps even vocabulary that can help people from different communities build a shared understanding. Boundary objects will be interpreted differently by different communities.

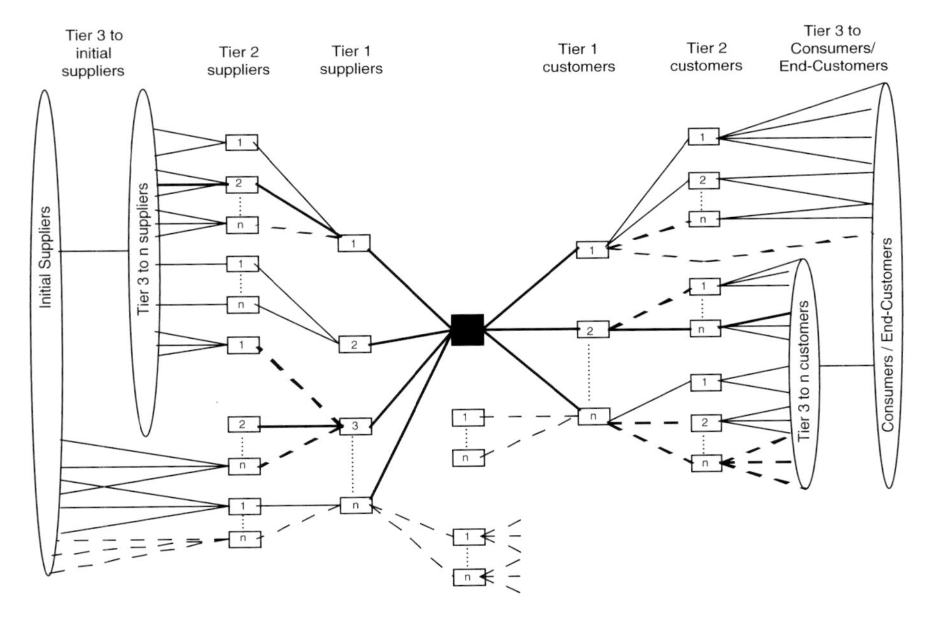

Fig. 3.1 Types of inter-company business process links—whole network perspective (from Lambert et al. 1998)

These dimensions have been used as design criteria in this research; note here that the dimensions do not solely refer to the structure and shape of the finished visualisation but to the medium used and the methods used to create it.

3.2.2 Representations of Multiple Organisations

Typically, representations of multiple organisations have been structured around concepts like supply chain, supply network, or value chain. They show flows of components, products and/or services, usually from left to right, taking a holistic view of an organisation—one box per company or company location, see Fig. 3.1. They do not, therefore, acknowledge that many organisations, especially those that are large, complex and hierarchic, do not have processes that fully integrate the behaviour of their sub-parts. This implies that a range of client and provider sub-organisations on which front line service delivery organisations depend upon must be represented.

Many methodologies have been developed to support representation of the more detailed flow of products, information and services, for example service blueprinting, BPMN,[2] COMET,[3] or IDEF.[4] These representations and methods

[2] http://www.bpmn.org/

[3] http://www.modelbased.net/comet/index.html

[4] http://www.idef.com/

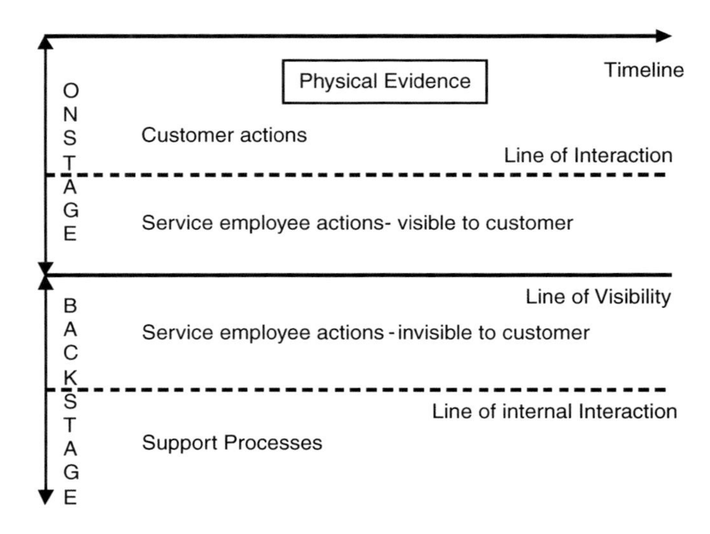

Fig. 3.2 Key elements of a service blueprint (adapted from Zeithaml et al. 2009)

may generate output that is overly complex and difficult to interpret. Our focus is on supplying senior management with a holistic appreciation of the enterprise of which they are part and that can represent:

- The scope of the enterprise
- Dependencies within and between their organisations
- Enterprise management issues

Why senior managers? Because attempting to manage complex through-life service at an enterprise level will require methods and behaviours that will lie outside the rules, methods and assumptions of normal single organisation business practice.

The most common service-specific representation is 'service blueprinting', a potentially useful representation of the interfaces between service providers and clients. There are a number of variants of service blueprinting described in the literature, for example, Lovelock and Wirtz (2004), Kingman-Brundage (1989, 1993) and Shostack (1984). They are all representations of the interactions between the client and manifestations of the service provider over time. A typical representation used for service blueprinting is shown in Fig. 3.2.

The main applications of this visualisation have been in the consumer markets of banking, retail and hospitality and have tended to focus on single value-adding processes, rather than on the value-adding enterprise itself. Clients interact with the provider face-to-face (within the line of visibility—Fig. 3.1), or remotely perhaps by telephone or the Internet. A useful concept from blueprinting is the notion of "backstage" and "onstage"—separate but coordinated sub-organisations within the provider organisation. While consumers do not possess a back office, complex businesses and public sector clients do. An important limitation of the

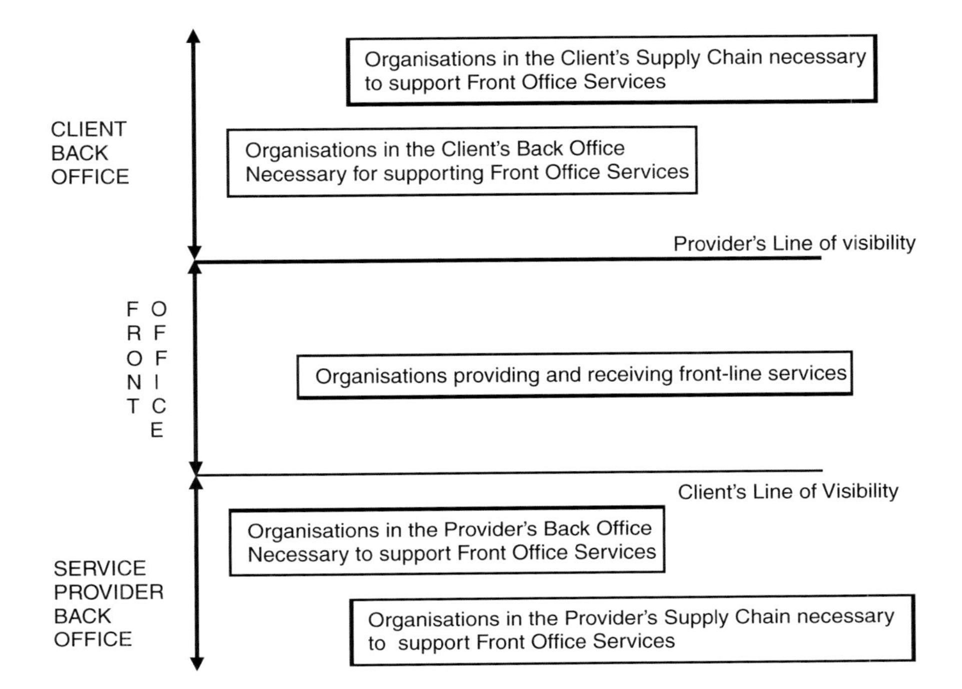

Fig. 3.3 Generic enterprise image

technique noted by Fließ and Kleinaltenkamp (2004) is its inability to capture dependencies between actions by the client and service provider. This is an important restriction since the client's role in obtaining effective and efficient service outcomes has long been known to be critical (Mills et al. 1983; Mills and Morris 1986). More recently, the term "co-creation of value" has been widely applied in the service literature (Vargo and Lusch 2006, 2008) and this research aims to increase awareness of value co-creation in client, provider, and third party organisations providing critical inputs. The importance of taking a sub-organisational perspective on co-creation is amplified by increasing complexity in the service task and when client and service provider are complex and hierarchical.

After some experimentation the key elements of an Enterprise Image were developed (see Fig. 3.3). The image differs from service blueprints in three important respects:

1. Both client and provider have back-offices.[5]
2. The time dimension is omitted since our aim is to represent the key contributors to service outcomes—not the detailed processes involved.
3. For simplicity, the line of internal interaction has been removed, again because we are taking an organisational rather than a process view.

[5] "Office" was preferred to "Stage" by this case community.

Note that the "front-office" represents the prime location of value co-creation, wherever, geographically, that happens to be.

3.3 Case Study Research Methods

An in-depth case study of the ATTAC programme enabled researchers to examine and interpret phenomena in situ and to understand the meanings actors bring to such phenomena. Case study research is also useful when the aim of the research is also to answer 'how' and 'why' questions (Yin 2003). This matches the wider aims of this research, to gain an understanding of how and why such complex service provision contracts actually materialise in practice. Though our overall focus was on understanding the obstacles and enablers to effectively implementing the service provision contract, our focus in this chapter is on the service enterprise issues raised and how these might be better and more widely understood.

The particular case study was chosen for two main reasons—it was the first of its scale and complexity between the Provider and the Client. Though both parties intended to continue to let and bid for such contracts, this first attempt was an opportunity for both parties to learn and this enabled the researchers to interview widely—six client and 22 provider interviews were conducted. These informants were classified into three groups—those involved in the design and implementation of the contract; those who supported implementation; and those who viewed the contract from a distance.

The interviews were conducted in 2008, were semi-structured, face-to-face, and took an average of 1.5 h. Interviews enable researchers to uncover how informants perceive and interpret situations and events (Bryman 2008). Themes covered were the scope of the contract; their role in implementation and the obstacles and enablers they met; their perceptions of issues in other areas of implementation; and the management structures and processes used. The case study analysis presented in the chapter has been rigorously validated through a series of presentations to key customer and provider contract and support functions. Written reports have also been made available for validation and feedback (Yin 2003).

The initial interviews used a diagram from the UK National Audit office report on Fast Jet Logistics Transformation as a starting point. This diagram focused on the hangar and supply chain activities but omitted much of the surrounding dependencies on the availability contract. A key question centred on what other parts of the interviewee's organisation and/or other organisations influenced the successful delivery of the ATTAC programme. Further organisational elements key to delivering ATTAC were thus identified and later ascribed 'front office or back office' status. These elements were raised by a series of respondents from relevant organisations—BAE Systems, Air Command and Defence Equipment & Support (DE&S). The initial focus in building the image was on volume, collecting descriptors for a wide range of organisational elements within each partner organisation and in the wider supply chain. Following initial outline development,

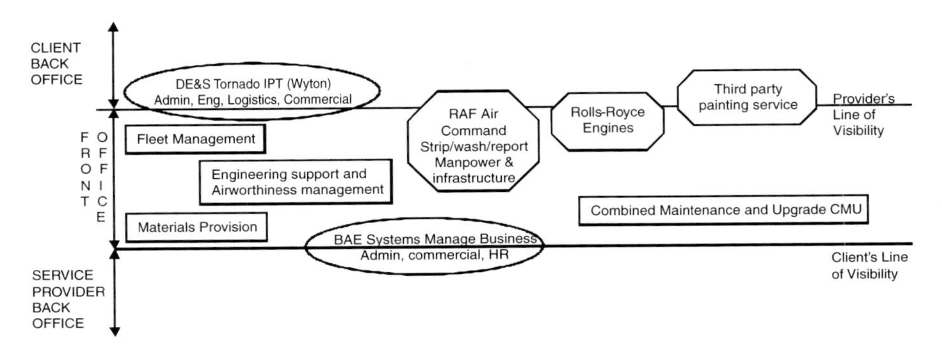

Fig. 3.4 Enterprise image—ATTAC showing partnered and independent direct service delivery organisations plus specific contract focused organisations

the Enterprise Image was presented to all interviewee constituencies to assess the validity and utility of the image.

3.4 Case Analysis

Since the case is complex it is inevitable that this section contains substantial details; however, the key outputs from this section are six defined organisation categories (Fig. 3.4). The six organisation categories are based on their location, reporting lines and role in providing the overall service.

3.4.1 Partnered Direct Service Delivery Organisations (Key: Rectangle)

These organisations are managed by the prime Service Provider, located where the operational services are delivered and staffed by Provider and Client staff. These organisations are fully visible to the Client and Provider and are placed firmly in the front office. In this case there are four such organisations:

- Combined Maintenance and Upgrade (CMU): Covering the main hangar activities that carry out depth maintenance tasks that result in aircraft with increased available flying hours.
- Fleet management: Providing the strategic and operational planning activities that translate the Forward Squadron requirements for Tornadoes of particular configurations into the schedule of aircraft through CMU.
- Materials provision: Covering spare parts and repair requirements planning and expediting to supply CMU and Forward squadrons.

- Engineering Support and Airworthiness management: Mainly located at RAF Marham, but represented at two other remote RAF bases in the UK and a forward base in Afghanistan. This activity resolves technical queries and safety issues.

3.4.2 Independent Direct Service Delivery Organisations (Key: Octagon)

These organisations are not managed by or are responsible to the prime Service Provider, but provide significant inputs to the support provision task of performing aircraft maintenance. They are located in positions that overlap front and back offices, corresponding to the degree to which they are visible to main clients and providers.

In ATTAC there are three main Independent Direct Service Organisations:

- Rolls-Royce, who manage the repair and overhaul of Tornado engines via the RB199 Operational Contract for Engine Transformation, otherwise known as ROCET, a contract between Rolls-Royce and the MOD
- RAF Air Command, who retain management of several key areas of depth maintenance:
  - The wash process plus all work connected with ejector seats, weapons and pylons.
  - Responsibility for the hangars themselves—the supply of electric/hydraulic power etc. and Information Technology infrastructure.
  - The supply of technicians, engineers and management personnel to the Partnered Direct Service Delivery Organisations
- SERCO, another third party company providing a painting service, one of the later, inline processes in the delivery of maintained aircraft and, therefore, a significant dependency.

3.4.3 Specific Contract Focused Organisations (Key: Ellipse)

These organisations are managed by the Prime Service provider or the Client and are focused solely on the focal contract. They are located in positions that overlap front and back offices corresponding to the degree to which they are visible to main clients and providers.

In this case there are two Specific Contract Focused Organisations:

- The "Manage Business" organisation is controlled by BAE Systems and operates on-base. It covers the commercial, financial, and Human Resource needs of the contract and operates principally in the front office, handling new contractual requirements, the acquisition of skilled industrial staff and technicians.

- The Tornado IPT (Integrated Project Team) is controlled by the MOD via DE & S and contains staff covering financial, engineering, logistics and commercial support of ATTAC. It is located at RAF Wyton—at the time the contract was signed the wide expectation was that much of this organisation would move to RAF Marham.

3.4.4 Back Office Support Organisations (Key: Parallelogram)

These organisations, within the Prime Service Provider or the Client, are focused more broadly than on the focal contract. Their support, however, is vital in providing the overall service effectively and is critical to service improvement. Typically they are located in "back offices" within the Client or the Prime Service Provider.

In this case we have identified four such back office organisations in the MOD and three in BAE Systems. They are placed in the appropriate back offices:

- Defence Estates manage the MOD's estate as a whole, providing advice and services on all property matters. Any change to structures on-base need support from this organisation
- Defence Storage and Distribution Agency (DSDA) is the sole provider of transport and off-base storage of Tornado parts
- Twenty Equipment/Commodity IPTs are dealt with here as a single class of organisation. The Tornado IPT has Service Level Agreements with these organisations for a wide range of equipment (e.g., ejector seats, munitions) that, in their centralised role, they provide to a range of other defence platforms
- Human Resources supervise HR plans and thus influence the supply of Engineering and supervisory RAF staff to the partnered organisations
- BAE Systems Central Purchasing at Salmesbury, Lancashire
- BAE Systems' Tornado Engineering support at Warton, Lancashire providing in-depth technical back-up
- BAE Systems' Human Resources at Warton, supplying appropriate management resources and oversight of human resource development

3.4.5 Key Supply Chain Organisations (Key: Rhombus)

These third party suppliers are key to the provision of services not already included. They are suppliers to either the Prime Support organisation, Client or both. The case study shows supply may consist of tangible goods, human resources, advice or opportunities to reduce costs. These are placed in the appropriate back office(s).

In this case we identified a key supply chain agency used by both Client and Provider and several key suppliers to BAE Systems. Only two are shown in Fig. 3.5:

- Panavia, based in Germany, is an organisation jointly owned by Alenia, BAE Systems and EADS. It arose from the extensive and complex supply chain for parts, repairs and engineering support resulting from design and manufacture work shares agreed between the UK, Germany and Italy who jointly funded Tornado. Designed more for manufacture than support, it remains an important agency for exchanging technical information and bulking up orders for spares from the fleet of over 900 Tornado aircraft built. Both the Tornado IPT and BAE Systems interact with Panavia over technical and supply matters.
- Morsons is the key supplier of contract technicians into the CMU.
- XChanging is an important supplier of outsourced HR services, particularly recruitment.

3.4.6 Governance Organisations (Key: Triangle)

These are functional organisations that determine how the rest of the organisations operate—for example their reporting rules, performance indicators, levels of authority, mandatory processes and policies. They are placed in the appropriate back office.

BAE Systems has policies set at corporate level within its strong functional structure and the MOD has Civil Service rules to work to.

3.5 Critique and Discussion of the Validity and Utility of the Image

When fed back to interviewees from all constituencies the Enterprise Image was revealed, organisation type by organisation type using Powerpoint.[6] Each organisation type was colour-coded as well as represented by a shape. This step-by-step building of the Enterprise Image was important to aid visual impact and clarity—two of the criteria suggested by Bresciani et al. (2008).

There were several feedback sessions, firstly to the BAE Systems senior executive of the ATTAC programme with related, Warton-based Human Resource Executives. This led to several identical presentations to the BAE Systems on-base management team with members of the RAF Air Command. This was followed by a presentation to a joint group of the two most experienced and senior representatives of the DE&S Tornado IPT and senior members of the BAE Systems on-base management team.

[6] Microsoft Corporation

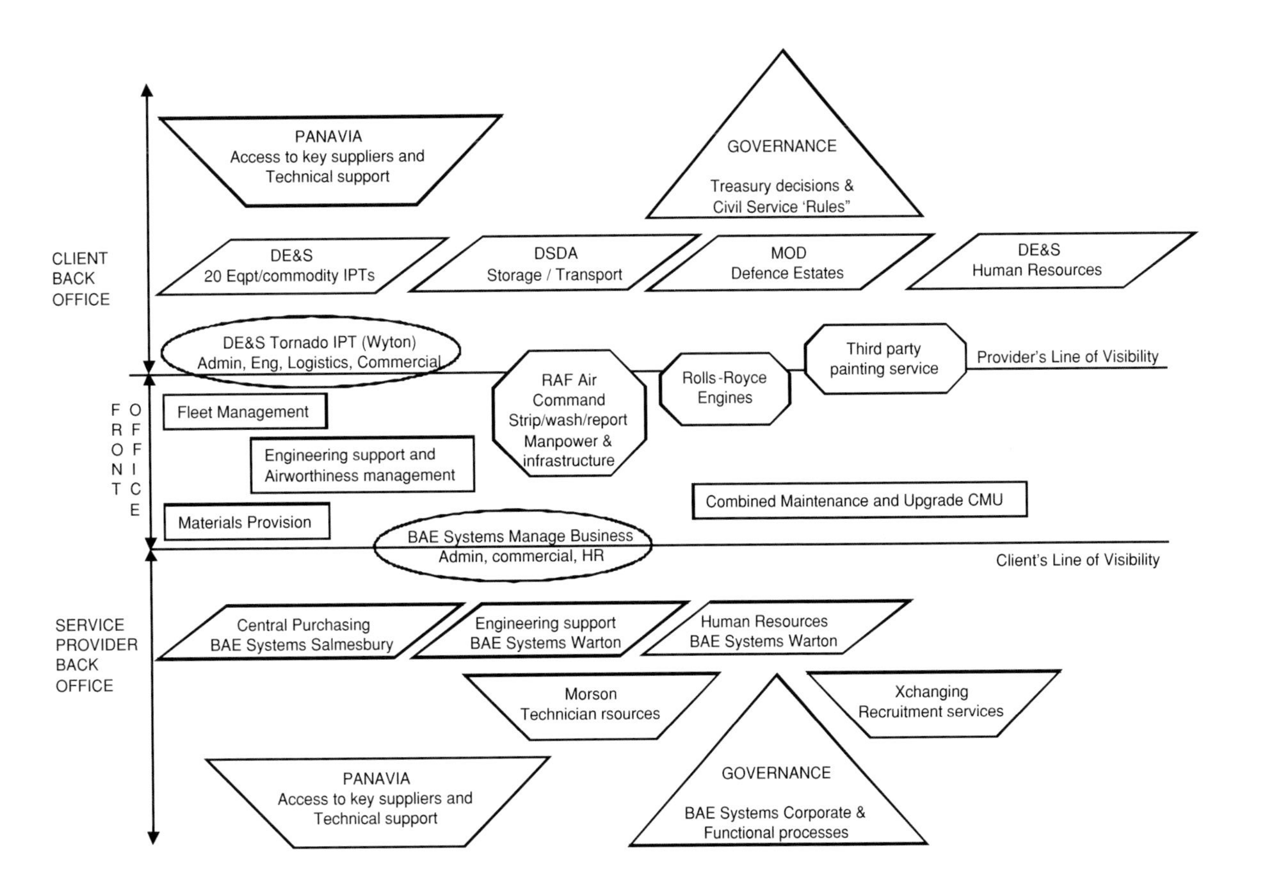

Fig. 3.5 Enterprise image ATTAC case study

The reactions of those involved in the front office could be described as "Yes—that's what we face! All those organisations need to adapt to this new situation." The Enterprise Image, drawn by researchers, appeared to be a valid representation of the ATTAC Enterprise with minimal suggestions for change. When the researchers shared their conclusion that little or no Enterprise-level management had been found, both Client and Provider agreed that each organisation managed the ATTAC contract from their own perspective and this was not ideal.[7]

For those more peripherally involved in back offices the Image appeared to provide, for the first time, an appreciation of the scope and complexity of the interdependencies within the ATTAC Enterprise. There was genuine surprise at their extent and the questions it raised for all were "How might this be managed more effectively?" and "What are the key areas for Enterprise management?" It was appreciated that Enterprise management raised major questions like "Who would lead?" and "How could such processes be managed in a fair manner?" The management of Service Improvement came to the fore. Though the need for ongoing cost reduction and service improvement led by the main Provider was clear in terms of performance improvement metrics, it was not made explicit in terms of the duties of all partners, including the Client. All partners need to be prepared to change methods, invest in training and other implementation aspects of cost reduction and service improvement if Enterprise improvement was to be achieved. This was clearly an important aspect of Enterprise management. Another major insight concerned the Image's ability to convey the dual challenges of on-base, operational enterprise level organisational design and management, and the need for supporting changes in the operation and structuring of parent organisations' back offices. The need for structural change arose from displaying four Enterprise Images representing all the fast jet support contracts in play and planned: Tornado, Harrier, Typhoon and the Joint Strike Fighter fast jet platforms—all platforms would require support from client and provider back-offices.

Thus far the Enterprise Image appears to work well in forming a picture these Providers and Clients can recognise and promotes discussion on the need for Enterprise management in both front and back offices. However, these preliminary conclusions have not been arrived at scientifically and further research is required.

3.6 Further Research

Our preliminary findings have provided encouragement for further research to refine the categories defined in the Tornado case and explore the Image's ability to encourage "Enterprise management" understanding and development in other

[7] A meeting 2 weeks earlier between BAE Systems, Rolls-Royce, DE&S and senior Air Command was mentioned—it had shown that an Enterprise perspective was required and useful and both BAE Systems and IPT leaders hoped to continue that dialogue.

sectors involving public sector clients and/or complex multi-organisational networks delivering ongoing service provision. The fact that the Enterprise Image was built by researchers enabled an initial classification of organisations to be defined, but such Images are then 'owned' by researchers rather than stakeholders. Further case studies may uncover further organisation types and/or an improved classification. However, we expect much of the practical utility of these Images to arise from Enterprise stakeholders who co-create their own Image. During that process, the position of the sub-organisations would be debated. A particular focus might fall on the degree to which the operation of the provider and/or client sub-organisations was visible and supportive to the aims of the Enterprise. It is, perhaps, around differences in perception on these matters that shared understanding may be built. While that can be attempted using the framework and categories in this paper, a more structured, facilitated tool used by principal Client and Provider stakeholders could be developed.

Research to refine a tool would use the perceptions of practitioners to evaluate the Image's efficacy in promoting collaboration based on Bresciani et al.'s (2008) "Collaborative Dimensions Framework" and the tool itself would be assessed using the "Feasibility, Usability and Utility" dimensions proposed by Platts (1993).

The purpose of the Enterprise Image and its creation process is to:

- help organisational partners understand the boundaries and interdependencies of the service enterprise;
- encourage them to take an enterprise perspective;
- operate as an initiator for discussion;
- provide a common reference point for multiple interpretations;
- and provide a basis for co-creating holistic Enterprise management processes.

Care is, however, needed in developing such visualisations, for while pictorial approaches may be attractive many pictures of the same phenomenon could be drawn. Each picture due to its size, shape and the methods used to create it will emphasise different aspects, reflect the perspective of the artist(s) and even be affected by the materials used. What are the potential biases in a particular picture of a multi-organisational enterprise? Should the image be put on its side with neither client nor provider at the top? Would the image have other useful contributions if drawn before the contract is signed? These reflections are of both academic and practical interest and will play a part in further research.

3.7 Chapter Summary Questions

The chapter has highlighted the challenges faced by multi-organisational service enterprises in gaining a shared understanding of the scope of their operation, shared aims, and the need for enterprise-level management processes. We have found that the Enterprise Image assists this process by providing an 'as-is' visualisation of the Enterprise but also it raises further questions:

- To what extent can an Enterprise Image be used to inform future scenarios where further services are incorporated and/or the service contract moves a level higher, for example the provider is asked to provide support for an airfield and the multiple aircraft based there?
- Can the Image be extended to show the "front offices" between providers and between providers and client constituencies?
- To what extent can such visualisations enable increased understanding and awareness of the need for Enterprise-level management processes and be converted into action? What further understanding is necessary?

Acknowledgements The authors would like to acknowledge and thank their sponsors within BAE Systems, Louise Wallwork and Jenny Cridland, and especially John Barrie for his support and direction. We would also like to thank the Tornado IPT, RAF Marham and the BAE Systems ATTAC management team for their engagement with our work. This research was supported by BAE Systems and the UK Engineering and Physical Sciences Research Council via the S4T project.

References

M.K. Ahuja, K.M. Corley, Network structure in virtual organizations. Org. Sci. **10**(6), 741–757 (1999)

S. Bresciani, A.F. Blackwell, M. Eppler, A collaborative dimensions framework: Understanding the mediating role of conceptual visualizations in collaborative knowledge work. Proceedings of 41st Hawaii International Conference on System Sciences (HICCS 08): (2008), pp. 180–189

A. Bryman, *Social research methods*, 3rd edn. (Oxford University Press, Oxford, 2008)

P. Carlile, A pragmatic view of knowledge and boundaries: Boundary objects in new product development. Org. Sci. **13**(4), 442–455 (2002)

J.H. Dyer, *Collaborative advantage: Winning through extended enterprise supplier networks* (Oxford University Press, Oxford, 2000)

S. Fleiß, M. Kleinaltenkamp, Blueprinting the service company: managing service processes efficiently. J. Bus. Res. **57**, 392–404 (2004)

T.R.G. Green, in *People and Computers V*, ed. by A. Sutcliffe, L. Macaulay. *Cognitive dimensions of notations* (CUP, 1989)

C. Hundhausen, Using end user visualization environments to mediate conversations: A 'Communicative Dimensions' frame-work. J. Vis. Lang. Comp. **16**(3), 153–185 (2005)

J. Kingman-Brundage, The ABC's of service system blueprinting, in *Designing a winning service strategy*, ed. by M.J. Britner, L.A. Cosby (American Marketing Association, Chicago, 1989), pp. 30–43

J. Kingman-Brundage, Service mapping: Gaining a concrete perspective on service system design, in *The service quality handbook*, ed. by E.S. Eberhard, W.F. Christopher (Amacon, New York, 1993), pp. 148–163

D.M. Lambert, M.C. Cooper, J.D, Pagh, Supply chain management: Implementation issues and research opportunities. Int J of Logist Manag **9**(2), 1–19 (1998)

C. Lovelock, J. Wirtz, *Services marketing: People, technology, strategy*, 5th edn. (Prentice-Hall, NJ, 2004)

A.D. Meyer, Visual data in organizational research. Org. Sci. **2**(2), 218–236 (1991)

P.K. Mills, J.H. Morris, Clients as 'partial' employees: role development in client participation. Acad. Manag. Rev. **11**(4), 726–735 (1986)

P.K. Mills, R.B. Chase, N. Margulies, Motivating the client/employee system as a service production strategy. Acad. Manag. Rev. **8**(2), 301–310 (1983)

J.F. Mills, A.D. Neely, K.W. Platts, M.J. Gregory, Manufacturing strategy: A pictorial representation. Int. J. Oper. Prod. Manag. **18**(11), 1067–1085 (1998)

K.W. Platts, A process approach to researching manufacturing strategy. Int. J. Oper. Prod. Manag. **13**(8), 4–17 (1993)

L. Shostack, Designing services that deliver. Harv. Bus. Rev. (Jan-Feb): (1984), pp. 133–139

R.E. Spekman, E.W. Davies, Risky business: Expanding the discussion on risk and the extended enterprise. Int. J. Phys. Distrib. Logist. Manag. **34**(5), 414–443 (2004)

S.L. Star, J.R. Griesemer, Institutional ecology, 'translations' and boundary objects: Amateurs and professionals in Berkeley's Museum of Vertebrate Zoology, 1907–39. Soc. Stud. Sci. **19**(3), 387–420 (1989)

S.L. Vargo, R.F. Lusch, Service-dominant logic: What it is, what it is not, what it might be, in *The service-dominant logic of marketing: Dialog, debate and directions*, ed. by R.F. Lusch, S.L. Vargo (ME Sharpe, Armonk, 2006), pp. 43–56

S.L. Vargo, R.F.L. Lusch, Service-dominant logic: Continuing the evolution. J. Acad. Mark. Sci. **36**(1), 1–10 (2008)

R.K. Yin, *Case study research: Design and methods*, 3rd edn. (Thousand Oaks, Sage, 2003)

V.A. Zeithamel, M.J. Bitner, D.D. Gremier, *Services Marketing: Integrating customer focus across the firm*, 5th edn. (McGraw-Hill, NY, 2009)

Chapter 4
Complexity Management

Glenn Parry, Valerie Purchase and John Mills

Abstract This chapter explores the nature of complexity that arises in high value contracts between large organisations. To develop a framework, a detailed case study was undertaken to identify the factors that create complexity. The case studied was the availability contract to provide depth maintenance and upgrade on Tornado aircraft between BAE Systems and the MOD. The contract, named ATTAC, is worth ~£1.3bn and the MOD engaged with BAE Systems precisely to enable them to more cost effectively manage the complex enterprise of over 22 organisations or business units that deliver this service. The work explores the operation from a range of perspectives, interviewing managers from across the organisations involved. The factors contributing to complexity are described in context and a framework is presented which clusters them into six key areas. It is proposed that this framework may then be used as a tool for analysis and management of complexity.

4.1 Introduction

Service providers seek to enhance their value proposition and their competitiveness through improving customer's experience (Fitzsimmons and Fitzsimmons 2005). Value is realised not through the resources which create a value proposition, but by the value in use jointly created by the client and provider (Sandström 2008; Prahalad and Ramaswamy 2003). To deliver value in use, customers' experience becomes integral to the offering (Sampson and Froehle 2006; Heineke and Davis 2007). Servitised companies are faced with increased complexity as

G. Parry (✉) · V. Purchase · J. Mills
Bristol Business School, Frenchay Campus, University of the West of England, Bristol, UK
e-mail: glenn.parry@uwe.ac.uk

I. Ng et al. (eds.), *Complex Engineering Service Systems*, Decision Engineering,
DOI: 10.1007/978-0-85729-189-9_4,

Main Complexity factors	Product Groups	Process	Contracting	Organisation	Finance	People
Contributing Complexity factors	Variant	Dependency	Contract completeness	Enterprise boundaries	Financial regulation	Cultures
	Fleet	Recovery	Objectives	Geographic dispersion	Financial reporting	Customer
	Resource	Reporting	Risk	Structures	Cost	Leadership
	Technical	Governance	HR legal	Stakeholders		Line management
	Cross platform	Security	Regulations	Politics		Roles
				Partnering Constraints		Learning
						Transition management
Factor Level	Emergent or Deterministic?					

Fig. 4.1 Framework for complexity factors perceived by enterprise managers

customers move from deriving value from products to value derived from service experience (Kates and Galbraith 2007). The competence to deliver these experiences comes from skilful co-ordination of complex combinations of resources (Vargo and Lusch 2008). Yet complexity is an inhibitor to process improvement and hence the realisation of this value (Bateman and Rich 2003). We propose that recognition and identification of the drivers of complexity within a service enterprise will better enable managers to realise value, reducing complexity where possible.

As a first step, an approach to complexity identification and classification has been developed that forms the focus of the work undertaken for this chapter. The research is based on a case study of the availability contract to provide depth maintenance and upgrade service for Tornado aircraft between BAE Systems and the MOD, named ATTAC. Interviews with senior management were used to identify factors that contribute to complexity. A broad range of complexity elements were identified in the interview data. The elements that were identified as contributing to the complexity of the service were placed within a framework, Fig. 4.1.

The factors can all influence each other and cases could be made for most sub-factors to be attributed under any lead heading. However, for the purposes of this analysis, the contributing factors were placed under the main complexity factor that best reflected the context being described in the specific interviews where that factor was raised. In some cases the context described spanned multiple main factors. We have grouped the contributing factors beneath the core factor that appeared most appropriate. Categorisation was pragmatic and has been developed to provide a starting point for organisational analysis and discussion. The factor level examines the unit of analysis—for example a product, process or service—displayed emergent behaviour, as a proxy for complexity, or was complicated but gave outputs that were predetermined and thus not in itself complex. For example, operation of a single product may provide pre-determined output, but the operation of a number of products together may give rise to emergent complexity. This thinking is expanded within this chapter.

Whilst we can make no claim for generalisation, we would expect that the framework may be suitable as an initiator of analysis of complexity for many enterprises.

The chapter will be structured as follows:

- A brief overview of management complexity theory from the organisational and contractual perspectives
- A description of the case study approach used
- Identification of the elements of complexity for this service engineering support enterprise
- Discussion of potential application of the framework and opportunities for future work.

4.2 Complexity

What makes a business 'complex'? Complexity is difficult to define and measure (Foley 1996; Murmann 1994; Pighin 1998; Kim and Wilemon 2003; Schlick et al. 2007) and there appears often to be a resistance to provide clarification if it involves simplification (Elliot and Kiel 1997; Cilliers 1998). There are many definitions and perspectives, but here we seek only to present a condensed overview to provide understanding. Simple systems with small numbers of components, such as a pendulum, are well understood. Complicated systems, such as a jet aeroplane have many systems that interact through predefined rules (Perrow 1999). Complex systems have many systems that autonomously interact through rule sets that only emerge over time (Amaral and Uzzi 2007). Businesses are frequently managed as though they are either simple or complicated systems, being predictable, linear, measureable and controllable (Boulton and Allen 2007). The concept that businesses are somehow mechanical by nature has underpinned much of the approach to strategy and management development (Mintzberg 2002). This view is reinforced by the approaches and tools employed by managers e.g., standardisation, metrics, value stream maps. If businesses were so coherent in nature the role of the manager could be automated and growth could be assured. The experience of managers informs us that process outcomes are far from assured, as systems of people are usually complex and are only mechanical in nature under specific, stable circumstances (Boulton and Allen 2007). So how are complex systems characterised?

A complex system may be described as one made of a large number of interdependent parts which together, make up a whole that is interdependent within some larger environment with which resources are exchanged (Anderson 1999).

Complex systems are not deterministic; as part of natural phenomenon it is not possible to see the final outcome at the start as there are infinite possibilities (Kao 1997). Emergent phenomena develop as local interactions between agents alter their behaviour in response to each other (Bonabeau 2003). Cumulative local interactions build and shape group behaviour, but group behaviour may be separate and distinct from the behaviours of individuals. As such, systems that exhibit emergence may respond in counterintuitive ways (Youn et al. 2008). Emergence is formed in a 'bottom up' process and is used as a characteristic to classify systems as complex. It is proposed that complex systems can be modelled from the bottom up, simulating systems level behaviour by defining the distinct behaviours of numerous individual agents (Bonabeau 2003).

Boulton and Allen (2007) provide a set of key principles for complexity: there is more than one possible future; tipping points may occur, when systems tip due to seemingly unimportant events into new forms with potentially radically different characteristics; systems are interconnected and diverse and it is the diversity that permits change and creativity; local variation is a prerequisite for novelty as change happens within the micro-diversity, unpredicted from the situational 'average', yet potentially impacting the global enterprise; systems are emergent and co-evolving, with characteristics constantly changing with time; complexity does not mean chaos and helplessness, but managers must both plan for clear action based on best information and recognise that the best laid plans may give rise to unintended consequences, unexpected outcomes and remain flexible enough to react accordingly.

Pascale (1999) further divides complex systems into groupings and posits a subset of complex systems named complex adaptive systems [CAS] for which are provided four tests. First, there must be many agents acting in parallel. Second, there are multiple levels of organisation. Third, the system is subject to the laws of thermodynamics and must be replenished with energy to prevent it from slowing down. Fourth, pattern recognition is employed by the system to predict the future and learn—thus CAS can change or learn from its past (Pascale 1999). Beinhocker (1997) proposes that CAS have three characteristics: they are open and dynamic systems; they are made up of interacting agents; they exhibit emergence and self-organisation. CAS transition between points of equilibrium occurs through self organisation and environmental adaption. Order is said to be emergent, not predetermined, and mechanisms for controlling it must also be emergent (Dooley 1996). Many systems are complex (they meet some of the criteria), but not all are adaptive, meeting all of the criteria. Inherent in this approach is the concept of grouping different complex systems and the idea of some being more or less complex.

Applying complexity to business processes or project management raises challenges (Whitty and Maylor 2009). Managers must understand the system under control (Taylor and Tofts 2009), developing the ability to cope with both environmental change and changes in the task/customer requirement which requires organisational structures that are sufficient to provide control and flexible enough to adapt to local needs (Schuh et al. 2008).

Attempts have been made to understand the degree of complexity and design organisations to function accordingly in their environment. From an organisational theory perspective, Daft (2007) states that the more external factors that regularly influence an organisation and the greater the number of organisations in the domain, the greater the complexity. In the context of a production process it is proposed that complexity would be measured along three dimensions, giving three factors (Daft 1992). The vertical axis shows the number of levels in an organisation, on the horizontal is the number of departments or job titles and on the third axis is the spatial complexity, perhaps of different geographical locations. A second two-axis model proposed by Kates and Galbraith (2007) explores manufacture and integrated product and service providers, linking organisational form to complexity and defining the space for organisational design. Their model follows Galbraith's (2002) rationale of matching an organisation's design to the complexity of its environment. The X-axis of Kates and Galbriath's model is titled 'Product and Service Integration' and shows a spread offering from standalone product at the low end to an 'integrated system' at the high end. They describe integrated solutions with reference to an IBM offering, the design and installation of a new call centre comprising a range of hardware, software and staff training. The Y-axis of their model is labelled 'Complexity'. Their dimensions of complexity include the number of products, variety of products and customers, customer geographic distribution and degree of customisation offered. They provide a rule of thumb for product number; four or fewer do not add complexity, whereas 12 or more can add significant complexity.

Care must be taken over viewing the delivery of service from a single business unit perspective. A clear trend towards increasing specialisation among firms (Mills et al. 2004) has inevitably been accompanied by the emergence of notions like "Extended Enterprise" (Dyer 2000) and "service value networks" (Basole and Rouse 2008). While firms have always been a part of multiple networks, their dependence on other network members and hence their inability to fully control their output, has grown alongside or as a consequence of the narrowing scope of their competences. It may be salient to view the firm as a nexus of contracts that together delivers a service (Jensen and Meckling 1976) and explores how the nature of the contracts to deliver service may be an indicator of complexity.

The case study presented is based upon an availability contract (Ng and Ding 2010) for the servicing of fast jets, providing depth maintenance and upgrade, and is between two parties. The literature on contracting and economics discusses complexity and similarly presents factors that contribute to complexity. The degree of contract complexity might be measured by the degree of customisation of formal contracts and the requirement for considerable legal work (Macneil 1978). Additionally, a simple measure of contract complexity is indicated by the length of the contract (in pages), which previous work applied as an indicator of contract complexity (Joskow 1988; Popo and Zenger 2002). Eggleston et al. (2000) discuss contract complexity as a continuum and identify the three 'standard' factors contributing to contract complexity as: (1) rich in the expected number of payoff-relevant contingencies specified in the contract i.e., the contract assigns numerous different contingencies against future possible outcomes. This covers obligations and entitlements; (2) variability in the magnitude of payoffs contracted to flow between parties i.e., contracts become more complex as the variability of payments increases under the specified contingency; and (3) severe in the cognitive load necessary to understand the contract i.e., the contract is difficult to understand and follow as the previous factors create a great number of permutations of potential outcomes, often spread across numerous annexes making prediction of future effects difficult. A contract is deemed 'complete', termed perfectly complete or 'p-complete' when it differentiates between all relevant future states. However, Eggleston et al. identify that this state is mainly theoretical and in most real world situations, due to factors such as a lack of trust, communication or unverifiable valuation, contracts are 'p-incomplete'. They may best achieve a state of being 'functionally complete', in which they are able to distinguish between states to the satisfaction of parties involved in a court. Faced with complex operations where futures are uncertain, contracts that provide that detail may fail to achieve even a functionally complete state. Longer contracts are more complete, but by necessity more complex. Eggleston et al. (2000) conclusions suggest detail may be sacrificed for the sake of simplicity, and suggest courts pay attention to the complexity of a contract when interpreting them.

From the research perspectives on complexity explored, we notice that findings and definitions from simple to complex are presented at a level of system aggregation selected by the authors to make their point (Perrow 1999; Cilliers 2000; Amaral and Uzzi 2007). All of these systems may be subject to further reductionism, yet the authors proposing them have taken a system, or system of systems approach and have taken the viewpoint of a pragmatic holist, even if they are of an in-principle reductionist viewpoint (Simon 1981), as they have chosen a level of abstraction that best illustrates the point they are wishing to make. In a search for understanding or solutions to a complex problem, the alternatives may be numerous holist approaches or reductionist approaches to make the problem faced simpler; the choice is made to best increase the chance of delivering a successful solution (Edmonds 1999).

To further explore how complexity may manifest and be usefully understood in relation to practice, we have undertaken case study research within an engineering service enterprise with the aim of identifying factors that contribute towards complexity.

4.3 Research Method

Pragmatically, we will follow Edmonds (1999) and argue that managers who are faced with complexity are fully able to identify it from their practical, experiential level of abstraction. Models of complexity are built upon the actions at the local level (Bonabeau 2003) and we propose that it is from the actor level that we may usefully identify characteristic drivers of complexity and explore the question of managing complexity in an enterprise.

We have chosen to proceed through a case study analysis (Yin 2003) and have focused upon the BAE Systems/MOD ATTAC contract which provides an availability service for the RAF Tornado aircraft. The particular case study was chosen for two main reasons—it was the first of its scale and complexity between the Provider and the Client. Second, though both parties intended to continue to let and bid for such contracts this first attempt was an opportunity for both parties to learn, which enabled the researchers to gain access to both client and provider and interview widely at the senior management level. Interviews were conducted face-to-face; 6 Client and 22 Provider interviews were undertaken with key managers from across the enterprise. Interviews were recorded then transcribed. The interviews were semi-structured at an average of 1.5 h per interview though some were considerably longer. They were conducted in 2008.

The interview transcripts were initially codified against the questionnaire and analysed to identify common responses and provide credence to findings whilst qualitative ethnographic analysis was employed to explore content (Bryman 2004). Further data content analysis followed a category scheme, deductively derived from the answers provided (Weber 1990). Three coders were used to ensure a degree of agreement among coders (Weber 1990). Participants were anonymous during analysis in an attempt to remove any bias and at the request of participants themselves (Seale et al. 2004). The anonymous results were presented back to managers who validated the findings.

The analysis of the interviews by the three coders identified 258 different quotations that referred to elements that made the service enterprise complex. The coders then examined these quotations and extracted the specific factor(s) that created the complexity. This analysis requires some interpretation but was repeated for all the quotations using three coders, meeting during workshops over 6 months to provide consensus in analysis and agreement of the emergent inductively derived categories. The resultant sets of data were analysed for duplication of factors. Thirty-eight different complexity factors were identified. Analysis of the factors and the context from which they were derived allowed them to be grouped into 6 core categories with 31 sub-categories and a factor level challenge to test if the unit of analysis displayed emergent behaviour or was complicated but predetermined. It was recognised that there was opportunity to place many of the sub-categories under a number of core categories, but pragmatically they are placed under a single core category which was found to best illustrate their context. The analysis presented in the chapter has been rigorously validated through a series of presentations to key customer and provider contract and support functions. Written reports have also been made available for validation and feedback (Yin 2003).

We shall now present the core and supporting complexities factors as perceived and described by the interviewees.

4.4 Factors Contributing to Complexity

Each factor as perceived by interviewees has been identified and coded during analysis. These factors are presented here in context of their description.

4.4.1 *Product Groups*

In the case example the product being studied is a fast jet, in particular the RAF Tornado aircraft. This is a complicated product as it is made up of a large number of other products. Individual components may form units, which combine to form systems. The authors contend that similar variant examples of the aircraft when new could not be described as complex as they were repeatable, predictable and any failure of the higher systems is subject to diagnosis by analysis through reduction. As the individual aircraft is no-longer new and has been serviced and modified numerous times, each aircraft has become unique with regard to its operation and service history and its future performance and service requirement is likely to vary accordingly. Aircrafts have increased variance which may cause issues for servicing operations when making changes to systems or components, perhaps giving rise to emergent behaviour, but individually each aircraft product could still not be described as complex. When observed as a group or operating as a fleet, we may see this individuality leading to complexity.

The product originally existed in a small number of different production variants, with later modifications made to individual aircraft according to intended use. There were originally three basic product variants: a fighter bomber, a reconnaissance aircraft and an interceptor. The product variants were subject to mid-life upgrade, which modified the entire fleet. As the aircraft were bought by four different nations, each nation also had its own variants. These factors add to the product variance contributing to potential fleet complexity.

The RAF Tornado aircraft are described as a fleet, but they are operated as numerous ‘fleets within fleets’. The 139 strong RAF Tornado fleet is currently modified to a number of varying standards, giving rise to 20 different aircraft clusters which have similar attributes that make up seven fleets (National Audit Office pp. 31, 2007). The various operational aircraft are equipped for their required mission. A number are ‘front line’ aircraft and may at any one point be equipped with a number of different weapon and sensor equipment. Some are trainer aircraft and will be modified according to the current training requirement.

The resources that were employed to construct the aircraft are geographically diverse, both nationally and internationally as the aircraft was manufactured by Panavia, a tri-national consortium (UK, Germany and Italy) formed and owned by BAE Systems, MBB (Now EADS) and Alenia. Each country made a significant contribution to the manufacture of major parts of the aircraft. In addition, numerous sub-contracts were given to companies to produce systems, sub-systems and components.

A swing-wing fast jet such as the Tornado is demanding and complicated as a piece of engineering. The multitude of interfaces, where the integration of the numerous sub-systems from different suppliers, coupled with the varying states of modification for each aircraft, may introduce technical complexity.

To reduce the possibility of complexity becoming a significant factor for the aircraft as a product group, an upgrade programme is currently underway to reset the fleet configuration to seven common clusters. More broadly a number of components and systems are common across a number of platforms (aircraft and other military equipment including ground vehicles etc.). This adds to its robustness and supportability during operation.

4.4.2 Process

The process of supporting the availability of the product is challenging. The prime contractor, BAE Systems, is a large manufacturing company and is able to operate successfully because of its process rigour. To deliver availability in this contract, the high level processes are partnered between the RAF and BAE Systems—two large organisations which operate many processes that need to integrate at various levels to deliver the availability levels required in the ATTAC contract. This can lead to situations where processes are duplicated, incompatible or require sign off from multiple parties which introduces bottlenecks. Also, many processes operate deep within the organisations that impact upon the ATTAC contract, but do not have line of sight to the operations. There may be many interfaces between the process and front line operations and this loss of line of sight was highlighted as leading to process complexity.

Many different process dependencies and interdependencies exist. The process integration across multiple organisations in the enterprise created complexity due to the dependency they had upon each other for the successful delivery of the contracted service. The value that each individual placed upon a particular aspect of the process frequently differed according to the value the party realised from the transaction, creating imbalances in the dependencies.

Process rigour and standard operating procedures were also difficult to stabilise due to emergent issues in the service requirement. Instances of emergent processes included 'unplanned parts requirements' and the need for completely new processes to be developed as some parts "*simply had never been bought before as spares*". These issues introduced new, unknown lead times to processes, placing a demand on all parties within the service enterprise to have process flexibility.

Such flexibility leads to a requirement for processes of recovery that may well be constructed 'ad hoc' locally by individuals. Recovery consumes resource and is an additional activity to the main planned enterprise processes. Whilst some recovery actions may be planned for, through its nature it remains fundamentally unplanned and may drive complexity as a factor affecting processes.

Data reporting processes e.g., performance metrics, costing etc. in multi-organisational enterprises may have been developed from different contexts and different requirements. This may lead to a multiplicity of data requirements from processes in operation. Some of these requirements, in this case due to both the military and aviation contexts, may be legally binding and require specific methods or signatures. Others may add no true value to the business and should be halted.

Many are fed into corporate or public governance structures as proxy-monitors of system performance. Compliance with governance requirements, public sector and corporate in this case, took up a great deal of time, added to the workload and complexity of management. Governance reporting may include data that are legally required and both understanding the requirement and legal standing of a request increased the complexity of process management.

Data and equipment that flows through the processes may also mean additional security measures are required. Restricted data such as software code, product detail or the service record for an aircraft may not be freely transmitted or displayed in a multi-organisational enterprise where not all the actors have sufficient security clearance. Weapon systems or the aircraft canopy and the ejector seats which contain explosive charges also place security restriction and impact upon the processes. This restriction on the free flow of data between parties significantly complicates processes. It also creates additional issues where it may not be possible to share data that would otherwise be useful. These restrictions to communication drive local behaviours which may give rise to unintended systems level outcomes, a core element of complexity.

4.4.3 Contracting

ATTAC, as multi-organisational enterprise that has a contract to deliver fast jet availability, has a complex contract that is between the primary contractor, BAE Systems and the client, MOD. Availability contracts deliver an agreed number of aircraft or engines at an agreed level of capability over the length of the deal. The ATTAC contract for Tornado is for availability of the aircraft, not the engines, though these are obviously a major dependence. This is a separate contract named ROCET and is held by Rolls-Royce Plc. Whilst the contract will have a number of companies named within it, there will also be a number of companies that have process dependencies that are not named on the contract (Mills et al. 2010).

The ATTAC contract is described as "*three and a half feet high*", which, allowing for colloquial exaggeration, would indicate that it is complex by the definitions of contract complexity (Macneil 1978; Joskow 1988; Popo and Zenger 2002; Eggleston et al. 2000).

The contract length is a direct result of a desire by the government to cover all the possible eventualities and create a functionally (f)-complete contract. However, due to the nature of service offered, approaching this task in this way is not feasible at the outset and could not be achieved at least until the enterprise has been in operation for a number of years and the processes and outcomes are established. Therefore the current contract may be described as not (n)-complete and complex by the measures of Eggleston et al. (2000).

A key driver of the contract length is risk. Each party had the desire to minimise risk to their institution and ensure that they are compensated for the risk that they take on. The organisations also tried to ensure that risk was bounded so that any liabilities will be limited. It was also important for the MOD not to transfer too much risk and jeopardise the future of the commercial organisations.

Within this partnered enterprise there is a minimum requirement for 50% of the personnel on the production line to be RAF and the other 50% may be from industry. This creates a situation where two people working together on the same job may be on completely different contracts with differing incentive structures. Local individual agent interaction may lead this to be a cause tension, potentially driving behaviours that create complexity systems level. It also makes the management of HR contracting more difficult as individuals seek to interpret their terms against others and equalise their conditions.

During interviews, the contract was described as being "*owned by lots of different people with different objectives*". The objectives written into the contract for the prime contractor are many but the enterprise partners may change their objectives over time, leading to either a conflict between contract and desired objectives or a requirement to change the contract. The number of contract changes made was described by one interviewee as "*absolutely enormous*". The multitude of objectives and changing needs also creates a tension for the prime contractor between the need to be customer focused and the need to be delivery focused.

4.4.4 Organisation

Whilst ATTAC was a novel contract which created a new way of interacting between partners, the larger surrounding organisational constraints were not changed to facilitate its operation. The enterprise that delivers the ATTAC contract is a complex organisation existing of over 30 different entities, both public and private, on and off base. Whilst the contract represents a significant sum of tax payer's money, approximately £1.3 billion, it is only a small percentage of the total financial streams managed by the two main parties, MOD and BAE Systems. These organisations co-create the value for ATTAC through servicing Tornado aircraft on base, primarily at Marham, but both have vast 'back office' support structures that directly and indirectly affect the operation.

This multi-organisational service enterprise creates a challenge in the identification and definition of a functional organisational boundary. Some form of definition is necessary for the organisational managers to understand and operate within, though it is recognised that this boundary will need be flexible and likely to constantly change.

The clarity of organisation structure that delivers the contract is lacking as it is difficult to conceptualise the numerous different entities that work together to deliver the service and there are currently few suitable tools (Mills et al. 2009a). This makes communication of organisational structure difficult and hence new managers are challenged to understand the structures that they are faced with managing.

Both main signatories to the contract also make extensive use of acronyms in their organisational structures. Such acronyms refer to semi-autonomous business units within both provider and customer organisations. Many struggled to remember what each acronym stood for. Adding further complication to gaining understanding is the often misleading business names that the various companies, business units or operations have. Even when known, both organisations are constantly in a state of flux and the individual units moved, renamed or shut down completely. This is a significant issue particularly for this organisation, as much organisational knowledge is lost because there is a frequent turnover of staff due to the 3-year rotation of military staff through the base.

Geographic dispersion adds complexity to the enterprise. The RAF has Tornado service at both Marham and smaller operations at Lossiemouth. The RAF Tornado support management team is located in RAF Wyton whilst MOD's procurement arm within DE&S is located in Bristol, the Defence Storage and Distribution Agency [DSDA] holds some of the parts in Germany, Panavia is an international organisation that was originally formed to build the aircraft and is a co-ordinator for many of the aircraft spares, and the RAF Head Quarters is in High Wycombe. BAE Systems have a similar operational spread.

Within and operating across the various organisations are management structures. During the initial transition to ATTAC implementation, many of these structures were mirrored MOD/BAE Systems structures to facilitate communication, though this can lead to many agents acting in parallel. There are further complications with the organisational and managing structures as we explored its supply chain management. The RAF operates a number of different aircraft that have commonality, if not directly in line replaceable units, then in the companies that supply them. A number of RAF integrated product teams [IPTs] were created to manage these parts or systems across the various platforms operated. Many of these were 'wrapped up' when the ATTAC project began. However, a number remain and the parts supply management for Tornado may be a very small part of their activity. Since they are not a signatory for the contract, this can potentially create further management challenges through diverse and potentially conflicting intra- and inter-organisational priorities.

This challenge also raises the issue of stakeholder complexity. The operational squadrons are the primary stakeholders in so far as they are the aircraft users. However, they are not the budget holders. Moving away from base there are numerous other public sector stakeholder organisations within the RAF such as Air Command—the RAF HQ, but also non-RAF organisations such as Defence Estates who provide the hangars, Defence Equipment and Support (DE&S) which equips and supports the UK's armed forces, the Treasury and ultimately Parliament. Within the private sector the contract is primarily run by BAE Systems MAS

[Military Air Solutions], who rely upon numerous BAE Systems organisations such as HR, purchasing and engineering support functions and are ultimately accountable to the BAE Systems board of directors and shareholders, many of whom do not have 'line of sight' to the service operation.

Political challenges arise as these stakeholders do not represent simple relationships. Key decision makers are frequently far removed from the front line operations. Decisions made in the UK Parliament around Treasury spending on the armed forces or decisions to go to war directly impact upon the service operation.

The partnering aspect of the contract was recognised as a key success factor. The nature of relationships between the partners, traditionally adversarial, has changed to a more co-operative service nature and this is influencing the organisation.

A key element is the shared resources. BAE Systems operates its service operation in an RAF hangar on an RAF base with 50% of their staff working on the aircraft. As stated before, there are a large number of other contributing organisations. However, legally the RAF cannot be a partner with business and they must 'own' the aircraft. There are also few options for finding alternative customers or suppliers. All these issues create organisational constraints.

4.4.5 Finance

The finances for the enterprise were organised to be, at least notionally, simpler through the use of an availability contract as opposed to a list of spares provision. However, the finances are difficult to manage due to the complexity of the enterprise, which in turn makes modelling and forecasting the cost of service difficult. Budgeted risk cost is available to facilitate planning, but is very difficult to access. It is also difficult for managers to put money aside in their budgets as deferred spend since there is a potential double penalty as unspent monies may be taken back and the overall budget may be reduced by that amount the following year. The budgets are controlled by a number of top level budget holders who are influential but many of whom may not be part of the contract, or even aware of the detail of the ATTAC contract.

The complexity of financial reporting is affected by contributing factors from process, contract and organisation. The challenge of appropriate enterprise integration goes beyond the budgeting and into their financial systems. Organisational units within the ATTAC enterprise use different operational and financial software systems which 'do not talk to each other'. They often have different processes and stakeholders.

Internally, the company has to attribute overheads to their staff which makes them appear expensive in comparison to subcontract labour. However, in 'real' terms, this is not the case.

Financial reporting regulations also add to the complexity. As the government is the customer of the service, there is no national taxation to pay. However, as the work is undertaken by industry, the government customer pays tax which it must then claim back. This creates additional financial processes which are non-value adding activity.

These processes all take up the time of the managers engaged in delivering the service and create a time lag between requirement and funding. To deliver the required service to the squadron in a timely manner the managers on both sides had to carry short-term financial risk, effectively outside of contract and without commercial cover, in order to ensure that aircraft were modified or serviced correctly and available.

Financial reporting from the partnered organisation is therefore complex and made more challenging by potential political pressures applied if the budgets are either under or overspent. This influences behaviour across and beyond the enterprise, though the understanding remains that the service should be delivering best value to the taxpayer.

The service is constantly under cost pressure. The availability contracts, including ATTAC were created as part of a cost reduction initiative by the MOD and Defence Logistics Organisation.[1] There is an underlying assumption that there will be a cost reduction year on year from the service operation—though this seems to be in contradiction to the increased demand being placed upon personnel due to the number of conflicts that the UK is involved in.

4.4.6 People

A large number of people are directly and indirectly involved in the ATTAC contract and it is the people who enable the complex service described so far to operate successfully. Collaborative process design by all stakeholders has helped reduce complexity.

Organisational cultures are strong across all the partners and this influences agents, process management and outcomes. The RAF develops specific cultures as part of its operations and BAE Systems has a distinct business culture. The various operational sites for all the partners also have specific sub-cultures. The cultural aspects of the business were described in both positive and negative terms. In the early stages of contract implementation, when problems were encountered the groups 'retreated' to their positions, making cross-cultural working difficult. Problems were attributed to 'working with civilians' or to customers failing to deliver necessary support. At later stages of contract implementation, the 'on base' partnered service has developed its own culture, including a strong sense of having a shared objective and this has enabled the contract to become functional.

---

[1] From 1 April 2007, the Defence Logistics Organisation ceased to exist and Defence Equipment and Support was established to manage all equipment throughout its life, from acquisition to disposal.

The ‘customer’ that the enterprise creates value for is difficult to define as there are so many different stakeholders. The final customer is the pilot of the aircraft, but they are not budget holders. They are part of a squadron, another customer, whose operational activity is controlled by fleet planning, RAF Air Command, the MOD, Treasury and Government respectively. But the contract was between DLO, subsequently reorganised into DE&S, and the service and operation of the aircraft is directed by ILOC, the Tornado and Logistics Operating Centre, all of whom may also be recognised as a primary customer. There is an additional complexity in that, as this is a partnered arrangement where value is co-created, BAE Systems and the other providers of government assets may also be seen as ‘customers’. In this context the term ‘customer’ begins to lose its meaning, though it is still very much in use. The authors highlight that great care needs to be exercised when using or engaging in discussion of customer and clarity should be sought on what exactly is meant.

The ATTAC contract has been discussed in terms of a partnered enterprise, but is made up of individual organisational units, staffed by individuals from one or possibly from a number of differing organisational cultures. For example, the hangar activity is undertaken by a mixture of civil and military personnel with very different cultural employer perspectives shaping their behaviour. With regard to leadership, we found that each of the organisations had leaders for ATTAC located within their management teams. The leaders managed their organisations contribution and also managed their interactions with the other organisations. Consistency of leadership can therefore be a problem both in terms of diverse objectives and more literally as personnel change over time. Managing the programme requires close relationships among the organisational leaders which is difficult to maintain due to rotation of staff.

The blurring of the organisational boundaries, with people from different organisations working side by side on the same job on site means that line management is more challenging. Military personnel may find themselves reporting to a business manager and vice versa. This may also reinforce cultural complexities as traditionally the RAF managers play a much greater role in the life and career management of their personnel, ensuring that they have broad experience but also that their families are looked after. This is in contrast to private enterprise where individuals play a much more active role in planning and managing their own career progression and would not expect any management interest in their family circumstances.

Roles that both the organisations and the people in the enterprise must undertake to deliver the service have changed considerably due to the nature of the ATTAC enterprise. Some of the roles are different as BAE Systems no longer sell the RAF an aircraft, but instead work alongside them to deliver a service to the front line squadrons. This changes the nature of some of the managers roles as they are intimately engaged with operations and constantly faced with the ‘customers’ personnel. Their role requires greater flexibility as much of their work requirement will be emergent as the value is co-created in the repair hangar. However, further back in the organisations, away from the customer/supplier contact points many of the managers' work requirements will appear to be the same. Functional managers and staff who provide support for service contracts may therefore be unaware of the changing business model within which they operate. This can also create added complexities as behaviours and decision making may not align with new business models.

Learning and recognition of change is required in many areas as the enterprise transforms from a product-driven buyer-to-supplier relationship to an enterprise that understands how complex engineering services are co-managed and co-delivered. The differences and similarities between products and services and their management need to be clearly understood by all parties. Likewise all enterprise parties need to learn how to most effectively implement the new requirements in this interdependent environment. The lessons learnt during the development and adoption of the ATTAC contract must also be captured and disseminated so that when future contracts are available, the enterprise can more rapidly and cost effectively develop.

Complexities were also attributed to the learning process necessary in contract implementation. The learning process will be ongoing; will require constant reinforcement and need to be an embedded part of the structure due to the rotation of staff, both in the private and public sector.

Transition Management is a constant challenge within this case study. The number of different businesses and departments and the seemingly constant change or reorganisation they undergo is a significant complexity. The move from product- to service-based contracts also brings in a significant number of new concepts to these groups, not least 'partnering to co-create value'. The ATTAC contract is only one of many different programmes that drive change initiatives which impact upon the enterprise including site closures, relocation and the introduction of Lean process improvement techniques.

The ATTAC contract is the latest contract development from the MOD and it is expected that in the future an increasing number of higher value contracts will be offered for partnered delivery. It is critical therefore that the service enterprise partners learn to effectively manage the organisational transition process. It should be noted that whilst transition may reach a point of equilibrium, there will be no recognised end to the process.

4.4.7 Emergent or Pre-determined

From the findings it would appear that the level of analysis may define whether a factor is complex and may exhibit emergent behaviour, or complicated with predetermined output that potentially contributes to higher level emergence. The behaviour of local agents shapes the behaviour of the whole system, but that systems level may be independent of individual action. For example, a single service operation undertaken on the aircraft may be described as complicated, but it would be difficult to make any case for it being complex. All of the different service operations have been detailed and captured in text, along with approximate times in which they may be completed, leading to single operations seemingly having pre-determined outcomes. However, further issues may arise when undertaking service and upgrade involving multiple processes that may necessarily create changes in priority and hence create changes to planned operations. It is the behaviour of local agents appropriately interacting and reacting when dealing with arising requirements that combines to create emergent complexity. Management are to question if a unit of analysis (process, system etc. being examined) exhibits complexity as the outputs are emergent or is complicated representing output of a predetermined system.

4.5 Using the Complexity Factor Framework

The ATTAC case has been presented as an example of a complex enterprise. Following interviews and detailed analysis the factors identified as potentially contributing towards complexity were placed within the framework shown in Fig. 4.1. Six core categories, 31 sub-categories and a factor level challenge were identified. Sub-categories were pragmatically placed beneath the core category that best reflected the context in which it was raised—though it is recognised that many subcategories may be relevant to a number of core categories.

It was widely recognised during interviews that many of the factors that add to complexity were not a result of customer requirement, but rather bureaucracy or historic precedence. As such, the generated complexity introduces time and cost. The method for application of the framework forms part of ongoing research with the enterprise and is currently being assessed as a basis for analysis of organisational change through reduction in complexity.

Each factor within the framework may be discussed with reference to a specific context using a three-factor analysis of complexity value, based on the groupings for waste from the Toyota Production System (Ohno 1988). First, those factors that add to the complexity of the systems that are non-value adding and can be immediately eliminated. Second, those factors that are currently embedded in process, equipment, contract or other legal frameworks and must be identified and strategies developed to remove or minimise their impact over a managed time period. Third and perhaps most important, those factors of complexity that create value and profit, embodying the reason that the company has been engaged to act due to its competency in managing these factors.

Each factor is discussed by a management team within their context—does the factor manifest in their operations and how does the factor affect them? Is there sufficient evidence to support their responses and if not, how may this be gathered? What plans are in place to act? Should the factor be applied to sub-processes and each explored separately? These factors are assigned a red/amber/green status in the framework to facilitate the identification and application of scarce resource upon the issues that are deemed in need of most direct attention. At the time of writing, practical application of this framework is ongoing.

4.6 Conclusions and Future Work

Pragmatically, we have taken complexity of a service enterprise to be defined from the actor's perspective, identifying factors that may lead to complexity via a qualitative case study using semi-structured interviews. The set of complexity factors that result are a product of the respondent's interpretations of their context. Thus the generated complexity and complex adaption described use a set of criteria defined by the interviewees, with complexity taken to be from their proposed level of abstraction (enterprise, system, component etc.). Whilst the identified six core, 31 supporting complexity factors and factor level challenge significantly exceed the three factors for complexity of an organisation proposed by Daft (1992), we find that the factors within the framework are supported by the general meanings of complexity found in the literature.

The case study draws upon the complexity of the enterprise that delivers the ATTAC contract, explored through the 'perceived' complexities identified in interviews with managers across the enterprise. The case is presented as an example of a complex service contract and the MOD engaged with BAE Systems precisely to enable them to better manage the complex enterprise which delivers the service. The meanings of the answers given to the question of complexity have been interpreted by the researchers, but by using three different researchers we have sought to minimise any bias. The factors presented represent a substantial analysis of our findings, but we make no claim for identification of a comprehensive set of factors that represent the true scale of complexity of the case study contract. By grouping into a factor framework we have further proposed a tool for managers to analyse their enterprise based on complexity, such that time and cost may be reduced. This tool is developmental at the time of writing. However, knowledge and understanding of the complex service delivered by the enterprise is still developing and it is expected that there will be further changes to the factor framework and guidance on its application as a tool for complexity management.

4.7 Chapter Summary Questions

Complexity obfuscates and introduces cost and may exceed that which is inherently needed to perform the service task. We provide a framework of complexities identified by managers that may be used to discuss and identify strategies for complexity management and cost reduction.

- Is our assumption that costs relate to complexity valid in your experience?
- What are the inherent complexities which are part of the job and what are created unnecessarily?
- Are plans in place to reduce unnecessary complexity—both over the short and longer term?

Acknowledgement. The authors would like to acknowledge and thank their sponsors within BAE Systems, Louise Wallwork and Jenny Cridland, and especially John Barrie for his support and direction. We would also like to thank the ATTAC team from both BAE Systems and RAF/MOD for their full engagement with our work. This research was supported by BAE Systems and the UK Engineering and Physical Sciences Research Council via the S4T project.

References

L.A.N. Amaral, B. Uzzi, A new paradigm for the integrative study of management. Physical and technological systems. Manag. Sci. **53**(7), 1033–1035 (2007)

P. Anderson, Applications of complexity theory to organizational science. Org. Sci. **10**(3), 216–232 (1999)

R.C. Basole, W.B. Rouse, Complexity of service value networks: conceptualization and empirical investigation. IBM Syst. J. **47**(1), 53–70 (2008)

N. Bateman, N. Rich, Companies' perceptions of inhibitors and enablers for process improvement activities. Int. J. Oper. Prod. Manag. **23**(2), 185–199 (2003)

E.D. Beinhocker, Strategy at the edge of chaos. McKinsey Q. **1**, 25–39 (1997)

E. Bonabeau, Predicting the Unpredictable. Harv Bus Rev **80**(3), 109–116 (2002)

J. Boulton, P. Allen, Complexity perspective, in *Advanced strategic management: A multi-perspective approach*, vol. 14, 2nd edn., ed. by M. Jenkins, V. Ambrosini, N. Collier (Palgrave Macmillan, Hampshire, 2007), pp. 215–234

A. Bryman, *Social research methods*, 2nd edn. (Oxford University Press, Oxford, 2004)

P. Cilliers, *Complexity and postmodernism: Understanding complex systems* (Routledge, London, 1998)

P. Cilliers, What can we learn from a theory of complexity? Emergence **2**(1), 23–33 (2000)

R.L. Daft, *Organisational theory and design* (West Publishing, St. Paul, 1992)

R.L. Daft, *Understanding the theory and design of organisations* (Thomson South-Western, Mason, 2007)

K. Dooley, A nominal definition of complex adaptive systems. Chaos Netw **8**(1), 2–3 (1996)

J.H. Dyer, *Collaborative advantage: Winning through extended enterprise supplier networks* (Oxford University Press, Oxford, 2000)

B. Edmonds, Pragmatic holism (or pragmatic reductionism). Found. Sci. **4**, 57–82 (1999)

K. Eggleston, E.A. Posner, R.J. Zeckhauser, Simplicity and complexity in contracts. John M. Olin Program in Law and Economics Working Paper No. 93, University of Chicago Law School. doi: 10.2139/ssrn.10.2139/ssrn.205391, 2000

E. Elliot, L.D. Kiel, Nonlinear dynamics, complexity and public policy: Use, misuse, and applicability, in *Chaos, complexity and sociology: Myths models and theories*, ed. by R.A. Eve, S. Horsfall, E.L. Lee (Sage, London, 1997)

J. Fitzsimmons, M. Fitzsimmons, *Service management: Operations, strategy, and information technology* (McGraw-Hill/Irwin, New York, 2005)

D.K. Foley, *Barriers and bounds to rationality: Essays on economic complexity and dynamics in interactive systems* (Princeton University Press, New Jersey, 1996)

J. Galbraith, *Designing complex organisations* (Addison Wesley, Reading, 2002)

J. Heineke, M. Davis, The emergence of service operations management as an academic discipline. J. Oper. Manag. **25**, 364–374 (2007)

M.C. Jensen, W.H. Meckling, Theory of the firm: Managerial behavior, agency costs and ownership structure. J. Financial Econ. **3**(4), 305–360 (1976)

P. Joskow, Asset specificity and the structure of vertical relationships: Empirical evidence. J. Law Econ. Organ **4**(1), 95–117 (1988)

J. Kao, *Jamming: The art and discipline of business creativity* (Harper Collins, New York, 1997)

A. Kates, J.R. Galbraith, *Designing your organization* (Wiley, New York, 2007)

J. Kim, D. Wilemon, Sources and assessment of complexity in NPD projects. R&D Manag. **33**(1), 15–30 (2003)

I.R. Macneil, Contracts: Adjustment of long-term economic relations under classical, neoclassical and relational contract law. Northwest. Univ. Law Rev. **72**, 854–905 (1978)

J.F. Mills, J. Schmitz, G.D.M. Frizelle, A strategic review of supply networks. Int. J. Oper. Prod. Manag. **24**(10), 1012–1036 (2004)

J. Mills, V. Crute, G. Parry, Enterprise imaging: Visualising the scope and complexity of large scale services. QUIS 11—The service conference, June 11–14, Wolfsburg Germany, 2009

J. Mills, G. Parry, V. Purchase, Public sector out-sourcing: Understanding the client's aspirations and fears. Proceedings of the 11th International La Londe Conference in Service Management, May 25–28 (The Institut d'Administration des Entreprises (IAE), La Londe Les Maures, France, 2010)

H. Mintzberg, *The strategy safari* (Prentice Hall, Harlow, 2002)

P.A. Murmann, Expected development time reductions in the German mechanical engineering industry. J. Prod. Innov. Manag. **11**(3), 236–252 (1994)

National Audit Office, Transforming logistics support for fast jets. Available at: http://www.nao.org.uk/publications/nao_reports/06-07/0607825.pdf. Accessed Feb 2010 (2007)

Ng I, Ding X, Outcome based contract performance and co-production in B2B maintenance and repair service. Dept of Management Discussion Paper Series, University of Exeter, 2010

T. Ohno, *Toyota production system: Beyond large-scale production* (Productivity Press, New York, 1988)

R.T. Pascale, Surfing the edge of chaos. Sloan Manag. Rev. **40**(3), 83–94 (1999)

C. Perrow, *Normal accidents: Living with high risk technologies* (Princeton University Press, NJ, 1999)

M. Pighin, An empirical quality measure based on complexity values. Inf. Softw. Technol. **40**(14), 861–864 (1998)

L. Poppo, T. Zegner, Do formal contracts and relational governance function as substitutes or complements? Strateg. Manag. J. **23**(8), 707–725 (2002)

C.K. Prahalad, V. Ramaswamy, The new frontier of experience innovation. MIT Sloan Manag. Rev. **44**(4), 12–18 (2003)

S. Sandström, B. Edvardsson, P. Kristensson, P. Magnusson, Value in use through service experience. Manag Serv Qual 18(2), 112-126 (2008)

S.E. Sampson, C.M. Froehle, Foundations and implications of a proposed unified services theory. Prod. Oper. Manag. **15**(2), 329–343 (2006)

C. Schlick, E. Beutner, S. Duckwitz, T. Licht, A complexity measure for new product development projects. Proceedings of the 19th international engineering management conference 2007 (IEEE Publishing, Austin, 2007), pp. 143–150

G. Schuh, A. Sauer, S. Doering, Managing complexity in industrial collaborations. Int. J. Prod. Res. **46**(9), 2485–2498 (2008)

C. Seale, G. Gobo, J.F. Gubrium, D. Siverman, *Qualitative research practice* (Sage, Thousand Oaks, 2004)

H.A. Simon, in *The architecture of complexity*. The sciences of the artificial (MIT Press, Cambridge, 1981), pp. 192–229

R. Taylor, C. Tofts, *Managing complex service systems* (Springer, New York, 2009)

S.L. Vargo, R.F.L. Lusch, Service-dominant logic: Continuing the evolution. J. Acad. Mark. Sci. **36**(1), 1–10 (2008)

R.P. Weber, *Basic content analysis* (Newbury Park, CA, 1990)

J. Whitty, H. Maylor, And then came complex project management (revisited). Int. J. Proj. Manag. **27**(3), 304–310 (2009)

R.K. Yin, *Case study research: Design and methods*, 3rd edn. (Sage, Thousand Oaks, 2003)

H. Youn, H. Jeong, M. Gastner, The price of anarchy in transportation networks: Efficiency and optimality control. Phys. Rev. Lett. 101(12). doi: 10.1103/PhysRevLett.101.128701, 2008

Chapter 5
Towards Understanding the Value of the Client's Aspirations and Fears in Complex, Long-term Service Contracts

John Mills, Glenn Parry and Valerie Purchase

Abstract This chapter focuses on the translation of public sector client aspirations and fears into a specification of the services necessary for a complex, long-term service availability contract. The contract is complex in many senses including that many independent organisations must work together to deliver contracted service outcomes and long-term being in excess of 10 years. These factors imply the need for enterprise level management processes in addition to stakeholder centric management. The alignment between the contracted services and the client's needs is investigated and the implications of partial mismatches are discussed. Particular issues raised are the effect on behaviours around contract operation; potentially missed opportunities to co-create value and build trust; and challenges to the achievement of enterprise-wide management processes. The research highlights the potential role of evolving and explicitly shared Client and Provider aspirations and fears as a basis for enterprise-wide management.

5.1 Introduction and Case Background

Normative first steps in taking an enterprise (Binder and Clegg 2007) perspective of a complex availability contract are: first, to define the boundaries of the enterprise—what is included and what is not? And second, to identify the interests and value propositions of the enterprise as a whole and its constituent organisations. Research described in Chap. 3 developed visualisation of a complex support enterprise, a first step in developing the generic visualisation capable of improving understanding of the interfaces, leadership and management challenges in complex multi-organisational contracts between the Business and the Public sector (Mills et al. 2009).

J. Mills (✉) · G. Parry · V. Purchase
Institute for Manufacturing, University of Cambridge, Cambridge, UK
e-mail: jfm@eng.cam.ac.uk

I. Ng et al. (eds.), *Complex Engineering Service Systems*, Decision Engineering,
DOI: 10.1007/978-0-85729-189-9_5,

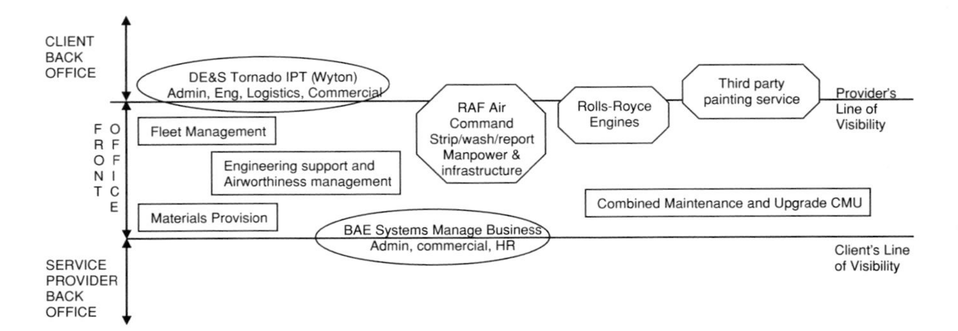

Fig. 5.1 Enterprise image of on-base ATTAC organisations

The enterprise environment is inherently complex and Chap. 4 has begun to systematically identify the complexity factors present. However, the focus of this chapter is on the role of client and provider value aspirations and partial mis-matches between these aspirations and the realities of the contract. An unexpectedly ambiguous environment was one result for front-line industrial providers where the lead providers were surprised to find that changes to plans were frequent; sudden, unexpected additions to the contract were the norm, and working outside the contract seemed essential to satisfy the on-base client.

A military fast jet, through-life availability contract[1] provides the context for the study. The support front office is located in an RAF (Royal Air Force) base in the UK and is supported by client, provider and third party on- and off-base organisations. A more complete view is detailed in Chap. 3. This chapter focuses on those organisations with an on-base (front office) presence; see Fig. 5.1.

The principals involved in the ATTAC contract are:

- BAE Systems as the lead provider
- DE&S (Defence Equipment and Support) as the original providers of the support service and the negotiators of the contract on behalf of the ultimate client
- The MoD (Ministry of Defence) as the ultimate client
- RAF Air Command as the users of the Tornado aircraft and who direct the activities of squadrons and aircrew.

On-base, several types of organisation were found (Mills et al. 2009):

1. *Partnered Direct Service Delivery Organisations*: They are managed by the lead provider (BAE Systems), located where the operational services are delivered and composed of BAE Systems and Air Command staff:

[1] ATTAC (Availability Transformation: Tornado Aircraft Contracts) is a long-term, whole-aircraft availability contract where BAE Systems take prime responsibility to provide Tornado aircraft with depth support and upgrades, incentivised to achieve defined levels of available aircraft, spares and technical support at a target cost.

- Combined Maintenance and Upgrade (CMU) includes most of the main hangar activities that carry out depth maintenance and upgrade activities resulting in aircraft with increased available flying hours.
- Fleet management provides planning activities that translate the Forward Squadron requirements into the schedule of the aircraft through CMU.
- Engineering Support and Airworthiness management resolves technical queries and safety issues.
- Materials provision plans and expedites spares and repair requirements to supply CMU and Forward squadrons

2. *Independent Direct Service Delivery Organisations*. These are organisations that are not managed by or responsible to BAE Systems, but are critical dependencies on the delivery of the service. In this case there are three such organisations:

   - Rolls-Royce manages the repair and overhaul of Tornado engines via a contract between themselves and the MoD
   - SERCO provides a painting service via a contract between themselves and MoD
   - RAF Air Command retained management of several key areas of depth maintenance: the strip and wash process, strip report and all work connected with ejector seats and weapons. They also manage the hangars' upkeep, power supplies and information technology infrastructure and provide technicians, engineers and management personnel to the Partnered Direct Service Delivery Organisations

3. *Specific Contract Focused Organisations*. These organisations are managed by BAE Systems or the Client and are focused solely on the focal contract. They may be located with the operational service provision, remotely or spread between them. In this case there are two such organisations:

   - The "Manage Business" organisation is controlled by BAE Systems and operates mainly on-base. It covers the commercial, administrative and human resource needs of the contract.
   - The Tornado IPT (Integrated Project Team) is controlled by the MoD via DE & S and contains staff covering administration, engineering, logistics and commercial support for ATTAC.

Given this necessary introduction to the case context the chapter proceeds as follows:

- A literature review of research in the area of outsourcing complex services and the extra issues involved with outsourcing publicly funded services to industry
- A brief description of the research methodology used
- Case analysis leading to the identification of the aspirations and motivations of Providers and Clients and a summary set of services required to deliver these aspirations
- A discussion of the issues raised
- A summary of findings and directions for future research.

5.2 Literature Review

With a clear trend towards increasing specialisation among firms (Mills et al. 2004) as they develop and concentrate upon their core competence has come an inevitable emergence of notions like "Knowledge-Sharing Network" (Dyer and Nobeoka 2000)—sets of firms with complementary competences that collaborate to deliver products and/or services or "Virtual Organisations" (Ahuja and Corley 1999)—a form of extended firm suited to the delivery of products and services that are competence-based. While firms have always been a part of multiple networks, their dependence on other network members and hence their inability to fully control their output, has grown alongside or as a consequence of the narrowing scope of their competences. Thus, calls for the need to take a wider "Enterprise" or "Network" perspective have grown in parallel with the need to understand and articulate competence in terms of an individual firm's value proposition.

The service enterprise presented here is composed of multiple independent organisations, as shown in Fig. 5.1. The services are highly customer specific, related to the particular requirements of the product in the client's context. Mathieu (2001) deliberately used the term 'client' instead of customer to emphasise a major change in the relationship necessary when transforming from equipment to service provider. 'Client' implied a professional, expert, service provider capable of delivering confidential advice, attention and support. The technical quality of the product might even become a hygiene factor in some contexts, for the client is looking for a 'solutions provider' (Galbraith 2002; Davies 2004; Davies et al. 2007; Windahl et al. 2004). The provider charges a fixed price to provide specified services over a set period rather than charging for each service event (e.g., breakdown or upgrade). The provider takes on the risk of equipment failure, establishing contracts that offer a set level of operational availability, often combined with a specified response time in the event of failure (Oliva and Kallenborg 2003). The notion of *availability* enables the client to evaluate the value or worth of the provider's offer compared to their current internal and external costs of ownership. The profitability of an individual contract is largely dependent on the provider's assessment of failure risk and the combined ability of the provider and the client to co-produce (Ramirez 1999) and co-create (Vargo and Lusch 2006, 2008) improved returns from this new arrangement. The challenges to making this transition are wide in scope and time consuming to achieve for both the main provider and the client. In addition to responding to new sources of profit and cost (Markeset and Kumar 2004), new capabilities are required in four domains (Windahl et al. 2004). Of these, partnering and networking skills are a particular focus here. Researchers are increasingly realising that client–provider partnering is a very limited view of the partnerships a solution provider must enter. Windahl and Lakemond (2006) emphasise the importance of partnerships within different departments in the solution provider, within the client and with other independent organisations necessary to produce the solution. Another difficulty is the need to share closely held design and/or financial data with partners and be honest about one another's performance (Foote et al. 2001). To illustrate the nature of the solution provider–client relationship, Helander and Moller (2007) assert that the solution supplier's senior management and client peers must interact over the sensitive out-sourcing of key functions and co-development and management of solutions over a long term. In other words the strategic direction of each partner must be shared. It becomes very clear that the "co-creation" of value (Vargo and Lusch 2006, 2008) requires 'co-management' by the client and the provider and that developing and managing enterprise-wide management processes may be central to the enterprise success.

Jost et al. (2005) study of private consortia taking over services previously provided by the UK public sector describes factors that help build successful relationships in this context. Jost et al. (2005) findings focus on three key activities at different levels of organisation—individual (assuring continuity), group (team-building) and organisation (reconciliation of objectives)—underpinned by the concept of trust (Vangen and Huxham 2003). The study also drew attention to the uncertainty in such contracts—all eventualities cannot be predicted at the outset and activities to reconcile objectives occur repeatedly in a cycle of negotiate, commit and execute (Smith Ring and Van de Ven 1992, 1994). While the co-creation of value remains central in these contracts, the cultural differences between the public sector and the commercial constituencies may add another challenge. For example the confusion created for military partners, used to visible symbols of rank and authority, by an indistinguishable set of suited industrial managers was significant—who *was* in charge? This is a fascinating area but outside the scope of this chapter.

The research explores challenges predominantly at the organisation level concerned with reconciliation of objectives viewed from the translation of predominantly client aspirations and fears into the services required in ATTAC.

5.3 Research Methodology

Case study research is useful when the aim of research is to answer 'how' and 'why' questions (Yin 1984). This matches our wider aim of improving understanding of how and why such complex service provision contracts actually materialise in practice, as perceived by involved (and uninvolved) actors in the provider and the client. Though our overall focus was on understanding the obstacles and enablers to effectively implementing the service provision contract, our focus in this chapter is on the service enterprise management issues raised and how these might be better and more widely understood.

This particular case study was chosen for two main reasons—it was the first of its scale and complexity between the Provider and the Client. Since both parties intended to continue to let and bid for such contracts, this first attempt was an opportunity for both parties to learn. This interest in learning from experience enabled the researchers to gain greater access to partner organisations—six Client and 22 Provider interviews were conducted. In this chapter we differentiate between informants from the industrial on-base providers, Air Command (the product user) and DE&S (the negotiators of the contract on behalf of the MoD). For reasons of confidentiality, detailed roles are not divulged.

Key questions for this chapter concerned the client motivations to involve industry in the provision of Tornado support and the motivations of industry to accept this opportunity. What value was expected for each stakeholder and had this materialised? These questions also provoked discussion from some client constituencies on what value had been put at risk by the contract and from the provider on what value was at risk through unanticipated requirements and dependencies in the task.

All interviews were recorded and transcribed. 'The case study analysis presented in the chapter has been rigorously validated through a series of presentations to key customer and provider contract and support functions. Written reports have also been made available for validation and feedback (Yin 2003). A full description of the case study analysis methods is described in Chap. 3.

5.4 Case Analysis

The analysis will examine the ATTAC case in terms of value aspirations and fears for each of the key stakeholders taken from relevant documents and case study interviews. The research findings will be presented in two sections: lead provider strategic motivation, and client motivations and requirements from the outsourcing including MoD, DE & S and Air Command perspectives.

5.4.1 Provider Motivations and Aspirations

The provider motivation for entering this contract was based on two major factors, first the Joint Strike Fighter contract funded predominantly by the USA and the UK appeared to be the last manned defence platform that would be developed for a considerable time. Defence air programmes were now much more likely to involve Unmanned Aerial Vehicles (UAVs). Second, the UK government (Defence Industrial Strategy 2005) was, and remains keen to move towards partnering on aircraft maintenance and support, providing alternative income for Defence OEMs (Original Equipment Manufacturers) that could help maintain the capabilities necessary for upgrading manned aircraft.

BAE Systems expected to continue support for Tornado and develop support contracts for two more fast jet platforms—Harrier and the Eurofighter (Typhoon). Since opportunities for OEMs and others to partner with governments in Defence support are increasing rapidly in the UK, the USA and elsewhere, the support business was seen as an important new market for Defence OEMs. This market was expected to provide steady ongoing profits from long-term contracts.

This motivation was widely recognised by the service provider and the client respondents. Better procedures for delivering the contract were expected by both parties to lead to increased profits for BAE Systems and ultimately in gain share for the MoD. There were, however, tensions and a lack of trust in the commercial values that were evident at the early stages of the contract implementation. Clients were suspicious of provider motivations who may 'still see it as a game that they have to go out and win'. There were also fears that having given this type of contract to industry, this may become a monopoly and result in price increases.

There was evidence, however, that such concerns reduced with experience of contract delivery. Significantly, on-base provider respondents did not tend to discuss their commercial motivations, focusing instead on the client's motivations and how the service provider might support the achievement of these objectives. This may have been due to an awareness of potentially conflicting values or may have been evidence of reconciling objectives through adopting the client's objectives. As the first major contract of its type, BAE Systems and the UK MoD were keen for it to be a success.

5.4.2 Client Motivations, Aspirations and Fears

The Defence Industrial Strategy White Paper (Ministry of Defence 2005) set out the MoD/UK government perspective on partnering with industry. It was widely accepted by all interviewees that the principal motivation for outsourcing Tornado support (and subsequently support of other Defence assets) was to reduce the cost per flying hour. The ATTAC contract offered savings of £510 million over 10 years. These savings arose from reductions in RAF and civilian-related personnel and the improvements a commercial organisation was expected to bring to task.

From the DE&S perspective, the organisation previously tasked with Tornado support, their task became one of negotiating a reduction in the price per flying hour and helping to implement the contract.

Further reductions were also expected from a gain/share agreement within the contract—with an open book partnership, savings made in excess of the target would be shared between the client and the provider. Particular areas such as supply chain management were seen as ripe for cost reduction.

There was thus a strong requirement that cost reduction would be ongoing. The need for a cost reduction service is clear with a lead from the lead provider. One major source of cost reduction concerns the problems that arise in the forward squadrons that are corrected without provider input. Sometimes, mis-diagnosis leads to a repair cycle costing over £30 K and a result of "no fault found". Good data on these events can be difficult to obtain and the provider could justly ask that the Air Command cooperates to provide an improved data service on these events.

However, reducing costs could not be achieved without other effects. The manpower reductions were felt to limit the flexibility of an organisation whose purpose called for very fast response. The old organisational arrangements were considered to be 'almost infinitely flexible' and thus could cope with 'surges and unexpected events'.

There is a strong requirement that the provider will be as flexible as possible. The importance of this aspect should not be under-estimated for the level of flexibility potentially required by a military defence client is difficult, if not impossible, to cover in a contract.

However, this requirement was fully recognised by the on-base industrial providers. They recognised their customers' concerns; that retaining flexibility was 'hugely important'; and that 'politicians will change their mind' resulting in new operational requirements and the need to change plans.

The manpower reduction had further consequences—potentially reducing the RAF's engineering knowledge and capability and implying changes in career structures. As industry assumes management of such contracts, there were concerns that there may be fewer career routes for service personnel who would be moving beyond operational duties and would have traditionally played a role within Integrated Project Teams. Such roles gave service personnel important knowledge and understanding of the system and skills sets working with industry.

Having fewer graduate engineers implied a change in career pattern to establish and maintain a new balance of scope and depth of knowledge and skills.

The Air Command perspective was not covered in depth in our interviews; however two potential losses in value from the new arrangements needed to be guarded against. The first concerned the skills of the technicians and whether they would be as well trained as before in a hangar managed by an industrial provider. In-depth skills give RAF technicians an understanding of the aircraft that can be used in the front line and the loss of such skills would be detrimental. Having fewer service technicians implied an adaptation of training procedures to maintain new net skill levels.

The second potential danger was a cultural factor. The defence customer emphasised that 'respite' from operational duties was an important factor for service personnel involved in delivering the ATTAC contract. In relation to this issue, it was important that personnel should not leave their squadron where they felt pride in their defence role, to then feel that they had been 'abandoned for a few years in a civilian organisation'. There was therefore a necessity to maintain the 'military culture' by ensuring that such personnel were not a minority in a largely civilian community.

5.4.3 Sample of Services Required to Deliver Client and Provider Aspirations and Fears

Having understood the context and the value propositions, it becomes possible to express the value propositions that have been contracted for and express them in terms of the service bundles offered (Vargo and Lusch 2008). Table 5.1 summarises the operational services within the contract, those support services implicit in the contract and un-contracted services.

We do not claim that this collection of services is comprehensive; they do however provide examples of client aspirations and fears translated into services

Table 5.1 Samples of contracted, implicit and un-contracted services

Contracted operational services	Lead
Depth maintenance service (CMU)	BAE Systems
Fleet management service	BAE Systems
Engineering support and Airworthiness data management service	BAE Systems
Materials provision service to CMU and forward squadrons	BAE Systems
Support services implicit in the contract	
Training service for industrial technicians and engineers	BAE Systems
Training service for RAF technicians and engineers	Air Command
Cost reduction and improvement service	BAE Systems
Forward data provision service on faults	Air Command
Service to support and develop the hangar infrastructure	Air Command
Service to provide strip, wash, ejector seat and weapons aspects of depth maintenance	Air Command
Service to assist integration with contracts outside ATTAC e.g., Engines and Painting	Air Command
Commercial service providing open book data, quotations etc.	BAE Systems
Un-contracted services	
Highly responsive service on all operational services	BAE Systems
Skills maintenance and development service for RAF technicians	BAE Systems
Skills maintenance and development service for RAF graduate engineers	BAE Systems
Respite provision service that maintains the RAF ethos	BAE Systems

required by the client whether they have been explicitly recorded in the contract or contracted for at any level.

5.5 Discussion of Issues Raised

As in many other outsourcing decisions, reductions in cost for the client and the prospect of profit for the provider lie at the heart of the decision. A cost focus on the client's part invariably leads to losses in value—increased dependence on the provider, a potentially slower response to emergencies and effects on careers in the client organisation. In this section we discuss un-contracted services and implicit services.

5.5.1 Un-Contracted Services

5.5.1.1 Flexibility and Responsiveness

On-base provider personnel are well aware of the client's requirement for fast response; indeed it is not advisable to refer to the contract when a new requirement arises, for the on-base client interprets that behaviour as lack of flexibility.

The need to be highly flexible and responsive creates challenges for the service provider. It was suggested that it would be 'very rare that we would say 'No'. It was often necessary to meet the defence customers' needs and agree afterwards how this should be handled in terms of the contracted agreements. This would mean that the provider needed to rely on, for example, 'minutes of meeting for their authorisation to undertake work'. This process of meeting requirements and negotiating contractual impact afterwards relied heavily on mutual trust between the provider and the customer, and presented challenges for the provider.

The client's behaviour when fast response is needed is to ignore costs. They do not expect to discuss how much the new requirement may cost—they expect action, costs can be discussed later. Commercial providers try hard to avoid spending money on tasks that are barely defined and for which no formal order exists. Responsive behaviours that please the client are not always understood in the provider's back office.

Nevertheless it can be no surprise that the client has not fully[2] put their flexibility and responsiveness needs in the contract since commercially, the provider would be taking huge risks and the contract price would rise substantially. The client, therefore, chooses to pay separately for each new requirement for fast response. The on-base providers had become reconciled to the client's need for flexibility and fast response. However, there was less acceptance of the provider's commercial requirements, with little progress on alternative co-created administration processes to handle emergencies and help on-base providers satisfy their back office managers and accountants.

5.5.1.2 Skills maintenance

The fear of loss of product capability and knowledge is likely to remain important since in forward positions, the squadrons need good current knowledge of their aircraft. They do not outsource the whole of Tornado support and remain involved in depth support for sound strategic reasons. Thus with lower numbers of RAF engineers and fewer opportunities for advancement, the development and maintenance of past levels of RAF capability will need active co-operation from industrial partners. For example, in planning actual movements into the roles necessary to equip RAF engineers for more senior roles, the RAF no longer has full control over each role; the provider's staff occupy many of these roles. Thus the two services in Table 5.1 that provide manpower from the Provider and the Client into the Partnered Direct Service organisations will need to be strongly linked. This issue, a problem for the RAF and an opportunity for the provider, is not unanimously appreciated. It may be regarded as a chore rather than a critical

---

[2] The contract does address "surge" requirements using the key metrics, however many requirements for flexibility are completely new.

piece of value add for the client since the provider can feel that 'by
trained them up, they're thinking of the next posting'.

The provider and client have an opportunity to co-create an im
this area. A review and modification of both technician and engi
education may enable training to be delivered more swiftly. These trai
neers and technicians benefit the provider since they are more capable of carrying out depth support; they benefit the client by enabling technicians and engineers to gain and maintain more capability than otherwise; and finally they benefit the client and provider by maintaining and developing the client's technical and commercial intelligence.

5.5.1.3 Respite

The provider is implicitly required to provide a respite service that maintains an RAF ethos embedded in the partnered organisations they manage. There is already some sensitivity to this on base among BAE Systems managers in terms of what can and can't be done with RAF technicians. This is a semi-reactive stance but articulated explicitly as here, this service could provide a trigger for ideas to further improve mutual understanding and respect in provider–client relationships. This is another significant opportunity for value co-creation between the client and provider.

5.5.2 Implicit Services

Most of the implicit services in Table 5.1 concern dependencies for the effective provision of contracted operational services. These are generally well represented in the contract and are an acknowledgement of the need for the client and provider to co-operate to deliver the operational metrics specified in the contract. The cost reduction and improvement service, however, is less well described. As shown in Sect. 5.4.2, ongoing cost reduction is expected by the client yet the contract appears to say little of how a cost reduction service might work. Compared to the other implicit services, a cost reduction service can involve any of the stakeholders in the client and provider organisations and extend into the supply chain. It is, therefore, a key "enterprise level" service, requiring involvement, co-operation and effort to implement new processes and ways of working across multiple stakeholders. It could also reasonably generate new implicit service requirements. For example, the forward data provision service on faults; see Table 5.1.

In summary, un-contracted services based on client aspirations and fears provide opportunities for service improvement and value creation. If they are articulated as services that need structure, management and review—perhaps to—they potentially promote trust in the provider by demonstrating that client requirements are understood, respected and being actively managed. Paying attention to more complex, enterprise level implicit services may be important to:

a. ensure that a collaborative approach is taken to complex inter-dependent tasks;
b. jointly generate proposals for future services rather than react to the client requests; and
c. signal the need for enterprise management of complex support organisations.

5.6 Preliminary Findings, Limitations and Future Research

This section is divided into two main parts. First, a discussion on findings with respect to previous research; new findings that are particular to military availability contracts and new findings worthy of wider recognition. Second, a critique of the research methods used, discussion of the paper's contribution and proposals for further research.

5.6.1 Discussion of Findings

This research supports Jost et al. (2005) study on the importance of reconciling partner objectives. In this case, there was evidence to suggest that the nature of partner motivations, requirements and, therefore, objectives are complex and interdependent. Partners in the ATTAC contract had diverse initial objectives that were recognised but little discussed. The extent to which objectives were reconciled was not clear, management processes that spanned the enterprise were little developed, however, the need for such enterprise management processes appeared to be emerging as we reported our findings jointly to providers and clients. It is clear, however, that a joint overall objective for ATTAC was not explicitly articulated. In our minds this would have been concerned with supplying an aspect of the UK's Defence capability effectively, safely and economically.

Jost et al. (2005) also asserted that complex service contracts were inherently uncertain since all eventualities could not be predicted at the outset. Thus reconciliation of objectives required regular discussion and review of objectives to ensure that evolving partner requirements were understood and taken into account. This research suggests some of this evolution might be accelerated if the partners' requirements were more explicit at the outset.

Implications from the preliminary findings from this case will be discussed in three sections:

1. Translating requirements into a contract;
2. Findings particular to complex military service out-sources; and
3. Findings worthy of wider recognition in complex service provision.

5.6.1.1 Translating Requirements into a Contract

A series of stakeholder motivations and fears have been identified, some of which are being met through the service contract while others are currently neither met, nor contracted for, in a systematic way. It is interesting to note that the contracted services in the case are services aimed at the 'product', while the uncontracted services are what Mathieu (2001) would call services for the 'client'; for example, maintaining engineer and technician capability. Likewise only the operational support of Tornado aircraft drew attention in the development of performance indicators and targets. This suggests both the client and provider quite reasonably have much to learn on the meaning of "service" in their context, as both took an enormous step from their 'product dominant' experience base and traditions in contracting for ATTAC. It is also possible that in not fully articulating their requirements in the contract, they obtained a more competitive contract price than otherwise. While the inexperienced lead service provider concentrated on tangible "product" aspects of the contract, the number of new requirements the client would need was significantly underestimated. The client must rely on provider goodwill for un-contracted services, especially in terms of avoiding steps that might oppose their long-term aspirations and needs.

The contract was created around the need for cost reduction rather than joint value creation and this will inevitably bias the nature of the partnership and slow the development of trust. Nevertheless, there is evidence that trust and mutual respect and understanding are developing in this contract, especially on base. However, the need for cost reduction and service improvement led by the main provider, though clear in the contract in terms of performance improvement metrics and a gain/share agreement, was not made explicit in terms of the duties of all partners, including the client. All partners need to be prepared to change methods, invest in training and other implementation aspects of cost reduction and service improvement if holistic enterprise improvement is to be achieved. At the time of ATTAC contract development this was an unpopular move, as far as many in the RAF were concerned, and this did not provide an environment where the duties of the Air Command in cost reduction activity could be discussed. Relationships have developed and now may be the time to discuss these aspects. The economic situation of the client has significantly worsened since contract signature, thus the cost reduction aspiration of the DE&S client and, we suggest, any UK public sector service client needs diligent attention. Ongoing service improvement, however defined, is a key potential advantage of public sector outsourcing to industry since the complementary capabilities and mindsets of industry towards cost reduction and improvement in general need to be fully exercised.

5.6.1.2 Lessons for Complex Military Support Outsourcing

The specific lessons for military outsourcing concern the strategic necessity for them to be incomplete, and the linked necessity for the industrial provider to maintain a military ethos across the service operation. Military clients must be knowledgeable about their equipment to be effective in a war zone where limited civilian expertise is available. This is particularly the case for aircraft where safety considerations vis a vis the public (as well as the military) are high. First-line maintenance of warships, tanks and other weapons cannot be outsourced to civilians in a war zone. There are large differences between the knowledge required to conduct first line maintenance; for example, in the air sector, training for both engineers and technicians is a lengthy process and now takes place in an environment where civilian and military personnel work together.

5.6.1.3 Lessons for Other Complex Service Outsourcing

We see no reason why the following preliminary findings from the ATTAC case are not of wide interest to complex service outsourcing across the public sector. The particular lessons we would draw attention to are:

a. The need for over-arching enterprise objectives to help reconcile individual partner objectives;
b. Evolving requirements and their role in understanding that non-contracted, yet vital services can assist partnership development as well as providing opportunities for value co-creation; and
c. Making explicit "implicit service requirements" that require management processes involving all partners, is a vital step towards holistic management of a complex service enterprise. 'Service improvement' services would appear the key in this context.

5.6.2 Limitations and Further Research

This research has three key limitations; the research methods used, its client centric focus and the normative assumptions on which the research is based.

First, the interview methodology could not achieve a comprehensive set of client aspirations and fears. This could be addressed using a framework that tested informants' views across the documented pros and cons of service outsourcing. In defence, the original interviews were focused on gaining knowledge across the ATTAC case and the methods did much to ensure the data collected were reliable. It was then concluded that the data generated on client aspirations and fears were sufficiently interesting to explore and publish. Second, the study was clearly client-centric; we now believe the providers' perspective to be equally important. Providers have rights, and from our observations, it appears an explicit view of the client, provider and product services are required to organise and negotiate Enterprise level management and improvement as well as set the scene for contract development. We thus concur with Gummesson's (2008) appeal for "balanced" rather than "customer centricity" in the case of complex service outsources. Third and finally, a root assumption of the research is that all partners will co-operate to make their aspirations and fears explicit and enable a reconciliation of their needs with the needs of the enterprise at large. As a result of that knowledge, service improvement at the Enterprise level can begin, as partners can more clearly view the implications of change for other partners and thus negotiate change. However, they may not wish to do that for many reasons—they lack trust in one another; they are not used to working with each other in this way; their back offices and corporate governance will not let them; clients refuse to relinquish part of their power and so on. Having seen how in this case, clients and providers discussed these issues at our validation report-backs, the benefits are clear for those supplying service and their client contacts. The road is likely to be much more challenging in their back offices. It is our belief, however, that those clients and providers who do experiment in this way are more likely to learn and thrive in complex service contexts.

Given these criticisms, what is the contribution? Research of this kind is relatively rare; it exposes aspects of the empirical reality of complex service enterprise life and that is the contribution. There is much research in the services management, supply and value chain arena suggesting generalised advice, as yet there is little in-depth empirical data on how aspirations and fears are translated into contracts or the utility of those contracts to form a basis for enterprise management.

Further research is required to more comprehensively identify client and provider aspirations and fears and thus the operational, support and un-contracted services required from both perspectives. This should not be restricted to the lead provider, for example in ATTAC at least three "second tier" providers could be (or could have been) included. Also, much more needs to be known about the means of promoting and supporting value co-creation in public sector service outsourcing and the conditions in which it may thrive. There may be opportunities to investigate this area in the near future if governments see service outsourcing as a significant and realistic means of implementing budget reductions.

Overall the study confirms previous advice for service providers to fully understand the value propositions of their clients and amplifies the importance of this advice when dealing with a complex public/private sector service enterprise. It illustrates how requirements can be framed in terms of the additional services required to deliver the core services. The research also suggests that un-contracted aspirations can be translated into potential additional services from both the client and provider that, jointly recognised, can lead to improved mutual understanding, respect, opportunities for further value co-creation and increased recognition of the need for enterprise level management.

5.7 Chapter Summary Questions

The chapter has highlighted challenges faced in translating client aspirations and fears into a complex service support contract and the potential benefits of understanding the clients' full requirements. Even though they may be unaffordable or too difficult to contract, many questions remain. In a complex service partnership, given the roles of providers and clients are to co-produce the service and co-create value, it is essential that all stakeholders openly understand the requirements of the client and provider constituencies.

- How can a sufficiently comprehensive and shared understanding of client *and* stakeholder aspirations and fears be achieved?
- How might these aspirations and fears be converted into services required by an "ideal" contract for both provider and client communities?
- Since an 'ideal' contract will be unaffordable, how best might un-contracted service requirements be respected?
- How might service improvement requirements be implemented more successfully in such partnerships?

Acknowledgements The authors would like to acknowledge and thank their sponsors within BAE Systems, Louise Wallwork and Jenny Cridland, and especially John Barrie for his support and direction. We would also like to thank the Tornado IPT, RAF Marham and the BAE Systems ATTAC management team for their engagement with our work. This research was supported by BAE Systems and the UK Engineering and Physical Sciences Research Council via the S4T project.

References

M.K. Ahuja, K.M. Corley, Network structure in virtual organizations. Org. Sci. **10**(6), 741–757 (1999)

M. Binder, B. Clegg Enterprise Management: A new frontier for organisations. Int. J. Prod. Econ. **106**(2), 409–430 (2007)

A. Davies, Moving base into high-value integrated solutions: A value stream approach. Ind. Corp. Change **13**(5), 727–756 (2004)

A. Davies, T. Brady, M. Hobday, Organizing for solutions: Systems seller vs. systems integrator. Ind. Mark. Manag. **36**, 183–193 (2007)

J.H. Dyer, K. Nobeoka, Creating and managing a high-performance knowledge-sharing network: The Toyota case. Strateg. Manag. J. **21**(3), 345–367 (2000)

N.W. Foote, J. Galbraith, Q. Hope, D. Miller, Making solutions the answer. McKinsey Q. **3**, 84–93 (2001)

J. Galbraith, Organizing to deliver solutions. Org. Dynam. **31**(2), 194–207 (2002)

E. Gummesson, Extending the service-dominant logic: From customer centricity to balanced centricity. J. Acad. Mark. Sci. **36**, 15–17 (2008)

A. Helander, K. Moller, System supplier's customer strategy. Ind. Mark. Manag. **36**, 719–730 (2007)

G. Jost, M. Dawson, D. Shaw, Private sector consortia working for a public sector client: Factors that build successful relationships: Lessons from the UK. Eur. Manag. J. **23**(3), 336–350 (2005)

T. Markeset, U. Kumar, Study of product support strategy: Conventional versus functional products. J. Qual. Maint. Eng. **11**(1), 147–165 (2004)

V. Mathieu, Product services: From a service supporting the product to a service supporting the client. J. Bus. Ind. Mark. **16**(1), 39–58 (2001)

J.F. Mills, J. Schmitz, G.D.M. Frizelle, A strategic review of supply networks. Int. J. Oper. Prod. Manag. **24**(10), 1012–1036 (2004)

J. Mills, V. Crute, G Parry, Value co-creation in a UK air defence, service availability contract—the problem of multiple customer perspectives and diverse cultures. The European Institute for Advance Studies in Management (EIASM) Naples Forum on Services: Service-Dominant Logic, Service Science and Network Theory, Capri, 16–19 June 2009

Ministry of Defence (MoD), Defence industrial strategy: Defence white paper, command 6697. Presented to Parliament by the Secretary of State for Defence. UK, December 2005

R. Oliva, R. Kallenberg, Managing the transition from products to services. Int. J. Serv. Ind. Manag. **14**(2), 160–172 (2003)

R. Ramirez, Value co-production: Intellectual origins and implications for practice and research. Strateg. Manag. J. **20**, 49–65 (1999)

P. Smith Ring, A. van de Ven, Structuring cooperative relationships between organizations. Strateg. Manag. J. **13**, 483–498 (1992)

P. Smith Ring, A. van de Ven, Developmental processes of cooperative interorganizational relationships. Acad. Manag. Rev. **19**(1), 90–118 (1994)

S. Vangen, C. Huxham, Nurturing collaborative relationships: Building trust in interorganizational collaboration. J. Appl. Behav. Sci. **39**(1), 5–31 (2003)

S.L. Vargo, R.F. Lusch, Service-dominant logic: What it is, what it is not, what it might be, in *The service-dominant logic of marketing: Dialog, debate and directions.*, ed. by R.F. Lusch, S.L. Vargo (ME Sharpe, Armonk, 2006), pp. 43–56

S.L. Vargo, R.F.L. Lusch, Service-dominant logic: Continuing the evolution. J. Acad. Mark. Sci. **36**(1), 1–10 (2008)

C. Windahl, N. Lakemond, Developing integrated solutions: The importance of relationships within the network. Ind. Mark. Manag. **35**, 806–818 (2006)

C. Windahl, P. Andersson, C. Berggren, C. Nehler, Manufacturing firms and integrated solutions: Characteristics and implications. Eur. J. Innov. Manag. **7**(3), 218–228 (2004)

R. Yin, *Case study research: Design and methods*, 1st edn. (Sage, Beverly Hills, 1984)

R.K. Yin, *Case study research: Design and methods*, 3rd edn. (Sage, Thousand Oaks, 2003)

Part II
Delivering Service Contracts

Rajkumar Roy

Service contracts underpin a longer term and closer relationship between a firm and its customer. Delivering a service contract for complex engineering systems requires redefining the value proposition; dealing with additional uncertainties and therefore costs; exploring opportunities for additional revenue generation and developing a sustainable risk-sharing environment within the supply chain. The contracting for services could bring additional benefits in comparison with the traditional contracting for product sale for the customer, the firm and its supply chain. The benefits could come from reduction in the whole life cost of the service provision, stronger partnership within the supply chain, increasing the loyalty of the customer and new revenue generation opportunity for the firm and its suppliers. The service contracts play a major role in shaping an environment for the long-term relationship essential for the complex services. This section presents the key concepts and challenges involved in delivering a service contract.

Complex Engineering Service Systems delivers value-in-use in partnership with the customer. This is a shift from good dominant logic of business to a service dominant logic. The shift in business logic requires transformation in organisational capability for both the customer and the firm. This section identifies the capabilities required to co-create the value within a 'value-web', where a value-web represents integrated capabilities of the customer, the firm and its suppliers. There are seven major attributes of co-creation that impacts the organisational capabilities: complementary competencies, empowerment and control, behavioural transformation, process alignment, behavioural alignment, firm's expectations, customer's expectations. The chapter presents a capability matrix by mapping the attributes against six dimensions of the service capability. The matrix can be the basis to define service based organisational capability within a firm.

Development of service delivery contracts need to understand and reduce the uncertainties involved over the entire duration of the contract. Uncertainty is defined as the indefiniteness associated with the outcome of a situation. Understanding of the uncertainties is even more important for a service contract due to its long-term nature and complex dependencies among the members of a value-web. The service uncertainty is driven by the quality of information flow and knowledge across a given value-web over the duration of a contract, which creates challenges in arranging the operational requirements. Therefore, predicting cost of a service delivery will also depend on the understanding. Service cost estimation is more challenging than estimating the acquisition cost of a product. The service delivery process involves potential demand variability, intangible expectations from customers and relationship management within the value-web, and that brings additional costs to the contract. The chapter presents a mapping between the major classes of uncertainties and the cost drivers for service delivery. The chapter demonstrates the impact of the uncertainties on service cost estimates through a case study. The uncertainty based cost estimation framework could provide essential cost information for a contract development and negotiation, and reduce the whole life cycle cost.

Revenue generation for the firm and its supply chain is dependent on the relationship with the customer and the risk-sharing environment for the service contract. There are increased risks when a firm moves towards the servitization. Incentives are used to influence the motivation and relationships among the members of a value-web. An appropriate incentive mechanism could develop sustainable value-web and reduce the whole life cost. Incentives motivate a firm or a supplier to achieve higher performance than the one specified originally and reduce the service oriented risks. It is often win-win for the firm and its customer. The chapter identifies the impact of incentives on organisational behaviour within a service contract through a case study. One of the major attributes for the service contract incentive is to encourage flexibility and adaptability within the value-web. The adaptability requirement often contradicts the whole cost reduction objective. The performance of the service contract depends on the value co-creation with the customer and that means the incentive mechanism may include the performance of the customer.

Finally, the section argues that successful service delivery is dependent on community level relationship among the members of a value-web. Commercial contracts are often limited to capture this aspect of the service delivery in an objective manner. The relationship is developed through relational governance. The governance is the strength of socio-commercial relationship norms present in the service delivery. The chapter observes the transition from an interpersonal relationship to a community level one with the maturity of two contracts. The study also highlights the challenges in defining the performance measurements for service contracts. It is observed that the relational governance is more efficient than purely commercial contract governance in delivering the service based contracts. Due to the dynamic nature of the service contracts, it is often difficult to specify the boundaries and roles, and measure the value-in-use as the outcome of the service. It is necessary that the service delivery focuses on achieving the required value-in-use rather than just satisfying the key performance indicators of a contract. Interpersonal relationships among the key personnel across the value-web play a major role in co-creating the value, and therefore selection of a right team is even more important for the service delivery.

The section provides the foundation for service contract development and delivery. The four chapters address the key aspects of the service contracting and highlight the need for more flexible relationship based governance for the service delivery.

Chapter 6
Redefining Organisational Capability for Value Co-creation in Complex Engineering Service Systems

Irene Ng, Sai Nudurupati and Jason Williams

Abstract There is evidence that service transformation is bringing substantial benefits to traditional design and manufacturing organisations leading them to invest in transforming into service firms co-creating value with their customers. However there is lack of understanding in how these organisations can effectively and efficiently (re)design their service delivery to co-create value with customer to attain optimal benefits. This chapter explains the seven key attributes that are essential in value co-creation: complementary competencies, empowerment and control, behavioural alignment, process alignment, behavioural transformation, firm's expectations and customer's expectations. It describes how the seven attributes demand the need for organisational structural change. The chapter then describes how the six dimensions of organisational capability, i.e., competence, capacity, culture, structure, systems and infrastructure should be redefined for better value co-creation and proposes key actions organisations need to take to develop the capability for value co-creation. In doing so the chapter provides a starting point for organisations to understand and begin to plan how their organisational capability could be re-configured for enhanced co-creation of value.

I. Ng (✉) · J. Williams
University of Exeter Business School, Exeter, UK
e-mail: irene.ng@exeter.ac.uk

J. Williams
e-mail: Jason.Williams@harmonicltd.co.uk

S. Nudurupati
Manchester Metropolitan University Business School, Manchester, UK
e-mail: S.Nudurupati@mmu.ac.uk

I. Ng et al. (eds.), *Complex Engineering Service Systems*, Decision Engineering,
DOI: 10.1007/978-0-85729-189-9_6,

6.1 Introduction

Traditional academic literature focuses on value as exchange value (e.g., Marshall 1927; Thomas 1978). This notion of exchange value underpins the traditional customer–producer relationships, where each party exchanges one kind of value for another, with something in exchange for something else. With the advent of servitization (Anderson and Narus 1995; Neely 2007; IBM Research 2005), a value-centric approach with service dominant logic puts delivering value-in-use as the key to superior competitive advantage (Vargo and Lusch 2008; Ng 2009). Consequently, to achieve value-in-use, the firm has to ask how value is created and understand the role of the customer within that context (Lengnick-Hall 1996). The challenge lies in changing the nature of collaboration between the firm and its customers. This challenge is amplified and becomes more complex when the firm shares resources across multiple contracts with different degrees of involvement with the customer. This is prevalent in complex engineering service systems such as firms operating in the defence industry (aircraft manufacturing, maintenance, repair and overhaul services). Collaborative activities become increasingly complex as they cross organisational boundaries.

According to Marion and Bacon (2000), traditional organisations with a closed system approach of complexity limit their organisation's ability to adapt to its environment resulting in loss of control and opportunities. Hence cross-organisational activities should be managed with an open systems approach by developing flexible capabilities to continuously change and co-create value with customers (Brodbeck 2002). Firms can increase their effectiveness by achieving a good 'fit' between their structures and coordinating mechanisms and the context in which they operate (Drago 1998). To collaborate in this way, organisations must nurture flexibility within processes and procedures and encourage and empower employees to reactively self-organise as change occurs. Such procedures, modelled on complexity theory, would suggest a capability to adapt and provide an enduring fit between structure and context (Brodbeck 2002). Competitive advantage may be gained through creating the capability to continuously adapt and co-evolve within the complex environments created, embedding a system capable of undergoing continuous transformation in order to respond to a dynamic business environment (Brodbeck 2002; Lewin et al. 1999).

The objective of this chapter is to discuss the seven attributes of value co-creation presented by Ng et al. (2010) and their impact on the organisational capability required to deliver complex services. A new definition of capability is proposed that is configured to enable effective co-creation through a value-web (capability integration perspective), rather than delivering its value proposition through a value chain (vertical integration perspective). Initially, it presents the seven attributes from a more technical paper on the subject. Later on it presents the six dimensions of service capability identified from literature. By mapping the dimensions of capability onto the attributes of value co-creation, we propose a matrix that redefines organisational capability and suggest actions for firms to achieve that capability.

6.2 Attributes of Value Co-creation

In service delivery, the *value* of the service is embedded in the processes and interactions between the customer and the firm over time. Recent literature have discussed these interactions as those where the value is co-created between the firm and the customer. For example, maintaining and servicing equipment and parts on site, integrating systems, or training. Consequently, whether benefits to customers are attained through tangible goods or through the activities of firms, a customer-focused orientation would focus on achieving value-in-use, delivered by the outcomes rendered by the firm's value proposition of goods or activities (Vargo and Lusch 2004, 2008; Tuli et al. 2007).

In striving to achieve value-in-use the firm now becomes an essential part of the consumption process. According to this perspective the relationships, service delivery processes and interactions between firm and customer become crucial in determining the value created. Hence greater concern should be placed on post-purchase interactions as they directly impact on value creation and the likelihood of future contracts and revenues (Bolton et al. 2008).

6.2.1 Value Co-creation

In co-creating value, firms do not provide value, but merely propose value (Vargo and Lusch 2004) and it is the customer that determines the value by co-creating it with the firm. As Ballantyne and Varey (2006) puts it, a "customer's value-in-use begins with the enactment of value propositions" (p. 337). Hence, a firm's product offering be they goods or activities, are merely value unrealised. Through this logic the offering is a, "store of potential value" (Ballantyne and Varey 2006, p. 344) until the customer realises it through a process of co-creation and gains the proposed benefit. Value co-creation is therefore the customer realising the value proposition to obtain benefits (value-in-use).

Woodruff and Flint (2006) suggested that customers have an obligation to assess the provider's needs and to assess their resources to deliver these needs as part of the co-creation of value. In doing so, there is a need to understand the role of the customer in the firm's processes and systems, and vice versa. Payne et al. (2008) developed a process-based framework for co-creation in which they proposed customer value-creating processes, firm value-creating processes and encounter processes through which customers derive benefits from the firm's value propositions. Ulaga (2001) argued that suppliers and customer organise the service system where value is jointly co-created with superior value arising from the effective combination of core competency and relationships (Kothandaraman and Wilson 2001).

Thus, for co-creation to be understood in its fullest the customer's role in attaining benefits for themselves cannot be ignored and firms have to face the challenge of understanding customer consumption processes (Ballantyne and Varey 2006). While there is clearly a need to better understand the dynamics of how value is co-created, literature in this area is scarce. Most research has discussed value co-creation in terms such as interactions, relationships, reciprocity, and customer orientation. Value co-creation has also been described as, "spontaneous, collaborative and dialogical interactions" (Ballantyne and Varey 2006, p. 344). Whilst accurate, such descriptions are not useful for developing the organisational capability or for designing services that effectively co-create value.

Oliva and Kallenberg (2003) noted that the transition from a transaction-based business model to a relationship-based model requires the firm to develop the capability to co-create value. This in turn requires an evaluation of organisational principles, structures, and processes—a major managerial challenge. This is also echoed in redefining the value chain towards a 'web' model (Prahalad 2004) or 'value constellations' (Normann and Ramirez 1993; Ramirez 1999) that could enable more effective value co-creation. This is particularly important for organisations that deliver to outcome-based contracts (see Ng et al. 2009), where the focus on delivering to outcomes extend the boundaries of the organisation's responsibility, compelling the organisation to co-create value with the customer and embed value co-creation as an organisational capability.

Current literature places more emphasis on relationships and less on organisational or service design that could facilitate such relationships. Much of the research in value co-creation resides in the theoretical and conceptual domain with little empirical evidence (e.g., Prahalad and Ramaswamy 2004; Vargo and Lusch 2004, 2008; Lengnick-Hall 1996). It is still unclear *how* firms should design their service delivery to co-create value with customer to attain the highest benefits. In other words,

> What attributes should a service system exhibit that would enable co-creation of value with the customer to attain beneficial outcomes and how should such attributes impact on the organisation's capability?

A study to uncover the attributes was conducted and the results reported in a more technical paper (see Ng et al. 2010). The following section will present the key aspects of the study and following on, develop a framework on managerial implications for organisation capability.

6.2.2 The Seven Attributes of Value Co-creation

The seven attributes were initially discovered through qualitative study, with data collected through interviews, participant observation, analysis of texts and documents as well as recording and transcribing. The data was then analysed through a grounded theory approach (Strauss and Corbin 1990) to arrive at the seven attributes of value co-creation (AVCs). The study also operationalised these AVCs and internally validated them using Exploratory and Confirmatory Factor Analysis from the data obtained through an internal survey. It found that the role of the customer in achieving outcomes or value-in-use is dependent on use practices in different contexts which 'push back' into the organisation's system. This meant that organisations have to develop a capability to manage open systems and where even when the customer and the firm do the exact same thing each time, the context changes and together with it, benefits, satisfaction and costs. The seven AVCs derived from Ng et al.'s (2010) study provide a starting point towards changing the internal organisation to ensure more effective interfaces with the customer.

1. Complementary competencies	5. Behavioural transformation
2. Empowerment and control	6. Congruence of the customer's expectations
3. Behavioural alignment	7. Congruence of the firm's expectations
4. Process alignment	

The first of the AVCs is *complementary competencies,* described as both the customer and the firm employees having to provide the right competencies, in terms of expertise and judgment, and complementary resources. Organisations exhibiting such an attribute would benefit from improved planning with increased resource demand and cost predictability. When the customer shares complementary information, material and skills, the firm will have the opportunity to learn and develop new technologies, skills and behaviours necessary in delivering the availability of service (Ng and Nudurupati 2010).

The second AVC is *empowerment and perceived control. Empowerment* is described as "employees with suitable autonomy to make situational decisions as well as to implement new ideas". *Perceived control* is defined as "employees and customers ability to demonstrate their competency over the environment". During the course of service delivery changes in roles and responsibilities cause discomfort and disruptions resulting in a reduced sense of control and security within individuals. Hence empowerment of the employees in the firm allow them to turn problems into opportunities, exercise personal judgement for greater effectiveness would improve the service efficiency and effectiveness. Also, allowing customers sufficient visibility of service delivery information and processes renders employees of both organisations better perception of control.

The third AVC is the *behavioural alignment* between firm's and customer's personnel. Success in co-creation is highly dependent on personal relationships. Ensuring the right behaviours such as co-operation, teamwork, trust and open communication (of plans) is essential in delivering the required outcomes.

To facilitate behavioural alignment, the firm and the customer also need to (re)align their processes to enable exchange of information through emails, telephone, meetings, and seminars. The processes should also enable smooth flows of material/equipment between the firm and the customer to enable efficient service delivery. For these reasons *process alignment* was identified as the fourth AVC.

The fifth attribute of *behavioural transformation* is essential for delivering outcomes (value-in-use) as customers need to be educated on the best usage of the firm's assets and activities. Thus, firm employees have to transform the behaviours of customers to ensure better usage in achieving outcomes. Better usage results in lower costs of delivery and higher satisfaction.

If the organisation and the customer have overlapping skills and roles, it creates ambiguity such as to who should perform certain tasks, why it should be them and not us etc., which can lead to a mismatch in expectations. Hence the firm should understand and be clear of the *customer's expectations* and vice versa. The congruence of expectations between the firm and the customer of each other represents the sixth and seventh AVCs.

The seven AVCs accentuate the need for structural change in firms to enable knowledge sharing, communication, interaction and innovation (e.g., Sawhney and Prandelli 2000; Grönroos 2004). Achieving value-in-use clearly does not follow the typical value chain (Porter 1985) with interactions compartmentalised into marketing, HR, operations, supply chain and logistics. Instead, value co-creation transcends discipline, functional and organisational boundaries of both the customer and the firm, focused only on outcomes and value-in-use. Value is co-created through interactions at every level and with every resource be it equipment or people all co-existing in a common service system. Using the seven AVCs, we redefine the organisational capability required for value co-creation. The dimensions of organisational capability are explored and discussed in the next section.

6.3 Dimensions of Service Capability

Where service is defined as *the application of knowledge and skills for the benefit of another* (Vargo and Lusch 2008), service capability is defined as, *the ability to deliver beneficial outcomes to satisfy customers*. The development and effective execution of service capability is a key source of competitive advantage as it represents the true ability to continually and consistently satisfy customers.

An organisation is said to have service capability when the business model is explicitly focused on achieving the co-creation of value with stakeholders through the application of specialist (core) competencies designed to create benefit for customers. Particular emphasis is placed on the effective utilisation of operant resources rather than operand resources (Constantin and Lusch 1994). In order to explore the dimensions of capability, a brief literature review is presented in the following sub sections.

6.3.1 Capability in Service Transformation

In planning and implementing the service transformation of an organisation towards effective service capability it is essential to understand both the current operating state and that required by the future operating model. A clear understanding of both states will inform the planning process, allowing detailed activities to be structured around the transformation requirements necessary to develop service capability. Capability is used as the basis for assessment because the transformation aims to develop the service capability of the organisation. To build and manage service capability a generic capability basis is required.

A capability perspective assists the understanding, integration and application of capability to achieve common objectives. Adopting capability as the basis of transformation is appropriate because it transcends organisational functions in the same way that effective co-creation of value requires coordinated activity across multiple functions and indeed organisations. Managing from a capability perspective promotes the creation of innovative solutions focusing on the integrated management of interlinking functions (possibly across organisational boundaries) and activities in a strategic context.

6.3.2 Capability in Strategic Management

Capability has been a much-studied topic within strategic management (e.g., Helfat 2000). The growing volume of research on firm capabilities links capability with performance, an indication of the importance of capability in creating and sustaining competitive advantage. While the intricacies of the relationship between capability, performance and competitive advantage are widely debated, it is clearly recognised that a firm's ability to manage and develop its capabilities over time is crucially important and will only become more crucial as levels of competition continue to increase. Most literature would agree that capabilities that are critical to a business achieving competitive advantage contribute to the firm's core competency (Hamel and Prahalad 1990).

Traditionally, organisations have focused on financial, strategic and technological capabilities to gain competitive advantage (Ulrich and Lake 1991). However, it is beginning to emerge that such capability perspectives must be supplemented by another capability, that of the firm's ability to manage people. Such an organisational capability emphasises the strong link between effective people management, performance and competitiveness. Ulrich and Lake (1991) see organisational capability as the glue between the traditional financial, strategic and technological capabilities. *"Managers who are able to understand and integrate all four sources* [of capability] *are more likely to build competitive organisations"* (Ulrich and Lake 1991, p. 82). Today, organisations are increasingly interacting with their customers resulting in value being embedded in a system of interactions. This means that boundaries of what is value is fluid and the organisation's capability has to contend with the management of such a value proposition as well as dealing with customer use variety to realise the value delivered by the firm.

In establishing organisational capability the organisation must therefore become adaptive by establishing internal structures (*structure*) and processes (*systems*) that aid the creation of competences (*competence*). Competence is further nurtured through selective recruitment and importantly, "effective human resource practices" (Ulrich and Lake 1991, p. 77). Recruitment and personal development procedures allow an organisation to build a stable resource base providing the necessary *capacity* to compete in the marketplace. Capability development involves, "adopting principles and attitudes, which in turn determine and guide behaviour" (Ulrich and Lake 1991, p. 77), i.e., the *culture* of the organisation.

Ulrich and Lake define a capable organisation as consisting of four critical elements: (1) A shared mindset both internally and externally (*culture*); (2) make use of management practices to build a shared mindset (*structure* to build *culture*); (3) create *capacity* for change through understanding influence and managing organisational systems (*systems* influence *culture*); (4) empower all employees to think and act as leaders (*structure* and *systems* nurture *competence*). In addition to the five dimensions arising from Ulrich and Lake's definition of organisational capability, there is a sixth dimension, *infrastructure*, that cuts across all four of the capability types (organisational, financial, strategic, and technological) by providing the physical environment needed for the operation of an organisation. Infrastructure includes buildings, equipment, materials and IT systems, all of which facilitate the working and interaction of the other capability dimensions.

By abstracting from Ulrich and Lake's four types of interacting capability, we propose a generic definition of capability and its six constituent dimensions. Establishing a generic model allows the capability to be tailored or nurtured for different purposes. In the case of service transformation we are interested in developing *service* capability, a particular blend that focuses on co-creating customer benefits, and in doing so align with Ulrich and Lake's definition that competitive advantage is build on customer value and uniqueness.

6.3.3 Extracting the Six Dimensions of Capability

Six generic dimensions of capability have been identified based on the examination of extant literature:

1. *Competence*: The level and type of knowledge and skills which can be brought to bear (to an acceptable level).
2. *Capacity*: The level of output possible in a given time period with a predefined level of staffing, facilities and equipment.
3. *Culture*: Collective assumptions, behaviour and values of a group of people.
4. *Structure*: The structure and associated governance mechanism that controls activity.
5. *Systems*: Processes, procedures and tools used to transform inputs into outputs.
6. *Infrastructure*: The material, equipment and physical environment that supports operational activity.

The six dimensions of capability are mapped against the seven attributes of value co-creation and represented as a matrix that illustrates the capability for value co-creation, which according to current literature (Maglio and Spohrer 2008), is how we would define the firm's capability for *service*. The matrix provides a framework for identifying key actions that can be taken by the firm and/or customer to enhance co-creation of value through service delivery resulting in greater benefits. This matrix with key actions is presented and discussed in the next section.

6.4 Redefining the Firm's Capability for Value Co-creation

In Sect. 6.2 we identified the seven attributes of value co-creation and in Sect. 6.3 we identified the six dimensions of capability. In this section, we will discuss how to configure the seven attributes against the six dimensions and the actions that could be taken to enhance an organisation's capability to co-create value through effective service delivery resulting in greater benefit for the firm and the customer.

As outlined in Table 6.1, organisational *competence* to co-create value is defined by how well the firm's institutional and human knowledge and skills can be applied to deliver against the seven attributes. So for example, the competency to ensure complementary competencies is defined by how well the firm's institutional and human knowledge and skills are able to continually ensure a complementary fit between the knowledge and skills of the firm and the customer. This type of competency can be attained with policies that ensure customer and company share information on each other's competencies, to achieve complementarity; policies that are also assisted by joint training and evaluation of technology and assets that are complementary. Furthermore, the firm is required to build expertise throughout the organisation that allow it to understand, align with and adapt to changes in customer expectations, processes and behaviours. Specific skill sets are needed to ensure behaviours are transformed in the firm and customer domains, where necessary changes produce improved outcomes. This is not merely about training of employees but ensuring the roles within the firm are redefined for the execution, rather than impediment, of such behaviours.

Organisational *capacity* is defined as the level and effectiveness of output possible in a given time period with a predefined level of staffing, facilities and equipment (resources). However, outputs are determined by how such resources interact and the quality of the interactions and processes. Thus, the capacity to co-create value is defined by the firm's ability to deploy necessary resources (be it people, facilities and equipment) to facilitate the service delivery in line with the seven attributes. Hence the capacity for complementary competencies is defined by how well the firm is able to deploy the necessary resources to complement the customer's lack of resources to achieve outcomes. Higher levels of trust can improve capacity effectiveness and quality by reducing transaction

Table 6.1 Redefining organisational capability for value co-creation in service—value co-creation capability matrix

DoC/AVC*	Competence	Capacity	Culture	Systems	Structure	Infrastructure
Complementary competences	*Def* The institutional and human knowledge and skills to ensure a complementary fit between the capability of the company and the customer. *Actions* Training to develop analytical and organisational skills. Policies to ensure customer and company share competency information.	*Def* The ability to deploy the necessary resources to complement the customer's lack of resources to achieve particular outcomes. *Actions* Train sufficient number of staff in analytical and organisational skills. Both parties to commit resources to ensure complementarity is maintained.	*Def* The collective assumptions, behaviour and values of employees in the firm that strive for fit between the competencies of the company and the customer. *Actions* Reward examples of win/win situations realised through complementary interdependence between parties. Develop formal skills sharing and work shadowing mechanisms to encourage knowledge transfer and an appreciation of others skills and roles.	*Def* The processes, procedures and tools utilised to map competencies allowing analysis of complementary fit. *Actions* Instil competency assessment of both customer and firm as standard operating procedure during due diligence and subsequent service delivery. Develop tools to allow competency comparison and aid cross functional and cross organisational working.	*Def* A set of governance structures and activities that maintain a core of stability and ability to learn about multiple customer competencies and environments; as well as a flexible and agile interface to manage changes in customer competency. *Actions* Allow agility in resource usage and governance to cope with variety of customer environments and capabilities.	*Def* The utilisation of IT and communications technologies as well as the physical environment to effectively support communication and knowledge sharing amongst the stakeholder community. *Actions* Develop infrastructure that effectively supports collaboration and skills sharing.

(continued)

Table 6.1 (continued)

DoC/AVC*	Competence	Capacity	Culture	Systems	Structure	Infrastructure
Congruence of expectations	*Def* The institutional and human knowledge and skills to continually ensure expectations are aligned across the stakeholder community. *Actions* Training in stakeholder management and partnering. Policies for constant communication and updates between employees of firm and customer.	*Def* The identification and allocation of adequate resources required to manage expectations. *Actions* Train sufficient number of staff in stakeholder management and customer orientation. Policies of open communication and transparency.	*Def* The collective assumptions, behaviour and values of employees in the firm that strive congruence of expectations across the stakeholder community. *Actions* Create a value set that emphasises regular and open communication and the benefits of shared goals.	*Def* The processes, procedures and tools transparently developed with and communicated to the customer where boundaries are clearly mapped and enable effective ongoing communication between parties. *Actions* Develop robust processes for regular and thorough communication across the stakeholder community.	*Def* A set of governance structures and activities that maintains a core of stability and ability to learn about the customer competencies and environments; as well as a flexible and agile interface to manage changes in customer competency and to strategically decide and communicate what the organisation is or is not able to do within its structure. *Actions* Appropriate communications strategy. Governance mechanism to encompass all key stakeholders and sufficient regularity of reviews.	*Def* The utilisation of IT and communications technologies and the physical environment to effectively support communication amongst the stakeholder community. *Actions* Develop infrastructure that effectively supports communication between parties through various medium.

(continued)

Table 6.1 (continued)

DoC/AVC*	Competence	Capacity	Culture	Systems	Structure	Infrastructure
Process alignment	*Def* The institutional and human knowledge and skills required to continually ensure an appropriate degree of process alignment between the company and the customer by recognising and adapting to any changes in customer requirements or processes. *Actions* Training in business process management and stakeholder management.	*Def* The identification and allocation of resources required to ensure compatibility with customer processes for all transformations (information, people or material/ equipment) required to deliver value. *Actions* Train sufficient number of staff in business process management. Develop infrastructure with capacity to support timely process alignment.	*Def* The collective assumptions, behaviour and values of employees in the firm that actively seeks to ensure an appropriate degree of process alignment between the company and the customer. *Actions* Nurture culture of agility that is open to change by rewarding innovation.	*Def* The processes, procedures and tools utilised to transform inputs into outputs together with the customer with a willingness to change and integrate interface processes for more effective co-production. *Actions* Implement Business Process Management to ensure all interface processes are mapped and reviewed regularly. Develop approach for formal process change control to ensure synergy between interface and internal processes within the company. Develop approach for managing customers through process change.	*Def* A set of governance structures and activities that maintain a core of stability and ability to learn about the customer resources, assets and environment; as well as a flexible and agile interface to manage customer process changes. *Actions* Allow agility in resource configuration, equipment and assets use, and governance mechanisms to cope with changing customer requirements and processes.	*Def* The utilisation of IT and communications technologies capable of mapping and communicating interface processes, roles and responsibilities, as well as identifying any changes during the service delivery process. *Actions* Create IT and communications infrastructure capable of mapping and communicating interface processes, roles and responsibilities.

(continued)

Table 6.1 (continued)

DoC/AVC*	Competence	Capacity	Culture	Systems	Structure	Infrastructure
Behavioural alignment	*Def* The institutional and human knowledge and skills to continually ensure an appropriate degree of behavioural alignment between the company and the customer by recognising and adapting to any changes in customer behaviour. *Actions* Role redefinition and development of strong interpersonal skills of firm employees as well as openness to adaptation and flexibility. Develop an Understanding of behavioural psychology.	*Def* The identification and allocation of resources that could be deployed for absorbing changes in customer requirements and behaviours. *Actions* Develop sufficient number of staff with interpersonal and behavioural skills capable of absorbing changes in customer behaviours.	*Def* The collective assumptions, behaviour and values of employees in the firm that actively seeks to ensure an appropriate degree of behavioural alignment between the company and the customer. *Actions* Redefinition of roles and responsibilities to nurture and reward a culture of agility and customer orientation	*Def* The processes, procedures and tools utilised to transform inputs into outputs together with the customer with a formal approach for monitoring and recording changes in customer environment and behaviour to aid delivery. *Actions* Develop the process and tools to monitor customer behaviour in order to identify changes that might affect service delivery and outcomes. Develop approach for formally changing and recording behavioural change within the company's standard operating procedures. Develop robust processes for regular and thorough communication across the stakeholder community.	*Def* A set of governance structures and activities that maintain a core of stability and ability to learn about and manage customer behavioural changes. *Actions* Appropriate communications strategy. Agility in governance and structure to absorb changes in behaviour and process caused by changes in customer behaviour.	*Def* The utilisation of IT and communications technologies to monitor customer behaviour and identify any changes during the service delivery process. *Actions* Utilise communication technologies that effectively facilitate the communications strategy and processes.

(continued)

Table 6.1 (continued)

DoC/AVC*	Competence	Capacity	Culture	Systems	Structure	Infrastructure
Empowerment and Control	*Def* The level and type of institutional knowledge and skills which seeks to achieve suitable empowerment of its employees, and able to assure the customer of the firm's competency, providing the customer with sufficient artefacts for a sense of being 'in good hands' across various contexts of use. *Actions* Role redefinition of employees for greater empowerment when dealing with customer changes. Training in stakeholder management and communications skills. Provide personal and professional development opportunities; including training in appropriate skills areas.	*Def* The identification and allocation of resources to ensure stakeholders remain fully appraised of progress and provided with adequate resource and infrastructure. *Actions* Ensure sufficient number of staff have redefined roles and responsibilities with higher empowerment and control. Ensure sufficient numbers of staff are competent in stakeholder management. Ensure technology investment has the capacity to support the customer's need for control.	*Def* The collective assumptions, behaviour and values of employees in the firm that strive for customer centricity and shared values which work to ensure fair distribution of control and authority between firm and customer. *Actions* Redefinition of roles and responsibilities to empower employees and encourage personal leadership. Nurture a culture of trust and openness through transparency and delegation of authority.	*Def* The processes, procedures and tools utilised to transform inputs into outputs together with the customer with appropriate communication channels to keep the stakeholder community informed and allow their opinions to be captured and influence the service delivery. *Actions* Develop robust processes for regular and thorough communication across the stakeholder community. Allow flexibility in internal processes and systems where benefits in service delivery can be realised.	*Def* Appropriate organisational structure and governance mechanisms to allow effective dissemination of authority allowing service delivery to be controlled for better outcomes. *Actions* Delegation of power and authority to nurture a sense of empowerment in the company and the customer. Use of technology and information systems to ensure transparency and control	*Def* The materials, equipment, technologies and physical environment that provide stakeholders with the visibility of and access to the information, people, equipment, materials and facilities required for service delivery. *Actions* Create an infrastructure that provides employees and the customer with visibility and access to the people, information, equipment and facilities involved in service delivery.

(continued)

Table 6.1 (continued)

DoC/AVC*	Competence	Capacity	Culture	Systems	Structure	Infrastructure
Behavioural transformation	*Def* The level and type of institutional knowledge and skills which seeks to be able to influence and adapt customer behaviour for more effective co-production. *Actions* Training in leadership and people management for firm employees, joint governance and training policies with customer.	*Def* The identification and allocation of resources required to transform customer behaviours as part of the delivery processes. *Actions* Ensure sufficient numbers of staff are competent in leadership and people management. Ensure sufficient technology resources exist to support employees in transforming customers.	*Def* The collective assumptions, behaviour and values of employees in the firm that strive for openness to change and willingness to be involved in orchestrating improvements in service delivery by changing customer behaviour. *Actions* Nurture a culture that strives for continuous improvement through empowerment and reward incentives.	*Def* The processes, procedures and tools utilised to transform inputs into outputs together with the customer with a formal approach and tools for auditing customer behaviours as well as managing any necessary change activity. *Actions* Develop regular customer behaviour audit as a standard operating procedure. Develop an approach for managing customers through behavioural change.	*Def* A set of governance mechanisms and activities that include the transformation of the customer as part of the firm's due process. *Actions* Create governance mechanisms that encourage staff to influence change in customer behaviour where service delivery and/or service outcome benefits can be realised.	*Def* The utilisation of IT and communications technologies to effectively support the company in changing customer behaviour. *Actions* Develop facilities and technologies that provide the tools and means to support company staff in changing customer behaviour.

* *DoC* dimensions of capability, *AVC* attributes of value co-creation

and monitoring demands and costs. Similarly, the necessary resources should be identified and allocated to facilitate the behavioural transformation and process alignment, and where appropriate the relevant training should be provided to sufficient numbers of people to enhance capacity effectiveness while co-creating value with customer.

The organisational *culture* to co-create value is defined as, holding collective assumptions, behaviours and values that are customer focused to collectively co-create value across all co-creation attributes to achieve joint outcomes as shown in Table 6.1. The culture to co-create value largely reflects a partnering culture which encourages, through reward and communication, win/win situations realised through complementary interdependence between parties. A variety of approaches are required to nurture the necessary culture which include: creating a value set that emphasises regular and open communication; formal skills sharing mechanisms; encouraging agility by rewarding innovation; promoting trust and openness through transparency and delegation of authority; and nourishing continuous improvement through empowerment and reward incentives.

The behaviour of individuals within the firm is hugely influenced by the prevailing culture. It is the behaviour of employees, acting upon and within the firm's capability, which ultimately determines the effectiveness of the firm's role in co-creating value with the customer. Hence, nurturing the culture to effectively co-create value should be seen as a formal and critical activity and so prioritised accordingly within any improvement programme.

As shown in Table 6.1, the organisational *systems* to co-create value are defined as the processes, procedures and tools required to consistently manage service delivery exhibiting all the co-creation attributes. In order to achieve this, the firm should deploy the tools and processes to assess and map competencies allowing analysis of the complementary fit across the stakeholder community. To ensure congruence of expectations robust processes should be incorporated to ensure roles, responsibilities and boundaries are clearly mapped and enable effective ongoing communication between parties to clearly understand each other's expectations. To this end the implementation of tools and techniques such as business process management (BPM) or other process change management techniques to map, change and integrate interface processes for better process alignment and more effective co-production are warranted. A formal approach and tools are required to aid the sustainability of delivery through behavioural alignment. Many such tools arise from the areas of team dynamics and collaboration. In addition a formal approach and tools for auditing behaviours and managing any necessary change activity is required to ensure the continued effectiveness of service delivery. Empowerment and control through the use of systems and processes can be developed by allowing flexibility in internal processes and systems to ensure adaptability and agility.

The organisational *structure* to co-create value is defined as the use of structure and governance mechanisms to maintain a core of stability whilst providing the ability to address and adapt to the seven co-creation attributes encountered across customer environments. Consequently structure should provide the ability to learn about and adapt to a variety of customer environments as well as a flexible and agile interface to manage changes in customer capability and requirement as shown in Table 6.1. A sense of empowerment and control in the firm and the customer should be nurtured by delegating power and authority appropriately to levels where the impact of decisions is best understood. Finally, flexible governance mechanisms and dissemination of power should be implemented to encourage staff to influence change in customer behaviour where service outcome benefits can be jointly realised.

The organisational *infrastructure* to co-create value is defined as the material, equipment, IT and physical environment that effectively support service delivery in accordance with the seven attributes, as demonstrated in Table 6.1. The firm should strive to create an infrastructure that utilises IT and the physical environment to effectively support communication and data sharing amongst the stakeholder community which will in turn help ensure congruence of expectations and behavioural alignment. The IT and communications infrastructure should also be capable of mapping and communicating interface processes, roles and responsibilities thus strengthening the process alignment. The infrastructure should effectively support collaboration and skills sharing and thus strengthen complementary competencies. It is important to create an infrastructure that allows the customer visibility and access to the people, information, equipment and facilities involved in service delivery as this will promote empowerment and control within the customer.

The framework (Table 6.1) which maps the six dimensions of capability to the seven attributes of value co-creation provides a prescriptive tool with which practitioners can begin to assess and diagnose the firm's capability to co-create value. Whilst in its current form the framework does not provide detailed guidance or measures with which to interrogate the firm, it is sufficient to form an initial analysis. An assessment can be carried out by comparing the organisation to the definitions laid out in the framework; any significant differences being areas requiring attention in order to enhance the organisation's capability to co-create value. It is important to understand the assessment provides only a perceptual indication of the firm's capability as it is reliant on the opinions, observations and knowledge of individuals within the firm.

Moving beyond assessment, firms seeking to enhance their capability to co-create value can use the actions included within the framework to direct activity within and across the firm. As with its diagnostic abilities, the current framework falls short of providing detailed guidance on coordinating improvement activity but nonetheless provides an indication of how individual constructs (cells within the framework) can be enhanced. This section has demonstrated how the firm can reconfigure its six dimensions of capability to facilitate the seven attributes of value co-creation to maximise benefits for all parties. It has also highlighted some key actions that firms can adopt in order to better configure service design and delivery to enhance the benefits arising from the co-creation of value.

6.5 Conclusion

Service transformation is beginning to bring about substantial benefits to businesses, leading traditional design and manufacture firms to invest in transforming into service firms capable of managing the customer and the co-creation of value in new ways. However current literature places the emphasis more on relationships and less on organisational or service design that could facilitate such relationships. Hence the emergence of knowledge for firms to reconfigure their capability to effectively manage service delivery is still slow.

This chapter developed knowledge with which organisations can more effectively configure the organisation as well as design and deliver service in such a way that it promotes the effective co-creation of value with the customer. Through integrating the dimensions of capability with the attributes of value co-creation a new method of developing and enhancing service capability is proposed. This new approach to service transformation now provides organisations with a tangible framework for transforming organisational capability, service design and delivery, thus presenting a significant step forward in developing effective service capability. The framework is void of any bias towards a particular industry, organisational size, type or sector. In fact the approach to service transformation provides a means for any organisation to improve its ability to co-create value, no matter the nature of the value it seeks to co-create or its current ability to do so. In this respect it presents a significant contribution for organisations seeking to deal with the practical implications of value co-creation. The key revelations this chapter presents to practitioners are:

1. The firm cannot create value, instead it must be co-created through relationships and interactions within a complex service system that includes the customer within that system.
2. The ability to manage the service system for effective co-creation of value with the customer requires a particular organisational capability, *a service capability,* one that would provide a significant competitive advantage.
3. The service capability to effectively co-create value requires firms to be agile in adapting to changing circumstances of customer use contexts and the ability to work across functional and organisational boundaries.

The framework presented here is only the beginning; further research is needed to enable empirical assessment of the capability to co-create value (i.e., how do organisations compare to the framework) and to provide a method for subsequently enhancing capability in a coherent and coordinated manner (i.e., how organisation's can improve their capability in line with the framework). Research is underway to address these requirements and a toolkit is currently being developed to enable the practical application of the framework and its thinking.

6.6 Chapter Summary Questions

This chapter has proposed that an organisation needs to be configured in such a way that it can effectively co-create value. The framework presented above uses six dimensions of capability as the basis for organisational design and poses them against the six attributes required for effective co-creation of value. With these in mind we pose a number of questions to organisations:

- Does the organisation have the right type and level of skills to address the attributes of value co-creation?
- Does the organisation's culture reflect the need to address the attributes of value co-creation?
- Does the organisation have the processes and tools in place to address the attributes of value co-creation?
- Is the organisation structured (including governance and control mechanisms) to provide the flexibility required to address the attributes of value co-creation?
- Does the organisation have the right type and level of infrastructure to support it in addressing the attributes of value co-creation?

References

J.C. Anderson, J.A. Narus, Capturing the value of supplementary services. Harv. Bus. Rev. **73**(1), 75–83 (1995)

D. Ballantyne, R.J. Varey, Creating value-in-use through marketing interaction: The exchange logic of relating communicating and knowing. Mark. Theory **6**(3), 335–348 (2006)

R.N. Bolton, K.N. Lemon, P.C. Verhoef, Expanding business-to-business customer relationships: Modeling the customer's upgrade decision. J. Mark. **72(January)**, 46–64 (2008)

P.W. Brodbeck, Implications for organisation design: Teams as pockets of excellence. Team Perform. Manag. **8**(½), 21–28 (2002)

J. Constantin, R. Lusch, *Understanding Resource Management* (The Planning Forum, Oxford, 1994)

W.A. Drago, Mintzberg's 'pentagon' and organisational positioning. Manag. Res. News **21**(4/5), 30–40 (1998)

C. Grönroos, The relationship marketing process: Communication, interaction, dialogue, value. J. Bus. Ind. Mark. **19**(2), 99–113 (2004)

G. Hamel, C. Prahalad, The core competence of the corporation. Harv. Bus. Rev. **68**(3), 79–91 (1990)

C. Helfat, Guest editor's introduction to the special issue: The evolution of firm capabilities. Strateg. Manag. J. **21**, 955–959 (2000)

IBM Research, *Services science: A new academic discipline?* (IBM Research, Armonk, 2005)

P. Kothandaraman, D.T. Wilson, The future of competition—value creating networks. Ind. Mark. Manag. **30**(4), 379–389 (2001)

C.A. Lengnick-Hall, Customer contributions to quality: A different view of the customer-oriented firm. Acad. Manag. Rev. **21**(3), 791–824 (1996)

A.Y. Lewin, C.P. Long, T.N. Carroll, The coevolution of new organisational forms. Org. Sci. **10**(5), 535–550 (1999)

P.P. Maglio, J. Spohrer, Fundamentals of service science. J. Acad. Mark. Sci. **36**(1), 18–20 (2008)
R. Marion, J. Bacon, Organisational extinction and complex systems. Emergence **1**(4), 71–96 (2000)
A. Marshall, *Principles of economics (1890). Reprint* (Macmillan, London, 1927)
A. Neely, Servitization of manufacturing. 14th international annual EurOMA conference (Turkey, 2007)
I.C.L. Ng, A demand-based model for the advanced and spot pricing of services. J. Prod. Brand. Manag. **18**(7), 517–528 (2009)
I.C.L. Ng, R. Maull, N. Yip, Outcome-based contracts as a driver for systems thinking and service-dominant logic in service science: Evidence from the defence industry. Eur. Manag. J. **27**(6), 377–387 (2009)
I.C.L. Ng, S.S. Nudurupati, Challenges of outcome-based contracts in B2B service delivery. J. Serv. Manag. (in revision) (2010)
I.C.L. Ng, S. Nudurupati, P. Tasker, Value co-creation in the delivery of outcome-based contracts for business-to-business service. AIM working paper series, WP No 77—May 2010 http://www.aimresearch.org/index.php?page=wp-no-77. Under review at J. Serv. Res. (2010)
R. Normann, R. Ramirez, From value chain to value constellation: Designing interactive strategy. Harv. Bus. Rev. **71**(4), 65–77 (1993)
R. Oliva, R. Kallenberg, Managing the transition from products to services. Int. J. Serv. Ind. Manag. **14**(2), 160–172 (2003)
A. Payne, K. Storbacka, P. Frow, Managing the co-creation of value. J. Acad. Mark. Sci. **36**(1), 83–96 (2008)
M.E. Porter, *Competitive advantage: Creating and sustaining superior performance* (Free Press, New York, 1985)
C.K. Prahalad, *The fortune at the bottom of the pyramid: Eradicating poverty through profit* (Wharton School Publishing, Upper Saddle River, 2004)
C.K. Prahalad, V. Ramaswamy, *The future of competition: Co-creating unique value with customers* (Harvard Business School Press, Boston, 2004)
R. Ramirez, Value co-production: Intellectual origins and implications for practice and research. Strateg. Manag. J. **20**(1), 49–65 (1999)
M. Sawhney, E. Prandelli, Communities of creation: Managing distributed innovation in turbulent market. Calif. Manag. Rev. **42**(4), 24–54 (2000)
D.R.E. Thomas, Strategy is different in service businesses. Harv. Bus. Rev. **56**(3), 158–165(1978)
K.R. Tuli, A.K. Kohli, S.G. Bharadwaj, Rethinking customer solutions: From product bundles to relational processes. J. Mark. **71(July)**, 1–17 (2007)
A. Strauss, J. Corbin, *Basics of qualitative research: Grounded theory procedures and techniques* (Sage, Beverly Hills, 1990)
W. Ulaga, Customer value in business markets an agenda for inquiry. Ind. Mark. Manag. **30**(4), 315–319 (2001)
D. Ulrich, D. Lake, Organisational capability: Creating competitive advantage. Executive **5**, 77–92 (1991)
S.L. Vargo, R.F. Lusch, Evolving to a new dominant logic for marketing. J. Mark. **68**, 1–17 (2004)
S.L. Vargo, R.F. Lusch, Service-dominant logic: Continuing the evolution. J. Acad. Mark. Sci. **36**(1), 1–10 (2008)
R.B. Woodruff, D.J. Flint, Marketing's service-dominant logic and customer value, in *The service dominant logic of marketing: Dialog, debate and directions*, ed. by R.F. Lusch, S.L. Vargo (ME Sharpe, Armonk, 2006), pp. 183–195

Chapter 7
Service Uncertainty and Cost for Product Service Systems

**John Ahmet Erkoyuncu, Rajkumar Roy, Partha Datta,
Philip Wardle and Frank Murphy**

Abstract Service orientation in the defence and aerospace industries through availability contracts is creating challenges in cost estimation due to uncertainties. The chapter initially presents the topic of uncertainty in service delivery, which subsequently leads to a classification of uncertainties. Then, the current uncertainty-based service cost estimation processes are explained. By identifying the existing challenges, an improved framework to estimate costs is highlighted. This firstly involves enhanced understanding of the influence of uncertainties and cost estimation capabilities through the analytical hierarchy process and the Numeral, Unit, Spread, Assessment and Pedigree (NUSAP) Matrix approach. Secondly, it involves using agent-based modelling to represent the dynamism in service cost estimates. The chapter ends with a case study application of the approach.

J. A. Erkoyuncu (✉) · R. Roy
Manufacturing Department, Cranfield University, Cranfield, Bedfordshire,
MK 43 0AL, UK
e-mail: j.a.erkoyuncu@cranfield.ac.uk

R. Roy
e-mail: r.roy@cranfield.ac.uk

P. Datta
Indian Institute of Management Calcutta, Kolkata, India
e-mail: ppdatta@iimcal.ac.in

P. Wardle · F. Murphy
BAE Systems, Hampshire, UK
e-mail: phil.wardle@baesystems.com

F. Murphy
e-mail: frank.murphy@baesystems.com

I. Ng et al. (eds.), *Complex Engineering Service Systems*, Decision Engineering,
DOI: 10.1007/978-0-85729-189-9_7,

7.1 Introduction

The defence and aerospace industries are increasingly becoming service oriented through performance-or outcome-based contracts such as Contracting for Availability (CfA), which seeks to sustain a system or capability at an agreed level of readiness, over an extended period of time, by building a partnering arrangement between Industry and their customers. This shift in paradigm has been studied in the servitization literature, where the focus of research has been on explaining the process of manufacturing organisations adopting through-life support for their products (Vandermerwe and Rada 1988; Oliva and Kallenberg 2003; Vargo and Lusch 2004, 2008; Aurich et al. 2006; Roy and Cheruvu 2009). A Product-Service System (PSS) is considered to be a special case of servitization, which provides value by integrating products and services (Baines et al. 2007) where CfA serves as an example approach. The concept value in use has received much interest, as it refers to the benefits that the customer obtains by operating a PSS (Ng et al. 2009). This means that the shift has enhanced the importance of outcomes, which are achieved by integrating goods and activities. As a result responsibilities across the supply network (customer, Industry, suppliers) have been redistributed, whilst focusing on the functional outcome (Alonso-Rasgado and Thompson 2006). Furthermore, the importance of developing an efficient service support system capable of keeping the hardware, software and other assets (e.g., process, people) in good working condition has become essential. This means that all associated parties need to understand the process, competencies, and assets required to deliver the customer's demand (Alonso-Rasgado et al. 2004).

In CfA the shift in the business model is associated with the shift in the level of the service content, where service tends to be considered as processes and is experienced, created or participated in, while its production and consumption are simultaneous (Lovelock 1983). Furthermore, the CfA context promotes an integrated view of the supply chain because of the need to attain the performance requirements of the customer and inability to provide all required support operations (Kim et al. 2007). Driven by this, in order to manage the reliance on external sources, incentives and/or penalties form a type of measure to improve the performance (e.g., availability) over the Contract duration across the supply network. Narayanan and Raman (2004) highlight that in order to achieve high overall performance incentives are necessary across the supply chain. This involves a consideration of a fair distribution of risks, costs and rewards of doing business. Moreover, various aspects, such as information and knowledge sharing and trust-based mechanisms are important to design well-aligned incentive schemes. Thus, in order to achieve the desired outcomes of CfA an alignment between objectives needs to be established. Also, for complex contracts in the bidding stage, various target-cost methods, known as pain-share/gain-share mechanisms could be employed to distribute the risk between buyers and sellers (Hughes et al. 2007).

Some of the main challenges in transforming the business model into CfA include tackling the cultural change to shift to an availability-based context, issues in establishing boundaries between organisational activities, challenges in establishing aligned objectives across the service network and increased complexity and unpredictability in costs (Ng et al. 2009). The challenge in cost estimation (e.g., for maintenance or spares) for Industry is driven by uncertainties, such as equipment usage rate, failure rates, repair turnaround time, beyond economical repair, no fault found rate, obsolescence rate and labour efficiency as well as financial measures, such as exchange and inflation/deflation rate. Considering that Industry traditionally was not responsible for most of these engineering tasks, capturing the dynamism of these uncertainties has created challenges. Along with this factor, the need to consider costs much before compared to the traditional context causes cost estimation challenges. Considering that at the bidding stage driven by cost estimates major agreements with large financial burden are agreed, there is a need to apply rigorous steps to take account of the influence of uncertainty on the cost in CfA.

There is an expectation that availability type contracting will grow in the defence and aerospace industries due to the national defence budget constraints and downturn in the airline industry, competition and growth of the customer's willingness to transfer responsibilities. For instance, in the Availability, Transformation; Tornado Aircraft Contract (ATTAC) service provided by BAE Systems the proposed solution involves the delivery of Tornado fleet availability in the configurations currently foreseen. This partnership between Industry and the customer enables an affordable solution as for instance costs are reduced through lesser arisings and 'No fault found' and saving the United Kingdom Ministry of Defence (UK MoD) some £500 million over the initial 10 years (BAE Systems 2008). Expected growth of CfA further increases the necessity of research that aims to enhance rigour in cost estimation.

The chapter combines literature and findings from real world case studies in complex engineering service systems and proposes methods to improve incorporation of uncertainties in the cost estimation processes. The sections below are organised as follows. Section 7.2 explains the sources of uncertainty in complex engineering services from a value in use perspective for the defence and aerospace industries. This section introduces the role of uncertainty in service delivery, while also classifying uncertainties for PSS. Section 7.3 explains the service cost estimation process in PSS, which includes a prioritisation scheme and the steps to conduct service cost estimation. This section also explains a methodology to reflect the influence of uncertainties in cost estimation. Section 7.4, presents a case study, which consists of results from the applied approaches in identifying uncertainty, and in service cost estimation through agent based modelling (ABM).

7.2 Uncertainty in Complex Engineering Services

Uncertainty involves the indefiniteness of the outcome of the situation and risk is considered to be the chance of loss or injury (Knight 1921; Bernstein 1998). For the bidding stage of CfA service uncertainty refers to uncertainties that arise in predicting the process of service delivery and associated costs. Service uncertainty is driven by the quality of information flow and knowledge across a given service network over the duration of a CfA, which requires alignment of operational activities. At this stage cost estimation commonly suffers from the unavailability of useful data to assist cost predictions, inability to identify the exact sources of uncertainty and link them to cost drivers. The move towards CfA changes the sources of uncertainties for Industry, which includes (Erkoyuncu et al. 2009a):

- Additional uncertainties arising from the in-service phase of the life cycle (e.g., The Concept, Assessment, Demonstration, Manufacture, In-service, Disposal cycle has been used by the UK MoD since 1999, CADMID). Under traditional arrangements these additional uncertainties were managed under follow-on contracts signed towards completion of the manufacturing phase
- Availability contracts also demand 'left shift' of the point-in-time at which some uncertainties are addressed yet the information needed to resolve some of them may not have been developed at the bid time, before the actual delivery begins

As well as the type of contract that is agreed, the scope of services to be offered also affects the types of uncertainties that are experienced (Roy and Cheruvu 2009; Armistead and Clark 1992). The service types that are delivered vary among organisations; however, some of the commonly provided services include spares, maintenance and training (Brax 2005; Oliva and Kallenberg 2003). Industry also provides services, such as supply and support chain management, integrated logistics support, asset management, equipment health monitoring, and reliability trend analysis, which require enhanced know-how. Uncertainty in cost of provision arises because the industry is no longer able to charge for provision on a per-transaction basis, but rather as a bundled and concurrent service at a fixed price.

The main problem in cost estimation arises from the lack of ability for both industry and customers to predict the demand rate and the cost of delivery over future decades. Furthermore, a match between demand and resources enables desired service quality standards to be achieved while also maintaining resource productivity rates (Slack 2010). This hypothetical balance does not tend to occur, which causes fluctuating service quality (Armistead and Clark 1992). In order to rectify this, it is necessary to identify and study the processes involved in service delivery (Johansson and Olhager 2004). Additionally, in the defence and aerospace industries, the issue is more to do with the co-creation of value, which requires enhanced collaboration across the supply network.

7.2.1 Service and the Role of Uncertainty

Service uncertainty is considered in relation to the success in achieving the cost, schedule or performance targets and knowing what the customer targets will be, and how the performance fluctuates over time (Slack 2010; Clark and Armistead 1991). From a project context processes refer to the work breakdown structure

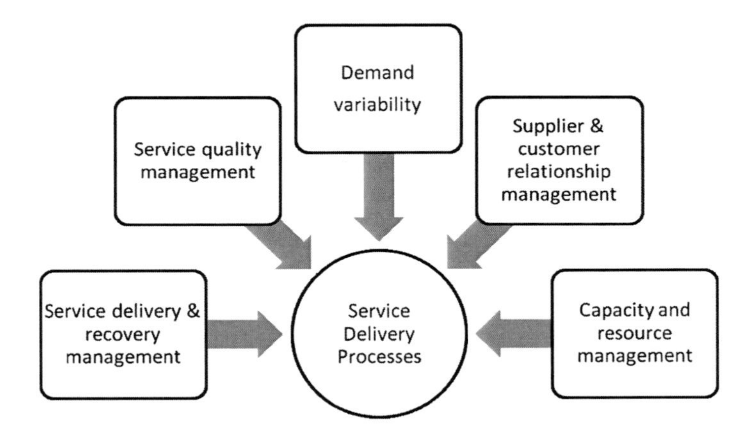

Fig. 7.1 Service delivery processes

(WBS). A WBS should structure the data required for various activities involved in the contract (DoD 2005). The selection of a set of performance measures or key performance indicators (KPIs) is a key step to define the performance assessment of activities in service delivery (MoD 2010). The equipment performance is driven by design for reliability, maintainability, and supportability (MoD 2010), where the significance of these metrics in CfA has advanced. Compared to the traditional model, these metrics are continuously contributing to the availability level and are measured throughout the life of the contract. Figure 7.1 represents the major service delivery processes that have been mentioned in literature. The processes involved in the provision of a service include demand variability, capacity and resources, customer relationship, supplier relationship, service delivery, service quality and service recovery (Cohen et al. 2006).

The role of uncertainty in service delivery is particularly driven by the agreed contract type, while different payment models are available, two of which have received more interest (Ng et al. 2009). Firstly, through a fixed price contract KPIs may be defined, in association with pain and gain share mechanisms. Furthermore, in the fixed contract, unless severe conditions arise, which are notified in the contract, uncertainty is owned by the Industry. In this case, inability to meet targets may result in payments being withheld or reduced. The second approach associates maintenance requirement with the equipment usage level. Within this model the customer brings in uncertainty by varying the level of equipment usage.

7.2.2 Classification of Uncertainty in Product-Service Systems

The types of uncertainties vary across projects. For instance, projects with innovative designs face a larger set of uncertainty, due to lack of experience, e.g., it is not possible to extrapolate from the experience of a prior "spares and repairs" contract. The types of uncertainties are driven by the scope of the support (e.g., contracting approach or service delivery model), which customer demand specifies. Classification of uncertainties based on a systematic process serves the purpose of eliminating or reducing personal bias (Ward and Chapman 2003). This in turn, by better assessment of uncertainty, offers performance improvements in support delivery and cost estimation. Through interaction with companies in the defence and aerospace industries, uncertainties have been categorised into commercial, affordability, performance, training, operations and engineering areas, where the list has been referred to as CAPTOE (Erkoyuncu et al. 2011):

- Commercial uncertainty considers factors that could affect the contractual agreement ("the Contract")
- Affordability uncertainty considers factors that affect the ability to predict the customers budget constraints (e.g., ability to pay) and perception of value for money (e.g., willingness to pay)
- Performance uncertainty considers factors that affect meeting the performance requirements in the Contract (e.g., KPIs)
- Training uncertainty considers factors that affect meeting the delivery of training requirements in the Contract (e.g., trainer availability)
- Operation uncertainty considers factors that are associated to providing the direct activities in maintenance requirement (e.g., in the naval domain onshore activities, fixing breakdown) in the Contract
- Engineering uncertainty considers factors that are associated to providing the indirect activities in maintenance requirements (e.g., in the naval domain offshore activities, obsolescence management) in the Contract

Figure 7.2 represents a classification of uncertainties in alignment with Roy and Cheruvu (2009), which defines a competitive framework for industrial PSS. Whilst the focus in a CfA is performance, this factor is influenced by all the other uncertainty categories. Furthermore, the uncertainty originating from the operating environment is covered under the engineering and operating uncertainty categories. Thus, the list of uncertainties can be considered to be a checklist that supports in achieving cost targets, which may relate to affordability for customer, profitability for Industry and sustainability for supply network. The study was conducted within the defence and aerospace industries, but is applicable on a cross sector basis. Each of the uncertainty categories contains a number of uncertainties, which have been classified into internal and external sources. Internal sources focus on Industry's capabilities that drive uncertainties. External sources focus on uncertainties that originate outside of the Industry (e.g., the customer or suppliers). Some of the internal uncertainties affecting the PSS delivery include performance against KPIs, rate of systems integration, skill level of maintainers, availability of trainers, bid success rate and cost estimation process (Erkoyuncu et al. 2011). Some of the external factors affecting the PSS delivery include rate of capability upgrade, quality of components, trainee skill level, customer ability to spend, and customer misuse (Erkoyuncu et al. 2011).

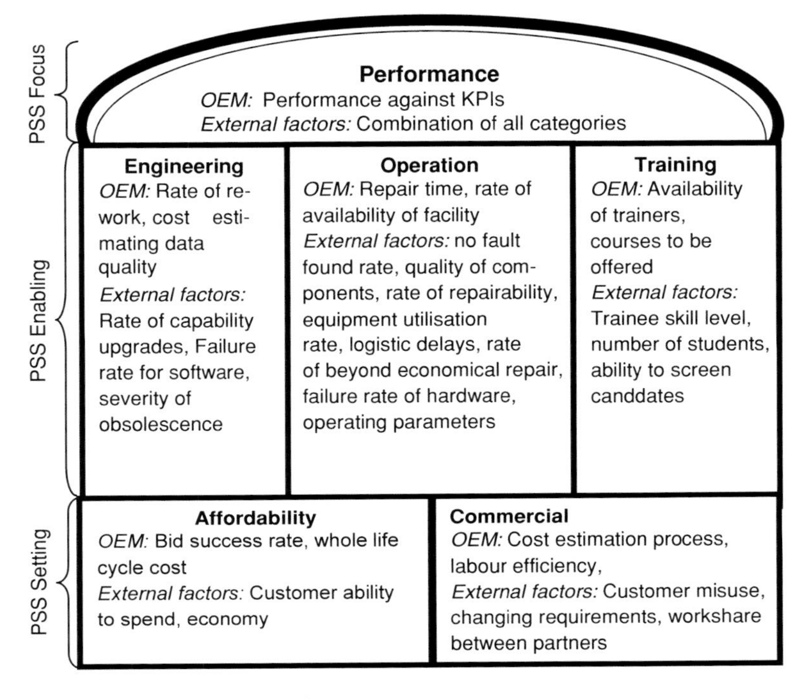

Fig. 7.2 Classification of uncertainties

Table 7.1 Example cost and uncertainty linkage

Uncertainty	Failure rate	Turnaround time	LRU cost	Repair cost	Labour availability
Cost driver					
Customer misuse	x		x		
Rate of emergent work		x		x	
Performance against KPIs		x			x
Capability upgrades	x		x	x	
Quality of engineers		x		x	
Rate of fault investigation				x	x
Repair time		x		x	x

Evidence from the study showed that each uncertainty may influence one or more cost drivers as illustrated in the matrix at Table 7.1. The matrix takes into account the typical cost drivers that arise at the bidding stage of a CfA, such as failure rate, turnaround time, repair cost and labour availability. On the other hand, the uncertainties follow the list presented in Fig. 7.2 along with additional uncertainties. The aim of the matrix is to provide a structured approach to manage the influence of uncertainty over cost drivers. This means that the reasons for the variation in a specific type of cost can be better understood. Although, providing such a list and pre-defined associations between costs and uncertainties may cause a bounded mind set, it particularly suits the bidding stage to enhance effectiveness in considering uncertainties. Furthermore, it also enables a basis for communication across projects to enhance the contents considered, while enabling better overall uncertainty management. Also, the classification of uncertainties into internal (OEM driven) and external factors aims to allocate the responsibility of managing uncertainties to the suitable stakeholders.

7.3 Service Cost Estimation

Cost estimation refers to the process of collecting and analysing historical data or creating analogy (e.g., when there is lack of data) and applying quantitative models, techniques, tools and databases to predict the future cost of an item, product, program or task (Boussabaine and Kirkham 2004; Curran et al. 2004). A cost estimate is ultimately developed to be translated into budget requirements (Roy 2003). Furthermore, a cost estimate also enables to determine and communicate a realistic view of the likely cost outcome, which is based on the considerations in the WBS (NATO 2009).

In the bidding process, Industry is challenged to establish sufficient confidence in a cost estimate given the immaturity of the design solutions for both product and service. Immaturity may arise for technical reasons (e.g., the means for co-creation of value with customer or supplier have not yet been explored), funding reasons (e.g., budget constraints), or timescale reasons (e.g., bid response deadline). Also, both Industry and the customer need to be confident that the cost estimate has not been inflated due to excessive uncertainty budgets or management reserves, which may affect the selling price. Concurrently, customers are challenged in establishing sufficient confidence in their "should-cost" estimates to be able to evaluate industries' bids objectively (Datta and Roy 2009). Under these circumstances it is necessary to focus resources on identifying and mitigating the largest sources of uncertainty. A prioritisation scheme needs to be utilised in order to facilitate this process (Chapman and Ward 2000). Methods, such as probability impact grids and uncertainty rating schemes have commonly been utilised to achieve this (Ward and Chapman 2003).

The Numeral, Unit, Spread, Assessment and Pedigree (NUSAP) Matrix approach has recently received much interest due to its ability to turn qualitative information gathered from expert opinion into quantitative results (Van der Sluijs et al. 2005). The approach consists of five qualifiers that are used to qualify quantities, including numeral, unit, spread, assessment and pedigree. The approach is capable of combining qualitative analysis (numeral, unit, spread) and the systematic multi-criteria evaluation of a given knowledge base. Pedigree is expressed by means of a set of criteria to assess these different types of uncertainties. Although, subjective input is necessary, explicit definition of each score enables a

Table 7.2 Pedigree Matrix: Scoring of uncertainties

Score	Basis of estimate	Rigor in assessment	Level of validation
1	Best possible data, large sample, use of historical field data, validated tools and independently verified data	Best practice in well-established discipline	Best available, independent validation within domain, full coverage of models and processes
3	Small sample of historical data, parametric estimates, some experience in the area, internally verified data	Sufficiently experienced and benchmarked internal processes with consensus on results	Internally validated with sufficient coverage of models, processes and verified data. Limited independent validation
5	Incomplete data, small sample, educated guesses, indirect approximate rule of thumb estimate	Limited experience of applied process with lack of consensus on results	Limited internal validation, no independent validation
7	No experience in the area	No established assessment processes	No validation

reduction of subjectivity as the understanding of concepts and scores is established (Erkoyuncu et al. 2009b). The pedigree criteria covered include basis of estimate, rigour in assessment and level of validation, as explained below:

- *Basis of estimate*. Refers to the degree to which direct observations are used to estimate cost. This measure focuses on the availability of data.
- *Rigour in assessment*. Refers to the methods used to collect, improve and analyse data.
- *Level of validation*. Refers to the degree to which efforts have been made to cross-check the data against independent sources.

Table 7.2 provides the definitions for each of the scores. Results from the approach enable us to classify uncertainties into high, medium and low level of uncertainty for services (e.g., maintenance). Subsequently an adequate process is needed to reflect the influence of uncertainty in cost estimation. Some of the challenges that require additional focus include ability to capture technical scope of service requirement (e.g., modelling of reliability or arising), inflation/productivity expectations and inadequate cost estimation methodology.

7.3.1 Uncertainty Based Cost Estimation Process

Cost estimates are subject to varying degrees of uncertainty. The Ministry of Defence (MoD) specifies nine key steps in cost estimation (MoD 2009), which depict the traditional approach to integrate uncertainty to cost estimation. Firstly, determine the scope of the estimate, by covering the phases of life cycle and the technical elements. Secondly develop a programme baseline, which involves capturing the rules and assumptions and provides the basis for the cost estimate. Thirdly, develop a WBS, which involves structuring of hardware, software, services, data and facilities at agreed and appropriate level of detail for the programme to structure the cost estimation. Fourthly, develop a cost-estimating methodology, which involves consideration of the phase of the life cycle and requirements of the cost estimate. The fifth step involves collection of data, which develops a data collection plan. The sixth step is concerned with data analysis, while applying statistical analysis of data to generate mean values and identify trends and outliers. The seventh step compiles and validates the results by selecting the cost model to develop cost estimates, which also needs to be validated with historical data and trends for similar projects. The eighth step undertakes risk and uncertainty analysis, by integrating risk and uncertainty to the generated cost estimate. A sensitivity analysis tends to be conducted to determine the cost impact of uncertainties. When quantifying uncertainty proves impractical, a qualitative assessment (e.g., using expert judgment) can be made. The ninth step is concerned with generating results, which develops a report on the cost results.

In the context of a PSS, support costs are as much a matter of user choice (patterns of use bounded by equipment capacity) as they are of equipment type and supplier efficiency. The modelling of support costs is distinguished between recurring and non-recurring costs (Roy et al. 2009). Recurring costs include labour payments, spares, repairs and tangible consumable costs. On the other hand, non-recurring costs include one-off costs, such as obsolescence management effort. The estimation of non-recurring costs is harder, due to the stochastic nature of variables (Curran et al. 2004). Adoption of service-oriented business models necessitates enhancement of modelling capability to reflect the dynamic nature of service delivery. Thus, approaches such as agent-based modelling have become popular for their ability to capture dynamism. Some of the main advantages of adopting this approach include (Datta 2007):

- Increased robustness against unpredictability and the dynamic nature of uncertainties that arise in service delivery
- With complicated phenomenon there is necessity to understand the influence of each uncertainty
- An agent-based model can represent many actors in particular their intentions, internal decision rules and interactions

To sum up, there are a number of challenges that are creating the need for an improved framework to consider uncertainty in service cost estimation. The scope of the improved framework in Sect. 7.3.2 is defined by some of these challenges including:

- The need for improving the prediction of uncertainties, such as equipment reliability or failure rates, repair time, demand rate for spares, obsolescence, and technology refreshes over cost estimates
- Difficulties that derive from the lack of useful data and poor timeliness of its availability, which enhances the need for expert opinion
- Limited time that is available to build uncertainty-based cost estimates

- Service delivery particularly depends on the service supply chain, where challenges arise from lack of homogeneity in supplier characteristics (e.g., cost, profit, sustainability)
- Difficulties in systematic representation of uncertainty driven by the be-spoke nature of offerings.

7.3.2 Improved Framework to Integrate Uncertainty with Service Cost Estimation

The improved framework integrates uncertainty to service cost estimation using intelligent agents. In this chapter to capture the dynamic nature of the cost impacts of incentive mechanisms and different risk sharing techniques a simple high level model is constructed with customer, Industry and two suppliers—one supplying spares and another supplying resources represented as agents. The model has the ability to replicate the evolution of maintenance costs for various time scales including the life cycle of the PSS delivery as a whole, a specific mid-term, or for the short-term, however, the minimum time scale of the model is a year. Inside each agent there are many parameters and decision rules that can be configured by the user in order to capture the proposed scenario. Broadly, the agents can be designed to represent any context (e.g., producing, consuming or selling) by assigning appropriate rules to define their behaviour. The parameters and variables serve the purpose of defining characteristics for each agent and they can be adjusted to perform what-if analysis.

Incentive mechanisms play a critical role in designing availability. Since CfA promotes reliance on the supply chain, the presented model considers various uncertainty sharing schemes between the spares, supplier and Industry. The three scenarios include (1) uncertainty with Industry, (2) uncertainty with supplier, (3) Industry and supplier share uncertainty. The complexity increases as the number of stakeholders rises. In order to arrange the incentives the TPPI (Target Price Performance Incentive) type of arrangement between customer and Industry is considered. The key features of this incentive mechanism are:

- It gives a price sufficiently stable at contract signature to allow internal customer and Industry approvals
- A price which easily changes with varying levels of equipment usage
- Fixed unit prices with Supply Chain providers
- A price which keeps uncertainty allowances to a minimum by understanding and allocating risks as appropriate
- Provides a financial motivation (and simple share-out mechanism) for both sides to want to work together to reduce ‘overall’ programme costs, not just contract costs.

The structure of the TPPI is shown in Fig. 7.3. The concepts are based on the air domain; however, the principles apply to other domains as well. A differentiation

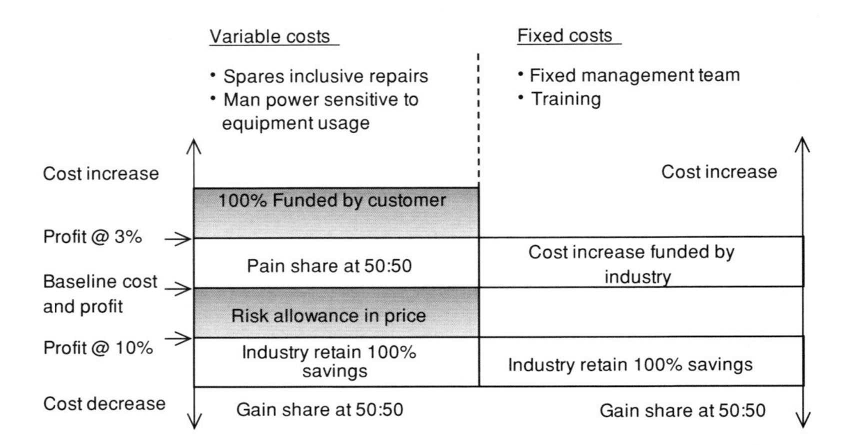

Fig. 7.3 Target price performance incentive mechanism

between variable costs and fixed costs is considered, as different measures are taken between stochastic and fixed costs in the pain and gain share mechanism.

The model assumes that initially a price is set for the pre-defined equipment usage (e.g., flying hours) levels. The price and payment for variable costs are adjusted annually from baseline if projected equipment usage differs from contract assumptions. Furthermore, annual adjustment also compares baseline cost/uncertainty against actual cost/uncertainty spend to calculate either pain share or gain share. Annual adjustment compares baseline cost/uncertainty against actual cost/uncertainty spend to calculate gain share. Industry is expected to attain profits between three and ten percent and any deviation from this range causes implications for Industry and the customer by means of savings or additional costs. The variable costs are subject to 50:50 gain and pain share keeping aside a certain percentage for profit. The assumptions of the model are:

- The model has been developed for peace time
- The customer and Industry have TPPI arrangement
- The spare consumption rate is assumed to be stochastic and expressed by a probability distribution attaining values from 1 onwards
- No cannibalisation is assumed
- Spare consumption rate is independent of equipment usage (e.g., flying hours) and may increase even if equipment usage reduces
- A certain amount of technical investment is necessary to reduce spare repair costs both in case of supplier and industry

Compared to the traditional way of integrating uncertainty to cost estimation, Fig. 7.4 presents the improved framework to integrate uncertainty to cost estimation for a PSS context. The improvement that the framework proposes originates from two sources. Firstly, rigour in assessing the impact of uncertainty on cost is enhanced. Secondly, the framework enables to represent cost in a dynamic manner. The process involves an iterative process, which initially begins by

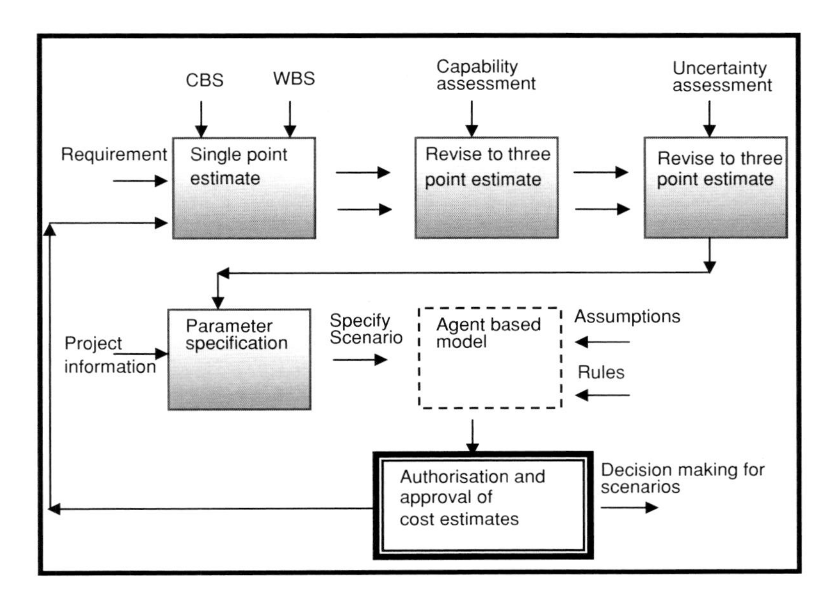

Fig. 7.4 Improved framework to integrate uncertainty to cost estimation

inputting a single-point estimate that has been generated from an alternative cost estimation approach. This estimate is revised by conducting a capability assessment for cost estimation and an uncertainty assessment. The categories and the process are influenced by the Capability Improvement Model for Cost Engineering (CECIM) Industry (Pickerin and Lewis 2001) and the NUSAP matrix. The assessment takes into account high level service cost estimation capabilities including platform specific capability, work and expenditure planning, data recording capability, service network cost, adaptation, analysis, assessment and control, and risk and opportunity assessment. As a result of this assessment a score is calculated by aggregating the level of capability for each mentioned aspect to define a project level cost estimating capability. This in turn is used to revise the uncertainty range using the guidelines provided by the Association of Advancement of Cost Engineering (AACE), which specifies classes for cost estimates based on confidence and suggests suitable ranges (AACE 2003). This revision process is also followed for the uncertainties.

The assessed uncertainties are linked to specific cost drivers, while calculating an uncertainty score for each cost driver. This then feeds into determining the suitable ranges for cost drivers. Assuming that there is a top-down estimating approach, as is typical early on at bidding, an initial total cost is required as an input. The user then ranks the significance of cost drivers that contribute to the considered total cost and through the analytic hierarchy process (AHP); the contribution of each cost driver is calculated. The AHP is a pairwise comparison method, which derives ratio scales of relative magnitudes for a set of elements by making paired comparisons (Saaty 2006).

Table 7.3 Cost driver significance assessment: AHP results

	Cost drivers					Initial total cost
Cost drivers	Failure rate	Turnaround time	Transport cost	Packaging cost	Repair cost	£10.000
Rank score	5	4	8	4	3	
AHP significance	0.04	0.02	0.15	0.02	0.02	
Uncertainty score & range	0.52 −20/+30%	0.62 −20/+30%	0.44 −15/+20%	0.26 −10/+15%	0.78 −30/+50	
Minimum cost	8.000	8.000	8.500	9.000	7.000	
Maximum cost	13.000	13.000	12.000	11.500	15.000	

Table 7.3 demonstrates indicative results from a participating project using the AHP analysis for cost drivers. The results represent an aggregate outcome of three participants, who have between 20 and 30 years of experience in service cost estimation. The results from the table show that the most important cost drivers are transport cost and failure rate. The uncertainty score is elicited through the NUSAP matrix and the range values are defined using the AACE guidelines. Based on the values the initial total cost estimate of £10.000 is refined for each cost driver to reflect the minimum and maximum values. These feed into the simulation model, along with input parameters including the percentage of costs accounted by the spares and resource suppliers, the fraction of gain/pain share by customer and for the Industry and for the uncertainty sharing scenario percentage of Industry responsibility. This process also requires the user to select the suitable scenario(s). This leads to the simulation model, where the interaction between the agents is triggered by variables, such as equipment usage and failure rate. The model output provides cost information regarding the specified scenario and supports in comparing between scenarios to support with the decision making.

In the cost estimation practice, the outputs from running a traditional model tends to field static results, which do not account for the element of time. On the other hand, ABM is time driven enabling to capture dynamic results distributed in time and space (Datta 2007). The simulation model, by assigning a distribution to represent spares and resource arising, enables to attach dynamism to events by triggering interaction between agents, which in turn causes the variation in cost. As a result the rules and the dynamism considered in the model enable to capture uncertainty more realistically (e.g., non-linear) compared to the traditional approach of cost uncertainty modelling.

7.4 Case Study

The case study section highlights an iterative process, which initially involved determination of major cost drivers and uncertainties, with a number of projects across the defence and aerospace industries. This subsequently led to the application of the developed agent-based model on a live project. Initially, a series of workshops were held with 20 Industry personnel involved in availability-type contracts (in areas of economic modelling, project management) to establish a generic list of cost drivers and associated uncertainties, which can be used to improve reliability while considering uncertainty in cost estimation. The discussions focused on establishing quantifiable cost drivers. Along these lines, inputs to economic and cost models were distinguished and the focus was put on value drivers, as they reflect the requirements of the customer and constitute the sources of costs. By scoring the influence of uncertainties for given cost drivers the study also involved ranking the cost drivers through AHP. The top level generic list of cost drivers based on functional breakdown along with components for each cost category includes, supply chain (e.g., stock level), engineering (e.g., query volume), maintenance (e.g., availability of labour), performance (e.g., customer actual usage), business management (e.g., incentive mechanism) and training (e.g., number of courses). Also, the study integrated the cost drivers with uncertainties, by defining associations, to better capture the sources of variation. For instance, for the supply chain cost category, the main cost drivers are arising rate, MTBF, purchase cost and repair cost. For each cost driver relevant uncertainties are defined as represented in Table 7.3.

The described simulation model was applied on a case study involving a very large military system-of-systems project in the naval domain with over 60 subsystems, largely procured from suppliers. The project is currently establishing the maintenance requirements with the customer, while the estimation of cost is creating challenges. In collaboration with three subject matter experts ABM was run for the case study, while input to the model followed the process covered in Fig. 7.4. The goal of the case study was to compare the cost impact of the scenarios that have been explained in Sect. 7.3.2, namely; (1) uncertainty with Industry, (2) uncertainty with supplier, (3) Industry and supplier share uncertainty. Across the three scenarios the inputs varied concerning the revised cost estimates and the definition of parameters. For instance, the influence of uncertainty on cost is the lowest when the Industry takes all risks (scenario 1). Also the influence of technical investment on cost varies across scenarios, affecting the actual cost across scenarios.

The plots presented in Fig. 7.5 compare indicative results for each of the scenarios considered in the case study. The results point to a consistent peak in all three plots, which could be described as initially less failures in the project, later more failures and replacement costs are high and at the end of the project the cost of replacing will be postponed, which means no cannibalisation. Based on comparison of the average cost levels across scenarios, risk with Industry constitutes the lowest whole-life-cycle cost. Thus, meaning that it would be better for the customer to pass on all the risks to Industry. In scenario 1, because the Industry takes ownership for its operations, it assumes that the uncertainty is minimised, which affects the revised cost estimates. In scenario 2, the Industry perceives high uncertainty arising from passing the risk to the supplier. Thus, with the difference between estimated maintenance costs and actual, technical investment costs

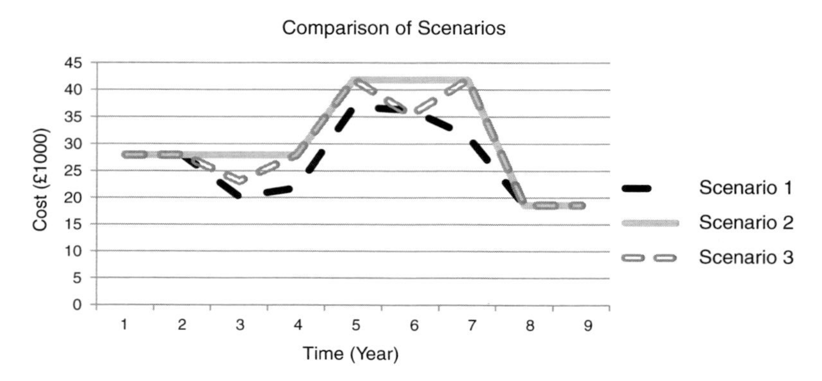

Fig. 7.5 Maintenance incentive cost impacts

contribute to Industry costs. Finally in scenario 3, by sharing the risk the technical investment is distributed, which reduces the technical cost for the Industry. The marginally small differences between the cost outcomes of the scenarios could be due to the relatively small cost figures considered in the case study. The conclusions drawn from the results are limited by the fidelity of the model (e.g., the model may not take account of practical difficulties such as being forced to pass risk to suppliers when they own unique Intellectual Property and will not reveal it to the prime contractor). Through validation it was suggested that some of the key benefits of the framework include costs that can be predicted for specified periods as well as for the long-term. Though, the framework and the ABM specifically suit the early stages of bidding where there is very limited information. Also, the framework enables intelligent management of the influence of uncertainty over cost early on in order to negotiate performance requirements against cost across the supply network.

7.5 Conclusion

One of the most distinctive characteristics of services is their process nature. Unlike physical goods, services are dynamic, unfolding over a period of time through a sequence of events and steps. Furthermore with manufacturing-based industries typically experiencing a shift towards service orientation major challenges have emerged due to the nature of services. Cost estimation, which has significant financial implications, has been challenged by various factors, such as determining the arising rate, obsolescence and technology refresh. The common theme across these complexities is the dynamic nature, which is increasingly challenging their representation through static models. To the contrary of static models, dynamic models (e.g., ABM) are able to reproduce the evolution of a system over time. Furthermore, it provides a suitable platform to create robust and accurate "what if" scenarios for decision making. Based on the case study it has been possible to recognise that taking a systematic approach to identify uncertainties in relation to cost drivers provides opportunities to enhance rigour in cost estimation. Furthermore, it also offers the prospect of understanding the root cause of uncertainty. Validation of the ABM model enabled to reflect that the rules and assumptions and the framework were suitable for the early stages of the bidding phase. Improvements to the presented ABM would need to take account of further complexities that arise from the supply chain. The importance of the ABM approach will increase as the complexity of the service solution grows (e.g., increase in the number of customers, suppliers or scope and complexity of maintenance).

7.6 Chapter Summary Questions

- What are the main uncertainties and cost drivers related to PSS?
- What are the main challenges in service cost estimation?
- What is the current uncertainty-based service cost estimation process?
- What improvements can be made to the uncertainty-based service cost estimation process?

References

AACE, International Recommended Practice No. 17R-97 Cost estimation classification system. TCM Framework: 7.3—Cost estimating and budgeting (2003)

T. Alonso-Rasgado, G. Thompson, A rapid design process for total care production. J. Eng. Des. **17**(6), 509–531 (2006)

T. Alonso-Rasgado, G. Thompson, B.O. Elfstrom, The design of functional (total care) products. J. Eng. Des. **15**(6), 515–540 (2004)

C. Armistead, G. Clark, Service quality: The role of capacity management (1992). Available at Cranfield Ceres. http://hdl.handle.net/1826/641. Accessed 12 April 2008

J.C. Aurich, C. Fuchs, C. Wagenknecht, Life cycle oriented design of technical product service systems. J. Clean. Prod. **14**, 1480–1494 (2006)

BAE Systems, Director's report: Business Review Strategy [Online] (2008). Accessed from: http://www.annualreport08.baesystems.com/financialstatements/~/media/Files/B/BAE-AR2009/pdf/directors-report-businessreview/strategy-action.pdf. Accessed 12 March 2010

T.S. Baines, H.W. Lightfoot, S. Evans, A. Neely, R. Greenough, J. Peppard, R. Roy, E. Shehab, A. Braganza, A. Tiwari, J.R. Alcock, J.P. Angus, M. Bastl, A. Cousens, P. Irving, M. Johnson, J. Kingston, H. Lockett, V. Martinez, P. Michele, D. Tranfield, I.M. Walton, H. Wilson, State-of-the-art in product service-systems. Proc. I Mech. E Part B J. Eng. Manuf. **221**(10), 1543–1552 (2007)

P. Bernstein, *Against the gods: The remarkable story of risk* (Wiley, New York, 1998)

A. Boussabaine, R. Kirkham, Whole Life-cycle costing: Risk and risk responses, 1st edn. (Oxford: Blackwell Publishing 2004)

S. Brax, A manufacturer becoming service provider–challenges and a paradox. Manag. Serv. Qual. **15**(2), 142–155 (2005)

C.B. Chapman, S.C. Ward, Estimation and evaluation of uncertainty: a minimalist first pass approach. Int. J. Proj. Manag. **18**(6), 369–383 (2000)

G. Clark, C.G. Armistead, Improving Service Delivery, Managing Service Quality, July Accessed from: Cranfield Ceres Access date: 12/04/2008 (1991)

M.A. Cohen, N. Agrawal, V. Agrawal, Winning in the aftermarket. Harv. Bus. Rev. **84**(5), 129–138 (2006)

R. Curran, S. Raghunathan, M. Price, Review of aerospace engineering cost modelling: The genetic causal approach. Prog. Aerosp. Sci. **40**, 487–534 (2004)

PP. Datta, A complex system, agent based model for studying and improving the resilience of production and distribution networks. (PhD Thesis, Cranfield University, 2007)

P. Datta, R. Roy, Cost modelling techniques for availability type service support contracts: a literature review and empirical study, CIRP J. Manuf. Sci. Technol. **3**(2), 142–157 (2010)

Department of Defense, DoD, Handbook on work breakdown structures for defense material items [Online] (2005). Available at: http://www.acq.osd.mil/pm/currentpolicy/wbs/MIL_HDBK881A/MILHDBK88A/WebHelp3/MILHDBK881A.htm. Accessed 11 March 2010

J.A. Erkoyuncu, R. Roy, E. Shehab, P. Wardle, Uncertainty challenges in service cost estimation for product-service systems in the aerospace and defence industries. Proceedings of the 1st CIRP IPS^2 Conference (Cranfield, UK, 2009a), pp. 200–207

J.A. Erkoyuncu, R. Roy, E. Shehab, K. Cheruvu, Managing uncertainty in service cost estimation for product service systems. Proceedings of the 7th International Conference on Manufacturing Research (ICMR 2009). (Warwick University, Warwick, UK, 2009b) 8–10 Sept

J.A. Erkoyuncu, R. Roy, E. Shehab, K. Cheruvu, Understanding service uncertainties in industrial product- service system cost estimation. J. Adv. Manuf. Technol. **52**, 1223–1238 (2011)

W.P. Hughes, I. Yohannes, J.B. Hillig, *Proceedings of CIB world building congress: Construction for development.*, ed. by T. Haupt, R. Milford. Incentives in construction contracts: Should we pay for performance? (Cape Town, South Africa, 14–17 May 2007) pp 2272–2283, Cape Town: Document Transformation Technologies. CD-ROM, ISBN 1-920-01704-6

P. Johansson, J. Olhager, Industrial service profiling: Matching service offerings and processes. Int. J. Prod. Econ. **89**, 309–320 (2004)

S.H. Kim, M.A. Cohen, S. Netessine, Performance contracting in after-sales service supply chains. Manag. Sci. **53**(12), 1843–1858 (2007)

F.H. Knight, *Risk, uncertainty and profit* (Houghton Mifflin, New York, 1921)

C.H. Lovelock, Classifying services to gain strategic marketing insights. J. Mark. **47**(3), 9–20 (1983)

Ministry of Defence, Cost estimation process (2009). Available via: http://www.aof.mod.uk/aofcontent/tactical/engineering/content/fe_costestprocess.htm. Accessed 11 Dec 2009

Ministry of Defence, What is reliability and maintainability? [Online] (2010). Accessed from: http://www.aof.mod.uk/aofcontent/tactical/randm/content/randm_whatis.htm. Accessed 08 March 2010

V.G. Narayanan, A. Raman, Aligning incentives in supply chains. Harv. Bus. Rev. Nov: (2004), pp. 94–102

I.C.L. Ng, J. Williams, A. Neely, Outcome-based contracting: Changing the boundaries of B2B customer relationships. Advanced Institute of Management (AIM) Research Executive Briefing Series. Accessed from http://www.aimresearch.org/index.php?page=alias-3. Accessed 09 March 2010 (2009)

NATO, RTO Technical Report, [Online]. NATO, Annex D, Pg D1. http://www.rta.nato.int/pubs/rdp.asp?RDP=RTO-TR-SAS-054 (Accessed 8th July 2009), (2007)

R. Oliva, R. Kallenberg, Managing the transition from products to services. Int. J. Serv. Ind. Manag. **14**(2), 160–172 (2003)

H. Pickerin, D. Lewis, A capability improvement model for cost engineering. Available at: http:/www.galorath.com/blogfiles/CECIM%20White%20Paper%20Version%20A-R4.pdf. Accessed 11 Dec 2009 (2001)

R. Roy, Cost engineering: *Why, what and how? Decision Engineering Report Series*, (Cranfield University, 2003) ISBN 1-861940-96-3

R. Roy, P. Datta, F.J. Romero J.A. Erkoyuncu, Cost of industrial product service systems (IPS^2). Keynote paper, 16th CIRP International Conference on Life Cycle Engineering (2009)

R. Roy, K.S. Cheruvu, A competitive framework for industrial product service systems. Int. J. Internet. Manuf. Serv. **2**(1), 4–29 (2009)

T.L. Saaty, *Fundamentals of decision making and priority theory with the Analytic Hierarchy Process. Vol VI of the AHP series* (RWS Publications, Pittsburgh, 2006)

N. Slack, *Operations management* (Financial Times Prentice Hall, Harlow, 2010)

S. Vandermerwe, J. Rada, Servitization of business: Adding value by adding services. Eur. Manag. J. **6**(4), 315–324 (1988)

J.P. Van der Sluijs, M. Craye, S.O. Funtowicz, P. Kloprogge, J. Ravetz, J. Risbey, Combining quantitative and qualitative measures of uncertainty in model based foresight studies: the NUSAP system. Risk Analysis, **25**(2), 481–492 (2005)

S.L. Vargo, R.F. Lusch, Evolving to a new dominant logic for marketing. J. Mark. **68**, 1–17 (2004)

S.L. Vargo, R.F. Lusch, Why 'Service'? J. Acad. Mark. Sci. **36**(1), 25–38 (2008)

S. Ward, C. Chapman, Transforming project risk management into project uncertainty management. Int. J. Proj. Manag. **21**(2), 97–105 (2002)

Chapter 8
Incentives and Contracting for Availability: Procuring Complex Performance

Nigel D. Caldwell and Vince Settle

Abstract Procuring complex performance where both the performance required and the infrastructure to support the requirement are complex, is the current frontier on procurement knowledge. Trends to bundle contracts for products and services together exemplify the challenge. These product-service bundles take the form of contracts for the use of the product (e.g., including long term maintenance and support) rather than just the product as an artefact. Such contracts bind a Prime contractor and the customer into complex long term agreements. Contractual incentive mechanisms have long been used to align the interests of customer and supplier in such projects. This chapter explores the use and role of incentives in complex engineering support environment, drawing on a case study from recent research on availability contracting to support fighter jets. The chapter presents the challenge to the conventional incentive mechanisms inherent in the combination of flexibility and cost control required in contracting for jet fighters availability.

8.1 Introduction

The aim of this chapter is to briefly outline the need for, and role of, incentives in complex support environments. It discusses the major types of contractual incentive, their advantages and disadvantages and how they relate to service support engineering, and links incentives to the S4T core integrative framework. The chapter provides a case study based on a composite of military fighter jet

N. D. Caldwell (✉)
School of Management, University of Bath, Bath, BA2 7AY, UK
e-mail: n.d.caldwell@bath.ac.uk

V. Settle
BAE Systems, London, UK
e-mail: vince.settle@baesystems.com

I. Ng et al. (eds.), *Complex Engineering Service Systems*, Decision Engineering,
DOI: 10.1007/978-0-85729-189-9_8,

contracts to explore detailed issues in contracting for availability and incentive mechanisms. The chapter ends with conclusions and comment on current trends and areas of future interest.

Contractual incentives for performance have a long history. Their importance can be illustrated through their use in the transportation of prisoners to Australia in the eighteenth and nineteenth centuries. Those ships incentivised on an outcome (i.e., number of prisoners delivered alive) were extraordinarily successful at keeping their cargo well through a hazardous and arduous journey. Other ships incentivised on inputs (speed, price and a share in any consumables such as food and drink 'left over' at the end of the journey) had terrible convict passenger attrition rates. So in this instance, giving those responsible for transporting the prisoners the incentive of being able to sell off any unused provisions for their own profit did not align with the intention of the contract. Following a scandal about the poor mortality rates and a public enquiry, the Home Department, the equivalent of the Home Office today, insisted that in all future shipments, the contractors should be paid for the number of convicts landed, rather than for the number embarked (for a full description of the history of incentives in UK government procurement, see Sturgess 2010).

This example from the transportation of convicts shows an eighteenth century use of an outcome-based contract for procuring complex performance. Given the state of technology at that time performance of the task was complex, and it was deemed necessary to incentivise the contractor. The example also illustrates how badly incentives can affect performance when they are not thought through, (as here) when some contractors were not incentivised on convicts delivered alive and well, this is often termed the problem of 'perverse incentives'. A perverse incentive creates an effect that was not intended by the customer.

Incentives include penalty incentives, the early days of individual train operating companies running the UK rail network provide an example of a penalty clause that became a perverse incentive. The penalty on running a train very late was so high as to incentivise the train operating companies to instead cancel the train entirely. Thus incurring a much smaller penalty (but at great disadvantage to rail users) an unintended consequence. Incentives are usually financial but can also include non financial incentives such as reputation by association and the carrot of future work.

Incentives have been formally defined by Her Majesty's Treasury (1991:1) as:

> A process by which a provider is motivated to achieve extra 'value added' services over those specified originally and which are of material benefit to the user. These should be assessable against pre-determined criteria. The process should benefit both parties.

The purpose of incentives has been defined by Broome (2002:111) as to:

> Align more closely the motivations of the contractor, consultant or supplier to those of the client, so that any of the participants, by working for the success of their organization, is indirectly working for the success of the project.

The key, as Broome states, is that an incentive will put more emphasis on achieving a client objective rather than a contractual obligation alone (2002:111).

Examples of financial incentives include profit sharing in cost-plus incentive contracts, bonus performance provisions attached to various lump sum and cost reimbursable contracts and multiple financial incentive mixes (Rose and Manley 2005). These detailed forms will be discussed below, the intention in this introduction is to show why incentives matter.

The chapter is structured as follows. Firstly, we introduce the notion of procuring complex performance (Sect. 8.2). Section 8.3 discusses contract strategy, building on the basic forms of contract to introduce contracting and contractual incentives for performance-based contracts such as availability. This section introduces the importance of identifying and allocating risk. Section 8.4 is a case study of incentive use in contracting for availability in military fighter jets. The case material provides a lead into the conclusions.

8.2 Procuring Complex Performance

It is becoming increasingly common for organisations to leverage service opportunities to deliver more value, to both the customer and the organisation itself. Cohen et al. (2006) suggest that this need to leverage service opportunities has been occurring since the early 1990s in Western Europe, the USA and Japan. As competition for manufactured products has increased and margins have become reduced, offering service solutions has been seen as maximising revenues and profit (Cohen et al. 2006). Many economies are increasingly servitized (Vandermerwe and Rada 1988). Advocates of the move to services cite downstream (i.e., the customers end of the value chain) markets as offering large potential revenues with higher margins, and also as requiring fewer assets than product manufacturing; and that due to their steady service-revenue streams, they can often be countercyclical (Wise and Baumgartner 1999). In complex engineering support, services typically range from routine and planned maintenance, to unscheduled repairs to inserting upgrades such as new technology into existing systems or platforms, through to disposal.

Vargo and Lusch suggest that there is a

> …need for refocusing substantial firm activity or transforming the entire firm orientation from producing output, primarily manufactured goods, to a concern with service(s)' (Vargo and Lusch 2008:254).

In the academic literature a concern with understanding these bundled offerings of products and services has led to an interest in the product service system (Mont 2002; Baines et al. 2007).

Industrial and public sector customers as business to business procurement professionals reflect these trends through increasingly purchasing (or in health settings commissioning) a combination of products and services. One example of this phenomenon is the blurring of traditional boundaries of ownership, design and post-construction performance in major construction projects. In combining the purchase of service with a product the buyer is procuring complex performance (Caldwell and Howard (2010); Lewis and Roehrich (2009); Caldwell et al. 2009) or what is increasingly known as Procuring Complex Performance (PCP). PCP can be defined as a combination of performance complexity (a function of the level of knowledge embedded in the performance and/or the level of customer interaction and infrastructural complexity. The latter involving substantial bespoke or highly customised hardware and software elements (Lewis and Roehrich 2009). Caldwell and Howard (2010) take a more relational approach to defining procuring complex performance writing of the need for a co-ordinated, relationship-focused approach to buying made necessary by the task being so composed of sub-elements that it cannot be achieved by the sequential or additive achievement of individual tasks or transactions.

This chapter addresses the contractual forms that are emerging to support the procuring of the complex performance inherent in newly 'servitized' business models and specifically contractual incentives in product service contracts where availability is the key customer requirement. Typical product service contracts must incentivise industry to provide new levels of service, for example, innovative environmental practices, ease of maintenance, flexibility once in use and ease of ultimate disposal. The client must in effect, procure complex performance (as opposed to a complex product or building); clients increasingly value the "in use value" of the product infrastructure over the tangible product. This theme is returned to in the discussion of the core integrative framework. However, the move towards a product service business model is not without risk. "In industries where excellence in product manufacturing and design form the key to uniqueness and hence power in the value network, diverting focus to an issue such as PSS [product service system] development is a recipe to lose rather than win the innovation battle" (Tukker and Tischner 2006: 1553). Incentive design can play a key role as our case study will suggest, in maximising the opportunities and minimising the risks in adopting a product service approach.

8.3 Contract Strategy

Although there are many variations, two contract payment systems have dominated product logic-dominant manufacturing and construction that is either price- or cost-based. Price-based contracts have a fixed price or rate, the customer does not know what costs are incurred. This can be particularly problematic when in large complex undertakings there is a time lag between the costs being incurred and the contractor being able to consolidate and report the costs incurred, potentially resulting in the customer facing a much larger bill than expected with litigation a real possibility. Under cost-based payment systems the actual costs incurred by the supplier are reimbursed together with a fee to cover overheads and profit (including target cost-based contracts). To avoid the problem of cost lag described above some form of cost transparency or 'open book' costing is

Table 8.1 Generic contract strategies

	Price based—fixed price contract	Cost based—cost plus contract	Outcome based—e.g., availability
Key advantage	Customer knows in advance the maximum cost	Customer can verify the basis of what they are being charged	Customer can specify the outcome they want High performance incentive for supplier
Key disadvantage	Supplier bears the risk of costs exceeding original estimates. Supplier incentivised to reduce costs but no performance incentive	Risk is shared but costs can escalate, supplier has no incentive to reduce costs, some performance incentive	Understanding and making risks transparent central to contract success

required. Table 8.1 lists some key features of both types but also introduces performance-based contracts, for example it might be a payment for the availability of beds in a hospital rather than a contract for the delivery of a facility.

Outcome or performance-based contracts are seen increasingly in the public and private sectors. Such outcome-based contracts are by nature long term: "In the case of resulted-oriented PSS [product-service systems], one actor becomes responsible for all costs of delivering a result, and hence has a great incentive to use materials and energy optimally" (Tukker and Tischner 2006, also Thierry et al. 1995). So for example for a traditional manufacturer in a complex engineering environment such as fighter jets, the scope of their business expands to include cost effective maintenance and support service (and after sales) in addition to the core product. Such manufacturers then face decisions about the shape of their business, and their business model.

The business model (Spring and Araujo 2009; Spring and Mason 2010) in defence, like many manufacturing models has historically have been able to make money out of initial product sales and the subsequent market for spares.

> And because manufacturing promised the greater returns through the provision of spare parts as well as the purchase of original equipment, the judgment of manufacturing primes about design tradeoffs and systems assignments had to be taken with some scepticism Sapolsky (2004:24).

The design, manufacture and sell spares business model in fact did not incentivise the defence OEM to produce reliability. Parts/systems that went wrong meant further sales, an old fashioned 'throw it over the wall' to the customer approach. Engineers could focus on new build and provision of spares and associated activity became a backwater (although a profitable one). Defence companies and their engineering talent were simply not incentivised to produce reliable easy to maintain components or systems. One consequence of this business model was an 'arms length' relationship between engineering and support, with no feedback loop from supporting maintenance (Davies 2004), and little 'design for supportability' (Goffin and New 2001).

Thus in addition to a manufacturer having to adapt to a new business of support the fundamental business model of the manufacturer has to change. In complex engineering environments the complexity of such contracts mandates incentive mechanisms to co-ordinate and align customer and contractor interests. The following case study now explores incentives in the complex procurement environment of fighter jets. The case study is based on over 30 interviews with managers and engineers and the personal experiences of one of the authors over a career of aircraft-related commercial management.

8.4 A Case Study of Military 'Fast Jet' Aircraft

When considering contracts for long term availability for military 'fast jet' aircraft, there are some key requirements unique to the military arena which make these contracts perhaps more difficult and challenging to construct than the ones in normal commercial environment between industrial companies or civilian public bodies. Incentivisation is the key to making such contracts work effectively and this case study examines some of the differences and difficulties in constructing incentivisation mechanisms.

It is generally accepted that military commanders, when in a war situation, demand the best products and services available, delivered when they want them and subject to constant changes as the threat changes. Whilst this could also be argued to be true of any public body, civilian or industrial contracts, the military are in a unique 'all or nothing' situation where being second best or being late is unacceptable if they are to win the war.

For these reasons, during the Cold War, the UK Government (and other NATO allies) adopted an approach whereby their Armed Forces and the civilian Ministry of Defence between them, enabled and controlled this required flexibility by carrying out the work 'in-house' (i.e., Royal Air Force (RAF) personnel servicing the aircraft) and having MoD staff purchase the necessary equipment and technical services to support this on a piecemeal basis from many contractors; i.e., setting up "enabling contracts" with a multitude of supplier companies to procure spares and repairs on an "as required" basis. For instance, the Tornado aircraft had some 350 separate contracts for the MoD team to manage. To add to this supply complexity, the capabilities of the aircraft need to be improved constantly as new technology becomes available and the capabilities of potential enemies also improve. Therefore, a key feature of the newly emerging CFA (Contracting for Availability) contracts is both to keep costs down and to keep aircraft availability up, such capability improvements must be embodied at the same time as routine servicing work. The use of servicing 'downtime' to fit upgrades maximises availability (when done well) and was previously not the case under in-house servicing. Any incentivisation mechanism, therefore, has to recognise that the technical standard of the aircraft will be enhanced again and again within the existing contract.

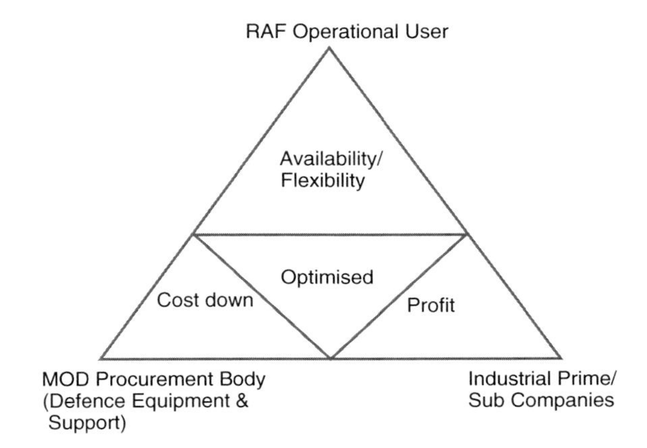

Fig. 8.1 The Benefits Triangle

In recent years, budget pressures have meant that the traditional in-house driven flexibility (i.e., in house servicing and separate upgrading) is unaffordable and although there have been some successes, recent studies have shown that partnering with industry on long term, output-based incentivised contracts is the way ahead to achieve demanded cost savings. Indeed recent Defence Support Review (Ministry of Defence and Deloitte 2010:7) states:

> incentivisation is key if the department [MoD] is to seek further savings in support costs once responsibilities are transferred to industry.

A particularly difficult issue remains though, in that the budget-holders demand for cost efficiencies should not be allowed to over-ride the military commanders' rightful expectation of demand-flexibility. They need to sit side by side.

Figure 8.1 illustrates the tensions within a typical availability contract for fighter jets (or other complex military procurements) between the operational user, here the Royal Air Force, the Treasury represented by the Defence and Equipment Support Agency and the civilian Prime contractor. As this case draws out the key requirements for the user are flexibility and availability. However the Treasury are concerned with minimising cost whilst the Prime contractor needs the stability to be able to plan for a sustainable business. At first view these three differing requirements can be seen as irreconcilable. So, how is this dilemma between cost efficient supply and maximum military flexibility resolved? It must be stated here that the solution of simply reducing cost by reducing quantity of products is not part of this paper (for example, if politicians decide that squadrons of aircraft will be taken out of the inventory). What is in discussion is trying to make the maximum use from the inventory available—and at the cheapest cost.

Incentivisation with Industry is seen as a key enabler in this. Examples of incentivised contracts are now available for the RAF's fast-jet fighter fleets, covering Tornado, Harrier and most recently, Typhoon. These contracts transfer the bulk of the servicing work [i.e., planned periodic servicing and capability embodiments] to Industry [and its' attendant contractual relationships with the layers of suppliers], through a Prime Contractor promising guaranteed outputs

Table 8.2 The different budget centres involved in providing output from an RAF base

Activity	Budget centre
RAF personnel, fuel, infrastructure services	RAF budget
Industrial personnel and supply chain Equipments	DE & S budget (Defence Equipment and Support) N.B. the prime contractors costs [and his supplier costs] are within this budget.
Buildings	Defence estates budget

(e.g., aircraft availability, flying hour levels, etc.). Put another way, 350 contracts can be merged into one contract, with the key feature that output is guaranteed, whereas, in the "enabling contract" regime before, there were no guarantees, only the best efforts of the RAF and civilian personnel managing the process.

A new pricing method has been developed by BAE Systems to introduce an incentivisation framework and attendant behavioural changes necessary to motivate all the parties to drive down costs. Referred to as Target Performance Price Incentive (TPPI), this method links technical/maintenance data and individual prices on specific equipments (repairs, spares) with an incentivised gainshare which rewards the parties fairly in accordance with the effort they have put into achieve savings through reliability improvements or servicing periodicity changes. So for example if the Prime Contractor can reduce the break downs on the aircraft which leads to 'repair arisings' at the supplier company on a particular component, then the TPPI gainshare mechanism allows for the reduction in through-life cost to be shared between customer, prime contractor and equipment supplier on an agreed pre-set basis. What is new is that TPPI can track savings on each equipment, rather than as previously where only total contract cost reductions would be known, with no audit trail as to why costs had come down.

However, data capture is essential to the process and this is an absolute must for the process to work. It is a difficult balancing act to manage this process, as within the industrial forum the prime contractor has to ensure engagement with his suppliers, who also have to be financially motivated to contribute their intellectual effort. A 3-way split of savings to include the supply base is much harder to achieve equitably than a 2-way split. A further issue is that the true end-to-end savings from any initiatives may be hard to determine, as the MoD budget process separates industrial spend controlled by MoD Defence Equipment and Support (D.E. & S) from other spend within the wider MoD. For example, the cost of providing an output from an RAF base will consist largely of the items listed in Table 8.2.

In seeking to carry out a cost benefit analysis of cost reduction initiatives and to construct an incentivisation method to reward all the parties according to savings effected, it is extremely difficult to assess the effect on all three military budgets in Table 8.2 without a long and cumbersome exercise, but without this any cost-benefit case is flawed. The current challenge is now to construct such end-to-end cost models and hence to establish the true cost of individual military platform outputs relative to their perceived military benefit.

Finally, it must not be forgotten that the possibility of cost overruns is always a threat (either for the customer, the contractor or both depending on the contract pricing method chosen) and commercial staff must always consider this in any risk assessment within the contract price. The issue here is determining firstly who is best placed to hold the risk, as risk always equates to price in a contract, and secondly how do we motivate the parties to work together to mitigate that risk? The methods to share the risk of cost overruns or the rewards of cost under run are now commonly referred to as Painshare or Gainshare mechanisms. The latest thinking is that the customer will accept a share of the pain if that means the datum contract price is lower, and thus more affordable. Although such sharing arrangements have been used in the past for example, large one-off development contracts, the new CFA Contracts are unique in that they can track the reasons for cost variances down to component level.

The challenge in constructing incentivisation methods in the military environment can, therefore, be summarised as:

- How to adequately motivate all parties to reduce costs but maintain the required flexibility?
- How to make the incentive mechanism simple to operate, yet take into account true end-to-end costs?
- How to make the mechanisms dynamic to change, as the products are operated in different ways, and updated, over time?

We now examine each of these in the context of fast jets.

The least developed of these challenge questions is that of flexibility for the military and it is that question which will cause the most difficulties for commercial staff who are trained to look for stability in contracts. The issue is how to construct a cost model and attendant contract conditions which accurately portray the increased price of risk (if any) of providing a far greater degree of output flexibility and with a rapid reaction time. If this is possible a reasoned judgment can be made by the customer's military staff and budget holders as to the price they are willing to pay for that flexibility.

Finally, questions are now being asked as to how to create the next generation of contracts and incentive mechanisms which move beyond single-aircraft contracts into multi-aircraft contracts, in order to give output flexibility across a range of aircraft in the RAF inventory but within one budget. Military Commanders now require a 'force-mix' approach to the deployment and use of their assets. For example, they may require a period of "surge" in flying for many months on one aircraft, as it has the capability to use a particular type of smart weapon. Yet, there is no increased budget available to pay for this higher usage.

The higher levels of MoD may, therefore, decide to "suppress" the flying levels on another aircraft in order to pay for this specific surge. Current individual aircraft contracts allow for a limited flexibility, and the force-mix approach would enhance this, but much improved budgeting and incentivisation models may be needed to allow the "budget balancing" as the cost of operation of each aircraft is different. What then is the way forward? What improved features should a second generation of CFA Contracts contain?

The challenge now is to be able to construct contracts and their incentive mechanisms which recognise force-mix, involving varying outputs as demanded, and at probably a short notice. However, they must be able to track savings and equitably reward the parties, when some stakeholders may be disadvantaged by reducing output on one aircraft to generate an increased output on another. This must be done whilst continuing to service the military need at the time, all within the original fixed budget! Taking this one [rather large] step further, it is possible to construct a scenario whereby force-mix applies to the full military inventory. Such an approach would allow budget trade-offs between aircraft, tanks, ships, etc. in terms of what assets are needed for particular missions, and the total potential cost.

Such a scenario would demand an ever-closer partnering relationship not just with the contracting customer (MoD D.E. & S) but with the Military as operators. Industrial companies, and particularly their commercial staff, have to understand how these products are used and deployed in-theatre to understand the flexibility required and be able to construct an incentive mechanism which will stand the test of time over perhaps 10–15 years in operation. Building upon the first generation contracts with output flexibility and an agreed risk-sharing mechanism, far more complex cost models are required to simulate the risk and cost involved in providing more flexibility and in shorter timescales. These must also recognise that any Prime Contractor must persuade his entire supply chain of the need to actively engage with this approach.

8.5 Discussion

Contracting for the availability of fighter jets is an example of procuring complex performance. Core to the contract is to enable instant flexibility to change outputs, e.g., primarily to increase or decrease the flying hours. However, additional complexity is added by the air force's need as a customer to have instant and unpredictable access to aircraft, and to be able when required to use the aircraft in unspecified ways. Given the unique 'bill of materials' of each fighter jet which will have already been through many major and minor maintenance and upgrade procedures the Prime contractor takes on huge risk which mandates the collection of robust data on usage and costs. Yet, there may at times be direct conflict between the Prime's need for reliable data for planning and cost management and military requirements for speed and flexibility. Current incentive mechanisms for gain and pain share rely on this detailed data collection.

The picture that emerges from the case is of the need for a strategic partnering relationship based on intense working together for collaborative advantage (Huxham and Vangen 2004). This relational approach is far removed from a transaction-based business model. Previous academic work in this area (Lamming 1993; Dyer and Singh 1998) has noted the need for new levels of information exchange in such relationships. Whilst such relational approaches do stress the need for mechanisms to enable inter-organisational dialogue the scale and complexity of the fighter jet Prime and user interface place huge stresses on traditional mechanisms such as early customer-contractor dialogue on risks and the requirement for a joint, not contractor only assessment log. These additional relational costs and the costs of the incentive mechanism itself have to be costed into the contract. From the interviews it was clear that the costs of collecting the information required to support certain incentive schemes were not insignificant both in financial terms and in the time involved.

In complex environment like fast jet support, companies do not just add value, they reinvent it according to Normann and Ramirez (1993:66). These authors suggest working with other economic actors like suppliers and customers to co-create value is more important than relying on a location within the value system or supply chain. What is significant about the customer role in value creation (and value destruction) is the ability the customer will have to ensure that the incentives are achieved/not achieved. There is a two part challenge, firstly to ensure that the customer (and all levels of the customer) has an understanding of how the contract is incentivised. Secondly that the customer value element to the work-mix is added in a way that is relevant, appropriate and sustainable. As the customer and end-user may not evaluate value delivered in a fixed or stable way it is critical that the incentive structure adapts to changes in the customer/end user, e.g., in military contexts these might be 'surges' when additional resources are required or slow periods when resource demand is reduced. This is a key problem in PFI hospital contracts where incentives support the expansion of hospital capacity by the contractor, for example rewarding the contractor by the number of beds the hospital supports. But such incentive structures do not support any moves by the end user to reduce capacity (for example through reallocating services around an area or by reducing demand). In effect one-way incentives are created that discourage the contractor from supporting strategic change by the user.

Earlier traditional manufacturing business models based on the volume supplied were discussed. With the pressure on the Prime to achieve reliability in availability contracting the supply chain faces lower volume requirements. A key challenge raised by the case study is how to effectively incentivise the supply chain of suppliers to the Prime to transition to a business model not based on volume. It is still relatively early in the development of availability contracting, and these first generation contracts have not addressed the issue to the depth that prolonged use of availability would require. The recommendation from the incentive literature would be the need to pay a premium over existing prices to compensate for lower volumes. But such 'largesse' may be restricted by the cost ceiling within which the Treasury require such contracts to be operated; this topic is significant and requires further exploration.

The Benefits Triangle (Fig. 8.1) was used to identify the tensions arising from the multiple parties within a fighter jet availability contract. So far this chapter has discussed the user and Prime perspectives. Turning to the MoD, DE & S or Treasury perspectives identifies a paradox at the centre of these contractual forms as often used in the UK. It is ironic that an outcome-based contract such as those discussed should be created through a focus on an input—a fixed or maximum contract value. For example, according to the National Audit Office:

> The Harrier Integrated Project Team's Joint Upgrade and Maintenance Programme cost model contained all the appropriate efficiency levers, but its starting point was the budget available and the estimate from the Prime Contractor on how much a contract may cost (2007:24).

It is often the case that incentive schemes are created within this maximum price, with the only option of cost reduction through gainshare. From the generic case study it would appear that the military are currently pushing at the frontier of how flexible a Prime can be under such a fixed-price (target price) contract. It is perhaps still not proven that where such contracts create a focus on cost reduction that innovative solutions (in innovation terms really new ideas or methods as opposed to incremental improvements) will be encouraged in the long term.

8.6 Conclusions and Trends in Incentive Design in Complex Support Environments

Contracting for a bundle of goods and services creates complexity, when these bundles are significant engineering capabilities such as availability in jet fighter contracts the procurement is for complex performance. Conventional incentive schemes including those with elements of shared pain/gain work to align customer and supplier/contractor interests. However, in the military environment of fighter jets support, the interests or requirements of the customer demand almost total flexibility from the Prime contractor, who is in turn reliant on robust data for their own planning purposes.

The Prime in such contracts needs to have the ability for example to accurately monitor contracts critically in major temporary cost overruns in periods of exceptional usage (e.g., flying hours). Yet it is at just such times that the military will have least time and sympathy for supporting data collection. In the case study it is clear that flexibility is the customers' key concern (within a tight budget envelope) which can then only be contracted for when translated into a business model that can model that risk, hence the repeated need for joint consideration and management of the risk registers. Any incentive scheme will only work well where it is cost effective and the contractor genuinely has control over performance.

Whilst it is easy to appreciate military needs for flexibility, advanced contracting for availability must also provide Primes with sustainable business models, which includes the capacity to plan internally, externally and financially. The increasing emergence of targets, Key Performance Indicators (KPIs) and performance criteria on the *customer* in such contracts are likely. Outcome-based contracts like availability in complex engineering support areas may require incentive schemes to include the performance of the customer.

Whilst such contracting does provide a reliable and consistent source of income for a Prime, there is an unanswered question over whether the Prime moving to increased flexibility of support provision is consistent with such Primes also coming up with truly innovative solutions. If, as the case study suggests, the next frontier for availability is contracts that support more than one platform, and even ultimately different platform bases (e.g., combining elements of land, sea and air platform availability) the supposed link between contracting for long term availability and innovation must be proven.

Finally it remains a challenge to incentivise the supply chains to achieve through put (volume) reductions as down times are reduced in better maintained availability contracts. Suppliers will need to migrate to new business models, some suppliers can become consolidators providing a bigger volume of parts, others may need higher prices to accept the change. There is little work in this area which will grow in importance as availability contracting matures.

8.7 Chapter Summary Questions

The chapter considers that the user/operator requires far more flexibility than the current Prime provision, and this drives the following questions:

- What processes can be taken to ensure the customers have articulated their real requirement?
- Is that requirement an assessable requirement (i.e., can it be measured cost effectively in some way to reflect performance against it?)
- Does the current incentive mechanism encourage behaviours on all sides that encourage collaboration?
- Is the distribution of benefit from the incentive scheme equitable over time?

References

T.S. Baines, H.W. Lightfoot, S. Evans, A. Neely, R. Greenough, J. Peppard, R. Roy, E. Shehab, A. Braganza, A. Tiwari, J.R. Alcock, J.P. Angus, M. Bastl, A. Cousens, P. Irving, M. Johnson, J. Kingston, H. Lockett, V. Martinez, P. Michele, D. Tranfield, I.M. Walton, H. Wilson, State-of-the-art in product service-systems. Proc. I MECH. E Part B J. Eng. Manuf. **221**(10), 1543–1552 (2007)

J. Broome, *Procurement routes for partnering* (Thomas Telford Publishing, London, 2002)

N.D. Caldwell, M. Howard (eds.), *Procuring complex performance: Studies of innovation in product-service management* (Routledge Taylor Francis, New York, 2010) (forthcoming)

N.D. Caldwell, J.K. Roehrich, A.C. Davies, Procuring complex performance: T5 and PFI compared. J. Purch. Supply. Manag. **15**(3), 178–186 (2009)

M.A. Cohen, N. Agrawal, V. Agrawal, Winning in the aftermarket. Harv. Bus. Rev. **84**(5), 129–138 (2006)

A. Davies, Moving base into high-value integrated solutions: A value stream approach. Ind. Corp. Change **13**(5), 727–756 (2004)

J. Dyer, H. Singh, The relational view: Cooperative strategy and sources of interorganizational competitive advantage. Acad. Manag. Rev. **23**(4), 660–679 (1998)

K Goffin, New C, Customer support and new product development & An exploratory Study. Int J of Oper & Prod Manag. **21**(3), 275–301 (2001)

C. Huxham, S. Vangen, Doing things collaboratively: Realizing the advantage or succumbing to inertia? Org. Dynam. **33**(2), 190–201 (2004)

R. Lamming, *Beyond partnership: Strategies for innovation and lean supply* (Prentice Hall, Basingstoke, 1993)

M.A. Lewis, J. Roehrich, Contracts, relationships and integration: Towards a model of the procurement of complex performance. Int. J. Proc. Manag **2**(2), 125–142 (2009)

Ministry of Defence, Deloitte, Defence support review: Phase 1 report (Refresh), 2010 (forthcoming)

O.K. Mont, Clarifying the concept of product-service system. J. Clean. Prod. **10**(3), 237–245 (2002)

R. Normann, R. Ramirez, From value chain to value constellation: Designing interactive strategy. Harv. Bus. Rev. **71**(4), 65–77 (1993)

T. Rose, K. Manley, A conceptual framework to investigate the optimisation of financial incentive mechanisms in construction projects. International symposium on procurement systems. The impact of cultural differences and systems on construction performance. Feb 7–10 (Las Vegas, USA, 2005)

H.M. Sapolsky, Inventing systems integration, in *The business of systems integration*, ed. by A. Prencipe, A. Davies, M. Hobday (Oxford University Press, Oxford, 2004), pp. 15–34

M. Spring, L. Araujo, Service, services and products: Rethinking operations strategy. Int. J. Oper. Prod. Manag. **29**(5), 444–467 (2009)

M. Spring, K. Mason, Business models for complex performance: Procuring aerospace engineering design services, in *Procuring complex performance*, ed. by N.D. Caldwell, M. Howard (Routledge Press, New York, 2010)

G. Sturgess, Commissions and concessions: A brief history of contracting for complexity in the public sector, in *Procuring complex performance*, ed. by N.D. Caldwell, M. Howard (Routledge Press, New York, 2010)

M. Thierry, M. Salomon, J.V. Nunen, L.V. Wassenhove, Strategic issues in product recovery management. Calif. Manag. Rev. **37**(2), 114–135 (1995)

H.M. Treasury, Guidance no. 58: Incentivisation (Her Majesty's Stationary Office, London, 1991)

A. Tukker, U. Tischner, Product-services as a research field: Past, present and future. Reflections from a decade of research. J. Clean. Prod. **14**, 1552–1556 (2006)

S. Vandermerwe, J. Rada, Servitization of business: Adding value by adding services. Eur. Manag. J. **6**(4), 315–324 (1988)

S.L Vargo, R.F Lusch, From goods to servive(s): Divergences and convergences of logics. Ind. Mark. Manag. **37**(3), 254–259 (2008)

R. Wise, P. Baumgartner, Go downstream—The new profit imperative in manufacturing. Harv. Bus. Rev. **77 (Sept–Oct)**, 133–141 (1999)

Chapter 9
Behaviour Transformation: An Examination of Relational Governance in Complex Engineering Service

Lei Guo and Irene Ng

Abstract In the present study, we investigated two outcome-based maintenance, repair and overhaul (MRO) service contracts in the attempt to better understand the effect of relational governance on firms' boundary-spanners behaviours in co-producing complex engineering service. Our field interviews indicate that managers are heavily dependent on interpersonal relationships to promote mutual cooperation in service delivery. The role of legal contracts in monitoring behaviours seems insignificant. Furthermore, we noted that with the development of interpersonal relationship, cooperation moved from reciprocal to communal. That is, at the early stage of relationship, both parties cooperate conditionally according to the norm of reciprocity. When the relationship becomes more mature, both parties share a common identity and work towards their collective goals. To that end, service performance would then be greatly enhanced. Through a theory-in-use methodology, our study maps the practices of behaviour transformation in complex engineering service systems, effected through interpersonal relationship development.

9.1 Introduction

The importance of social relations in supporting economic exchanges has been recognised across a wide range of theoretical perspectives (Eccles 1981; Granovetter 1985; Macaulay 1963). Most agree that relational governance, which is dominated by relationship norms, figures prominently in explaining the success and stability of inter-organisational exchanges. Yet, in equipment-based services, e.g., complex engineering service, the provider and the customer organisations often rely on legal contracts monitoring service activities. In contrast to people

L. Guo (✉) · I. Ng
University of Exeter Business School, Exeter, UK
e-mail: lg232@exeter.ac.uk

I. Ng et al. (eds.), *Complex Engineering Service Systems*, Decision Engineering,
DOI: 10.1007/978-0-85729-189-9_9,

centric services or 'pure' services, the significance of social relationships in equipment- based services is often overshadowed by a focus on the more tangible products attributes. Our study offered the first empirical investigation of relational governance in complex engineering service from an interpersonal relationship perspective. In the present study, two outcome-based maintenance, repair and overhaul (MRO) contracts of defence equipment provision have been examined. Unlike the conventional MRO contracts where the firms are paid on the basis of maintenance, repairs or spare parts used, the contracts were awarded on the basis of the availability of the equipment. The service provider relies heavily on the customer's operand (e.g., tangible equipment) and operant resources (e.g., knowledge and experiences) to deliver service outcomes. It is not possible to deliver an outcome without the customer co-producing the service with the firm. In day-to-day work, service performance is accomplished by individual employees, especially those who span the boundaries of the two organisations. Hence, there is a need to understand service co-production at individual levels.

Despite the importance of such perspective, there has been a lack of attention devoted to the study of the individual service provider and customer in complex engineering services literature. Nevertheless, our field work reveals that interpersonal relationships between them are essential to the overall service performance. Managers are often dependent on their relations with the customers to enhance inter-organisational cooperation. Furthermore, our interviews indicate that with the development of interpersonal relationships, the cooperation between the individual providers and customers can move from reciprocal to communal. That is, at the early stage of relationship, they cooperate conditionally based on the norm of reciprocity. When their relationship becomes more mature and both parties share a common identity, they will cooperate communally in the attainment of collective goals. To that end, service performance would then be greatly enhanced.

We begin with a literature review on equipment-based services, service co-production, and the interplay between relational governance and contract governance that will serve to illustrate the inadequacies of academic literature in providing insights of relational governance in equipment-based services. Following on, the method section outlines the details of the exploratory investigation of 19 in-depth interviews with individual providers and clients at all levels in both contracts. These interviews were coded and categorised through a theory-in-use methodology, supplemented by academic literature. We then close with a discussion on the implications of the findings, study limitations and directions for future research.

9.2 Literature Review

9.2.1 Service Co-production

The concept of equipment provision has been around since the start of industrial era. Indeed, as early as Adam Smith, the Wealth of Nations provided the dominant view that countries produce excess quantity goods to be exported and generate wealth. To date, the provision of equipment has become more complex as competition heightened, firms have felt the pressure to add value, predominantly through the provision of services. Research has shown that firms add on services integration with clients' capabilities. This provision has been commonly referred to as 'the servitization of manufacturing' (Baines et al. 2009). It has resulted in combinations of offerings to generate value from both products and services in bundled packages, frequently through providing through life support for their products by manufacturers (Vandermerwe and Rada 1988; Anderson and Narus 1995). Recently, the servitization of manufacturing has been reviewed and revisited. The emergent service-dominant logic (S-D Logic) (Vargo and Lusch 2004) provided a new paradigm that is rapidly gaining an established position in marketing literature. S-D Logic is captured in a list of Foundational Premises, such as 'the customer is always being a co-creator of value' with the firm. There is a distinction between value co-creation and service co-production though both imply the involvement of the customer and the firm. With S-D Logic, value is viewed as customer value which is proposed by the firm and unfolded by the customer over time. Nevertheless, service co-production is a process where the customer and the firm involve in delivering service outcomes. In the present study, our focus is service co-production.

Co-production represents a central construct in service literature (Zeithaml et al. 2006), such that the customer always plays an active role in the service offering. This conceptualisation derives from a specific characteristic of the service provision, namely, that the production phase cannot be disconnected from consumption activity (Lovelock and Wirtz 2004). Customer participation in service production has been defined as "the degree to which the customer is involved in producing and delivering the service" (Dabholkar 1990, p. 484). In this article, we adopt the view of co-production as "engaging customers as active participants in the organization's work" (Lengnick-Hall et al. 2000, p. 359). Therefore, we do not consider the situation of customer self-service (Meuter et al. 2000). When we refer to service co-production, we mean the joint production of service outcomes. Joint production is a situation in which both the customer and the firm's contact employees interact and participate in the production (Meuter and Bitner 1998).

The early work in co-production was largely firm-centric, the economic rationale provides the focus on cost benefits to the firm in terms of productivity gain, with customer labour substituting for employee labour (e.g., Fitzsimmons 1985; Lovelock and Young 1979; Mills et al. 1983; Mills and Morris 1986). The domain has then shifted to customer-centric, research in this theme has focused on when and why customers may be motivated to participate in service production as partial employees. This perspective has widely discussed topics such as technology readiness (Dabholkar 1996), provision of adequate training (Goodwin 1988) and identifying customer psychological responses (Bendapudi and Leone 2003). The concept of service co-production recently has been enhanced by S-D Logic (Vargo and Lusch 2004). That is, the personalized interaction between the firm and the customer thus becomes the locus of service co-production and resultant value creation. The new framework implies that all the points of customer–firm interaction are critical for creating value and value is co-created through their reciprocal and mutually beneficial relationship (Vargo et al. 2008). We argue that such relationship is grounded by interpersonal relationships between individual providers and customers. For example, when a customer talked about the relationship with the service provider, most likely, he/she was referring to the firm's employee, David or Susan. Therefore, co-production management need to place simultaneous emphasis on the role of interpersonal relationships in managing a successful inter-organisational relationship, and stimulating cooperative attitudes and actions. Although some authors argued that customer participation in service process is integral to service success, affecting overall performance and quality (e.g., Kelley et al. 1990; Schneider and Bowen 1995), interpersonal interactions between the individual providers and customers have been studied only in a limited fashion, especially in equipment-based services. This lack makes it impossible to draw strong conclusions about the effect of interpersonal relationships on service outcomes.

9.2.2 The Interplay of Relational Governance and Contract Governance

In inter-organisational exchanges, formal contracts are mechanisms that attempt to reduce risk and uncertainty (Lusch and Brown 1996). However, an undue reliance on the formal contract signifies a transaction oriented approach and an adversarial relationship (Gundlach and Achrol 1993). Williamson (1975) noted that contracts are incomplete because of parties' inability to write a comprehensive agreement that covers future contingencies, and Macneil (1980) recognised that the legal contract cannot explicitly state how potential situations will be handled in the future. It follows that strict adherence to the written contract may preclude the necessary flexibility in an exchange. Exchange performance can suffer when detailed contracts are used without a well developed social relationship (Cannon et al. 2000), and "the misuse of contracts could create irreconcilable conflict and other forms of dysfunctional behaviour that could ultimately harm channel member performance" (Lusch and Brown 1996, p. 19). Ghoshal and Moran (1996) argued that the use of rational, formal control has a pernicious effect on cooperation. They contend that for those parties being controlled, the use of rational control signals that they are not trusted to behave appropriately without such controls. Consistent with this logic, Bernheim and Whinston (1998) developed a formal model and showed that making contracts more explicit may encourage opportunistic behaviour surrounding actions that cannot be specified within contracts.

Research has long recognised that the importance of social relations in supporting inter-organisational exchanges. For example, Poppo and Zenger (2002) propose that for relationally-governed exchanges, the enforcement of obligations, promises, and expectations occurs through social processes that promote norms of flexibility, solidarity, and information exchange. Flexibility facilitates adaptation to unforeseeable events. Solidarity promotes a bilateral approach to problem solving, creating a commitment to joint action through mutual adjustment. Information sharing facilitates problem solving and adaptation because parties are willing to share private information with one another, including short- and long-term plans and goals. As the parties commit to such norms, mutuality and cooperation characterise the resultant behaviour. In addition, Poppo and Zenger (2002) also suggest relational governance may promote the refinement (and hence increased complexity) of formal contracts. As a close relationship is developed and sustained, lessons from the prior period are reflected in revisions of the contract. As a consequence, relational exchanges may gradually develop more complex formal contracts, as mutually agreed upon processes become formalised. In the present study, relational governance is defined as the strength of the relationship norms present in the service activities co-produced by the individual providers and customers.

However, researchers cannot agree on whether relational governance and contract governance function as substitutes or complements (see Poppo and Zenger 2002 for a full review). One stream of research has viewed them as substitutes. The underlying logic is if contracting parties trust each other, there is little need for contractually specifying actions. Informal social controls push these formal contractual controls to the background. In particular, Adler (2001) argued that trust can replace formal contracts by ‘handshakes’ in marketing exchanges. Dyer and Singh (1998) concluded that formal contracts are rather unimportant as informal self-enforcing agreements which rely on trust and reputation ‘often supplant’ the controls characteristic of formal contracts. Alternatively, other research contends the complementary relationship between relational governance and contract governance. In particular, in settings where hazards are severe, the combination of formal and informal safeguards will deliver greater exchange performance than either governance choice in isolation (Poppo et al. 2008). A common rationale is that the presence of clearly articulated contractual terms, remedies, and processes of dispute resolution as well as social controls can inspire confidence to cooperate with the counterparty.

In the execution of service contracts, relational governance becomes a necessary complement to the adaptive limits of contracts. It may heighten the probability that social relations will enhance trust and cooperation, and safeguard against hazards poorly protected by the contract. Typically, prevailing literature is mainly concerned with contractual and relational governance at the firm's level. Monitoring service activities at individual levels remains complex and troubling in management practices. Specifically, in outcome-based delivery, governance is complicated by the lack of specification in individual boundaries and roles. In the present study, we offer an empirical investigation on relational governance in complex engineering service from an interpersonal relationship perspective.

9.3 Context and Research Method

Our data was collected from two prime contractors working for the Ministry of Defence (MoD), UK. They provided the MRO service for defence equipment, e.g., fastjet and missiles. The contracts were awarded on the basis of the availability of the equipment. Both organisations were awarded for the MRO of whole operable life of the equipment till the out-of-service date. The availability of the equipment means that subject to certain conditions of how and where the equipment is used by the customer, the firm is obliged to deliver a set number of flying hours on the fighter jet and a fix percentage availability over a certain period of time (e.g., 95% availability) for the missile system. While the MRO service is outsourced, the MoD had a big role in the partnership which is to provide Government Furnished Materials (GFX) including supplying physical facilities, material, data, IT and manpower to facilitate the company in achieving its outcomes. Both parties were co-located at the service site and worked together for the contracts. Under this circumstance, interactions at the interpersonal level are intense. This context is best for the study of relational governance from an interpersonal relationships perspective.

The field study was carried out over a 2-year period from 2007 to 2008. Multiple qualitative methods had been employed to extract data for the purpose of understanding the dynamics arising from outcome-based service contracts. We had 19 interviews with team members from both sides, attended the contractors' orientation meetings, presentations and visited the service sites several times. Meanwhile, supplemental materials such as project briefing brochure and contracting documents (although not released to be reported in this article) were collected. The logic behind using multiple methods was to secure an in-depth understanding of the phenomenon in question. These interviews followed the phenomenological focus, that is, from the perspectives of the participants (Thompson et al. 1989). The primary aim of this type of qualitative investigation is to understand experience as closely as possible as its participants feel or perceive it. The depth-interview method offered the opportunity to gain insight into individuals' subjective experience with the contract. The course of the interview dialogue was set largely by the participants. The interviewer's questions were formulated in concert with the participants' reflections and were directed at bringing about more thorough descriptions of specific experiences. The primary objective was to allow the participants to articulate their own system of meanings (Thompson et al. 1989): the personalised meanings and meaning-based categories that constitute the individual's abstracted understanding of governance in service delivery.

The interviews included 11 managers from the service providers and 8 members from the clients in both contracts. The goal was not to compare between size and structure of the two joint delivery teams but to provide personnel at each level. The participants represented all levels of the teams (4 project directors [senior executives], 2 general managers [mid-level executives], 9 project managers [mid-level executives] and 4 engineers/technicians [staff]). These positions were included primarily because of significantly interpersonal interactions involved in their work. The interviewees were not selected in advance but by availability and also sequential recommendation from those who participated. Interviews were completed between 50 and 120 min. Prior to the interview, the participants were asked for their consent, which assured them and their firm of confidentiality regarding the interview texts and any other materials provided. The purpose of study was described to each participant as an exploration of outcome-based service contracts they were working on. The interview began by obtaining general background information on the participant (e.g., position, employment history and length of time on the contract). It then shifted to specific topics such as relational governance using the following probe: "How would you find the relationship between you and the customer?" or "Tell me about the relationship." The logic of this interview flow is straightforward: questions at the beginning of the interview provide the broader descriptions needed to contextualise the participant's specific experiences in the later part. Each interview was recorded and transcribed.

Data analysis proceeded, using discourse analysis to uncover the meanings and emerging themes of relational governance. The discourse analysis of this research was primarily focused on verbal discourse as opposed to written discourse. This has been termed the "talk as text" approach (Potter 1996). As talk and texts are parts of social practice, they 'construct' knowledge rather than 'describe' facts. The discourse analysis of the interview transcripts proceeded with the use of a hermeneutical procedure outlined by Thompson (1997). The hermeneutical circle is implemented by continually modifying the initial understandings of participant's personal experience with specific contracts, based on earlier passage with respect to the participant's general background. Earlier readings of the text inform later readings, and reciprocally, later readings were incorporated into the interpreting participant's broader life experience. As the analysis proceeds, the textual interpretations were broaden, with the resultant thematic structure reflecting the understanding of individual experience in the broadest context. Accordingly, readings on texts expressed by different participants allow the researcher to identify key patterns of relational governance across all the interviews and derive general managerial implications.

As demonstrated in the literature review, the divergence between practice and what the literature suggests is the backdrop against the reasons why a theory-in-use methodology seems warranted. In their book Theory Construction in Marketing: Some Thoughts on Thinking, Zaltmann et al. (1982, p. 113) illustrate the fundamental concept of a theory-in-use approach:

> Practitioners... are generally more concerned with informal theory based on everyday observations (versus controlled experiments), having less than precise concepts (versus explicit empirical referents), and being related to one another intuitively (versus in rigorous testable relationships). The informal theory built and maintained by practitioners in their everyday activities represents an important source of insight for the researcher concerned with formal theory. By mapping these informal theories and applying their own creativity, a researcher may gain insights into marketing phenomena which might not otherwise be obtained.

This methodology is therefore an exercise in reconstructed logic; mapping informal theory, linking with academic literature and developing a greater understanding of the phenomenon. Our purpose is to understand, formalise and document practitioners' experience of governance in service co-production as a contribution to academic literature.

With the assistance of qualitative software Nvivo 7, we started with open coding of the transcripts of the interviews, scrutinising the transcripts line by line in order to identify key words and phases that would give us insights into what was happening in the data. At this stage, coding was unstructured and hundreds of codes had been identified. Inevitably these need to be reduced as coding moves on to a more abstract level in the search for patterns and themes that suggest a relationship. We employed axial coding which clustered the open coded nodes in terms of their dynamic interrelationships. These categories were then re-evaluated and gradually subsumed into higher order categories which suggested the emergent themes of relational governance in equipment-based services. Meanwhile, memos were written to note ideas and reflections during data collection and analysis. The coding and memos had been constantly revisited and compared in searching for patterns of relational governance. The interviews had been conducted in an ongoing manner until no new insights arose from fresh data. The fresh data were compared with existing transcripts and were scrutinised for any new information. Whilst no new concepts emerged, the significance of those identified were reinforced and strengthened with further examples in different contexts. These concepts paved the way for categorisation.

9.4 Findings

9.4.1 Contract Governance

Our findings:

- Comprehensive contracts cannot provide complete safeguards against potential risks and uncertainties.
- The lengthy procedure of contract amendment has negative impact on service performance.
- Outcome-based contracting has difficulty in specifying boundaries and roles.
- Financial incentives failed to motivate interpersonal cooperation.
- Strictly working around the contract can lead to service failure.

Contract governance has previously been examined in terms of the existence of a detailed agreement (Cannon et al. 2000), its enforcement under conditions of violation (Antia and Frazier 2001), and the stipulation of expected behaviours and roles in the exchange (Lusch and Brown 1996). As far as concerned by our interviewees, contract governance has several limitations. First, even the most comprehensive contract cannot provide full safeguards against potential risks and uncertainties as the negotiating parties only write down clauses they agree on but remain silent on issues on which they have debate. One of the interviewees explained this practice to us:

> Well contracts tend to be tend to be silent on a lot of, on a lot of issues, you don't tend to may be because you haven't thought of them may be 'cos you didn't particularly want to broach that particular subject because you knew it would be difficult or whatever so sometimes you just don't bother to write them down and you will work your way through that once you get err once you get into the contract... Well inevitably, inevitably you never, you are never able to cover I don't think everything that you are going to have to be able to do. I mean first of all you would have a contract which was probably five times the size and you probably never actually reach a conclusion you have just got to assume that certain things kind of will happen. You are not, you never ever going to be able to tie everything down to you know a black and white in a contract it just isn't going to happen, it's not practical.

Secondly, the contract amendment procure is lengthy. In some situations, the foremost task is to solve the problem; otherwise, service performance will suffer from the bureaucratic process. Under this circumstance, the service provider and the customer would prefer to get things done rather than wait for the contract amendment. One interviewee described the lengthy procure of contract amendment:

> Now if you just follow the absolute letter of the contract then we would just stop, we wouldn't do anything right we would officially declare that we are not going to provide this manpower and we would go through a lengthy process of asking us for a quote, we would give them the quote and then get the money approved and the money finally turns up they put a contract amendment in and then we get to go out and we recruit sub contractors. That will take months and months and months.

Thirdly, outcome-based contracting has difficulty in specifying boundaries and roles. Individual service providers and customers did not sufficiently understand their roles in terms of the tasks and behaviours required in service delivery. As a result, the provider strictly adhered to the Key Performance Indicators (KPI) activities as defined in the contracts. On the other hand, the customer supervised rather than co-produced the service with the provider as they perceived that the provider was solely responsible for service outcomes.

Fourthly, contracting mechanisms such as 'gain share' and 'pain share' of financial incentives crafted to promote the firm-level cooperation failed to motivate interpersonal cooperation as individual customers (i.e., users) were more concerned with operational efficiency and effectiveness. The customer's perception that the provider was after profit rather than operational capability almost caused a difference in commitment. The service provider complained that the end user community was reluctant to get involved in service delivery. One interviewee shared his observation of the end users' ignorance:

> They (end users) don't know what's going on in that background and the user doesn't necessarily want to get involved in a lot of these contracts. Whether it be (end users),

> whether it be any (end users) you know if it came for (end users) or something like that, the user is just principally standing back going, well it's got nothing to do with me.

Finally, strictly working around the contract had caused insufficient communication and interactions at individual levels. Although there were formal communication channels such as regular meetings, interpersonal interactions were lacking. As observed by many interviewees, the provider and the customer would sit separately during the monthly meeting and not mingle with each other at all. As the customers recalled, at the beginning stage of the contract, a lack of interpersonal communication with the provider had left them in the dark. They had no idea why the provider had delivered the service differently from what had been done in-house. One customer told us the misunderstanding of the provider's practice,

> That's not to say they are doing them worse they are just doing them differently but they hadn't communicated that that's how they were going to do business. So therefore we were scratching our heads and not understanding why they were making some of the decisions that they were making but it was because they hadn't told us that their philosophy was that way or this way.

Consequently, service co-production was going to break. However, both parties cannot afford the contracts to fail, as the customer wanted to achieve service outcomes whereas the provider looked forward to business success because any failure would be harmful to the company's reputation. Therefore, they decided to work on relationships to rescue the service.

9.4.2 The Relationship Development

Our findings:

- Interpersonal relationships developed through informal interaction and socialisation.
- Interpersonal similarity is key to relationship development.

The individual provider and customer described their initial relationships as "an arranged marriage". The newly married couple has yet to find any common goals in life. The customer community treated the provider as an opportunist who was eager to make money from the marriage. Some even considered that working with the provider would be a nightmare. Likewise, salient cultural differences have discouraged interpersonal interactions. For example, many of the end users, in this case the soldiers, had problems in dealing with industrialists. Similarly, industry people also found it difficult in adapting to military culture:

> there is very much a focus on the task at the expense of anything else that is a bit of a different culture, and so that has for some individuals understandably who didn't join up to be military they have found it difficult to adjust to.

As stated early, in order to rescue the contracts, both organisations as well as their employees were making efforts to develop relationships. The firms were trying to put the right people at the boundary positions. For example, they recruited ex-military staff who could understand the protocols of the army and build rapport with the soldiers. The shared experience tended to break the ice of interpersonal interactions. On the other hand, individual providers and customers realized that informal events facilitating relationship development. Social activities such as golfing, outings and barbeques were organised to promote interpersonal interactions. The ongoing interaction did not develop in a vacuum but rather, it accounts for the social relationship bonds. The providers and customers became more joined up through socialising. Many relationships seemed to be developed outside of the work environment. One interviewee told us how they got to know the counterparty through social events:

> we had a night out with them so you get to know people that way socially and they've also got I think it's a, end of July like an open day barbeque and all that which I think most of us will be going to. Again you will get to know them that way.

In addition, we found that the influence of setting on formulating relationships was rather weak. Although co-located and work together on a daily basis, individual employees only developed close relationships with those who were similar to them.

9.4.3 Relational Governance

Our findings:

- Through developing interpersonal relationships, co-producing service became possible.
- When service provider–customer relationships developed into in-group relationships, cooperation moved from reciprocal to communal.

Cooperation and mutual adaptation provided the flexibility to cope with inevitable uncertainties that arise in the contracts. The expectation of continuity generated incentives to invest in relationships. As well, the expectation of longevity minimised the need for precise performance measurement in the short run as both parties expected that short- term inequities would be corrected in the long term.

Relational governance refers to a social institution that governs and guides exchange partners on the basis of cooperative norms and collaborative activities (Heide and John 1992; Macneil 1980; Zaheer and Venkatraman 1995). The mechanisms through which relational governance attenuates contract hazards are both economic and sociological in nature. Economists emphasise the rational, calculative origins of relational governance, emphasising particularly expectations of future exchanges that prompt cooperation in the present. Sociologists emphasise socially derived norms and social ties that have emerged from prior exchange (Uzzi 1997, p. 45). Trust is therefore considered a trait that becomes embedded in a particular exchange relation. Many conceptual (e.g., Bradach and Eccles 1989;

Dyer and Singh 1998; Uzzi 1997) and empirical (e.g., Artz and Brush 2000; Dyer and Chu 2003; Poppo and Zenger 2002) works have established the benefits of trust-based forms of governance. In essence, once an exchange partner is granted trustworthy status, they are expected to behave in a trustworthy fashion in the future. Williamson (1996, p. 97) suggests that the term trust is misleading, arguing that "because commercial relations are invariably calculative, the concept of calculated risk (rather than calculated trust) should be used to describe commercial transactions." Nonetheless, there is considerable overlap in the arguments of sociologists and economists surrounding trust and cooperation. Poppo and Zenger (2002) summarise that both sociologists and economists, for instance, argue that repeated exchange encourages effective exchange, and that repeated exchange provides information about the cooperative behaviour of exchange partners that may allow for informed choices of who to trust and who not to trust. While the mechanism may differ slightly, both economists and sociologists emphasise that reputations for trustworthy behaviour are rewarded and reputations for untrustworthy behaviour punished in the broader network of potential exchange partner. Consistent with the literature, our interviewees agreed that interpersonal trust is the cornerstone of relational governance. Following the economic logic, one interviewee told us the trustworthy status is conditional upon the benefits that accrue from trustworthy status over time contrasted with the benefits that accrue from self-interested moves that break from the trustworthy status. He said:

> we have to be nice it's okay we can trust them on this it's fine we trust them and you know the first time they ever let us down then that will destroy four or five years of relationship the same as if we ever let down that would destroy the relationships.

Amongst the advantages of relational governance, cooperation and resultant work efficiency and effectiveness were specially highlighted. As we discovered in our study, interpersonal relationships were considered as the glue that made all things work. The interviewees suggested that what was stated on paper (e.g., the contracts) may never happen but relationships helped make sure things get started and finished in a positive tone. Even when the process got stuck, having the right relationship could move it on. Otherwise, it would just stop there. As a result, interpersonal relationships enhanced efficiency and effectiveness in service co-production. For example, one interviewee thought social relationships were the solution of dispute:

> Now you can either spend two years having the fight and winging or if you have got the relationships you can just, it will get sorted out so it's things that you don't put financial fund it works it just makes everybody's life a lot easier and things just get done.

Noteworthily, our interviewees made it clear that relying on relationships does not mean shortcutting everything. Things can go wrong if one attempts to shortcut an established process in the contracts. It echoes the previous research that relational governance is complementary to contract governance. Next, we use examples to illustrate how relational governance was effected through relationship norms. Throughout the interviews, the norm of reciprocity was underlined. When one party gave favour to the other, he/she had the expectation that the counterparty would repay e.g., helping with the milestone task. Typically, the individual providers were developing personal relationships with key customers in exchange of resources, information and preferable treatment. Their cooperation was reciprocal, as illustrated by one interviewee, "they might bend the rules a bit and you might bend the rules a bit". We thus termed this relationship as the exchange relationship where relationship parties cooperate conditionally to facilitate the attainment of each other's goals.

Interestingly, we noticed the other type relationship norm which we refer to it as communal cooperation. Over time, some of the relationship development efforts gradually eliminated estrangements. The individual providers and customers discovered their interpersonal similarities and eventually moved from the 'them vs. us' split into a collective 'we'. Getting back to the marriage analogy, the marriage did not actually happen at the time they signed the contracts, but when both parties made the emotional commitment to act as one united team. After working on the contract for nearly one and half years, for example, we observed the 'we-ness' on some occasions. As a visitor, we could hardly distinguish between them when both parties were working at the service site. Our interviews also revealed, as a result of relationship development efforts, some of the individual providers and customers started to share a salient identity and they treated each other as 'one of us'. In some cases, the individual providers had been empowered to plan and solve problems for the customers with full knowledge of the goals. One interviewee shared his experience of being treated as an integral part of the team by the customer:

> just recently only a couple of months ago I went on a, on a military field trip with them into (the place) you know just another member of the team basically that's how I get treated.

He continued later,

> Oh yeah I am on the (the client's premises) now so I sit with (the client's team leader) and his team once a week and we go through they tell me things which are military, we go through military issues and they, they ask me for my industrial opinion. So they'll say err we are thinking about closing (the place) next week (name of the interviewee) for a day because we've had a really good err (the inspection) how does that affect output for us? I'll say well you'll lose a day's production you know, it'll cost us X number thousands of pounds but I am consulted the issue isn't the answer the issue is the consultation is the issue.

Social identity approach (Haslam 2004) specifies the circumstances under which individuals are likely to conceive themselves either as separate individuals or as part of a collective group. This in-group relationship is characterised by the norm of communal cooperation, that is, both parties cooperated for the attainment of collective goals. One individual provider told us he was supported by the customer's team leader when attempting to correct certain behaviours of the end users for the purpose of cost saving:

> Right you know right but that's the kind of relationship we've got. I can go in there and be very outspoken with them because I've built up that level of trust with them and you

know. Yeah but that's the kind of relationship we've got you know and (the client's team leader) sits there and he says I agree it's completely unacceptable and you know.

9.5 Discussion, Implications and Limitations

Our findings indicate that relational governance plays a very significant role in the delivery of outcome-based service contracts, especially in promoting service co-production at individual levels; whilst contract governance seemed unimportant. Although comprehensive contract mechanisms have been designed to safeguard service delivery, managers in fact rely heavily on relational governance to improve service performance. They tend to count on reciprocal or communal cooperation to weather through the various uncertainties in service delivery. In relationship development, we note that interpersonal similarities can foster close relationships. Likewise, informal interactions and socialising eliminated differences and misunderstanding. Accordingly, two types of individual provider–customer relationships were discovered. In exchange relationships, relationship parties cooperate based on the norm of reciprocity. As relationships develop into in-group relationships, both parties cooperate unconditionally and work towards collective group goals. A major contribution of this study is the advancement of an interpersonal relationship perspective of relational governance on service contracts through a theory-in-use methodology. It is our belief that this contribution provides pedagogical benefits and substantial relevance to the practice of service co-production and management thinking.

In this article, we have used our field study to structure inquiry into how relational governance works on service co-production at individual levels. Through our findings, we present an integrative view of contract governance and relational governance in service contracts. We not only bring together interpersonal relationships and relational governance but also relate cooperative behaviours to types of relationship norms. Furthermore, drawing on our field observations, we address the gap between theory and field research by introducing the idea that different types of relational norms, reciprocity and communality, can affect interpersonal cooperation in service co-production which has not been studied adequately in literature.

The striking factor in our study is that relational governance as a strategic resource to improve service performance. This seems particularly advantageous for outcome-based service contracts where environmental and behavioural uncertainties are a major obstacle in delivering outcomes, and where interactions between equipment, people and information increase the complexity. Our article is descriptive rather than normative in nature, and does not focus on performance implications of service delivery. Thus, generic and explicit recommendations on how organisations should manage contractual and relational governance cannot be drawn directly from our research. However, assuming that qualitative case studies provide better insights, we at least can make tentative statements of how our observations will guide managerial decisions:

- Outcome-based delivery in lack of specification in boundaries and roles, results in a greater dependency on relational governance.
- Delivering outcomes, rather than contracted performance measurements, is the primary focus of service co-production, with contracted performance measurements a secondary result arising from the primary focus.
- Interpersonal relationships oil the wheels of service co-production. Without proper relationships, the individual providers and customers won't co-produce anything and the service provided is not sustainable.

In addition, this relationship is a dyad, we urge the provider and the customer organisations to align their personnel arrangement at all levels with the counter side and relationship bonding activities are highly recommended.

The predominant emphasis of this study is to obtain in-depth understanding of the phenomenon of contractual and relational governance in equipment-based services. Although the investigation of two outcome-based contracts would still be relevant to the theory-in-use methodology (Zaltmann et al. 1982), multiple cases should be chosen as a way of discovering more patterns of relational governance. The study scope has limited our ability to generalise our findings.

It is interesting to note that what originally started as a study on relational governance of service contracts seems to have obtained quite unexpected results. Along the course of our study, we discovered two different types of interpersonal relationships between the service provider and the client: exchange and in-group. Specifically from the point of view of managing interpersonal cooperation, in-group relationships motivate collective actions which may hugely increase service output. At the best of our knowledge, this type of relationships has not attracted enough attention in academic research as much marketing literature is focused on the exchange type of relationships. Our study suggests a divergence between academic literature and practice in so far as the focus and approach towards individual provider–customer relationships are concerned. In light of this, there is a need for greater research attention towards the communal type of relationships.

9.6 Chapter Summary Questions

The chapter has raised the issue that relational governance is more important than contract governance in co-producing complex engineering service at individual levels, and hence it drives a few concluding questions:

- First, how can the firms train their boundary spanners to use relationship-building strategies?
- Second, can relationships be duplicated to other service contexts? i.e., another service contract?
- Third, what do the firms need to do to encourage the formation of a common identity in the service provider–customer joint work team?
- And finally, is there an issue if the key contact person leaves the team?

References

P. Adler, Market, hierarchy, and trust: The knowledge economy and the future of capitalism. Org. Sci. **12**(2), 214–234 (2001)

J.C. Anderson, J.A. Narus, Capturing the value of supplementary services. Harv. Bus. Rev. **73**(1), 75–83 (1995)

K.D. Antia, G.L. Frazier, The severity of contract enforcement in interfirm channel relationships? J. Mark. **65**(4), 67–81 (2001)

K. Artz, T. Brush, Asset specificity, uncertainty and relational norms: An examination of coordination costs in collaborative strategic alliances. J. Econ. Behav. Organ. **41**(4), 337–362 (2000)

T.S. Baines, H.W. Lightfoot, O. Benedettini, J.M. Kay, The servitization of manufacturing: A review of literature and reflection on future challenges. J. Manuf. Technol. Manag. **20**(5), 547–567 (2009)

N. Bendapudi, R.P. Leone, Psychological implications of customer participation in co-production. J. Mark. **67**(1), 14–28 (2003). January

B.D. Bernheim, M.D. Whinston, Incomplete contracts and strategic ambiguity. Am. Econ. Rev. **88**(4), 902–932 (1998)

J. Bradach, R. Eccles, Price, authority, and trust: From ideal types to plural forms. Annu. Rev. Sociol. **15**, 97–118 (1989)

J.P. Cannon, S.A. Ravi, T.G. Gregory, Contracts, norms, and plural form governance. J. Acad. Mark. Sci. **28**(2), 180–194 (2000)

P. Dabholkar, How to improve perceived service quality by improving customer participation, in *Developments in marketing science*, ed. by B.J. Dunlap (Academy of Marketing Science, Cullowhee, 1990)

P. Dabholkar, Consumer evaluations of new technology based self-service options. Intern. J. Res. Mark. **13**(1), 29–51 (1996)

J. Dyer, W. Chu, The role of trustworthiness in reducing transaction costs and improving performance: Empirical evidence from the United States, Japan, and Korea. Org. Sci. **14**, 57–68 (2003)

J. Dyer, H. Singh, The relational view: Cooperative strategy and sources of interorganizational competitive advantage. Acad. Manag. Rev. **23**(4), 660–679 (1998)

R.G. Eccles, The quasi-firm in the construction industry. J. Econ. Behav. Organ. **2**(4), 335–357 (1981)

J.A. Fitzsimmons, Consumer participation and productivity. Serv. Opera. Interfaces **15**(3), 60–67 (1985)

S. Ghoshal, P. Moran, Bad for practice: A critique of the transaction cost theory. Acad. Manag. Rev. **21**(1), 13–47 (1996)

C. Goodwin, I can do it myself: Training the service consumer to contribute. J. Serv. Mark. **2**(4), 71–78 (1988)

M. Granovetter, Economic action and social structure: The problem of embeddedness. Am. J. Sociol. **91**(3), 481–510 (1985)

G.T. Gundlach, R.S. Achrol, Governance in exchange: Contract law and its alternatives. J. Pub. Policy. Mark. **12**(2), 141–155 (1993)

A. Haslam, *Psychology in organizations: The social identity approach*, 2nd edn. (Sage, London, 2004)

J. Heide, G. John, Do norms matter in marketing relationships? J. Mark. **56**(2), 32–44 (1992)

S.W. Kelley, J.H. Donnelly, S.J. Skinner, Customer participation in service production and delivery. J. Retail. **66**(3), 315–335 (1990)

C.A. Lengnick-Hall, V. Claycomb, L.W. Inks, From recipient to contributor: Examining customer roles and experienced outcomes. Eur. J. Mark. **34**(3/4), 359–383 (2000)

C.H. Lovelock, J. Wirtz, *Services marketing: People, technology, strategy*, 5th edn. (Pearson/ Prentice Hall, Upper Saddle River, 2004)

C.H. Lovelock, R.F. Young, Look to consumers to increase productivity. Harv. Bus. Rev. **57**(3), 168–178 (1979)

R.E. Lusch, J.R. Brown, Interdependency, contracting, and relational behavior in marketing channels? J. Mark. **60**(4), 19–38 (1996)

S. Macaulay, Non-contractual relations in business: A preliminary study. Am. Sociol. Rev. **28**(1), 55–69 (1963)

I.R. Macneil, *The new social contract: An inquiry into modern contractual relations* (Yale University Press, New Haven, 1980)

M.L. Meuter, M.J. Bitner, Self-service technologies: Extending service frameworks and identifying Issues for research, in *AMA winter educators' conference*, ed. by D. Grewal, C. Pechmann (American Marketing Association, Chicago, 1998)

M.L. Meuter, A.L. Ostrom, R.I. Roundtree, M.J. Bitner, Self-service technologies: Understanding customer satisfaction with technology-based service encounters. J. Mark. **64**(3), 50–64 (2000)

P.K. Mills, J.H. Morris, Clients as 'partial' employees of service organizations: Role development in client participation. Acad. Manag. Rev. **11**(4), 726–735 (1986)

P.K. Mills, R.B. Chase, N. Margulies, Motivating the client/employee system as a service production strategy. Acad. Manag. Rev. **8**(2), 301–310 (1983)

L. Poppo, T.R. Zenger, Do formal contracts and relational governance function as substitutes or complements? Strateg. Manag. J. **23**(8), 707–725 (2002). August

L. Poppo, K.Z. Zhou, T.R. Zenger, Examining the conditional limits of relational governance: Specialized assets, performance ambiguity, and long-standing ties. J. Manag. Stud. **45**(7), 1195–1216 (2008). Nov

J. Potter, *Representing reality: Discourse, rhetoric and social construction* (Sage, London, 1996)

B. Schneider, D.E. Bowen, *Winning the service game* (Harvard Business School Press, Boston, 1995)

C.J. Thompson, Interpreting consumers: A hermeneutical framework for deriving marketing insights from the texts of consumers' consumption stories. J. Mark. Res. **34**(4), 438–455 (1997)

C.J. Thompson, W.B. Locander, H.R. Pollio, Putting consumer experience back into consumer research: The philosophy and method of existential-phenomenology. J. Consum. Res. **16**(2), 133–147 (1989)

B. Uzzi, Social structure and competition in interfirm networks: The paradox of embeddedness. Adm. Sci. Q. **42**, 35–67 (1997)

S. Vandermerwe, J. Rada, Servitization of business: Adding value by adding services. Eur. Manag. J. **6**(4), 315–324 (1988)

S.L. Vargo, R.F. Lusch, Evolving to a new dominant logic for marketing. J. Mark. **68**, 1–17 (2004)

S.L. Vargo, P.P. Maglio, M.A. Akaka, On value and value co-creation: A service systems and service logic perspective. Eur. Manag. J. **26**(3), 145–152 (2008)

O.E. Williamson, *Markets and hierarchies* (McGraw-Hill, New York, 1975)

O.E. Williamson, Transaction cost economics and organization theory, in *The handbook of economic sociology*, ed. by N.J. Smelser, R. Swedberg (Princeton University Press, Princeton, 1996)

A. Zaheer, N. Venkatraman, Relational governance as an interorganizational strategy: An empirical test of the role of trust in economic exchange. Strateg. Manag. J. **16**(5), 373–392 (1995)

G. Zaltmann, K. LeMasters, M. Heffring, *Theory construction in marketing: Some thoughts on thinking* (Wiley, New York, 1982)

V.A. Zeithaml, M.J. Bitner, D.D. Gremler, *Services marketing: Integrating customer focus across the firm*, 4th edn. (McGraw-Hill, New York, 2006)

Part III
Service Information Strategy

Alison McKay, Duncan McFarlane and Chris Pearson

The effective delivery of complex engineering services critically relies on the timely availability of appropriate information in forms suitable for use in a service operation. Additional complexity arises from the fact that service information typically has many owners and users dispersed across multiple organisations. As the trend for businesses to change from the delivery of goods and equipment to the delivery of complex services increases, so does the demand for improved understanding of how engineering information needs to change to support complex engineering services. Building this new understanding was one of the key challenges addressed by the S4T project.

The chapters in this section resulted from research carried out as part of the Service Information Strategy work package. Research in this work package examined the role and organisation of service information in the provision of availability and performance-based service contracts. The overall research question addressed was, "What are the most effective strategies for service information provision in support services solutions?" The three chapters provide a grounding for ways in which information and information systems might evolve as an organisation transfers from being a goods or equipment provider to a complex engineering service provider. Hence, with reference to the Core Integration Framework discussed in the Introduction, this section is primarily concerned with enabling appropriate information transformations which are central to delivering a complex engineering service.

As the research began it was recognised that the information requirements for complex engineering services are multifaceted and highly dependent on the nature of both the service offering and the underpinning service agreement. In parallel, a key challenge for industry lay in understanding how businesses might best transition from being product to service delivery enterprises. Early research highlighted the need for practitioners to understand information implications, risks and opportunities at different stages of the PSS (Product Service System) life-cycle. For example, when contracts are being agreed all parties in the co-creation process need to be able to appreciate the implications of their decisions for information through the whole life of the PSS whereas when service information systems are being built detailed information requirements need to be defined.

In the first chapter (Chap. 10), a framework for the identification of service information requirements, the so-called "12-box model", is introduced. The 12-box model combines aspects of service supply networks and service lifecycles. It was used to provide insights on information maturity in nine industrial PSSs through a traffic light approach where information requirements were labelled as being either green (completely addressable by the current solutions), amber (partially addressable by current solutions) or red (calls for new solutions). In addition, positive feedback from the BAE Systems Support Council led to the 12-box model being used within BAE Systems and mapped to current service development processes within the company.

An important challenge lies in understanding service affordability, feasibility and risk from the perspective of both supplier and customer, and the impact of performance-focused service requirements on service costs. In the second chapter (Chap. 11), an application of discrete event simulation to service systems for complex engineering products is described. The chapter uses a number of different service contract types and information availability scenarios to highlight the role that computer-based simulations might play in understanding trade-offs to be made between service affordability and performance when developing service contracts.

Having identified information requirements service developers need to understand their detail in order that appropriate service information systems can be designed and built. In the third chapter (Chap. 12), a service information blueprint is introduced as a means of defining service contracts and processes, and relationships between the two. By supporting both (a) the capture of process definitions (including process flows, steps and decompositions) and performance requirements related to both the contract and the parties in the co-creation process, and (b) the definition of relationships between process definitions and performance requirements, the service information blueprint provides a bridge from PSS to traditional engineering information system design and development methodologies.

Service industries today represent over 70% of employment in the USA and Europe. Increasingly, traditional manufacturers are exploring strategies for evolving their product offerings into product-service systems. This is a response to increasing purchase costs of capital equipment that means owners are using assets for longer and require suppliers to take a greater role in their support. In this section we have presented three contributions that can be used by organisations to build understanding of how their engineering information needs to change to support service offerings. The "12 box model" (Chap. 10) enables managers without specific IT experience to engage with information systems by identifying stages that information must pass through to fulfil a service contract. On this basis, the information blueprint (Chap. 12) can be used by service designers to define service processes and relate them to both the service contract and existing information resources, and service simulations (Chap. 11) provide means by which the operation of service processes can be visualised to provide insights on their likely performance with respect to the service contract.

Chapter 10
A Framework for Service Information Requirements

Rachel Cuthbert, Duncan McFarlane and Andy Neely

Abstract The increase in the percentage of revenue gained by nations as a result of growth in the service sector has been accompanied by the use of performance-based contracts to support such services. These performance-based contracts tend to be long-term in nature and often involve multiple parties in their service operation. In order to support such a service operation, information is required for its design, delivery and evaluation. This chapter provides a review of models and frameworks for the design and development of services. It focuses on providing details of a framework which has been developed for the determination of, specifically, service information requirements in order to design, deliver and evaluate services provided against engineering assets. It provides some general analyses of problems encountered within service providing organisations, and highlights areas of good practice. Further areas of application for the proposed framework are also presented.

10.1 Introduction

The proportion of revenue derived from different sectors of the economy has changed. In the UK, service, industry and agriculture account for 76, 23 and 1% of the nation's Gross Domestic Product (GDP), respectively (The World Factbook

R. Cuthbert (✉) · D. McFarlane · A. Neely
Institute for Manufacturing, University of Cambridge, Cambridge, UK
e-mail: rc443@cam.ac.uk

D. McFarlane
e-mail: dcm@cam.ac.uk

A. Neely
e-mail: adn1000@cam.ac.uk

I. Ng et al. (eds.), *Complex Engineering Service Systems*, Decision Engineering,
DOI: 10.1007/978-0-85729-189-9_10, © Springer-Verlag London Limited 2011

2008). This growth in the UK's service sector has been steady from around 55% in the 1970s (OECD 1996). The growth in the provision of services may be from traditional services, such as education and healthcare, through to those provided in an industrial context around complex engineering assets. While the split is not indicated by these statistics, they reinforce an apparent trend in the servitization of engineering organisations. Research by Neely indicates that the boundaries between manufacturing and service firms are breaking down across the globe. From a study of publicly listed engineering companies from around the world, 30% had been deemed to have servitized and 70% had remained as pure manufacturing firms (Neely 2009). Therefore, in consideration of the figures for the GDP above, a number of the service providing firms may potentially be derived from the industrial or service categories.

Alongside this rise in services there has been an increase in the use of contracts to support the delivery of such services. A number of types of contracts are used in such contexts ranging from a traditional context where these were generated on a case-by-case basis to a situation where contracts are now performance-based agreements, often involving multiple parties. Typically, four types of contracts are referred to within this chapter, namely, spares and repairs, spares inclusive, availability and capability. In the context of a spares and repairs (or discrete) contract, the industry is contracted for the supply of assets while the customer is responsible for the repair and overhaul of assets. Spares inclusive are similar to spares and repairs contracts in terms of the supply of parts, while joint customer–supplier teams undertake the maintenance and overhaul and share the risks. However, responsibility for the equipment remains with the customer (National Audit Office 2007). In the context of an availability contract the risk and responsibility is, again, shared between the customer and the supplier. The supplier is often the equipment design authority while both customer and supplier commit to contractual performance guarantees, or Service Level Agreements (SLAs) in order to enable a fit-for-purpose asset (National Audit Office 2007). The supplier/service provider guarantees availability of assets, at a pre-determined level, throughout the lifetime of the product (Cohen 2006; Kim et al. 2007). These agreements may last from months to tens of years (Gruneberg et al. 2007) and, as such, the relationship between the customer and supplier has lengthened. A capability contract places all risks and responsibilities on the supplier, who is responsible for providing a given capability without specification of the means by which this may happen (National Audit Office 2007).

As contracts move from traditional towards capability, the number of contracts provided by an organisation often reduces while their value increases. In addition to the financial and strategic incentives of higher revenue (Wise and Baumgartner 1999) with increased longevity, the change has also given rise to environmental benefits. This change has also impacted the nature of the risk associated with the service offering (Stremersch et al. 2001).

Information is critical to support the contract and, thereby, to furnish the service. While information about the asset location, condition and use is of importance to the service, additional information elements are needed in order to provide the service including, for example, information about the logistics, resources and finances of the service. Information has been described as the lifeblood of the organisation, the most valuable resource in industry today, a prerequisite to invoke the service and, despite this, an undervalued resource (Court 1995). Information is central to requirements determination for products and services. Given that the fundamental task of delivering a product or service is to ensure that the final outcome matches the specification (Levitt 1972), it is of utmost importance that information requirements are determined as accurately as possible. For the purposes of this chapter, information related terms will be defined as follows:

- *Information:* a combination of fact, context, meaning and relationships (as opposed to data which may be defined as that which is known or granted without a wider sense of meaning or interpretation).
- *Service Information:* information used to make decisions and take actions in a service environment. This will include, among others, information on maintenance, operations, resource, finance and engineering.
- *Information Requirements:* information needed to make decisions and take actions in a service environment (Cuthbert 2009).

These long-term performance-based contracts which are of a multi-organisational nature require an increased level of trust, openness and sharing in order for the service to be successful. A number of frameworks have been proposed to support the design and development of services, but information is not central to these. Situated within the Core Integrative Framework (CIF) is the transformation of information in the provision of services (Ng et al. Introduction chapter). The work described within this chapter is of a framework to enable the determination of information requirements for the design, delivery and evaluation of services. The framework may also be used to determine the information transformation required as services evolve.

10.2 Background

A significant amount of research has taken place over the years in the area of information requirements. However, relatively little research has been devoted to information requirements in a service context, and hence we predominantly draw here on research in relation to engineering design and information systems development.

In the context of engineering design, information provides a key resource at all stages of the product lifecycle (Hicks et al. 2002; Court 1995, 1997; Court et al. 1997; Lowe et al. 2004). Information influences beginning-of-life design and development decisions for new products (Hicks et al. 2006), middle-of-life product maintenance and design improvements, and end-of-life disposal and recycling actions. At each stage, information about the asset configuration and use is crucial in making informed decisions.

In the context of systems development the beginning-of-life phase or the requirements phase is recognised as the most important but also seen as a major problem (Davis 1982). Developers of systems tend to have limited understanding of the nature of the information needs while users may often be unable to accurately define what information they need (Wetherbe 1991; Watson and Frolick 1993). Furthermore, requirements can often be thought of as complete and static when, in fact, they will be continually evolving and differ between system users (Court et al. 1997; Flynn and Jazi 1998; Dearden 1964; Grudnitski 1984).

In a similar way to the engineering design and information systems development contexts, information within a service context also needs to be recognised as an essential input. A key element of services is that they receive and process customer information (Wathen and Anderson 1995). When compared with a manufacturing environment, information may be said to be "the raw material of service organisations and how it is processed will have a direct bearing on their productivity" (Wathen and Anderson 1995). Its quality, quantity and the process performed upon it will have a direct impact on the quality and productivity of the output operation. Services require the involvement of the customer in order to determine and understand their requirements (Fleiss and Kleinaltenkamp 2004) and, in circumstances where the service provided supports a complex engineering asset, a large amount of information about the use, components and configuration of assets is needed. Despite this, the design and implementation of services is not a well understood process (Tax and Stuart 1997).

In service models, information is dealt with to varying degrees, often focussing across the breadth of the service or in depth on other elements of the service. A review of a selection of these is provided in order to show their contribution in relation to information requirements determination for the design, delivery and evaluation of services.

A model for the development of new services was created by Johnson et al. It shows the new service development process as a continuous cycle (Johnson et al. 2000) and highlights some of the key enablers for the process which are shown as core elements within this model. However, information is excluded as one of these core enablers.

The design and delivery of service quality which reviews the design and delivery of the service was modelled by Ramaswamy (1996). A model for the design and management of services, reviewing service design and service management as two distinct domains, was also created. The design specification, its attributes and performance standards are developed within the design stage. From this, design concepts are generated and evaluated for subsequent development. Delivery and evaluation of the service is encompassed within the service management stage. The delivery is viewed as the implementation stage while the more significant evaluation stage comprises performance measurement, assessment of satisfaction and performance improvement.

Parasuraman et al.'s service quality gap model (Fig. 10.1) highlights four main gaps in the service process:

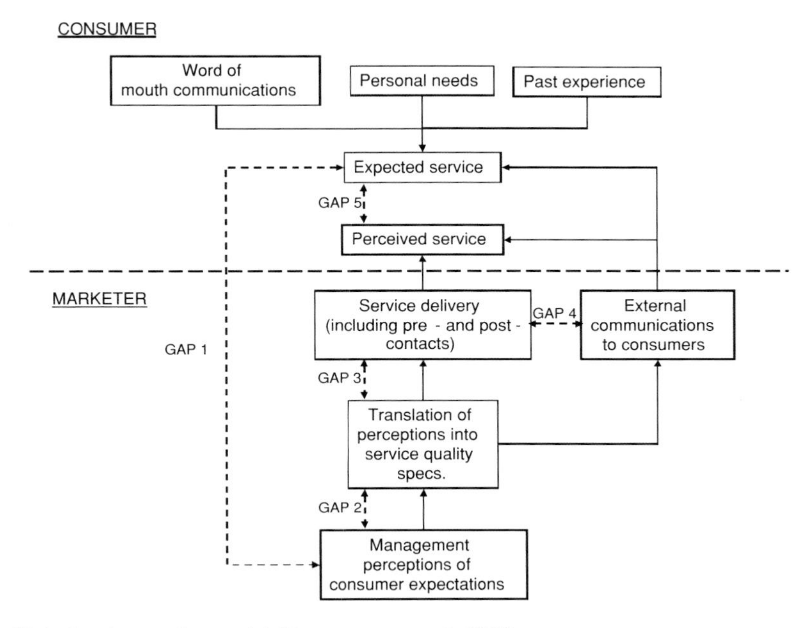

Fig. 10.1 Service quality model (Parasuraman et al. 1985)

- the gap between customer expectations of the service and the company understanding of those expectations,
- the gap between the company understanding of customer expectations and the development of customer-driven service designs and standards,
- the gap between the development of customer-driven service standards and the actual service performance by company employees, and,
- the gap between the service delivery and the service provider's external communications or what the supplier advertises to a third party that it is providing which may raise the customer's expectations are highlighted (Parasuraman et al. 1985; Zeithaml and Bitner 2000).

Sakao and Shimomura developed Service Explorer, a computer aided design (CAD) tool consisting of four sub-models. These sub-models are termed the flow, scope, view and scenario models and, respectively, represent the different players in the service supply network, the boundaries of the service network, the variables which impact the customer experience and the service receivers and their behaviours in receiving the service (Sakao and Shimomura 2007). This model does not deal with information requirements throughout the service.

Johnston and Clark describe a model which shows services as consisting of five main elements: the idea, experience, outcome, operation and value (Johnston and Clark 2005). This model provides a useful framework from which information may be highlighted or specified, but does not explicitly refer to information.

Industrial tools and models which are available tend to focus on the depth of information in a specific organisational function, whereas, the academic tools and models tend to focus on the breadth of service with scant consideration of information. There is a degree of overlap between areas of these frameworks in terms of the aspects of the service operation which these capture.

In most cases, the perspectives taken, the coverage provided by the framework or the enablers highlighted within the frameworks do not provide a broad information perspective across the service offering. What is evident overall is that these frameworks each deal with different issues relating to the design and modelling of service operations, covering service management, design, evaluation, quality and the enabling elements of services but without specific reference to information as one of these.

Breadth and depth of information throughout the design, delivery and evaluation of services is necessary in order that the desired service is developed by the provider, delivered to the customer and improved by the provider. Without key information throughout these stages of the process, the operation of such services to a desirable quality is likely to be challenging. With these other models in mind and their limited coverage with respect to information, an alternative framework, focussing on information requirements throughout the different stages of the service, offers real benefits which are critical to the effective design and delivery of the improved services.

10.3 A Model for Service Information

The previous section highlighted the lack of a framework to provide breadth and depth of information in a service providing environment. A framework which encompasses these requirements is proposed and described within this section (Cuthbert 2009). The evidence on which this framework is based is from close collaborations with a number of companies which have also led to a number of case studies (Cuthbert 2009).

The framework proposed in Fig. 10.2 deconstructs the service process from an information perspective. It is based on a representation of a notional four stage service supply chain. Each stage is represented by one of the four rectangles in the figure, namely, service need, service specification, service offering and service operation. These stages encompass the customer's domain (at the right hand side of Fig. 10.2) where a service need is generated. This is then jointly developed by the customer and supplier into a formal service specification. After this has been developed, the supplier progresses design of the service offering to meet the specification, and the design of the service operation to enable the service to be delivered. Traversing these service supply chain stages are underlying processes and information flows which are required to enable the service. Phase one is design information, used to develop the service and predominantly flows from the customer's domain to the supplier's domain. Phase two refers to information required

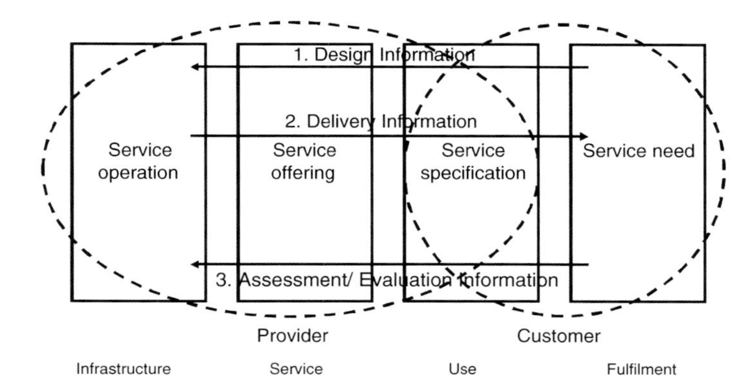

Fig. 10.2 Service information framework (developed from Cuthbert et al. 2008 by the addition of the terms infrastructure, service, use and fulfilment to correspond with the stages of the supply chain. The area within which provider and customer operate and have influence has also been updated to improve industrial relevance.)

in order to deliver and improve the user's interaction with the service. The third phase provides an indication of the performance and the effectiveness of the delivered service and a means by which the service may be measured. This may enable the provider to ensure that what is being delivered satisfies that which is required by the customer and also provides a mechanism from which improvements to the service may be made where necessary. While work on the nature of information services is evident (Berry and Parasuraman 1997; Lovelock 2001) the work reported within this chapter differentiates between information forming part, or all, of the service and information playing a role in the effective delivery of the service. This framework focuses on the latter. The role of three categories of information is considered, which form part of the overall service information requirement, namely, design, delivery and evaluation information (McFarlane et al. 2008).

The right hand side of the Service Information Framework represents the domain in which the customer has the most influence while the centre and left hand sections represent the domains in which the supplier has the most influence. Throughout the service supply chain are underlying process and information flows which correspond with information required to design, deliver and evaluate the service. The evaluation stage ensures that, as far as possible, the delivery is critiqued against the design to ensure that the customer's needs are satisfied. This ensures that the delivery is aligned with the design and the customer's need.

This framework was developed by dividing it into a three-by-four matrix as shown in Fig. 10.3. In this figure, the columns correspond with the stages of a nominal supply chain and the rows correspond with the underlying processes and information flows. This framework should be regarded as a closed loop system with feedback from the evaluation stages into both the design and delivery stages in order to constantly improve the service.

	Service Operation	Service Offering	Service Specification	Service Need
Design	4 Technical information to plan & develop the delivery of the offering	3 Technical/ architectural/ legal information to design offering	2 Information to formalise service contract	1 Conceptual information about customer requirement
Delivery	5 Technical info to run service/ infrastructure	6 System level functional information to fully supply service offering	7 Information with respect to service use	8 Information from provider enabling user to exploit service
Evaluation	12 Operational information on performance of service infrastructure and operations	11 Info relating to the effectiveness of service offering and its SLA metrics	10 Information to illustrate the perception/ expectation of the service – vs SLA	9 Information to determine fulfilment of customer need
	Infrastructure	Service	Use	Fulfilment

Fig. 10.3 Developed service information model; a three by four matrix of information types required in service development (developed from Cuthbert et al. 2008)

The framework presented in Fig. 10.2 provides a structure to inform data collection. In conducting case studies a key question is which firms, sectors and organisational characteristics would be appropriate? The criteria of this research require companies to display a core set of characteristics. This must include the provision of services against a contract monitored via some key performance indicators. Contracts should represent an increased level of risk to the organisation as their product or service offering is extended. There should also, ideally, be multiple partners involved in the provision of the service as this is likely to highlight some of the information sharing and ownership issues which the framework seeks to determine.

Before using the framework, a set of high level, generic information is gathered in relation to each of the case study organisations and the terms around the contract. The framework is then used as a guide in a semi-structured interview. Ideally, several interviews would be carried out for one particular contract to provide different perspectives of the same service and contract. The text detailed within each of the 12 boxes of this model represents a high-level generic description of the information type required at each stage. These are reviewed with the interviewees in numerical order, detailing the types of information needed within each box/stage. The information required for each box is recorded and these are reviewed to gain an understanding of which of the required information is actually available to the service provider. These are taken directly from the order and direction of the arrows, representing the underlying processes and information flows, in the original service information model of Fig. 10.2. A more detailed description of the information required in each of the twelve boxes follows.

10.3.1 Design Process

In Box 1, information is required around the service need. This is a set of requirements from the customer which may be conceptual as the customer may not fully understand what they require, despite believing that they have a clear set of requirements. Box 2 is the information required to determine the specification of the service design. This is information from Box 1 combined with other information from the supplier and other external sources which may provide a new perspective on, for example, the context of the service and how it may change in future. The information required at this stage will be formalised into a contract and subsequently signed off by both parties. Box 3 requires information around the service use to develop the service offering. This should use a full set of requirements drawn from the specification and is used to determine how the solution will look. The means by which the service will be measured needs to be designed into the process. Box 4 requires information around the infrastructure on the service operation. This is the space in which the technical solution is developed, and the hardware and software are specified. Information is required to help determine the tools, maintenance facilities, people and other resources.

10.3.2 Delivery Process

Boxes 5–8 correspond with the delivery process. For Box 5, information is required on the hardware and infrastructure issues or failures which prevent or impede the service delivery. Information on other infrastructure issues which may help with the planning of, or possibly impede the delivery of, the service is also required. For Box 6, system level functional information to supply the service is required. Information on how the service is actually delivered by the supplier is required. This may include information on the availability of the asset for maintenance to plan the delivery of the service, as well as usage of information and information on the asset relating to work carried out by the customer of Box 7 will include information with respect to the service use, information on how the assets have actually been used (against what they were specified for) and information on the actual consumption of the service (spares, hours used etc.) compared with what was specified. Box 8 requires information to enable the customer to best use/achieve fulfilment from the service. This may include information about the assets (provision of training) or information on the customer's use of the service compared with their stated requirements.

10.3.3 Evaluation Process

Boxes 9–12 correspond with the evaluation process. Box 9 requires information which evaluates the customer's fulfilment from the service delivered against the original need stipulated by the customer. This should indicate the customer's perception of their requirements against what was received or made available to the user. Box 10 requires information which provides an evaluation of the specified service agreed between supplier and customer. This will enable an evaluation of the perception versus expectation of the formally agreed requirements and will be measured against the contractual SLAs. Box 11 measures the effectiveness of the service offering and the SLA metrics and whether the SLAs skew the service. It seeks information to determine whether the service which has been delivered is equivalent to that which was designed, based on the specification. It is also used to determine whether the service is being used fully or whether elements of the service are not being used. Box 12 seeks information to determine whether the elements of the service (infrastructure) are performing as they should be to satisfy the service design/delivery.

10.3.4 Framework Application

The framework has been applied to a range of different services provided within traditional service organisations and also for services delivered around complex engineering assets involving different contract types. It has been used to provide insights on information maturity in nine industrial Product Service Systems (PSSs) through the use of a classification approach where information was categorised as being both required and available, inconsistently available or not available but required. Figure 10.4 summarises the general state of information availability for the three scenarios (S1, S2 and S3) of the washing machine case study of Chap. 11 highlighting areas where information is consistently available, inconsistently available or unavailable. In a full application of the model, each shaded box (e.g., the design stage of the service need) would be subject to a detailed iteration to determine the availability of specific types of information within that box. Examples within the category of the design stage of the service need might include information on the systems or capabilities required of the service, the customer's budget, or the likely geographical deployment of the asset.

Overall outcomes from the nine industrial PSSs indicate some general areas of problems encountered within service providing organisations as well as areas of good practice. These are discussed in the following section.

10.4 Analysis Using the Framework

In the first instance, this framework has been applied extensively to a number of services provided both in traditional service providing sectors as well as in services provided against complex engineering assets. In each case, the type of contract against which the service was provided differed from spares and repairs, spares

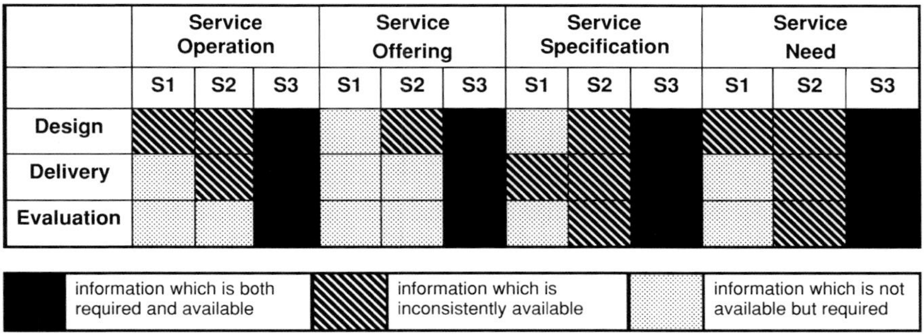

	Service Operation			Service Offering			Service Specification			Service Need		
	S1	S2	S3	S1	S2	S3	S1	S2	S3	S1	S2	S3
Design												
Delivery												
Evaluation												

information which is both required and available

information which is inconsistently available

information which is not available but required

Fig. 10.4 High level example application of the service information model to the washing machine case study of Chap. 11

inclusive, availability and capability (Cuthbert 2009). The type of contract is of significance as the information required, owned and available to the customer and supplier in different contractual scenarios is likely to differ due to the differing roles and responsibilities of each party. However, the framework was predominantly used to highlight differences in the information required and available for a number of services. From this, a gap in the information which was consistently required and available could be determined. Despite a number of case studies being carried out against services with different contract types, information problems re-emerge with similar themes across the case studies, and it is these common themes which are presented within this section.

In all instances of its application, the framework showed that a limited set of information tends to be available around the customer need. This may be shown as a deficiency due to the suppliers not being thorough enough in their requirements gathering processes or attributed to the customers not always understanding their own needs of the service, or the potential of the service which could be offered. In instances where the customer has a clear definition of their needs, these may still be difficult to articulate and specify.

Applications of the framework illustrated a consistently poor level of historic information available to service providers in relation to the complex engineering asset being serviced. This was often a result of the service being passed from one provider to another, and the information which was also passed between providers to enable the service provision. With such a deficit of information, the design and delivery of the service must be based upon models and estimates, rather than more factual information. This deficiency also includes information on external factors impacting the service such as information from third party providers.

The framework is designed in such a way that the design and delivery of services should always be compared through evaluation information. A dynamic approach to this is required to ensure that the services delivered satisfy the original needs. Furthermore, in carrying out this comparison any evolving changes in the nature of the service may be highlighted by the framework. The realities of such services are that the requirements are constantly evolving as customer requirements change and as new capabilities may be provided by suppliers. This indicates the need for a regular review of the service, which this framework may provide.

Insights were also provided by the framework into the information requirements for the different contracting arrangements of the different services. Availability contracts tended to show a better level of information at the design stage enabling a thorough understanding of customer's needs. During the delivery phase of the services, more information about the costs and requirements of the services tends to be available. In the evaluation stage, more information is available to enable analysis of services provided in order to populate performance metrics. In addition, information to forecast the likely use of assets and their failure is available.

Further to the application of the framework to understand differences in information required and available and the impact of different contract types, a number of other application areas may be considered in the areas of contracts, organisational function, sectors and projects. Firstly, the framework may be used to indicate the types of information required at different stages of new or existing systems or processes. It may also be used to enable an understanding of the gaps within existing (IT) systems, processes, information (required and available information). The framework may be used to highlight dependencies upon information shared between organisations. In such cases, weaknesses in information links/exchange between organisations and their vulnerability in terms of who owns what information may be highlighted. Finally, areas where improvement is required in terms of information may be illustrated.

10.5 Conclusions

A framework has been presented which has been used to determine information requirements for the design, delivery and evaluation of services. It has been demonstrated to provide a high level assessment of the information required for the service, and to understand the gap between this and the available information. Its application has been across a broad range of situations, including different contracts, organisational functions, projects and sectors.

This is useful in a number of situations. For organisations providing services against contracts, it may enable an understanding of the different information needs for different types of contracts and enable the organisation to upgrade its service provision, more easily, from one contract type to another.

From an organisational or functional (e.g., R&D, finance etc.) perspective, it may provide a gap analysis of information required and available in different areas of the service operation. This may enable the organisation to identify information-deficient areas within their organisation and to focus their efforts in order to improve their performance as a company and the service delivered to their customer.

For an assessment of different industries, the model may be applied to understand how advanced different sectors (e.g., Engineering (Commercial and

Defence), Information Technology, Finance, etc.) are in terms of information provision. This may enable an organisation to learn areas of good practice from a non-competing business and apply this within its own domain.

Areas where this framework has been applied successfully include cross-sector, -project and -contract scenarios. Within some of these areas, the framework could be developed in order to take the information assessment to a more detailed level.

10.6 Chapter Summary Questions

This chapter has raised a number of issues in relation to the information elements of service design, delivery and evaluation. Pertinent to this is understanding fully what information is required at each stage of the design, delivery and evaluation of the service and what information gaps are present within existing services. Of note within this work is the information required around the customer's need as well as an understanding of the information needs of the service provider. The type of contract will also impact the information availability to the service provider and present some dependencies and vulnerabilities within the information available. Some key questions are:

- How may required information be best identified, and made available within a service environment?
- What mechanisms may ensure that the customer fully understands their requirements, and the full capability on offer by the provider? How is this best articulated?
- What process may be used to ensure that the service remains dynamic and evolves with the customer's needs and the supplier's capabilities?

References

L. Berry, A. Parasuraman, Listening to the customer—the concept of a service quality information system. Sloan Manag. Rev. (Spring):65–76 (1997)

M. Cohen, Paying by the second. Serv. Manag. (April): 22–24 (2006)

A.W. Court, The modelling and classification of information for engineering designers. Ph.D. Thesis, University of Bath, 1995

A.W. Court, The relationship between information and personal knowledge in new product development. Int. J. Inf. Manag. **17**(2), 123–138 (1997)

A.W. Court, S.J. Culley, C.A. McMahon, The influence of information technology in new product development: Observations of an empirical study of the access of engineering design information. Int. J. Inf. Manag. **17**(5), 359–375 (1997)

R. Cuthbert, P. Pennesi, D. McFarlane, The impact of different support service contract models on provider information requirements. Proceedings of POMS 19th Annual Conference, La Jolla, California 2008

R.C. Cuthbert, The information requirements for service development. MPhil thesis, University of Cambridge, 2009

G. Davis, Strategies for information requirements determination. IBM Syst. J. **21**(1), 4–30 (1982)
J. Dearden, Can management information be automated? Harv. Bus. Rev. **42**(2), 128–135 (1964)
S. Fleiss, M. Kleinaltenkamp, Blueprinting the service company; managing service process efficiently. J. Bus. Res. **57**, 392–402 (2004)
D. Flynn, M.D. Jazi, Constructing user requirements: A social process for a social context. Inf. Syst. J. **8**, 53–83 (1998)
G. Grudnitski, Eliciting decision makers' information requirements: Application of the rep test methodology. J. Manag. Inf. Syst. **1**(1), 11–32 (1984)
S. Gruneberg, W. Hughes, D. Ancell, Risk under performance-based contracting in the UK construction sector. Constr. Manag. Econ. **25**, 691–699 (2007)
B.J. Hicks, S.J. Culley, R.D. Allen, G. Mullineux, A framework for the requirements of capturing, storing and reusing information and knowledge in engineering design. Int. J. Inf. Manag. **22**, 263–280 (2002)
B.J. Hicks, S.J. Culley, C.A. McMahon, A study of issues relating to information management across engineering SMEs. Int. J. Inf. Manag. **26**, 267–289 (2006)
S. Johnson, L. Menor, A. Roth, R. Chase, A critical evaluation of the new service development process; integrating service innovation and service design. Chapter 1 from Fitzsimmons and Fitzsimmons, 2000
R. Johnston, G. Clark, *Service operations management: Improving service delivery*, 2nd edn. (FT Prentice Hall, Harlow, 2005)
S. Kim, M. Cohen, S. Netessine, Performance contracting in after-sales service supply chains. Manag. Sci. **53**(12), 1843–1858 (2007)
T. Levitt, Production-line approach to service. Harv. Bus. Rev. Sept–Oct: 41–52 (1972)
C. Lovelock, *Services marketing: People, technology, strategy*, 4th edn. (Prentice Hall, New Jersey, 2001)
A. Lowe, C. McMahon, S. Culley, Characterising the requirements of engineering information systems. Int. J. Inf. Manag. **24**, 401–422 (2004)
D. McFarlane, R. Cuthbert, P. Pennesi, P. Johnson, Information requirements in service delivery. 15th international annual EurOMA conference, Groningen, June, 2008
National Audit Office, Transformation logistics support for fast jets (2007)
A. Neely, Exploring the financial consequences of the servitization of manufacturing. Aim Research Working Paper Series, 2009
Organisation for Economic Co-operation and Development (OECD), Int. Serv. Stat. (1996)
A. Parasuraman, V. Zeithaml, L. Berry, A conceptual model of service quality and its implications for future research. J. Mark. **49(Autumn)**, 41–50 (1985)
R. Ramaswamy, *Design and management of service processes: Keeping customers for life* (Addison-Wesley Publishing Company, MA, 1996)
T. Sakao, Y. Shimomura, Service engineering: A novel engineering discipline for producers to increase value combining service and product. J. Clean. Prod. **15**, 590–604 (2007)
S. Stremersch, S. Wuyts, R. Frambach, The purchasing of full service contracts: An exploratory study within the industrial maintenance market. Ind. Mark. Manag. **30**, 1–12 (2001)
S. Tax, I. Stuart, Designing and implementing new services: The challenges of integrating service systems. J. Retail. **73**(1), 105–134 (1997)
The World fact book, Central intelligence agency. Accessed March 2009 (2008)
S. Wathen, J. Anderson, Designing services: An information-processing approach. Int. J. Serv. Ind. Manag. **6**(1), 64–76 (1995)
H. Watson, M. Frolick, Determining information requirements for an EIS. MIS Q. Sep (1993)
J. Wetherbe, Executive information requirements: Getting it right. MIS Q. **15**(1), 51–65 (1991)
R. Wise, P. Baumgartner, Go downstream: The new profit imperative in manufacturing. Harv. Bus. Rev. **77**(5), 133–141 (1999)
V. Zeithaml, M. Bitner, *Services marketing: Integrating customer focus across the firm*, 2nd edn. (Irwin McGraw-Hill, Boston, 2000)

Chapter 11
Investigating the Role of Information on Service Strategies Using Discrete Event Simulation

Rachel Cuthbert, Ashutosh Tiwari, Peter D. Ball, Alan Thorne and Duncan McFarlane

Abstract This chapter details the work on a demonstration to illustrate the impact of information in the context of complex engineering services. The demonstration is achieved via a simulation model which illustrates factors, such as different service contracts, different levels of product condition information required by and available to the service provider, other constraints on the service system and different service performance levels achieved. The contribution of this work is showing, through simulating several scenarios, how support services may be improved as a result of providing better product condition information feedback to the service provider. In addition, the model factors in a number of other variables which have a significant impact on the level of the service provided. Results from the simulation models are presented, and a discussion of areas for further work is also provided. This discussion includes some suggested next steps and future information-related questions which the model may seek to answer.

R. Cuthbert (✉) · A. Thorne · D. McFarlane
Institute for Manufacturing, University of Cambridge, Cambridge, UK
e-mail: rc443@cam.ac.uk

A. Thorne
e-mail: ajt@emg.cam.ac.uk

D. McFarlane
e-mail: dcm@eng.cam.ac.uk

A. Tiwari · P. D. Ball
Manufacturing Department, Cranfield University, Cranfield, UK
e-mail: a.tiwari@cranfield.ac.uk

P. D. Ball
e-mail: p.d.ball@cranfield.ac.uk

I. Ng et al. (eds.), *Complex Engineering Service Systems*, Decision Engineering,
DOI: 10.1007/978-0-85729-189-9_11,

11.1 Introduction

The research has focused on understanding the information requirements for organisations which traditionally manufactured engineering products but which are now moving into the provision of services to support these products. Accompanying these services are support contracts which may take various forms and be of varying duration.

Much of the research to date has provided background on the state of information provision in relation to the delivery of services for such organisations (Tomiyama et al. 2004). This has highlighted some issues around the information required to deliver service contracts and issues around running multiple contracts where service resources must be prioritised against information available to the provider and the potential contract value or penalties.

Discrete Event Simulation (DES) involves the modelling of a system, as it evolves over time, by representing the changes as separate events. In discrete event simulation, the operation of a system is represented as a chronological sequence of events. Each event occurs at an instant in time and marks a change of state in the system. There are numerous applications that make use of DES particularly in manufacturing systems (Pidd 2004) but with potential beyond manufacturing.

Within this project, the authors have used discrete event simulation to develop models of various scenarios in the context of an engineering-based service environment. These models demonstrate how support services can be improved as a result of providing better product condition information feedback to the service provider. This is a novel application of discrete event simulation that has typically focused in the past on modelling manufacturing operations. Simulation has been applied to model some service provision but the role of product condition information feedback for different business models has not been explored.

Key questions which this work seeks to answer, at a high level, are to understand the role of product condition information, how this information can impact current performance and how the information can impact future contracts. More specifically, the questions which this work answers are:

- Does improved product condition information lead to improved service performance in terms of operational Key Performance Indicators (KPIs)?
- At what point does product condition information have no, little or most impact (bearing in mind other variables, bottlenecks in the process etc.)
- What other variables compete with such information provision?
- What is the impact of resources on meeting a service level in a given scenario?

11.2 Background

Although the field of simulation is well established, there is limited literature on the use of simulation for evaluating service solutions. Komoto and Tomiyama (2008) propose an Integrated Service CAD and Lifecycle simulator (ISCL). ISCL plays the role of CAD/CAE tools, used for product design, within a Product-Service Systems (PSSs) design context. In ISCL, the service CAD supports systematic generation of alternative PSSs based on service modelling and the use of a lifecycle simulator analyses their economic and environmental performances. They argue that the performance of service is influenced by market conditions (e.g., preference of the market for upgrading service) and technology conditions (e.g., interval of new function releases) (Tomiyama et al. 2004). They illustrate that a designer can systematically define and search PSSs as a combination of service contents, service channels and corresponding activities defined in the service CAD.

Shimomura et al. (2008) propose a method for evaluating service solutions based on the Quality Function Deployment (QFD) with some mathematical reinforcements. They argue that customers generally evaluate services and products concurrently rather than separately. They have shown that a specific function is the most important among a multistage structure of the target service. Shimomura et al. (2009) recently proposed a method for designing service activity and product concurrently and collaboratively during the early phase of product design. They provide a fundamental, unified representation scheme of human process and physical process in a service activity in order to increase customer value within services by extending service blueprints. They observe that the original blueprint, widely used to describe service activities, lacks design information and has insufficient normative notation to apply to actual services.

Hara et al. (2009) propose a design method to integrate products and service activities for total value. They argue that their design method differs from Traditional Engineering and Concurrent Engineering (CE) in that products and service activities are designed in parallel according to customer value. In addition, they claim that the CE design process fails to design for value or for the relationship between products and service activities. They replace product planning with value design and service content design. The service blueprint is argued to be unable to correlate customer value and service activity properly, and finds difficulties in the association of the described service process to the customer requirements. They propose that the blueprint of a service offering, such as PSS, should contain information concerning the product and its service behaviour as well as information on the human activity associated with the service. They extend the service blueprint to include the product and its behaviour.

There is potential for the engineering sector to learn from others on how information is modelled. The focus of simulation within the healthcare sector has been on scheduling resources within hospitals and other health-related facilities. For example, Davies and Davies (1986) discuss the use of simulation within the healthcare domain for the planning of health service resources. Further to this, a limited number of attempts have been made to use Discrete Event Simulation (DES) in a service design context. A Virtual Enterprise Architecture (VEA) has been modelled and simulated using DES for the development of an automated baggage handling system which can be used in airports (Møller et al. 2008). This involves an Automated Transport and Sorting system (ATS). A virtual enterprise is a network of several different organisations working to achieve a common goal. The goal in this instance is the provision of service to customers. A logistics service has been evaluated using DES modelling. In this case, the service is separated from its products and a framework has been generalised and validated for VEA based on DES (Møller et al. 2008).

The field of discrete event simulation and the commercial tools available (such as WITNESS from Lanner Group Limited, Arena from Rockwell Automation and Simul8 from Simul8 Corporation) have so far focused on modelling traditional manufacturing and service systems.

A relationship between service design and DES has been identified. Many other examples of DES can be cited but none showed the use of different levels of information on service provision. This research is a novel application of discrete event simulation to the domain of engineering-based service environments.

11.3 Demonstrator Design

11.3.1 Demonstration Requirements

The objective of this work is to scope and develop "a simple, illustrative demonstrator that will be developed to illustrate key differences between current and future information requirements and capture key elements of the blueprint proposed". Key areas which this would then simulate are:

- Information Structures
- Service Contracts
- Information requirements for different types of service contract
- Information Model
- Frequency/Availability of information
- Service performance levels achieved

The simulation should highlight the concurrent provision of services, by one provider, to a number of customers. The different customers may have different types of contracts provided with different performance levels and the provider may be privy to different levels of information at different frequencies and with different levels of quality/reliability. The owners and users of the information may vary between contracts and the level of demand, placed upon the supplier by the customer, is likely to vary.

The types of contracts which might be of concern in this context are indicated in Fig. 11.1.

Fig. 11.1 Types of contract (National Audit Office 2007)

Contract	What is the contract?	What is involved?
Spares & repairs	Contractor supplies spare parts	Customer responsible for repair/overhaul
Spares incl.	Spares and repairs plus overhaul and repair	Supplier and customer jointly responsible for repair/overhaul
Contracting for availability	Performance-based agreements	Supplier delivers "fit for purpose" equipment. Spares resource provision may be shared
Contracting for capability	Supplier responsible for delivery of a capability	Customer does not own the equipment. The risk & responsibility lies with supplier

11.3.2 Simulation

In order to illustrate the requirements and to highlight some of the key issues around the research to date, a number of scenarios were developed which would be part of the simulation. All scenarios are set up with standard elements:

- Three form contracts running in parallel to service three sets of washing machines. These are supported by one service provider but may have different performance requirements for each.
- Maintenance resource.
- Spares management (ordering system and inventory).

The details of each scenario are described in the following sections.

11.3.2.1 Scenario 1: No Information/Feedback Available

The first scenario will provide a traditional maintenance process provided to the assets in the order in which failures have occurred. Service personnel will arrive at machines as they become available for carrying out maintenance. No product information will be available to plan the maintenance ahead of time. The cause of the failures will only be diagnosed by the maintenance person during the first maintenance visit meaning that a high proportion of the initial maintenance visits to the assets within the fleet will result in the machine not being repaired.

If a machine needs a part, the maintenance person will check for spares in inventory. If the part is not available it will be ordered and, in the meantime, the machine then goes back to the end of the queue awaiting maintenance. Once the part is available it will be shipped to the machine. When available, the maintenance person will return to the machine to complete the repair.

Spare parts can be ordered as a one-off part but are generally controlled through an ordering system (e.g., periodic ordering with variable batch size). The assets are affected by two failure modes which will require different spare parts in order for them to be repaired. These will be part 1 with lead time 1 and part 2 with lead time 2.

Maintenance can be carried out by different maintenance groups as they are assigned to fleets. One or more maintenance groups may be assigned to a particular fleet for maintenance, or to visit a particular fleet first if maintenance is required, depending on the requirements of the contract and, subsequently, the priorities set within the model.

In this scenario the time requirements break down into one resource unit for the first visit, one spare lead time unit for the delivery of the part required and two time units for the diagnosis and repair of the asset.

11.3.2.2 Scenario 2: Component Failure Information Available

The second scenario illustrates a reactive maintenance process enhanced by machine information. Here machines fail, sensors diagnose the fault, communicate this with the service operation and maintenance is initiated once a replacement spare part has arrived at the machine (the maintenance activity is triggered by the dispatch of the spare part from warehouse). This spare part will be ordered by the machine on failure of the part concerned. The shipment of the parts will be direct to the machines. Following the arrival of the part at the machine followed by a possible delay (due to unavailability of personnel), the maintenance person will arrive at the machine to carry out replacement of the part. This will take place in the order in which the machine breakdown occurred. The two types of part have different frequencies of failure and different repair times.

Information about the failures on the machines will be available. This information will include information about which part has failed, when the spare part leaves the warehouse/reaches the machine and what the options are to re-order spares based on historic consumption. Key differences between scenarios 1 and 2 are:

- The elimination of the exploratory visit (there and back)
- The elimination of the ordering process
- The elimination of the possibility of a repair visit prior to part arriving
- Improved information quality on spares required

Resulting in:

- Reduced downtime
- Reduced repair time (resulting from the exploratory visit)

A limitation which remains with this scenario is that multiple part failures may not be diagnosed. A benefit over Scenario 1 is that the system provides the potential for reduced inventory levels at the supplier, for reduced lead times and lower staff loading.

Scenario 2 provides improved resource utilisation as it eliminates an initial visit to diagnose the failure. In this the time requirements break down into one resource unit for the first visit, one spare lead time unit for the delivery of the part required and one time unit for the repair of the asset.

11.3.2.3 Scenario 3: Component Failure and Machine Usage Information Available

In this third scenario a proactive maintenance process is modelled. The detection of impending faults triggers maintenance activity prior to machine failure through the availability of product condition information. In this scenario, the machine will order a spare part prior to its predicted breakdown. This part is shipped directly to the machine from the warehouse followed closely by the maintenance person.

The failure of the machine will be based on the same failure distribution for the particular part as used in earlier scenarios, meaning that the part may arrive before or after the breakdown has occurred. Ideally, the part should arrive before breakdown has occurred. Secondary faults, linked to the predicted failure of the part specified, will be allowed for within this scenario and reordered at the required time. The lead time of the spare part may also be factored into this scenario as the failure detection will be programmed to allow for this timeframe. Repairs within the model are currently prioritised statically. This is based on the priority assigned to an overall fleet within the scenario and this is fixed throughout a run. Priority is given to failed machines, followed by those still running but with a replacement part available as failure is imminent.

Cross-contract decisions need to be included to improve the performance of the service (and minimise loss of revenue) via KPIs across the fleet through adjustments in the:

- service resource required
- level of spares required
- fleet size required to achieve KPI

Information about the failures on the machines and across fleets will be available, including:

- Which component has failed
- When spare part leaves warehouse/reaches machine
- Status of all machines—idle/working/failed etc.
- Option to turn on spares reorder based on historic consumption

The maintenance in this scenario will largely be preventative based on the machine usage/remaining life. The scenario is currently configured for the same service personnel to carry out both non-scheduled and preventative maintenance. The time taken by each resource for repairs will only require one time unit to cover the asset repair time only.

Key differences between Scenarios 2 and 3 are:

- A further reduction in downtime prior to machine repair taking place
- The time when the machine is not working is predominantly the repair time. However, in a few cases, the machine may break down before the spare part arrives in which case some waiting time will occur prior to the repair taking place

- Prediction parameters can be varied so that the machine should not breakdown in Scenario 3

Within Scenario 3, resource may be optimised to show the benefits of the improvements of Scenario 3. Scenario three may also compensate for effects of variable lead time by adjusting the prediction parameters for part failures. However, this may lead to unnecessarily early replacement of parts and a higher overall level of spares consumption. This can be factored into the reorder time with minimal impact on the maintenance process.

11.3.2.4 Different Environments within Scenarios

Within each scenario, three fleets are specified. Within the context of washing machine service provision, different contracts simulated might correspond with a launderette, a hospital and a domestic user. These different environments are likely to expect different performance levels. They will also be subject to very different levels of usage, different users with different training levels and load/cycle requirements. For example, the launderette machines will be subject to mixed uses, mixed loads and cycles, and a range of commercial and domestic users who are not trained. The hospital machines will be subject to continuous use, heavy loads and heavily soiled items requiring high temperature cycles, and operated by multiple users who are likely to be trained. The domestic machine is likely to undergo less frequent use and light loads operated by a trained single user.

11.3.3 Parallels between the Demonstrator Scenarios and Engineering-Based Service Solutions

The fleets within scenarios have been set up to represent different possible scenarios within engineering-based service organisations. These might include:

- "fleets within fleets". For example, in the air world this might be BAE Systems Harrier with aircraft located in the UK and overseas where some of the fleet is sea-based rather than land-based. BAE Systems Tornado is also another example, where a priority fleet, standard fleet and a training fleet exist.
- Equivalent assets, operated by the same customer, being located in different locations around the world. Implications from this will be more significant as a result of travel time to and from the asset. Alternatively, support staff may be located near the assets, in which case there will be a resource cost in doing this.
- Equivalent assets, operated by different customers with different asset performance and service requirements (e.g., a civilian/commercial maintenance, repair and overhaul (MRO) provider).

The separation of these fleets within scenarios enables priorities to be given to customers which would be more costly in the event that a KPI is not met. Priority can be given which will impact the allocation of spares and resource in particular.

Scenario 1 might be likened to a traditional reactive spares and repairs approach to servicing such assets. There are also instances within this scenario where spare parts, ordered for one machine, may be stolen by another.

11.3.4 Simulation Modelling

This section will describe the modelling used for the scenarios described in Sect. 11.3, and provide further detail on the scenarios in the context of the model.

The combination of the availability of machines for incoming washing loads, the availability of spares and the availability of service groups leads to a complex situation where there is competition for resources. The most appropriate means for understanding the likely performance is via simulation.

Simulation modelling is a technique that enables abstract representations of systems in a software model on which "what if" experimentation can be performed. A specific type of this modelling technique is discrete event simulation which is able to model production systems, inventory and service systems over a period of time. As the model runs, particular events are modelled, such as the breakdown of machines, competition for resources and availability of maintenance personnel. Resources in the simulation can be modelled using probability distributions. This enables events, such as breakdowns, to be represented with the same unpredictability as it occurs in reality. The software selected to perform the modelling here was WITNESS (manufacturing edition) (lanner.com) which incorporates the capability to perform discrete event simulation. The modelling domain for this software is in manufacturing environments. It has the ability to model service systems and specific functionalities surrounding breakdowns and the replication of machines for fleets. Standard results are automatically generated by the software and additional results are collected by incorporating "additional results counters".

11.3.4.1 Description of Models with Screen Shots

The three scenario models created in WITNESS for this work incorporated three fleets which may be activated as required. Each fleet can contain different parameters for washing load arrivals, cycle times, breakdowns and repairs. The fleet sizes may also be easily adjusted in size. The model's layout is shown in Fig. 11.2.

As the time progresses in the model washing loads arrive within each fleet at fixed intervals. Each load will be randomly allocated to an available washing machine. Each washing machine processes the washing load according to a variable cycle time and then exits the model adding to the number of loads 'completed'. There is a small amount of buffering for incoming washing should no machines be available to process the washing. If the buffering is exceeded the washing loads are rejected and recorded in the results as 'lost'.

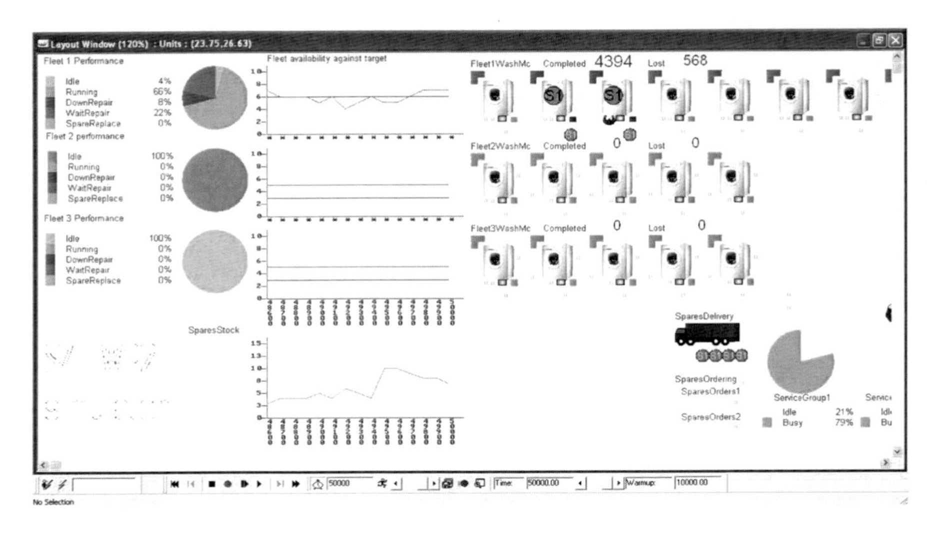

Fig. 11.2 Screen shot of the model after being run with fleet 1

Each washing machine has two variable breakdown patterns associated with the failure of two types of parts. The first part is likely to fail more often and the spare is quicker to replace and more predictable to source if stocks are unavailable. The second part associated with the second breakdown type is less likely to fail but when it does it takes longer to replace and is less predictable to source if stocks are unavailable. Spares arrive in batches at regular intervals to the spares delivery area. When spares are required but unavailable, one-off spares orders are placed. Service groups are able to attend to failed machines and there is flexibility around which fleets they cover. For simplicity, within the examples here, each service group is associated with one fleet; however, they can be associated with more than one fleet. For this purpose the fleets are prioritised. Should there be any competition for resources between fleets then the priority order between them, from highest to lowest, is fleet 1 then fleet 2 and then fleet 3.

The model of the fleets can be used in each of the three scenarios described earlier. One model is used for the three scenarios rather than splitting each into three models. This means that variations between scenarios are due to control logic rather than specific functionality unique to one particular model. The control logic uses standard WITNESS breakdown and repair functionality as well as considerable macros to support this unusual application. The scenarios were as follows:

- In the first scenario, the service group visits a failed machine, diagnoses and replaces the appropriate part with a spare if available. If no spare is available then the spare is ordered, the service group leaves the machine
- and returns once the spare becomes available. Ordered spares are not allocated to a fleet or a particular machine and are sometimes 'stolen' by other machines
- In the second scenario, the failed part is sensed by the failed machine. The appropriate spare part is then ordered and delivered to the machine. Only once

the spare has arrived at the failed machine will the service group attend and carry out the repair. This repair will then take the same time to carry out as the repairs in Scenario 1.

Scenario 3 has the same failure patterns and service group functionality as earlier scenarios along with functionality to predict the next failure type. Each machine monitors its run time and will order appropriate spares prior to the likely failure. Once the spare arrives at the machine the service group will attend, stop the machine, replace the spare and effectively cancel the impending breakdown. If a machine fails prior to prediction then the behaviour follows that of Scenario 2.

11.3.4.2 Test Areas for Models

Three scenarios as described above were modelled. The resulting parameters collected were:

- Washing loads. Completed and lost.
- Fleet utilisation. Average percentage times which machines spend idle, busy, down waiting repair and down being repaired. Availability of machines (count of machines not down). Number in fleet currently available.
- Spares stock.
- Service group utilisation. Average percentage times idle/busy.

Most of these results are illustrated as the model runs, see Fig. 11.2. Results shown were aggregated for each fleet and service group. Results were readily available at individual machine and service group operator level but were not considered valuable here.

Varying the input parameters described earlier (washing load arrival rates, fleet size, machine cycle times, breakdown frequencies, repair times, service group size, spares delivery quantities, spares lead times) impacts the above results. The work described here is focused on the different strategies for maintaining assets. Parameters, such as arrival rates and breakdowns, were not changed during experimentation. The focus was on the control logic in the scenarios and associated influences that a service provider has on areas such as spares availability and service group availability. In general, the comparisons focused on the different scenarios and the way in which spares and service group configurations impacted on fleet utilisation and washing loads completed and lost.

11.4 Results

11.4.1 Assessing the Simulations

This section discusses the three simulation scenarios which have been set up and run as well as some initial results from these. The results from these simulation scenarios were examined to assess the impact on the given KPI for the contract as a result of variations in the:

- size of fleet;
- service/maintenance resource level;
- spares availability; and,
- lead time.

These were also reviewed in the context of increasing information availability in each successive scenario. For the purposes of illustration only one fleet was modelled, as running multiple fleets introduces the complexity of competing for personnel and spares across fleets.

11.4.2 Simulation Outputs

The three scenarios were run on the base model for the first set of results. Simple changes were then made to understand differences in behaviour. For simplicity, the results shown are illustrative based on a single run rather than being an average of multiple runs. The warm-up time was 1 year ($\sim$10,000 h) to ensure that a minimum of three major breakdowns were experienced by every single machine. The run time for the results collection was just over a further 4 years (to 50,000 h). A typical run without graphics would take the order of 10 s with an average desktop PC. The input data are shown in Fig. 11.3. The results obtained from running the scenarios are summarised in Fig. 11.4.

11.4.3 Analysis of Results

Key points to gain from the results shown in Fig. 11.4 are detailed in the following section. Key overall results obtained were as expected:

- As the level of information available to the supplier increases (from Scenario 1 to Scenario 3) the number of washes achieved is higher. The number of washes which is lost due to machine non-availability decreases with increased information availability.
- Similarly, as the level of information available to the supplier increases (from Scenario 1 to Scenario 3) the utilisation of resource decreases at the same time as a performance increase.
- For Scenarios 1 and 2, the wait repair time is high due to the maintenance resource being a constraint. The wait-repair time is lower in Scenario 3 as the machines wait for the person once the part has arrived at the machine (usually ahead of the actual failure).
- Scenario 3 also has a higher running time and achieves a greater number of washes because the information available within the scenario means that more

Basics

Fleet	Size	washing int arr tm	priority
1	10	15	1
2	5	15	2
3	5	45	3

Washing machine cycle time

Min	Max
30	90

Time units hours

Failures

Same for all fleets, times base on machine busy time

Failure	Spare	Breakdown (MTTB) Min	Mean	Max	Repair (MTTR) Min	Mean	Max	Distribution Used	Predicting the failure Increment	Threshold
1	1	600	1200	1800	60	90	120	Uniform	60	1200
2	2	1200	2400	3600	120	180	240	Uniform	480	1200

incremented each cycle time
spare ordered once reaches threshold

Spares leadtime

Spare	Min	Mean	Max	distribution used
1	120	240	360	uniform
2	10	240	480	uniform

Fig. 11.3 A summary of input data used for the model

Scenario 1

Fleet	Fleet size	Washes Int arr rate (hrs)	Achieved (number)	Lost (number)	Achieved (%)	Fleet performance (%) Idle	Running	Repair	Wait repair	Service groups (%) Staff	Idle	Busy
1	10	1 to 15	3563	1394	72%	0	53	26	20	1	3	97
1	10	1 to 15	4353	606	88%	6	65	26	2	2	37	63
1	12	1 to 15	4748	214	96%	11	59	25	4	2	29	71
1	14	1 to 15	4957	2	100%	23	53	21	3	2	27	73

Scenario 2

Fleet	Fleet size	Washes Int arr rate (hrs)	Achieved (number)	Lost (number)	Achieved (%)	Fleet performance (%) Idle	Running	Repair	Wait repair	Service groups (%) Staff	Idle	Busy
1	10	1 to 15	4394	568	89%	4	66	8	22	1	21	79
1	12	1 to 15	4817	138	97%	11	60	7	21	1	13	87
1	14	1 to 15	4902	60	99%	18	52	6	23	1	10	90
1	10	1 to 15	4639	322	94%	10	70	8	12	2	59	41
1	12	1 to 15	4874	85	98%	21	61	7	11	2	56	44
1	14	1 to 15	4960	2	100%	31	53	6	9	2	55	45

Scenario 3

Fleet	Fleet size	Washes Int arr rate (hrs)	Achieved (number)	Lost (number)	Achieved (%)	Fleet performance (%) Idle	Running	Repair	Wait repair	Service groups (%) Staff	Idle	Busy
1	10	1 to 15	4886	75	98%	14	73	10	3	1	47	53

Warm up 10,000 hrs (~1yr)
Run time 50,000 hrs (~5yr)

Int arr rate = the time between arrival of the next washing load (input value)
Service groups are unique to fleets

Fig. 11.4 A summary of the base simulation runs and variants

machines are attended to prior to breakdown occurring. However, this has the impact that the consumption of spares is higher. If machines had been run to failure, the full life of the parts would have been realised and hence fewer spares consumed but the replacement of the part would have been reactive rather than scheduled. There is higher downtime but lower wait-repair time due to more frequent repairs but machines can carry on running whilst waiting for ‘repair’.

- Typical of any system, where the demand on available capacity is high, the idle time will be low and queues (or in this case rejected or ‘lost’ washing loads) will occur.
- Where fleet repair times are high the service group utilisation will be high.
- Where fleet idle and run times are low, the fleet availability (number of machines available at any one time) will be low.

Comparing Scenarios 1 and 2:

- A noticeable difference can be seen in the repair time of machines. This difference is due to the service group often not having the available spare on their first visit to a machine in Scenario 1. If the spare is not available, the service group leaves the machine in the broken down state and returns later when both maintenance personnel and spare part are available. Where the service group is highly utilised the delay in returning to the machine to carry out this repair can be significant.
- A lower repair time is evident in Scenario 2 compared with Scenario 1 giving a greater machine running time for Scenario 2. This is due to the lack of machine information in Scenario 1 meaning that the machines wait for spares to be delivered following a first visit from the maintenance personnel after which the part is ordered. Machines in Scenario 2 automatically re-order the spare part directly from the supplier meaning that the repair time is then just due to the delay of the maintenance personnel arriving and repairing the machine.
- The more proactive approach to fault diagnosis in Scenario 2 (through machine monitoring) can lead to a higher performance and, through more complete testing by changing the variables within the simulation, could require less investment in the number of assets required to achieve the same performance.

Figure 11.4 also shows the results for Scenarios 1 and 2 when the number of people in the service group and the fleet size are varied. As the constraints on both the number of people in the service group and the fleet size are relaxed, the benefits in Scenario 2 are marginal. There are, however, clear benefits in Scenario 1. It can be seen from Fig. 11.4 that when the system is 'under stress' (i.e., resource constrained) Scenario 2 shows better results than Scenario 1.

11.5 Discussion

11.5.1 Areas Illustrated by the Modelling

The main point shown by the simulation is that of improved machine performance and utilisation and reduced resource utilisation with increased product condition information availability for the different scenarios. An outcome of the later scenarios is that the level of manual data entry (for example, on part re-ordering) is reduced. This will enable improved product information quality and help to reduce errors and misinformation.

Within scenarios, benefit may be derived from understanding how contracts may drive decisions to prioritise which contracts or fleets should be serviced and in what order. This would minimise the level of likely penalties from each contract if the service level for one or more of the fleets is not attained. Such simulation would provide a useful insight into the business implications/costs of service (non)

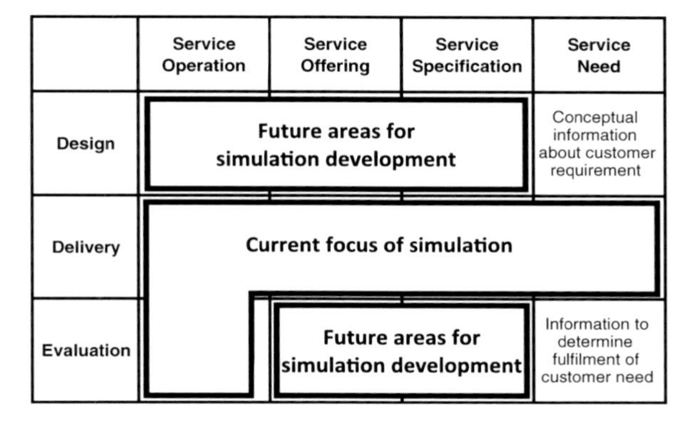

Fig. 11.5 Mapping of simulation coverage to 12-box model

provision. Hence simulation modelling of this type has the potential to assist with the business case for implementing component health monitoring equipment onto products.

Linking this work to the earlier research, in particular the 12-box model (Cuthbert et al. 2008), areas which the simulation currently and potentially could satisfy are illustrated in Fig. 11.5. The work here has focused on delivery and part of evaluation. There is potential to use simulation in the fundamental design to understand the implications of different design configurations and more subtle policy options as well as in the area of evaluation.

The Core Integrative Framework (CIF), proposed in this book, models value co-creation through three interrelated components (Materials and equipment, Information and People) in a partnered environment. The focus of the simulation work shows the impact of information on the transformation of materials and equipment, in the context of the supply chain, maintenance and repairs, and also how it impacts the transformation of people in terms of resource utilisation for value co-creation. The main emphasis is to demonstrate how the availability of appropriate information transforms service provision.

11.5.2 Limitations of the Modelling

The key limitations and assumptions of the modelling are as follows:

- Travel time for the service personnel is not considered.
- Transportation time for the spare parts is not considered.
- The location of spare parts is not considered.
- Input data used for the model (Fig. 11.3) are artificially generated.
- Simple distributions are used for the input data.
- Single, rather than multiple, runs have been used.
- The model has been developed (so far) for demonstration purposes only.
- The model needs to be populated with real data and then validated, before it can be used for decision making.

- Predicted failure is based on busy time of the machine that fails randomly rather than sensing actual failure in advance.

11.5.3 Future Developments of the Simulation

One area where it might be useful to extend this work is to separate the "diagnose" and "repair" times and to add a fixed time for part order administration which may help to highlight some of the improvements, in terms of resource utilisation, between scenarios. The impact of this would be evident in Scenario 1 in particular.

Another area which might be interesting to investigate would be to understand the costs around the service including the level of resources required, the amount of inventory held/owned by the customer or supplier at a given point in time. In addition to this understanding, for each fleet, the machine location and the travel time required for maintenance personnel to reach the machines should be known, as this will have an impact.

Extending the work, in Scenario 3 in particular, may lead to further interesting results. There are a number of ways in which this could be done, but examples of this might include allocation of dedicated maintenance personnel to either non-scheduled maintenance or preventative maintenance. Repairs might also be prioritised based on the requirements of other machines serviced under other contracts provided by the same provider. This prioritisation will take into account factors such as the number of other machines within the fleet of assets which are working, the cost of not repairing machines in other contracts (against KPI penalties) and the type of failure/time to repair/availability of spares/resource.

An increased benefit could be introduced within Scenario 3. In this scenario, the machine predicts which part will fail and anticipates this event by re-ordering the spare part prior to breakdown. However, it may be the case that the machine is better at diagnosing the failure (whether from its own "knowledge" or based on trends), compared with the maintenance personnel, and in so doing may pick up multiple possible failures which, under normal circumstances, may be masked by other apparent symptoms. For example, if a pump fails then the machine may know that the filter is likely to fail and should also be ordered at this point.

A further development, of relevance to Scenario 3 in particular, would be to show how variable lead time may be accommodated if the machine predicts failures far enough ahead. However, as the lead time of parts increases, parts may be replaced more often when they have more remaining life than would ideally be the case. Repairs may become more "scheduled" than dynamic/predictive and this would lead to an interesting problem around the cut-off point between types of maintenance schedules. The issues here would be to understand the:

- trade-off point for scheduled versus predictive maintenance policies, and,
- issues around the location/travel time to repair the machine.

The efficiency of the machine could be determined to provide a more dynamic assessment of remaining machine life throughout its life. This might then lead to developments around the level of preventative maintenance carried out which might strongly correlate to machine efficiency and reduce non-scheduled maintenance time. Factors impacting machine efficiency would include remaining life, usage and spares replacement as required.

Given the scalability of the scenarios, the final goal would be to explore the scenarios across multiple contracts requiring different KPIs, where the impact of this might be the greatest and the most important within Scenario 3.

11.6 Chapter Summary Questions

This chapter has introduced a simple, proof of concept study examining the way that simulation can be used to examine service information values. In the future, more detailed simulations can be conducted to answer the following key information questions:

- What is the impact of different information provision on different contract types?
- What is the band of operation performance within which product condition information does have an impact on service outcome?
- What financial (or other) information would help to provide service performance improvement for decisions around the trade-off between fleet size and resource level for different contracts?
- What might be the impact of real information and the associated information dimensions (timeliness, accuracy, completeness) on the performance of the service?
- What impact would improved accuracy (i.e., of stock levels) have on service?

Acknowledgments The authors acknowledge Dr Gokula Annamalai Vasantha and Mr Kostis Gkekas from Cranfield Innovative Manufacturing Research Centre for their support in preparing Sect. 11.2 (Literature) of this chapter.

References

R. Cuthbert, P. Pennesi, D. McFarlane, The impact of different support service contract models on provider information requirements. Proceedings of POMS 19th Annual Conference, La Jolla, California, 2008

R. Davies, T. Davies, Using simulation to plan health service resources: Discussion paper. J. R. Soc. Med. **79**, 154–157 (1986)

T. Hara, T. Arai, Y. Shimomura, T. Sakao, Service CAD system to integrate product and human activity for total value. CIRP J. Manuf. Sci. Tech. **1**(4), 262–271 (2009)

H. Komoto, T. Tomiyama, Integration of a service CAD and a life cycle simulator. CIRP Ann. Manuf. Tech. **57**, 9–12 (2008)

C. Møller, S.S. Chaudhry, B. Jørgensen, Complex service design: A virtual enterprise architecture for logistics service. Inf. Syst. Front. **10**(5), 503–518 (2008)

National Audit Office, Transformation logistics support for fast jets (2007)

M. Pidd, *Computer simulation in management science* (Wiley, Chichester, 2004)

Y. Shimomura, T. Hara, T. Arai, A service evaluation method using mathematical methodologies. CIRP Ann. Manuf. Tech. **57**, 437–440 (2008)

Y. Shimomura, T. Hara, T. Arai, A unified representation scheme for effective PSS development. CIRP Ann. Manuf. Tech. **58**, 379–382 (2009)

T. Tomiyama, Y. Shimomura, K. Watanabe, A note on service design methodology, Proceedings of DETC, ASME, Paper No. 57393 (2004)

Chapter 12
A Blueprint for Engineering Service Definition

Alison McKay and Saikat Kundu

Abstract Increasing numbers of businesses are moving from the supply of physical products to the delivery of product-service systems. The resulting need to support information related to both physical artefacts and associated services has a number of implications for the design of information systems used to support product-service systems through their lives. The focus of this chapter lies on service in the context of product service systems. Designers of service solutions need to be able to answer the question, "What information is needed in service design to enable the delivery of service excellence?" A key prerequisite to answering this question lies in understanding service elements that need to be supported, performance requirements of the service and how the service elements are related to the required performance. This chapter introduces a service information blueprint that has been designed to support service designers in gaining this understanding. The service information blueprint is a general purpose model for service definition that has been used both to define "as-is" and "to-be" services, service breakdown structures and service performance indicators, and to specify relationships between service processes and service requirements.

12.1 Introduction

Increasing numbers of businesses are moving from the supply of physical products to the delivery of product-service systems. The resulting need to support both physical artefacts and associated services has a number of implications for the

A. McKay (✉) · S. Kundu
School of Mechanical Engineering, University of Leeds, Leeds, UK
e-mail: a.mckay@leeds.ac.uk

S. Kundu
e-mail: s.kundu@leeds.ac.uk

I. Ng et al. (eds.), *Complex Engineering Service Systems*, Decision Engineering,
DOI: 10.1007/978-0-85729-189-9_12,

design of information systems used to support product-service systems through their lives. The focus of this chapter lies on service in the context of product service systems. Designers of service solutions need to be able to answer the question, "What information is needed in service design to enable the delivery of service excellence?" A key prerequisite to answering this question lies in understanding service elements that need to be supported, performance requirements of the service and how the service elements are related to the required performance.

A service information blueprint is introduced in this chapter. A review of literature that informed the theoretical framework on which service information blueprint is built is provided in Sect. 12.2. Three key questions are introduced at the end of Sect. 12.2 and answered through the remainder of the chapter. A case study that is used to illustrate both key features of the service information blueprint and answers to the questions posed at the end of Sect. 12.2 is introduced in Sect. 12.3. This leads into a discussion of the requirements for the service information blueprint (in Sect. 12.4) and a description of how the service information blueprint addresses these requirements (in Sect. 12.5). Finally, in Sect. 12.6, future trends in service development are anticipated and areas for further research outlined.

With respect to the CIF (Chap. 4) the service information blueprint focuses on information transformation.

12.2 Key Characteristics of Service

Means of understanding product information needs are well established and embedded within the development methods of product information exchange standards such as STEP (ISO10303-1 1994). Typically they begin by building an understanding of the activities and information flows to be supported and are followed by a detailed analysis of the information flows that are deemed to be in scope. This yields product information requirements. In product development projects the activities to be supported typically align with the product development processes for the products and organisations concerned (Ulrich and Eppinger 2004). Stage gated product development processes are important because they allow practitioners to focus on key decisions, and the information needed to support them. Product information system designers couple knowledge of key characteristics of physical products with product information requirements to develop information solutions.

Requirements for the service information blueprint were derived from two sources: the activities that service developers carry out and key characteristics of services. During service development activities, key issues that service designers need to be able to address lie in the trade-offs to be made between service affordability and performance. This leads to the requirement to be able to link service process definitions with the key performance indicators from service contracts. This is achieved in the service information blueprint by supporting both:

a. the capture of service processes (including process flows, steps and decompositions) and performance requirements related to both the contract and the business requirements of the service provider, and
b. the definition of relationships between process definitions and performance requirements.

In addition, for companies designing multiple service offerings, there is a desire to gain benefits of scale by enabling the reuse of service definition elements which requires capabilities to create catalogues of service parts and configure services using these "standard" service parts.

Given these high level requirements there is also a need to understand key characteristics of services and how they relate to characteristics of physical products. Johne and Storey (1998) provide a review of literature related to new service development. New service development includes the definition and delivery of services; the literature relates largely to traditional service products such as financial and hospitality services. Five key characteristics that distinguish service products from physical products are identified: intangibility, perishability, non-ownership, inseparability of production and consumption, and variability. As a result of these characteristics, the information structures applied to the definition of physical products cannot be applied directly to services. A key difference between the information content of a physical product definition and that of a service definition lies in the core structure around which it is built. Both have a product structure but physical products are defined around their physical parts (typically through a bill of materials (BoM)) whereas service products are defined around the processes that are used to deliver them (for example, through process flow charts and scripts). A key difference between these information cores is that the relationships in a BoM are part-whole relationships between part definitions whereas the relationships in a service are flows between process step definitions. Engineering information systems to support the lifecycles of product-service systems need to accommodate these distinctions without compromising the need to preserve commonalities between physical and service products.

Research on philosophy and engineering casts insights on relationships between technical artefacts and the processes and actions that surround them. A key tenet of the arguments put forward by Vermaas and Houkes (2006), (Houkes and Vermaas 2009) on the dual nature of technical artefacts is that technical artefacts have both designed physical structures and intended functional structures. On intended functional structures, Vermaas and Houkes, in their ICE (Intentionalist, Causal-role, Evolutionist) theory (2006), assert that when engineers ascribe functions to artefacts they have to consider explicitly the goals for which agents use artefacts and the actions that constitute their use; the agents' actions are captured in a "use plan". Mumford (2006) discussess distinctions between function, behaviour and capacity of physical artefacts and provides the following definitions for function and capacity:

- capacity is a property of an artefact that is understood according to what it can do or what function it can play in relation to other properties;
- function is a capacity plus the use plan that exploits it for an intended purpose.

Design rationale, as captured using tools such as the D-Red software tool (Bracewell et al. 2004), is a means by which designed physical structures might be related to intended functional structures. Design intent, for example as captured using advanced requirements management techniques (Agouridas et al. 2001), enables intended functional structures to be related to stakeholder intent and so aspects of what Vermaas and Houkes refer to as use plans. Illies and Meijers (2009) explore further relationships between artefacts and action schemes. The notion of an action scheme can be used in the context of product-service systems to include maintenance and support activities. As such, it can be argued that the service part of a product-service system is a kind of action scheme.

On designed physical structures, Simons (2003) uses mereology[1] to provide a theoretical basis for the definition of physical product structures, of which BoMs are a common manifestation. If services are regarded as products, or parts of product-service systems, then the following questions arise given the discussion in this section.

1. What are the intended functional structures of service products and how might they be represented?
2. What are the designed structures of service products and how might they be represented?
3. What kinds of service definition relationships are required to support the information needs of service development processes?

In the rest of this chapter we provide answers to these questions.

12.3 Case Study

A service deliverer's requirements arise from two sources: the requirements specified in the contract with the customer and internal requirements needed, for example, to ensure that delivery of the service offering is affordable and the business sustainable in the longer term. For example, efficiencies might be gained by sharing common service elements, akin to sharing standard parts in physical products, without detrimental effects on the quality of the delivered service. As product service system offerings develop, a number of different kinds of contract are being identified. The defence sector is one of the most advanced in identifying different kinds of service contract. The Logistics Coherence Information Architecture (LCIA)[2] is a joint Government (UK MoD) / industry framework developed to improve the way in which these different kinds of support service contract are

[1] "Mereology (from the Greek μερος, 'part') is the theory of parthood relations: ..." [http://plato.stanford.edu/entries/mereology/].

[2] The Logistics Coherence Information Architecture defines a structure for use in identifying information requirements for logistics support solutions [http://www.modinfomodel.co.uk/].

Table 12.1 Example service level agreements (SLAs) and key performance indicators (KPIs)

	Availability type contract	Spare only type contract
KPI 1	Call-to-Repair response time- 24 h (max)	Call-to-Repair response time- 48 h (max)
SLA 1	Provider to supply, install, maintain, repair and support spare parts and the whole machine	Provider to supply and install spare parts for the Coffee Maker
SLA 2	Service package includes planned preventive and predictive maintenance, and unplanned breakdown maintenance	Provider to repair back to safe working and/or operating conditions for the Coffee Maker
SLA 3	Provider is responsible for customer training and to demonstrate the use of the Coffee Maker	Provider responsible for customer training and to demonstrate the use and/or operation of the Coffee Maker
SLA 4	Provider to supply user manuals/training materials to the customer	Provider to supply user manuals/training materials to the customer
SLA 5	Customer to pay annual fixed price to the provider for availability of the product (Coffee Maker) and services	Customer to pay separate price for each of the new or replaced spare parts supplied by the provider
SLA 6	The price includes both spares and services (i.e., complete availability and ready for use)	Customer to pay annual fixed price to the provider for repair services that cover for a whole calendar year
SLA 7	24 months minimum contract period	12 months minimum contract period
SLA 8	Services are provided on both as-planned and on-demand basis for the duration of the contract	Services are provided on-demand basis for the duration of the contract

achieved; LCIA identifies a range of contracts from the provision of spares through the delivery of equipment availability to the delivery of capability. The case study used in this chapter is a fictitious maintenance and repair service for a coffee making machine. Two kinds of contracts are considered: availability and spares only type contracts. In the availability type contract the manufacturer supplies coffee making machines to its customers and takes responsibility for ensuring that the machine is available for use for an agreed proportion of the time. For example, an availability of 95% would mean that the machine must be available to use for 95% of the time. On the other hand, for the spares only type contract, the manufacturer supplies spare parts for the coffee making machine and carries out all repair activities; however, the customer pays separately for the spare parts. Typical service level agreements (SLAs) and key performance indicators (KPIs) for these two kinds of contract are shown in Table 12.1. Commonalities and similarities in the processes needed to deliver services under the two contract types are summarised in Table 12.2.

The service provider delivers services to a number of customers, some on a spares only basis and others as availability contracts, for the coffee making machine in this case study. To improve efficiency, the company maximises the sharing of common processes across different services, products and customers. As such, when designing services a goal is to minimise service elements that are specific to the product being serviced for the customer.

Table 12.2 Process commonalities and differences across the two contract types

Availability type contract	Spare only type contract
Service provider carries out planned preventive and predictive maintenance to an agreed schedule	Customer takes responsibility for the maintenance and repair of the machine
If there is a breakdown the customer contacts (by telephone) the service provider, reports the problem and schedules a service engineer's visit	
The engineer is expected to visit the customer within 24 h of the problem having been reported.	The engineer is expected to visit the customer within 48 h of the problem having been reported
The service engineer diagnoses the cause of the breakdown and tries to resolve the problem on-site. If the problem cannot be resolved on-site the machine is taken from the customer's premises for repair. During the repair, if necessary, the service provider either purchases replacement parts from its suppliers or uses parts that have been manufactured in-house	
An alternative machine is provided to the customer for temporary use	–
The service provider pays for any new or replacement parts.	The customer pays for any new or replacement parts.
The faulty machine is then repaired by the service provider's maintenance and repair team	
Other planned, preventive and predictive maintenance might be carried out	–
The service provider's call centre staff contact the customer to arrange the return of the repaired coffee making machine	
The standby coffee making machine is returned to the service provider	An invoice for the cost of the repair is sent to the customer

With respect to the questions posed at the end of Sect. 12.2, the following answers can be provided.

1. *What are the intended functional structures of service products?*

The intended functional structures of services come from the service contract and business strategies of the organisations involved in the delivery of the service (both customers and suppliers). In the case study these are the KPIs and SLAs from the service contract, and the company goal to maximise sharing of service elements across products, services and customers.

2. *What are the designed structures of service products?*

The designed structures of services are the means by which the intended functional structures are realised. For this case study they are the processes used to deliver the service, such as those that can be derived from the descriptions given in Table 12.2. In addition, the designed structure is likely to include references to standardised process elements.

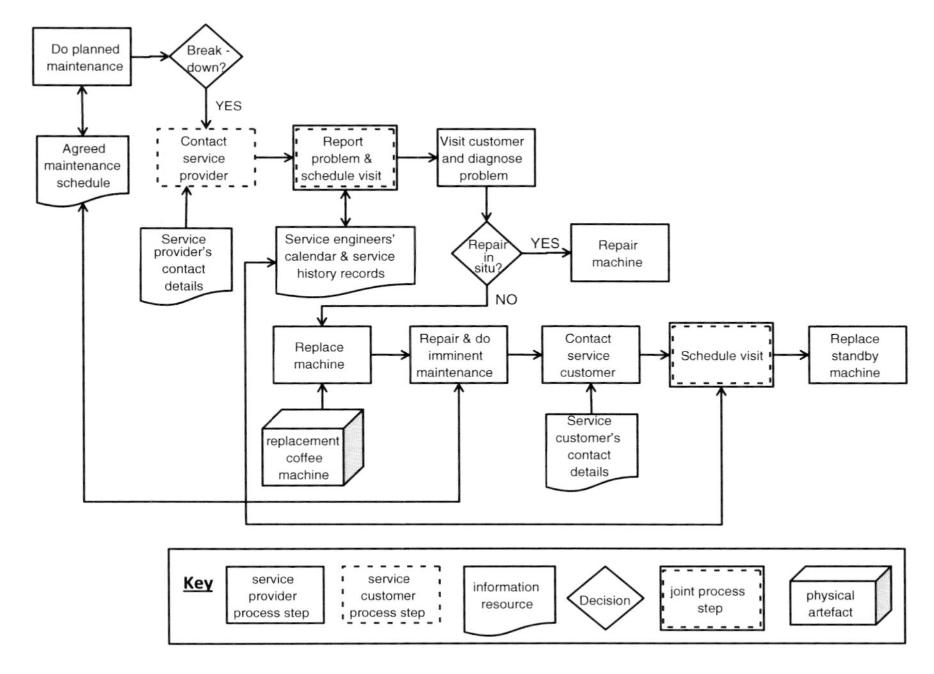

Fig. 12.1 Flow chart of the Availability Contract service case study

12.4 Requirements for a Service Information Blueprint

Requirements for the service information blueprint came from two sources: the information needed to support downstream applications such as service simulations and the needs of the service development process itself. These can be boiled down into two questions.

- What do service developers need to be able to say? and
- What do service developers need to be able to do?

12.4.1 What do service developers needs to be able to say?

The flow chart given in Fig. 12.1 is the result of an analysis of the availability contract example given in Table 12.2; an edited version showing the differences in process flows and information uses for the spares only option is given in the Appendix. The key in Fig. 12.1 is instructive in answering our question. It can be seen that there was a need to capture three kinds of service process: those carried out by the customer, those carried out by the service provider and those carried out by both parties collaboratively. The spares only contract flow chart in the

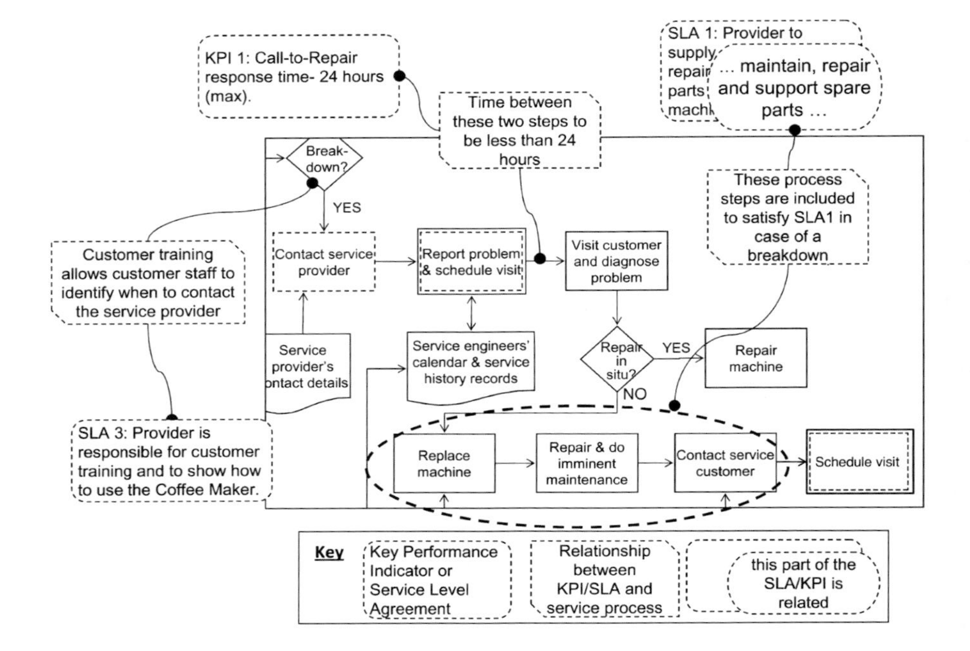

Fig. 12.2 Relationships between process steps and SLAs and KPIs

Appendix shows how these process steps change across different kinds of service contract. Real-world situations are often more complicated than this because services are frequently delivered and consumed by networks of stakeholders. Common examples are package delivery services where the customer experience (and so service quality) is influenced by the goods provider, the goods deliverer and the customer; and industrial services such as those delivered to the armed forces where the service is co-delivered and consumed by a number of agencies who are related to each other in different ways. These kinds of service are beyond the scope of the service information blueprint presented in this chapter.

There was also a need to capture information used in the service process; typically these would be held in business information systems (for example, contract and customer contact details) and engineering information systems (for example, service histories and information about replacement products and parts). The example used in this chapter oversimplifies these information systems; a key role for the service information blueprint lies in supporting service developers in identifying the detailed information requirements of their service; this is covered in more detail later.

Finally, although not shown in the key to Fig. 12.1, it is important to capture relationships between process steps and the service requirements to which they contribute. For this reason, a third kind of relationship to capture is between the process steps shown in Table 12.2 and the service level agreements (SLAs) and key performance indicators (KPIs) given in Table 12.1. Examples of these kinds of relationship, for the availability contract are shown in Fig. 12.2.

Again, the key in Fig. 12.2 is instructive in identifying the kinds of relationships that need to be supported. It highlights the need to be able to relate aspects of the service process to parts of and whole KPIs and SLAs. This requirement is mirrored in references to the service process where there is a need to refer to:

- individual process steps, such as in the case of SLA3,
- relationships between process steps, such as in the case of KPI1, and
- collections of process steps that are not necessarily defined as a self-contained group within the service definition, such as in the case of SLA1.

In addition, although not shown in Fig. 12.2, there is a need to be able to refer to parts of process steps, such as the *Repair* part of the *Repair & do imminent maintenance* process step.

12.4.2 What do the service developers needs to be able to do?

Discussions with service design and development personnel led to the identification of basic requirements for service developers. In essence they need to be able to indentify implications of decisions made during contracting, especially on affordability and performance. A key issue lies in enabling both the supplier and customer to understand the affordability of requirements defined in the contract, that is, what will it cost the supplier to deliver the requirement and is the added benefit worth enough to the customer for them to pay for it?

Hence the basic needs of service developers were defined as being:

- to capture service definition processes in ways that can support evaluation through, for example, simulation, risk assessment and life cycle cost models;
- to understand how the process elements relate to the contract; and
- to understand information and other resource requirements both at contract definition and during service development.

With respect to the questions posed in Sect. 12.2, the following answers can be provided.

1. *How might the intended functional structures of service products be represented?*

Service level agreements and key performance indicators in the service contract need to be captured in such a way that relationships between them and process elements of the service can be represented and visualised.

2. *How might the designed structures of service products be represented?*

From the case study, the designed structure of a service is its process structure. Both flows and part-whole relationships between process steps are required.

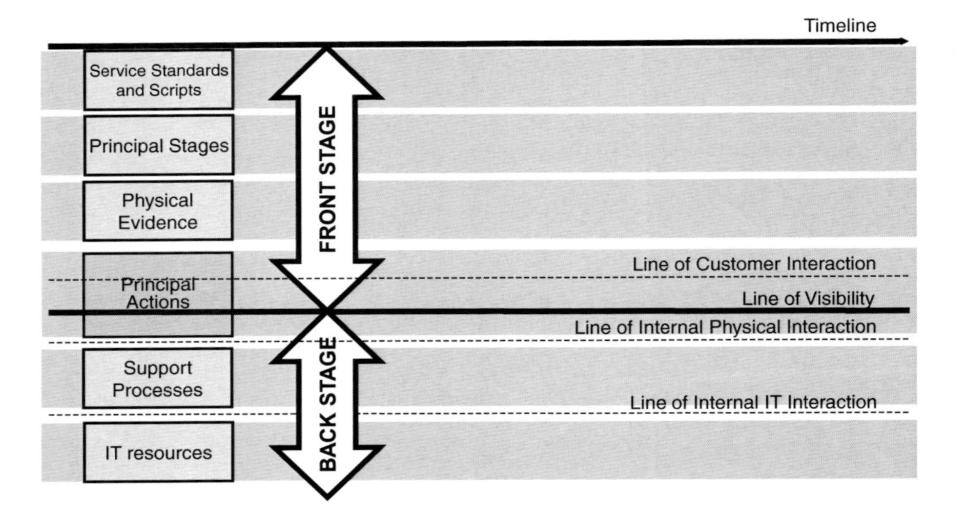

Fig. 12.3 The overall architecture of the service information blueprint

3. *What kinds of service definition relationships are required to support service development processes?*

An analysis of the example given in Table 12.2 highlights the following relationships that need to be supported:

- flow relationships between service process steps;
- references to standardised service elements;
- information on who carries out which process steps;
- relationships between elements of the functional structure and elements of the designed structure;
- relationships to product design and configuration data; and
- relationships to information generated through the service delivery process.

12.5 The Service Information Blueprint

Approaches to the definition and visualisation of service systems have emerged from a number of disciplines in recent years, including social and behavioural sciences, business, design and information technology. They are often tailored to different aspects of service definition processes; summaries of a range of these tools and techniques are available (Tassi 2008, 2009; Engine Group 2009). Service blueprinting approaches have traditionally been used to capture services, such as those in the hospitality and financial sectors, that are not associated with complex products. The service information blueprint introduced in this chapter is an

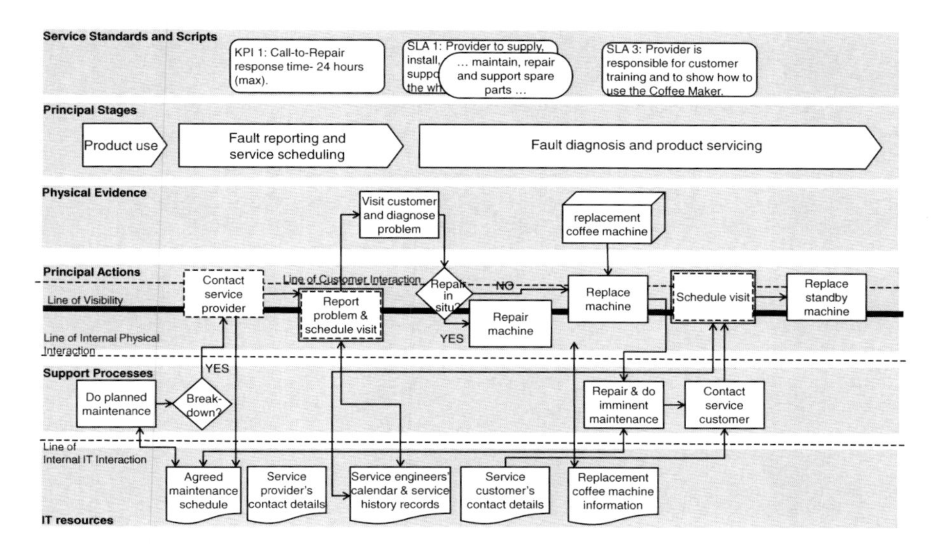

Fig. 12.4 Application of the service information blueprint to the coffee making machine availability contract

adaptation of the traditional service blueprinting technique that suits the needs of complex service systems intended to meet requirements set out in a contractual agreement with a customer and, potentially, other stakeholders. The overall architecture of the service information blueprint is shown in Fig. 12.3 and an application to the coffee making machine availability contract is shown in Fig. 12.4.

It can be seen that the service information blueprint is constructed around a collection of "swim lanes"; these are shown by the horizontal grey bars in Fig. 12.3. The idea of swim lanes is adopted from the Business Process Modelling Notation OMG (1997–2010). However, the content of the swim lanes is based on the core elements of a service blueprint.

It can be seen from Fig. 12.3 that the swim lanes are labelled as either Front Stage or Back Stage. Front stage lanes can be seen, or experienced, by the service customer whereas back stage lanes are not seen or directly experienced by the service customer. In addition there are a number of lines superimposed on the swim lanes in Fig. 12.3. The timeline represents the chronology of actions conducted by the service customer and provider. A line of visibility defines the service processes that are visible to the customer during service delivery. If the enactment of a service blueprint is seen as a simultaneous production and consumption of the service then this line governs which parts of the delivery of the service are visible to the customer. Finally there are three lines of interaction: contact person (visible actions), internal physical interacti

- Service Standards and Scripts specify target performance levels for the service. Examples from the coffee maker example would be the KPIs and SLAs in Table 12.1, as shown in Fig. 12.4.
- Principal Stages are the key process steps as seen by the customer. Examples from the case study include *Product use*, *Fault reporting and service scheduling* and *Fault diagnosis and product servicing*.
- Physical evidence addresses the intangibility of the service itself. Service activities are essentially intangible but they often leave tangible or physical evidence that they have been carried out; this evidence often informs customers' perceptions of the quality of the service. In the coffee maker example the replacement machine, the tone of voice of staff who arrange the engineer's visit and the visit by the service engineer are examples of physical evidence.
- Two kinds of process are captured.
  - Principal Actions can be of three types.
    - onstage principal actions by the customer, for example, *Contact service provider* in the coffee maker example;
    - onstage principal actions by the service provider, for example, *Replace machine* in the coffee maker example; and
    - backstage principal actions by the service provider, for example, *Schedule visit* in the coffee maker example.
- Support Processes are backstage processes that are carried out by the service provider. *Repair & do imminent maintenance* is an example from the coffee maker example.
- IT resources are the information systems, both engineering and management information systems, used to support the service delivery. In the coffee maker example the service history of the machine and contact details for the customer are examples.on and internal IT interaction. The swim lanes contain the service definition itself.

Product service systems are often designed and developed to improve performance such as reducing time and cost, improving quality and responsiveness, and maintaining a sustainable business. Effective delivery of services with improved performance demands access to high quality service information (i.e., complete, correct, minimal and available to the right people at the right time). This is pertinent throughout the lifecycle (which typically includes definition, development, delivery and evaluation) of the product-service system. To address this there is a need to capture service information requirements while defining services. The service information blueprint has information requirements associated with each service activity (i.e., the principal actions and support processes). These are intended to allow service designers to answer the following questions in the service definition:

- What information is needed by this activity?
- Does the information exist?

- If the information exists, where is it?
- If the information does not exist, where might it come from?

For each of the service activities, three types of service information are considered: input information, process information and output information. This builds on the service information classification scheme proposed by Berkeley and Gupta (1995). For the service information blueprint, input information refers to the information that is needed before the service activity can be performed; process information is the information required by the actor while performing the activity; and output information refers to information that is available as a result of the performance of the service activity.

In addition, the service information blueprint captures information associated with relationships:

- between two elements within a swim lane of a service blueprint, such as the flows between process steps are represented by arrows in Fig. 12.4; [3] and
- across the swim lanes of a service blueprint, for example, the relationships shown in Fig. 12.2.

To date the service information blueprint has been used to aid the understanding of as-is and to-be services. The blueprint can be used to provide a structure for service definition workshops and a software prototype that can be used to create digital service definitions is available.

With respect to the questions posed in Sect. 12.2, the following answers can be provided:

1. *How might the intended functional structures of service products be represented?*

In the service information blueprint SLAs and KPIs are currently represented as text that can be structured to enable parts of requirements to be identified. For example, this allows "...maintain, repair and support spare parts ..." in Fig. 12.2, which is a part of SLA1, to be identified and then referenced by the relationship to the relevant service process steps.

2. *How might the designed structures of service products be represented?*

In the case study used in this chapter, the service information blueprint represents service processes as process steps related by flow relationships. Although not included in this chapter, the service information blueprint also supports the definition of part-whole relationships between process steps; this is a key area of functionality needed to support the definition of service breakdown structures.

3. *What kinds of service definition relationships are required to support the information need of service development processes?*

In the case study introduced in this chapter, the service information blueprint is used to capture the following relationship types identified at the end of Sect. 12.4:

[3] Relationships within a given swim lane can be one of two types: connection and composition,

- flow relationships between service process steps;
- information on who carries out which process steps; and
- relationships between elements of the functional structure and elements of the designed structure.

Although not included in this chapter, the service information blueprint also supports the definition of references to standardised service elements (using hypertext links) and, through its underlying information framework [defined in McKay and de Pennington (2001)], it offers the potential to define relationships to product design and configuration data.

12.6 Discussion

The research that resulted in the service information blueprint concluded with a road mapping activity where future trends in service information strategies were anticipated. It was recognised that current practice in service design and development tends to focus on bespoke services created to meet the needs of individual contracts. A key area for improvement, for example in reducing costs and improving the quality and reliability of service offerings, lies in standardisation across services. A key priority in this respect lies in the use of common information system solutions which, in turn, leads to a need to use standardised service processes. In the medium term it was anticipated that service design and development activities will become more like product development activities where new offerings are built using standard parts. This would change the nature of service offering from being bespoke-to-the-customer to including standard-to-the-supplier elements and, in the longer term, variants of services that are standard to the supplier. As commonality across service elements increases opportunities for significant savings in service delivery and the use of common IT/IS services for common service functionality becomes a real possibility. Ultimately these could be used to enable service providers to manage fleets of services in a similar way to which fleets of products are managed today.

A number of drivers for product service systems development were identified and could be used to inform future generations of service information blueprint.

- *Need for more defined/accurate information at start of process to define service* Although desirable, and what a lot of designers and stakeholders might like to have, this may not be feasible given the constantly changing landscapes for and within which services are designed and delivered. In addition, the process of designing and delivering a service might change such information. Thus, an alternative need could be to build better capability of working with processes in addition to physical products within organisations that are moving to the delivery of services. A number of discussions during the research centred on advanced business process modelling approaches and techniques such as ARIS (Scheer 2000). The service information blueprint complements these systems by enabling the association of process steps with information requirements and elements of the contract.

- *Service cataloguing and reconfigurability*
  An analysis of current practice with colleagues in BAE Systems led to the conclusion that typical service development processes involve the design of bespoke services that respond to the needs of individual customers. Key issues lie in the lack of consistency across resulting service offerings. For example, opportunities for substantial savings from using the same process for a given task could be identified in terms, for example, of avoiding the need for bespoke information solutions and overly specialised training of personnel, and of increasing transferability of people and information across different service solutions. This led to the identification of a need to support the creation of service elements through the configuration of "standard" service elements, akin to the use of standard parts in physical products: service cataloguing and reconfigurability. Ideas on service configuration can be drawn from a large body of work in product configuration. Product configuration involves linking physical elements together to form new products; part-whole relationships are used to define product breakdown structures and connection relationships are used to define how parts within a product breakdown structure relate to each other to deliver intended functionality. Different kinds of connection relationship occur in product breakdown structures: for example, mating conditions in assemblies and functional interactions in functional definitions. These ideas can be transferred to service products in the service information blueprint because it is based on an underlying meta-model that treats physical and service products in a consistent way. However, whilst it is possible to use the service information blueprint to define service breakdown structures, their use requires understanding of issues that were beyond the scope of this research.
- *Reusability of service elements*
  This can occur within and across service information blueprint swim-lanes. However, more understanding is needed of how to (a) select service parts and (b) connect them to each other to create new services that will behave in intended ways. Given a service catalogue, from which service elements can be selected, how will service parts be found and then re-used? What are the equivalents of, for physical parts, bearing calculations for use in the selection and "assembly" of service elements?
- *Platform-independent services*
  Two kinds of platform agnostic service were identified: services that can be plugged into multiple platforms and assume that certain data from the platform is available and services that use minimal data from the platform and so can be applied in multiple contexts. The service information blueprint could inform research and development in this area, for example, by supporting the identification of relationships between the service definition and physical product.

12.7 Concluding Remarks

In this chapter we have introduced a service information blueprint that allows service definitions to be created and used to identify both information requirements for service delivery and relationships between service activities and service requirements. A number of companies are using process mapping techniques. The service information blueprint complements these techniques by supporting the definition of relationships between service processes and service requirements in addition to service processes. The way in which the service processes are represented make the service definitions well-suited to support some of the longer term aspirations outlined in Sect. 12.6. For example, they include information needed to support service simulations and the kinds of relationship supported could be used to support the configuration of new services from standard service elements. Further research is needed to demonstrate this potential and build understanding of the additional functionality that would be needed.

We began this chapter with a question faced by designers of service solutions, namely, "What information is needed in service design to enable the delivery of service excellence?" The research upon which the chapter is based focused on the first half of this question and information implications of alternative service options in the context of product service systems. In delivering service excellence, the service design needs to be operationalised. This is typically done through networks of organisations (so-called "enterprise networks") and, within each organisation, the people who deliver and receive the service. Further research is needed to understand both social and technical aspects of service excellence and the information needed to, for example, capture customers' experiences of a service both quantitatively and qualitatively.

12.8 Chapter Summary Questions

This chapter introduced a service information blueprint that embodies answers to the following questions.

1. What are the intended functional structures of service products and how might they be represented?
2. What are the designed structures of service products and how might they be represented?
3. What kinds of service definition relationships are required to support the information needs of service development processes?

Acknowledgements The authors thank the industrial and academic members of Work Package 2 in the S4T project (BAE SYSTEMS, MBDA, Institute for Manufacturing at University of Cambridge, and Decision Engineering Centre at Cranfield University) for their input to project workshops and Peter Dawson from the Keyworth Institute at the University of Leeds who supported the work through the establishment of software prototypes. Both informed the development of the service information blueprint reported in this chapter.

Appendix

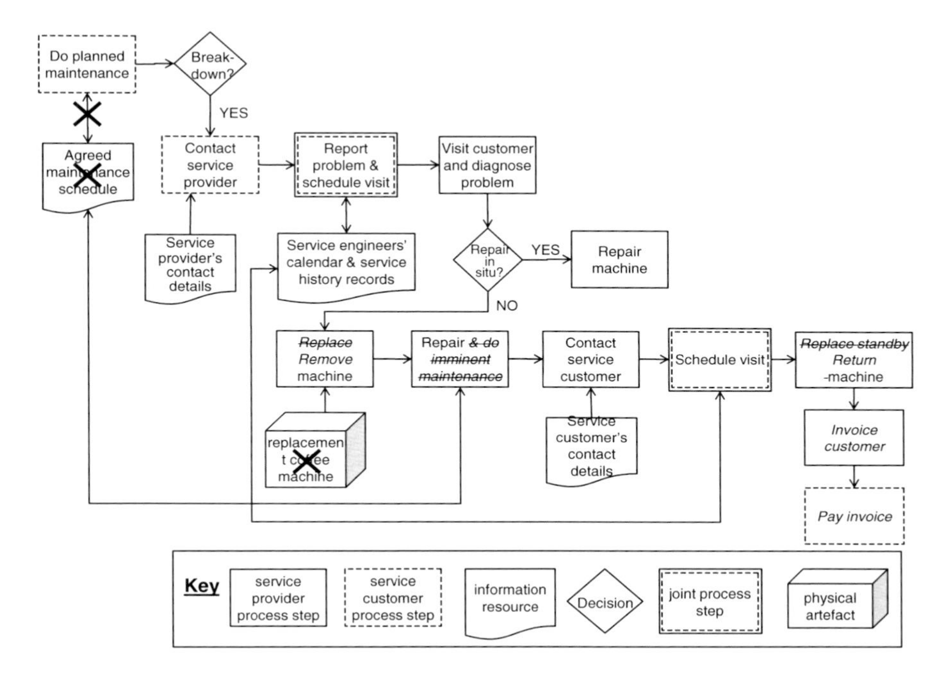

Fig. 12.5 Flow chart showing changes to the Availability contract for the spares only contract service case study

References

V. Agouridas, J.E. Baxter, A. McKay, A. de Pennington, On defining product requirements: A case study in the UK health care sector, in *13th international conference on design theory and methodology* (ASME, Pittsburgh, Pennsylvania, 2001)

B.J. Berkeley, A. Gupta, Identifying the information requirements to deliver quality service. Int. J. Serv. Ind. Manag. **6**(5), 16–35 (1995)

R.H. Bracewell, S. Ahmed, K.M. Wallace, DRED and design folders, a way of capturing, storing and passing on, knowledge generated during design projects. in *DETC'04, ASME design automation conference*. Salt Lake City, Utah, 2004

Engine Group, Engine service design. Available at: www.enginegroup.co.uk. Accessed Nov 2009

W. Houkes, P.E. Vermaas, Contemporary engineering and the metaphysics of artefacts: Beyond the artisan model. Monist **92**(3), 403–419 (2009)

C. Illies, A. Meijers, Artefacts without agency. Monist **92**(3), 420–440 (2009)

ISO10303-1, Industrial automation systems and integration–Product data representations and exchange–Part 1: Overview and fundamental principles (1994)
A. Johne, C. Storey, New service development: A review of the literature and annotated bibliography. Eur. J. Mark. **32**(3/4), 184–251 (1998)
A. McKay, A. de Pennington, Towards an integrated description of product, process supply chain. Int. J. Technol. Manag. **21**(3/4), 203–220 (2001)
S. Mumford, Function, structure, capacity. Stud. Hist. Philos. Sci. **37**, 76–80 (2006)
R. Tassi, Design della comunicazione e design dei servizi: Progetto della comunicazione per la fase di implementazione (communication design and service design: Implementing services through communication artifacts) Politecnico di Milano, Italy (2008)
R. Tassi, Service design tools: Communication methods supporting design processes. Available at: www.servicedesigntools.org. Accessed Nov 2009
A.W. Scheer, *ARIS: Business process modeling*, 3rd edn. (Springer, Berlin, 2000), p. 218
P. Simons, *Parts: A study in ontology* (Clarendon Press, Oxford, 2003)
K. Ulrich, S. Eppinger, *Product design and development* (McGraw-Hill, New York, 2004)
P.E. Vermaas, W. Houkes, Technical functions: A drawbridge between the intentional and structural natures of technical artefacts. Stud. Hist. Philos. Sci. **37**, 5–18 (2006)

Part IV
Complex Product Integration

Svetan Ratchev and Emma Kelly

Service systems may be defined under a variety of contexts. At its barest level, a service involves the application of competences for the benefit of another. This thereby infers that a service may be a form of action, performance or business requirement that is used as a type of value exchange between provider and client. Complex engineering service systems invariably employ complex technological solutions to meet availability, reliability and capability demands, as prescribed by the client. These solutions are intrinsically intertwined with both the needs of product life cycle management and the delivery of through life support.

The key behind the success of a service system is its interdisciplinary qualities. Focus is aimed not on one aspect, but on service as a system of interacting parts. The science behind a service system builds on ideas from existing disciplines and the transformations required between materials and equipment, information and people in delivering value to the client. The introduction to this book highlights the Core Integrative Framework (CIF) and the mapping of the three aforementioned transformations. The chapters within this section concentrate on work undertaken on the S4T programme in maintenance and capability enhancement, as well parallel investigations into contracting for capability and availability. The more knowledge intensive and customised the service, the more it depends on the interaction between two key transformations identified in the CIF; transforming materials and equipment and transforming information. This section addresses the many technological challenges associated with delivering a service system.

The first chapter introduces the service support solutions currently being used and adapted in the UK Defence Environment to deliver Contracts for Availability and Contracts for Capability. These contracts aim to provide operational availability through an integrated and effective support solution, The academic theory behind availability and capability is considered alongside the practical delivery of availability through comparing different support solutions. The over-arching concepts behind the two contracts aid in delivering a support solution that provides improved operational availability. The challenge is to re-emphasise and reiterate the engineering inputs and connections with reliability, maintainability and logistic support. To this end, Health and Usage Monitoring Systems,

Reliability Centred Maintenance, Condition-Based Maintenance (CBM) and Prognostic Health Management are considered enablers for engineering and support planning. The nature of these techniques and the challenges for their adoption are investigated.

Subsequent chapters in the section address CBM. Innovative approaches are needed to improve maintenance and reliability as well as to create unique opportunities for value co-creation between stakeholders. The chapter on 'Modelling Techniques' focuses on investigating the applicability and implementation of combining prognostic modelling with Condition Based Maintenance (CBM) and its role in providing improved service provision for the repair and maintenance of complex systems. Staying within the theme of CBM, the following chapter addresses the operational requirement for component replacement decision analysis. The underlying issue of 'why is residual life prediction important?' is both addressed and calculated through utilising available condition monitoring data and incorporating prognostic modelling options, which are optimised to estimate maximum availability whilst preventing component failures.

The chapters in the second half of this section address methodologies that have been developed to assist in the planning, scheduling and outsourcing of tasks associated with maintenance, platform modernisation and support solutions of complex engineering systems. The first proposed methodology is based on the delay time modelling concept and may be used to optimise maintenance service intervals with respect to specific criteria such as system cost, downtime and reliability. The concept may also determine the optimal inspection interval for multiple types or levels of maintenance and service. Simulation methods for the development of optimal service support procedures using an integrated product, process and resource structure are also proposed. Significant cost drivers can be identified and quantified through the virtual development of work breakdown structures for service support processes, the results of which have system level significance in reducing risk in service provision through the support of value co-creation. The provision of simulation output data for higher level cost modelling can also inform the strategy required for supply chain engagement through incentivisation.

The services aspect of delivering a support solution includes the use of new technologies and methods for managing technical products over their life cycle and ensuring that client's required capability and availability demands are met. This imposes new challenges on subsequent maintenance, repair and capability enhancement procedures. A framework for the development of a Maintenance Dashboard is proposed to address this, the underlying purpose being to establish an approach that supports the decision–making process on whether to maintain, repair, upgrade or update a given asset. Through incorporating maintenance and capability enhancement, both facets are considered in their entirety as opposed to in isolation. Technology insertion provides the means to proactively sustain and enhance the functionality and associated performance levels of legacy product platforms. The final chapter in this section proposes a transformation mapping approach that brings these three groups together for both vision setting and activity planning.

The chapters have been developed by a community working in the field of support solutions, with contributions from academics, service providers and service clients alike. The views presented have arisen from both the research conducted and test case applications, which provide a variety of insights from the different perspectives of those involved.

Chapter 13
Contracting for Availability and Capability in the Defence Environment

Christopher J. Hockley, Jeremy C. Smith and Laura J. Lacey

Abstract This chapter discusses the extensive support solutions being used and adapted in the UK Defence Environment to deliver what are described as Contracts for Availability (CfA) and Contracts for Capability (CfC). They aim to provide operational availability through an integrated and effective support solution, generally by industry, or a combination of industry and the Ministry of Defence (MoD). The chapter considers the academic theory for availability and capability before reviewing how the defence environment currently delivers CfA and CfC. It identifies the issues and diversity of current contracts in the military environment where operational availability and capability are vital yet are required in variable ways with periods of both dormancy and high intensity. The chapter analyses the practical delivery of availability and capability using defence examples which compare and contrast the different support solutions currently being delivered.

13.1 Introduction

Availability of equipment, facility or service is vitally important in the UK military environment and probably more so than in the usually profit driven commercial sector. Availability of defence equipment could significantly affect

C. J. Hockley (✉) · J. C. Smith · L. J. Lacey
Department of Engineering Systems and Management, Defence Academy of the UK, Cranfield University, Shrivenham, UK SN6 8LA,
e-mail: c.hockley@cranfield.ac.uk

J. C. Smith
e-mail: j.c.smith@cranfield.ac.uk

L. J. Lacey
e-mail: l.lacey@cranfield.ac.uk

I. Ng et al. (eds.), *Complex Engineering Service Systems*, Decision Engineering,
DOI: 10.1007/978-0-85729-189-9_13,

how operations can be conducted and lack of availability during military operations will almost certainly put military personnel in additional harm's way. With the ever increasing cost and complexity of military equipment, the defence budget is always under pressure to reduce costs and to out-source much of the activity that is not directly involved in the delivery of military effect. Thus, traditional support activities, such as maintenance and repair have been candidates for civilianisation in many solutions under the general banner of 'Contractor Logistics Support' (CLS). Initially this merely gave contractors responsibility for holding inventory and supplying spares. This was soon expanded into Private Finance Initiative (PFI) solutions where contractors provided everything from the facilities to the manpower, examples being the Joint Services Command and Staff College at Shrivenham and aircraft simulator services. The MoD observed the resultant benefits such as having the cost per year defined, make long-term budgeting easier. Whether it is in fact cheaper is an ongoing question as it is a complex equation that does not compare like with like. Two other terms came into common usage: Contracts for Availability (CfA) and Contracts for Capability (CfC). Under a CfA contract industry is required to deliver outcomes defined in terms of availability. Initially defined in terms of platform (ship, aircraft, vehicle etc.) availability has evolved into outcomes more clearly linked to the MoD's operational requirements such as available flying hours. Under a CfC contracting model industry is required to deliver a complete capability which by definition would include the operators, maintainers and all the support. This has proved difficult as there are few defence capabilities in which it would be acceptable for civilians to replace serving military crews and support personnel. A contractor would also find it difficult to deliver a contract without full control over all of its aspects and this situation would exist if military personnel were an integral part of the capability. A few MoD contracts are described as and justify the label CfC, namely, the provision of aircraft simulators. Others, for example the new air-to-air refuelling system for the Royal Air Force (RAF), do not really justify it, as it is planned to include uniformed RAF aircrew and possibly some of the groundcrew. CfA has been an effective solution for the MoD so far but a problem with the CfA model is that it has exposed a naivety about the make up of availability and what can be contracted for successfully. It is thus essential that we first understand the theory and make-up of availability.

13.2 Availability Theory and Practice in the Military Environment

First we need a reasonably simple definition so that we can analyse availability and what is important; the following is an adaptation of those in various standards (British Standards Institution 1991 and Defence Standard 00-49; Ministry of Defence 2008).

> The probability that the system or equipment used under stated conditions will be in an operable and committable state at any given time (Hockley 2009).

When contracting for availability does this simple definition describe whether the equipment is in the right condition or role for the task? For example the C130 Hercules aircraft has two distinct variants in its fleet, the C130K and C130J. Within the C130K fleet there is a "stretched" aircraft which can carry more. There is also a set number of aircraft with specialist equipment fitted for the "Special Forces". Does a CfA contract need to deal with the ever-changing daily operational requirements in order to deliver the availability the user actually wants for specific roles and tasks on a specific day? For this reason we need to understand what is going to influence availability before a contract can be created.

The availability definition deals firstly with *probability*. This will tell us the level of uncertainty and inevitably there will be statistics involved which need to be understood and appreciated. Means and averages can be stated but meaningful boundaries that capture what the user really wants are needed. When considering probability there are certain considerations which must be evaluated. They are: the operational consequence of failure, whether there are any reversionary modes, the required mission duration, the logistic consequence of failure and what is realistic and achievable. The next important aspect of the definition is *under stated conditions*. This needs to define the function and how and in what environment the equipment is being used and in what environment it was designed to be used. It is primarily about environmental conditions but there can be many facets, such as purely geographic (hot and dusty or cold and wet) the environment seen if fitted to different platforms (helicopter versus road vehicle) and the maintenance, storage, handling and transport environments. The word *operable*, the next important aspect of the definition, implies that the equipment can be committed to operations and will start its mission successfully. Finally *time* is highlighted. Here we identify that at any particular time, the equipment will be able to operate (to start its mission). However, because we have simplified the definition it says nothing about how long it should operate for and whether the equipment will complete the mission successfully, but perhaps it should! Availability is therefore, synonymous with what the military calls "Operational Readiness" and is clearly a function of both operating time which depends on reliability, and downtime, which depends on maintainability and supportability. Downtime will be determined firstly by how easy and quick it is to perform the maintenance or repair—its maintainability features—and secondly by the time taken to supply the support resources, such as facilities, personnel and spares.

Time is also an important aspect of the definition and indicates the ratio of time it is available—or its "uptime"—to the total time that it could be available and the total time will be made up of both the uptime and downtime. Similarly we can also think of uptime as being equivalent to the time it is working between the times when it is not, or in other words when it has failed. Therefore, if uptime is equivalent to mean time between failures (MTBF) in that case the downtime will be equivalent to mean time to repair (MTTR).

$$A = \frac{\text{Uptime}}{\text{Total Time}} = \frac{\text{Uptime}}{\text{Uptime} + \text{Downtime}} = \frac{\text{MTBF}}{\text{MTBF} + \text{MTTR}}$$

Availability has to depend on the factors that make up both elements of the equation but is surely more dependant on the support environment and resources which will affect the downtime. Yet the first aspect we must consider in what will affect the downtime is what the designer has actually designed into the equipment because only he can influence the repair time with good maintainability features. It is for this reason, and if we assume we have ideal support conditions, that we need to designate this availability as inherent or intrinsic availability A_i.

In other words everything that is needed for repair and support is available. The time to repair is then the active repair time so must not be designated merely mean time to repair, MTTR; rather it must be designated Mean **Active** Repair Time, MART. So if availability is defined as being in ideal support conditions, it must be called intrinsic or inherent availability with the suffix i (as A_i) MTTR must always be written as MART to ensure it means only the active time spent on maintenance and repair. Sometimes maintenance instead of repair is mentioned and then it is designated mean active corrective maintenance time (MACMT). However, there is no provision made here for scheduled or preventive maintenance and this is dealt with by *achieved availability* A_a which also defines availability in an ideal support environment. Consequently mean active preventive maintenance time (MAPMT) will describe the active time spent on scheduled maintenance. The downtime is then both the active corrective and active preventive maintenance times.

$$A_i = \frac{\text{Uptime}}{\text{Uptime} + \text{Active Repair Time}} = \frac{\text{MTBF}}{\text{MTBF} + \text{Mean } \textbf{Active} \text{ Repair Time}}$$

However, it is operational factors and actual operational conditions, not ideal conditions, which are really important. What needs to be considered is what actually affects availability in an operational environment. There is also no consideration of standby or dormant periods. Most importantly there is no consideration of the downtime occurring from logistic and administrative delays such as delays for spares, manpower, tools or test equipment. This is what determines *operational availability* A_o which needs to include standby time in the uptime and all the factors included that contribute to the downtime. Only in this way will there be a realistic definition and measure of availability for equipment so that commanders will have an accurate assessment of availability for operational planning purposes. So:

$$A_o = \frac{\text{OT} + \text{ST}}{\text{OT} + \text{ST} + \text{TPM} + \text{TCM} + \text{ALDT}}$$

where OT—Operating Time, ST—Standby Time, TPM—Total Preventive Maintenance Time, TCM—Total Corrective Maintenance Time, ALDT—Administrative & Logistic Delay Time.

Only by using operational availability, A_o, can we include the often lengthy standby or dormant periods that most military equipment experiences between periods of high intensity use. Moreover, only then can all the many variables that affect actual repairs such as availability of resources be properly assessed. A_o is therefore dependent on both the amount and the cost of logistic support supplied; the greater the level of logistic support, the greater the operational availability that might be delivered. However, having sufficient spares to hand and manpower for every eventuality increases the cost of support. As the logistic support in terms of management and cost is increased, inevitably the operational availability delivered will tend towards the designed-in limit i.e., towards A_i. In reality of course there will always be some margin lost to delays and to essential active maintenance.

The MoD, however, can no longer afford the luxury of allocating vast resources to support its operations and is thus seeking ever greater efficiencies through CfA; it is often used to buy a total support package where the contractor owns the spares, provides the labour and is responsible for the administrative delay time. Setting requirements for CfA contracts is going to be difficult particularly so if it is operational availability that is required. If the contractor does not control all the support systems, including facilities, manpower, spares and the entire support chain, he cannot be held accountable for A_o as some parts of the support are entirely outside his control. Contracting for third party responsibility for such things as the efficiency and success of the supply chain is possible, but can be fraught with problems. Contracting for downtime is possible and traditional requirements might state *x% availability with maximum unscheduled downtime* $=$ *t minutes*. However, this is imprecise; Numerous factors can affect downtime if it is operational availability that is required. For example, maintainability requirements also need to be set and should use suitable metrics for both corrective and preventive maintenance; a metric such as mean maintenance man-hours per flying hour might be specified, but this needs to be defined at each appropriate level of servicing and also for scheduled and unscheduled maintenance. Alternatively it might be specified that the removal of a single component should take no longer than X hours or Y minutes. Those components which must be capable of being removed within a set period could be stated. Regarding manpower and tools, the number of maintainers might be stated and the use of standard or specialist tools specified. Yet these are not the only considerations when embarking on a strategy to reduce support effort with CfA. A variety of support factors need to be considered, such as maintenance policies, philosophies and processes being used by the customer and whether they are content to transfer them all to a contractor. If so, must the contractor comply with the user's policies? The geographical location in which the maintenance and repairs will be conducted and whether this will potentially put the support service in harm's way is a factor of rising significance and concern. Other considerations include: where will the spares need to be positioned and what effect will the communication system and transport have on the delivery of availability? A contractor seeking to deliver any CfA contract, and particularly one for operational availability, will have many significant investment factors to consider, for example: what investment is required for the provision of support resources, spares, facilities, tools and manpower and how will this affect the price of the proposal? How much investment will be needed? Lastly there are organisational factors to be considered. How will the flow of information be managed between the user and the support solution provider and will the arrangement cause delays in the effective delivery of the required availability?

To answer these questions shows the importance of understanding the large number of factors in the availability equation and the need to define exactly what sort of availability each party is actually talking about. Simply talking about MTTR, might have two very different meanings depending on whether it means total time including any delays, or just the time actively undertaking the repair. It must be a vital step to achieve a clear and common understanding of exactly what is included and what is not, otherwise there is no chance of getting anywhere near delivering proper CfA that meets both users' and suppliers' expectations. High reliability is ultimately vital to mission success but working to improve a system's reliability can be more expensive than implementing more simple measures to reduce delays in the area of ALDT. Therefore, when contracting for availability, it is essential to be very clear about who is responsible for the logistic downtime which will critically affect the delivery of operational availability to the user. All too often assumptions are made for availability with neither a clear and common understanding of the terminology and definitions, nor of the actual availability required and the possibilities that various support solutions could deliver.

The MoD user's aspiration, as one of their prime availability requirements, is a reduced logistic footprint. This will include all the aspects of Logistic Support, such as: the lines and levels of repair and the repair location, manpower—who and how many, where the spares will be kept and how many are needed, repair versus replacement which may affect turnaround times, and the physical location for the logistic support. There may be differences in peacetime and training support compared to the operational environment. Considering delay times, how these are measured needs to be established and what should be included: debriefing, planning, diagnostic and administration times and time awaiting manpower or other resources. Other factors must now also be considered and defined, such as the mission length or the period over which the availability is required. Thought should be given to what is the minimum, the maximum, the average or indeed the range of values that the availability must be kept within? The definition of the availability period might be time, distance or some other mission parameter such as number of rounds fired. Similarly there must be a clear understanding of who the user is; it may be a single user or a number of users. The boundary needs to be defined as to whether the availability refers to the fleet, an individual system, or a sub-system. Importantly, a definition of availability needs to be explicit as to whether it is inherent or operational availability, or whether it is some user-defined availability between these two extremes which will then need its own suffix. When determining availability requirements the frequency and value of the data points can have a major effect though on the perception of overall availability. For instance if there is a monthly availability return, the values might still provide the required average, yet at a particular moment during the month they might not be providing the minimum level required. Other aspects that must be determined are what frequency of measure is needed and what activity will be measured: a Battlefield Mission or Battlefield Day.[1] How the required data are captured and measured and how to decide what data are important to capture, is vital.

Once the correct definition of the required availability has been decided, there are a few more considerations to be dealt with for CfA contracts. Any CfA arrangement must benefit and incentivise both parties: MoD and contractor. It is unlikely that a successful partnership will be achieved without mutual trust and openness and it will be necessary to understand where responsibilities and contractual boundaries lie and how they will work in particular environments and locations. There are other questions to resolve. What is going to be measurable and will it deliver the information to demonstrate satisfaction and success? Will one solution such as "power by the hour" which works for one fleet, work for another type of equipment? Are the definitions for availability able to be clarified without any chance of misunderstanding? Who will be responsible for every aspect of the delivery of the availability requirement and in particular the supportability aspects? Will the military user need to retain maintenance experience and currency for the inevitable periods of operational commitment for which contractual support may not be achievable or desirable and for which organic, uniformed military capability will be necessary? Will just contracting for spares provision provide enough incentives for a contractor to design in reliability?

If all these questions can be answered satisfactorily, then requirements may have been developed which will have forced the risks to be examined in some detail. There are nevertheless still problems to be faced such as dealing with incomplete systems, the need for clear contracts, renegotiation points, breakpoints, incentives, gain-share options and longevity of the contract to provide sufficient incentive for suppliers to make the necessary initial investment.

13.3 MoD and US DoD Policy for Availability Contracting

MoD is being driven to be more efficient, not by competition but by budgetary reality, its *public service agreement* responsibilities to the taxpayer, political pressures and social pressures. A powerful driver for the outsourcing of support services and the adoption of CfA contracts in the MoD was the Strategic Target placed on the Defence Logistics Organisation (DLO) in 2000 to reduce the cost of its outputs by 20% over a 5-year period. However, demonstrating the risks this engendered, Charles Haddon-Cave QC, in his wide-ranging review following the

[1] In essence this is a description of the activities which a given combat element will execute and experience in achieving a mission or during a day of operations. It will include, for example, distance travelled (tracks and wheels), rounds fired, terrain covered.

crash of the Nimrod (XV230) in Afghanistan in 2006 stated, "There is no doubt, in my view, that dealing with the waves of organisational change and the cuts and savings stemming from the SDR and the 'Strategic Goal', proved a major distraction for many in the DLO, particularly during the period 2000–2006. The overriding imperative during this period was to deliver the cuts and change required" (Haddon-Cave 2009). In the case of the Nimrod crash it was a contributory factor in the sequence of events that led to the crash.

MoD and industry would maintain that CfA contracts enable agile and responsive acquisition, facilitating incremental acquisition, technology insertion, upgrades and updates of military equipment because industry has more control of the equipment when undergoing maintenance and repair. The arguments for CfA also include the benefits of long duration contracts giving industry the commitment and confidence to invest and rationalise. There are also mutual benefits to be gained from incentives and gainshare of any profit and efficiencies. Industry makes a persuasive case that they are best placed to incorporate new technologies, many of which may have a cost but which are more likely to be adopted to drive improved supportability. Industry is more likely than the MoD to invest in technology because it is less prone to have to redistribute money within an annual budget from its planned support expenditure to pressing operational imperatives. When industry gains experience in what the support costs are any savings it can make should be returned in additional profit and gainshare.

A large contributor to support costs, and inevitably availability, is inventory. It generates significant storage, transport and personnel costs and incurs cost of capital and depreciation charges; yet a large stock of spares in the right place will help deliver better availability. Having just what is required in the right place is obviously the aim and having complete control over inventory is an essential aspect of delivering better availability. Similarly a more intelligent approach to the whole aspect of support and maintenance by an improvement in engineering and asset management is required to reduce the downtime equipment many experience. Many approaches are available, including better diagnostic systems, Prognostics & Health Management, design improvements so there is an understanding of the *physics of failure*, and Health and Usage Monitoring Systems (HUMS). The MoD has been slow in adopting these techniques and processes, a likely cause being its response to the imperative for substantial financial efficiencies in the short term, which shaped its early focus on CfA initiatives. In the USA the experience has been markedly different, yet it faced essentially the same cost reduction goals and the aim to improve equipment availability. In 2001 the DoD identified Performance Based Logistics (PBL) as their preferred support strategy for military systems. PBL initially arose from efforts to reduce support costs; however, published policy and practice actually placed performance first. The US Defense Planning Guidance 2001–2005 (Department of Defense 1999) advanced cost reduction intent by setting a goal for each department to reduce operation and support costs of fielded equipment by 20% by financial year 2005. An area identified for cost savings was reducing demand on the supply chain through improving reliability and maintainability (R&M) of equipment. Arguably the DoD had been progressing this goal for many years under various initiatives. However, it is instructive to see how the USA has kept the aim of improving levels of R&M at the top of their requirements. Other areas identified were to reduce supply chain response time and increase competitive sourcing of product support. In the DoD Directive 5000.1 (Department of Defense 2003), they stated their requirement to implement PBL strategies *that optimise total system availability while minimising cost and logistics footprint*. The aim of PBL was to purchase *performance* by delivering operational availability, lowest cost per unit of usage, operational reliability, minimal logistics footprint and lowest logistics response time. They made the distinction between owning and managing inventory. It was asserted that if Defence retains ownership and contracts industry only to manage inventory, industry is not inclined to invest to improve R&M because, instead, it just buys more inventory which Defence has to pay for. Concurrently the UK experience was coming to the same conclusions. Inventory optimisation had progressed as far as industry could go without taking full responsibility for it. R&M had improved but only where realistic and affordable metrics had been specified. Alas, MoD did not understand the support cost drivers as it owned few good systems for collecting and managing the performance and usage data. Indeed in many cases the data just did not exist.

13.4 Support Contracting for Availability

The traditional support contracting model usually encapsulated three separate contracts: a repairs contract, a spares contract and a Post-Design Services (PDS) contract. The repairs contract, typically a firm-price arrangement, negotiated to cover just a few years, was based on past equipment failure rates adjusted for future predictions based on modifications, usage rates and known deployments. The contractor's cost breakdown structure would be assessed to ensure they were not making excessive profit. Equipment was fed in on a *per-arising* basis, to a contractually agreed maximum. As the contractor was paid per-arising there was no incentive to reduce the number of arisings. MoD, under obligation to reduce costs, had little leverage that could be applied other than postponing the inevitable or, for example, delaying in-year work and paying for it later thus creating a backlog. The contractor, solely responsible for the repair, left all other ownership of the supply chain falling to MoD. There was no incentive for MoD to reduce stocks of spare parts because it was money already spent and the stock served as an insurance against items becoming obsolete. The spares contract provided the spares that would be used to do the repairs. It drew on historical demand data. Spares were typically bought in batches to smooth spending profiles in-year and to attract a discount for quantity. Often suppliers would only agree to produce whatever was deemed to be an economic quantity, so MoD would end up with more than it required anyway. Platform or equipment specific spares would be bought by the Integrated Project Team (IPT) and common use items, which were

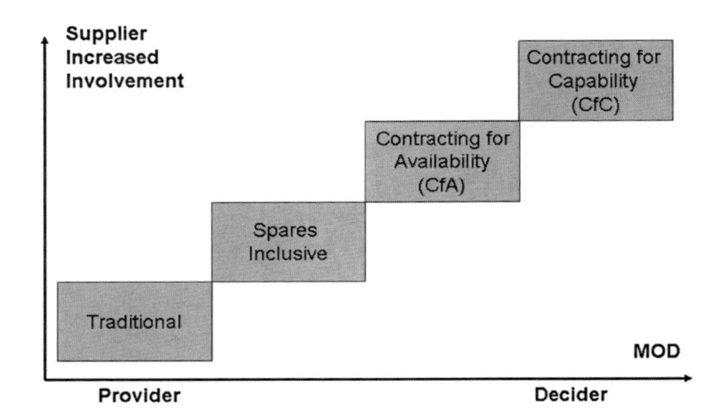

Fig. 13.1 The MoD transformation staircase (Ministry of Defence 2005a)

generally more readily available, would be bought by a commodity IPT. Under the spares contract, suppliers would be under no obligation to reduce spares utilisation, to manage obsolescence proactively, or to develop alternatives. As MoD was responsible for spares procurement, any non-availability gave the contractor a good excuse for failure to meet a repair schedule or milestone. The final contract was for PDS. As it was always difficult to predict the demand for post-design activities, PDS contracts were something of an insurance policy, usually an expensive one, against the unexpected or unforeseen. They ensured that the design authority (DA) retained up-to-date drawings and specifications thus enabling the MoD to task them with defect investigations, repair schemes, technical improvements and modifications. PDS contracts were often used to cover a multitude of other requirements which were not funded elsewhere and so were a very useful enabling tool.

Having three such contracts for each platform or major equipment inevitably led to fragmentation and incoherence. There was lack of visibility of the whole support chain and the various dependencies and linkages, the cost drivers and how to manage them. There was also a culture of lack of responsibility as it was always someone else's fault or problem. In trying to illustrate the need to evolve from these traditional support arrangements to models which gave greater involvement to industry and grew MoD's role as "Decider rather than Provider", the DLO proposed a graphic, the transformation staircase which illustrated, simplistically, four contractual positions (Fig. 13.1) (Ministry of Defence 2005a). Some IPTs interpreted this, incorrectly, as a directive to move up the staircase towards Contracting for Capability rather than to be *in the best place* contractually at a given point in the lifecycle. This misunderstanding persuaded the DLO that this graphical illustration should no longer be endorsed. However, it is important to understand the basic differences between the four broad arrangements it describes. In the traditional model, industry delivers assets to MoD. In turn MoD delivers support through life using its own organic capabilities. Industry sells spares to MoD but industry is not capable of delivering any benefit through more extensive ownership of the support chain and its cost and performance drivers. In the spares inclusive model, MoD outsources the ownership and management of spares. MoD

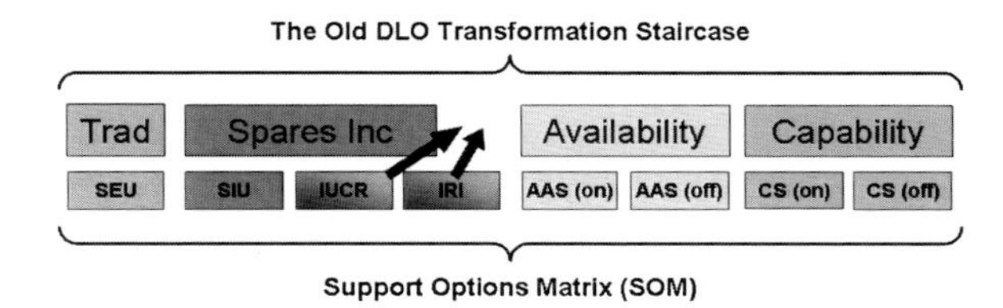

Fig. 13.2 The MoD support options matrix

then benefits from industry's ability to buy spares at lower cost and its ability to plan spares requirements more effectively than MoD. MoD saves on inventory costs and charges by holding fewer spares and on the administration costs of spares provisioning. It also enables incentives to be agreed for reducing turnaround times and opportunities for capital spares and asset disposals or purchase avoidance of spares not required. CfA is the next model, where the supplier is responsible for delivering platforms and equipment to an agreed performance and output standard. It is usually described in terms of availability of equipment ready for operation, but might also be described in a number of hours or days of availability. CfC is the most comprehensive model where the supplier is responsible for providing the whole capability and outputs to agreed performance standards. Conceptually this model presents real challenges, primarily around how capability is defined and how realistic it is to expect the contractor to be responsible for all of its constituent parts and to be contracted to be so. A more sophisticated model than the transformation staircase, and one which is in favour and still valid, is the Support Options Matrix (SOM) (Smith 2009). Its purpose is to:

- present the range of support options with more granularity,
- provide a clear rationale behind each support option,
- help a project team to make the right decisions for a given support chain,
- provide clarity of requirement to Industrial partners.

As such, the SOM is intended to help an IPT move to the *best place* in support contracting. It provides more options—as illustrated at Fig. 13.2 and explained in more detail below—and a project team leader is able to appreciate the nature of support cost drivers and who is best placed to manage them—MoD or industry. Figure 13.2 also shows how the simple transformation staircase and the SOM can be matched together with many more options now available. These are as follows:

- Spares Exclusive Upkeep (SEU). MoD defines the "upkeep" policy, the requirements and plans. Industry completes "depth" upkeep which MoD pays for. MoD pays for all spares used, including those consumed in depth, and pays for industry's labour and overheads,
- Spares Inclusive Upkeep (SIU). Industry develops a spares provisioning capability, plans and procures spares for depth "upkeep". MoD should benefit from industry's ability to plan spares more effectively and buy them at a lower cost,
- Incentivised Upkeep Cost Reduction (IUCR). Industry is incentivised and given the freedom to change the support, repair and test regime to reduce the cost of

spares. This would require a longer term contract and precise MoD forecasting to assess demand,
- Incentivised Reliability Improvement (IRI). Industry is incentivised to develop and implement improvements to the reliability of the equipment,
- Asset Availability Service (on Balance Sheet) (AAS (on)). Industry is responsible for providing MoD with a specified level of equipment availability on assets that remain on the MoD's balance sheet,
- Asset Availability Service (off Balance Sheet) (AAS (off)). Industry is still responsible for providing MoD with a specified level of availability but capital assets are not on the MoD's balance sheet,
- Capability Service (on Balance Sheet) (CS (on)). MoD contracts with industry to provide a defined capability and has ownership of defining capability upgrades. MoD is responsible for ensuring compliance with safety legislation,
- Capability Service (off Balance Sheet) (CS (off)). The MoD contracts with industry for the provision of a defined capability which, because there are third party users, requires that the capital assets are not on the MoD's balance sheet.

Given the much better granularity of the SOM, there are some generalisations and issues to be addressed. MoD is moving to a position of being only interested in buying performance outcomes. The contractor then receives payment in return for agreed levels of performance; this might be numbers of platforms available to the military for tasking or the completion of maintenance actions within specified time frames. MoD and the contractor need to agree to a common understanding of what is required; performance standards, therefore, need to be carefully specified. Undoubtedly the contractor is made more accountable with CfA, whereas before in the traditional contracts it was a transactional relationship, where MoD initiated and managed the transaction, telling the contractor "how it was to be done".

CfA is all about strategic partnerships, as espoused by MoD in the Defence Industrial Strategy (Ministry of Defence 2005b), demanding the long-term commitment from both parties that enable strategic investment, rationalisation and decisions which both must make to be more efficient and effective. Generally, it takes a long time to reach a baseline from which a CfA arrangement can be negotiated and activated and it must be allowed to evolve and mature. Many of the contracts have employed a "soft-start" where risk and the amount of work are gradually transferred to the supplier over a period, typically a year or two. Similarly, the gain-share arrangements are often not a simple straight-line graph, but have thresholds for changes in gradient or step-changes and also some absolute limits. Things may, however, go wrong and when they do the first reaction should not be to reach for the contract but, rather, to work together to find a solution. Successful partnerships will deliver the aspirations of both the user and the support provider but require a code of partnering behaviours to be developed and agreed; effectively there should be an ethos of shared values, information and vision—"need to share" replacing "need to know".

The basic model for CfA is for MoD to have fewer, more rationalised, contracts. The prime contractor is left to manage relationships with his suppliers and performance measurements and management arrangements are agreed and applied by both parties. CfA and CfC only really attract those with broad shoulders who are adept at providing integrated solutions. Pricing structures will also be more complex, perhaps with a non-variable element which is a guarantee of income to the contractor, together with a variable element for performance above the requirement specified. A lesson from all the current CfA arrangements is that incentives are required to motivate successful delivery. In many CfA arrangements MoD and contractors are working together, physically co-located and following common processes, sharing information and taking a shared approach to problem solving. In many respects with CfA, risk is transferred to industry and there is a financial cost attached to this; it should never be forgotten, however, that ultimately operational risk lies with the MoD!

13.5 Comparison of Contract Options: Fixed Wing

The first significant CfA contracts originated in the air environment and have grown in number, encapsulating fast jets, transport aircraft and helicopters. Aircraft simulators have been highlighted where commercial operators took over the complete provision of the simulator service and its support. The RAF Simulator technician trade-group was disbanded and many technicians were immediately re-employed by the commercial operators. Also aircrew instructors were recruited from ex-RAF aircrew. These contracts were not initially labelled as CfC although they do in fact provide the complete capability which is implicit in the term.

In the fixed wing area MoD/BAE Systems/Rolls Royce negotiated the ATTAC contract and ROCET, the parallel contract for the RB199 engine (which powers the Tornado). The Harrier Platform Availability Contract (HPAC) valued at £574M was signed in April 2009 and incorporates the Harrier Joint Upgrade & Maintenance Programme (JUMP) which finished in late 2009. The Typhoon Availability Service (TAS) contract, valued at £450M, was signed 4 March 2009 (BAE Systems 2009; National Audit Office 2007).

Before these arrangements, all the corrective maintenance and most of the preventive maintenance was done by the RAF at each operating base; the only exception was the Major, or largest of the scheduled servicings, often including large modifications or upgrades, which was usually done by industry in their facilities. A major change occurred in the period 2001–2006 with these contracts requiring a major rationalisation of RAF engineering and support and leading to industry taking over much of this work at regional centres, or hubs, with RAF personnel concentrated with the operational squadrons. LEAN[2] manufacturing principles were applied, in which aircraft are "pulsed" along a line incorporating

[2] The RAF have adopted LEAN principles in its maintenance operations as a way of reviewing the way it operates in order to drive out waste in all its processes and maintenance practices.

on-platform and off-platform maintenance and upgrade tasks. Reporting in July 2007, the NAO noted the savings and improvements the ATTAC and ROCET initiatives had generated (National Audit Office 2007):

- The pulse line at RAF Cottesmore reduced time for a Minor servicing by 19%,
- The JUMP programme (combining the Minor and Major upgrade programme) had reduced the time to convert Harrier GR7 to GR9 standard by 43%,
- Improved repair processes on the pulse line for Harrier Pegasus Engine have reduced turn-round times by 59%,
- The Tornado GR4 pulse line reduced Minor servicing repair times by 37% compared with previous times at the Defence Aviation and Repair Agency,
- ROCET has delivered 100% availability of the Tornado RB199 engine, reduced rejection rates and reduced numbers of engines in depth repair which generated savings of approx £12M in its first year, shared on a gainshare basis,
- Tornado and Harrier Integrated Project Teams' costs have reduced from £711M in 2001–2020 to £328M in 2006–2007, a cumulative saving of £1.4Bn.

The NAO also highlighted some of the difficulties associated with the fast jet CfA arrangements. The contracts had all retained some elements of RAF manpower providing either some specialist services or some skills. It was difficult to establish why this had occurred but the belief was that it was to retain the specialist technical skills available for operational deployments. At the time of the Report, however, MoD had not been able to meet its contractual commitments for military manpower for either the Tornado or the Harrier pulse lines. The skill and experience mix of the workforce, therefore, did not match the plan or the requirement. It was also probable that MoD did not have sufficient commercial, cost modelling and project management skills to have developed wholly successful and commercially viable support solutions in the first place. There had inevitably been an aversion to risk and concern about losing direct RAF manpower involvement in some of the depth tasks and this attitude shaped some of the contractual arrangements. Whether MoD also had the skills and expertise necessary to negotiate ever more complicated support arrangements and contracts, given the increasing complexity and likely volume of industrial logistics support it needed, is an open question. MoD has since made some effort to up-skill its personnel in these areas but still lacks sufficient personnel with the relevant experience and skill levels to equal their industry counterparts (National Audit Office 2007).

There are many supply chain issues which CfA brings to the fore. The NAO commented on pulse line delays at both RAF Marham (Tornado) and RAF Cottesmore (Harrier) being attributed largely to failure to deliver spares on time with on-time averages for Tornado at 81% and for Harrier at 72%. For sourcing of spares, the NAO noted that the Harrier IPT controlled 77% of spares required by the JUMP pulse line, but had limited influence over activities of commodity IPTs (National Audit Office 2007). These commodity IPTs and their suppliers, further down the supply chain, can have a serious impact on spares availability. Under CfA, industry should ideally take on full responsibility for all spares provision and the question needs to be addressed as to whether it matters to MoD if it loses this hitherto direct relationship with its commodity suppliers. The key issue is the risk involved in delegating responsibility to a prime and how confident MoD can be that the prime will be able to manage its lower tier suppliers satisfactorily.

13.6 Comparison of Contract Options: Rotary Wing

The other air capability area that has made significant progress in availability contracting is rotary wing. All helicopters are now supported via a form of CfA partnered contract using a generic model called Integrated Operational Support (IOS)—the Sea King Integrated Operational Support (SKIOS) and Integrated Merlin Operational Support (IMOS) being two examples. MoD and AgustaWestland (AW) have established a Strategic Partnering Arrangement (SPA); the term *arrangement* adds substance to the principle that what matters is how the partners work together to resolve difficulties rather than resorting to contractual, legally binding, mechanisms. It covers 10 years' support for all rotary wing assets for which AW is the Design Authority (DA) but is reviewed every 2 years. Focused on business transformation as a key enabler, it is encapsulated in a MoD/AW Business Transformation Incentivisation Agreement (BTIA) where Key Measures of Success include: operational serviceability, performance, cost and time improvement, relationship maturity, DA effectiveness, health and sustainment (Pay and Collins-Bent 2008). The dynamics of partnering are laid down together with the process of setting appropriate milestones to drive transformation; this covers the top-level requirements for aircraft availability, schedule adherence, responsiveness, and sustainable on-shore DA skills. The BTIA also covers the essential details of cost, establishes where there are clear dependencies, the cost retentions and how rewards are balanced at a 3 years review. AW maintains that the SPA and BTIA have been developed to overcome the challenges of protecting military flexibility whilst improving industrial predictability (Pay and Collins-Bent 2008).

In practice the IMOS and SKIOS contracts are priced with an element linked to flying hours. There are non-variable and variable elements of pounds per flying hour (££s/fg h). For IMOS, the price, from contract initiation is valid between 70 and 120% of the MoD target of flying hours. With this arrangement the more the Merlin Mk1 and Mk3 helicopters fly, the more turnover and contract value that industry achieves.

With the payment and profit mechanism based on ££s/fg h, the incentivisation mechanism works on both a *penalty and a bonus* mechanism. If hours are under 70% then there is a penalty and if hours are between 70 and 120%, there is an increasing incentive. There are also Key Performance Indicators (KPIs) for the operational fleet size and the Merlin Mk1 training system availability, but the principal one is for aircraft serviceability. In general the IOS contracts rely on good management information systems, meaningful KPIs, incentivisation mechanisms and partnering or "teaming" together with other management arrangements. They provide for significant value leverage and have delivered recognised

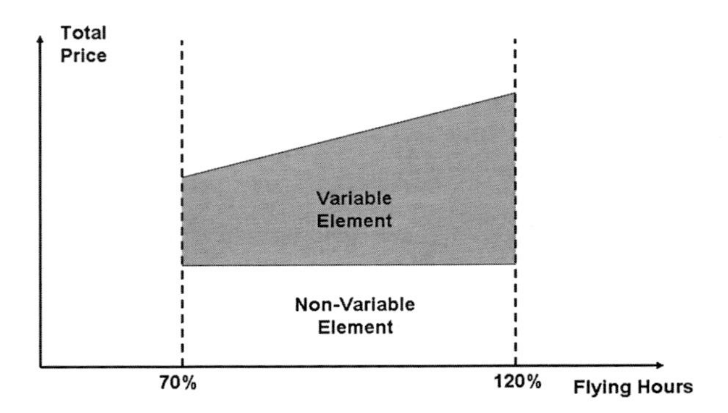

Fig. 13.3 Merlin contract price for flying hours (Pay and Collins-Bent 2008)

savings over previous arrangements. Some of this has undoubtedly been as a result of the re-organisation of Royal Navy (RN) and RAF maintenance arrangements and the centralisation of depth servicing. The fixed price contract with incentives has also encouraged the contractor to generate and incorporate modifications or system enhancements to improve reliability and to achieve contracted fleet availability whilst reducing maintenance and recovery effort by the Prime and sub-contractors (Fig. 13.3).

Under the SKIOS contract AW have gone further for the Search and Rescue (SAR) fleet by taking on responsibility for providing 1st line maintenance activities for the SAR Sea Kings at 8 bases (7 in UK, 1 at the Falkland Islands). This departure from normal operational support arrangements, where the RAF carry out the 1st line support, means this arrangement is tending towards CfC although not actually reaching genuine CfC status because the Sea Kings are flown by RAF aircrew. It is possible because the SAR units are fixed to provide 24 h SAR cover and are, therefore, not subject to the same need for flexible deployments.

Many lessons are being learned from the AW experience. There are joint teams at all levels of management in both technical and commercial areas allowing the contractor to be able to see and understand how and why the customer needs to "move the goal posts" as operational needs change. The customer must also have sight of the contractor's problems in meeting availability and be aware of the activities that the contractor may propose to limit his liabilities, for example modifying equipment to improve reliability in order to reduce the contractor's repair costs within the fixed price being paid. There needs to be due diligence in the technical issues based on a mutual and shared understanding of the risks and benefits. Likewise there needs to be logistics due diligence, which means a shared understanding of the repair and supply loops for the contractor's support strategy.

13.7 Comparison of Contract Options: Maritime

The RN used to have five Island Class Offshore Patrol Vessels (OPVs) which were commissioned 1976–1979 with the role of fishery protection and oil and gas field protection. Additional roles were added by the late 1990s for assisting HM

Customs and Excise, scientific and environmental roles and assisting vessels in distress. It was soon realised that the Island Class vessels did not have the capabilities to perform all these roles effectively and they were also proving costly to maintain and support. In December 2000 an Invitation to Tender (ITT) for the replacement OPV capability was issued to eight UK shipbuilders. MoD was looking for innovative ways of delivering the capability. In March 2001 Vosper Thorneycroft (VT) was selected as the preferred bidder and the contract was placed in May 2001. The vessels entered service during 2003. The acquisition period, just 3 years, was remarkably quick, particularly so for ships, and there were several reasons. Firstly, the Island Class ships had become too expensive to maintain and there was considerable pressure to reduce the short-term support costs. Gapping the capability was not an option so a new, cheaper solution was urgently required. Senior MoD management were under pressure to deliver early savings to prove the worth of the new Smart Acquisition process. The OPV fleet provided an opportunity to deliver clear short-term savings and demonstrate to politicians and taxpayers that Smart Acquisition was a success. In December 2000 VT had been forced to issue redundancy notices to much of its workforce and there was consequently considerable political pressure to ensure that VT won the contract with an order for at least three ships to maintain these jobs and to keep its shipyard open. As a result of winning the contract, VT was able to reduce job losses to 120.

The acquisition was a breakthrough in terms of CfA for MoD as it involved VT leasing three ships to the RN with support and maintenance provided through a Contractor Logistics Support (CLS) arrangement. VT retained ownership of the vessels and chartered them to the MoD for 5 years, with a daily charge for complete logistic support. VT also guarantees 960 days availability per year across the three ships. Only when those ships are available for MoD tasking is the MoD liable for CLS costs. In the first year of operation, the three River Class OPVs achieved 97.5% availability, compared with a maximum of 82% for the five, less reliable, Island Class they replaced. There are other advantages as well. The new vessels are more capable; they are also 30% larger. They are more fuel efficient and have a smaller crew, 30 compared with 35 on the Island Class. Total RN manpower involved in their operation has consequently reduced. Based on this success, a further vessel, similar in design, was acquired as the Falkland Islands Patrol Vessel. At the end of its initial 5-year lease the MoD decided to extend the lease for a further 5 years, out to 2013, and with the same CLS arrangements.

The fact that uniformed RN crews are part of the capability and are provided and paid for by MoD, puts this arrangement into the CfA contracting model. There is no doubt, however, that it has been delivered successfully and has generated savings for MoD. Such contracts would seem to be most suited to relatively simple, commercially available technologies used in routine, predictable operations such as fishery protection, rather than front-line warship capabilities. The contract incentivises the company to meet and exceed the capability and availability requirements. Leasing may not always be the cheapest option however, but it does provide smooth, predictable yearly budget requirements where the risk is transferred to industry. It enables a "try before you buy" approach that also reduces risk significantly. The counter argument of course is that support costs are usually significantly lower in the first few years of operation and should MoD then transfer ownership of the OPV class, it is in the next 20 years that support costs could be expected to increase. Another advantage that had encouraged the leasing option was that under the Resource Accounting and Budgeting processes which MoD operates as a public sector department, it was beneficial not to hold the ships on its own balance sheet. There has been some debate, however, as to whether it is acceptable to hold what are classed as National War Fighting Assets off balance sheet. If it becomes policy that this is not acceptable and MoD then has to buy them the balance of financial benefits will change.

13.8 Comparison of Contract Options: Land

The provision of availability of vehicles and systems in the land environment presents different challenges and less progress has been made than either air or maritime. The MoD examined how to combine its concept of Whole Fleet Management with a whole fleet support arrangement for their armoured fighting vehicle (AFV) fleet. There are compelling advantages to retaining a UK industrial AFV capability and a whole fleet support solution would underpin that capability by providing a maintenance and upgrade capability for current and future equipment both in peacetime and operational theatres. The project, known as Armoured Vehicle Support Transformation (AVST), aimed to move land vehicles towards CfA solutions type by type and progressively towards the optimal solution for each vehicle fleet. One arrangement, already in place and informing development of AVST, was the AS90 Equipment Support Agreement (ESA). This was a single 5-year CLS contract with BAE Systems to supply spares and repairs on demand over the 5 years, together with technical support, for the AS90 155 mm Self-Propelled Artillery System. AVST's intended benefits were to develop an engineering logic for hull/turret structural integrity and improve materiel availability by aligning the industrial supply chains. It was also intended to reduce costs by collecting and utilising data to reduce failure rates, improve demand planning, improve inventory and optimisation. AVST sought to apply continuous improvement by learning from experience and applied a structured approach to obsolescence management which is a critical issue for many military equipment. The changes expected would certainly deliver platform efficiencies by having a common requirement, developing a single solution and using it multiple times for the many different vehicle fleets. The significant number of vehicles in each AFV type makes the need for cohesive and workable solutions very important. AVST has yet to deliver on the challenging requirement to provide a coherent service through-life whilst providing improved vehicle availability and faster response times to the Front-line Command.

13.9 Conclusion

By dealing with the theoretical issues of availability first it has been shown that great care must be exercised when defining the exact availability required and by whom. There is much potential for confusion in what exactly can be contracted for in availability due to the downtime being misunderstood by both the customer and support solution provider. Nevertheless, CfA and CfC are the most recent popular service support solutions championed in the Defence environment. CfA has developed from what in the past was known in its more limited form as Contractor Logistics Support and a number of variants of the CfA model have been developed, the IOS model in the rotary wing domain being one example. What these new over-arching concepts do is to bring the whole support solution together in a partnership for the delivery of better operational availability. Yet what is unclear is the real connection with reliability, maintainability and logistic support. The engineering inputs and connections to these disciplines need to be re-emphasised and re-iterated. They each effect operational availability but it is unclear to many who require the simple delivery of operational availability just what these effects are and how important each one is. MoD policy is to re-evaluate projects and select support solutions that deliver the required availability and the most effective support rather than a progression from traditional arrangements to CfA and CfC. The SOM provides a greater range of options and provides a process for projects to determine the best option. In a number of examples it has been shown that there are already many ways of delivering CfA and CfC although few are perhaps true CfC. Indeed the debate as to what defines the whole capability is one that justifies a separate chapter in its own right. Nonetheless, it has been possible to show some of the pitfalls that must be avoided and issues that must be resolved to deliver successful and innovative support solutions which are meaningful and useful to the user. This would be a beneficial area for further research both to encapsulate the many and varied options, but also to expand the understanding of the SOM and to provide more definitive guidance to MoD and commercial organisations alike.

Despite the potential pitfalls and lack of a clear understanding of the real user requirements for availability, the move to CfA and CfC has been shown to be beneficial for the user, the wider MoD and for companies who provide support solutions in the defence environment. For the MoD, lower overheads and clarity of in-year budgeting are but two of the value creations. For industry, increased revenue from increased yet efficiently run business will generate more profit.

13.10 Chapter Summary Questions

The chapter has identified the many misunderstandings in trying to define availability and capability contracts in the Defence environment. It has compared what is happening in practice across the air, sea and land environments and shown there is no single format that is successful. It raises the following questions:

- How do we re-connect the essential attributes of reliability, maintainability and logistic support with the delivery of operational availability?
- What are the essential attributes and principles for a successful availability contract and an innovative support solution?
- Can contracts for capability be realistically delivered for defence equipment?

References

BAE Systems, BAE Systems welcomes new £574 million Harrier availability contract. BAE Systems News Release. Ref 067/2009, 23 April (2009)

British Standards Institution, BS 4778-3.1 1991. Quality vocabulary Part 3—Availability, reliability & maintainability terms (1991)

Department of Defense, Defense planning guidance update for fiscal years 2001–2005. Washington, Apr 1999

Department of Defense, DoD Directive 5000.01 [Online] Available from: http://www.med.navy.mil/sites/nmrc/documents/dodd_5000_1.pdf. Accessed 1 Dec 2009 (2003)

C. Haddon-Cave, The Nimrod review. Page 397, Para 13.174. ISBN: 9780102962659 (2009)

C.J. Hockley, Availability [Course notes], College of Management and Technology (2009)

Ministry of Defence (MoD), Defence logistic organisation, Plan 2005 (2005a)

Ministry of Defence (MoD), Defence industrial strategy. Defence white paper (CM6697). (Her Majesty's Stationery Office, London, 2005b)

Ministry of Defence (MoD), Defence standard 00-49 guide to R&M terminology used in requirements, Issue 2, June (2008)

National Audit Office, Transforming logistics support for fast jets. ISBN 9780102947304 (2007)

G. Pay, N. Collins-Bent, A partnered approach to contracting for availability—Lessons learned by MOD/AgustaWestland on IMOS & SKIOS. [Presentation] Contracting for availability and capability symposium, Shrivenham, June 17–18 (2008)

J.C. Smith, Maintenance Support Strategies [Course notes], College of Management and Technology (2009)

Chapter 14
Enabling Support Solutions in the Defence Environment

Christopher J. Hockley, Adam T. Zagorecki and Laura J. Lacey

Abstract Health and Usage Monitoring Systems (HUMS), Reliability Centred Maintenance (RCM), Condition Based Maintenance (CBM) and Prognostic Health Management (PHM) are enablers for engineering and support planning and are not being exploited to their full potential in the military environment. This chapter explores the nature of the techniques and the challenges for their adoption in the military environment. It shows that there is a connection not only between engineering solutions that involve one or more of these techniques which aim to provide effective support solutions, but that there is also a compelling case for their adoption to improve operational availability to benefit both the user and those who provide support solutions. The chapter first reviews the nature of failure and the consequential need for maintenance. It then reviews the techniques of RCM and CBM before looking at the processes of HUMS and PHM. Operational availability and its constituent parts and enablers are not commonly understood by either the user community or the support solution provider. Consequently HUMS and Prognostics are not yet generally recognised as being able to improve operational availability and make support solutions more effective. The benefits of RCM and CBM on in-service equipments are likewise not being exploited fully.

C. J. Hockley (✉) · A. T. Zagorecki · L. J. Lacey
Department of Engineering Systems and Management, Defence Academy of the UK, Cranfield University, SN6 8LA, Shrivenham, UK
e-mail: c.hockley@cranfield.ac.uk

A. T. Zagorecki
e-mail: a.zagorecki@cranfield.ac.uk

L. J. Lacey
e-mail: l.lacey@cranfield.ac.uk

I. Ng et al. (eds.), *Complex Engineering Service Systems*, Decision Engineering,
DOI: 10.1007/978-0-85729-189-9_14,

14.1 Introduction

In this chapter the state of the art for predictive maintenance and repair in the UK Armed Forces is reviewed. The chapter looks at the nature of failure before reviewing the philosophies and capabilities of Reliability Centred Maintenance (RCM), Health and Usage Monitoring Systems (HUMS), Condition Monitoring (CM), Condition Based Maintenance (CBM), and Prognostic Health Management (PHM). The discussion focuses mainly on the Army, Royal Navy (RN) and Royal Air Force (RAF) with references to civilian industry and US armed forces where necessary. Current practices for maintenance within the UK Armed Forces and opportunities to improve these practices through technology are identified. Which technologies should be applied to increase the effectiveness of all forms of maintenance are investigated. The statement *technology can help,* is not taken for granted, rather it is considered as a proposition leading to the investigation of how it can be achieved.

14.2 Reliability Practice and Modelling Failures

The approaches to maintenance have varied greatly through the evolution of systems and their increasing complexity. Until approximately the 1950s the approach to maintenance was very simple, whenever something failed it was repaired. In the 1950s a new strategy became dominant, pre-planned maintenance which incorporated the idea of preventive maintenance. In the 1980s new strategies started to be developed; their aim was to increase reliability and reduce the maintenance burden in terms of cost and time. The new techniques included: failure modes and effects analysis (FMEA), usage and condition monitoring and design for maintainability. Two factors played a critical role in this evolution: constantly increasing complexity of systems resulting in a growing difficulty in supporting maintenance; and increased understanding of the nature of failure—both the physics of failure and the statistical aspects. The increased understanding of the nature of failure also additionally led to an understanding of the linkage to the safety and economic consequences of failure.

In recent decades information technology has offered a new array of tools and processes to support maintenance and logistics. The new developments include automated monitoring systems that utilise a wide range of sensors, and modelling techniques that allow understanding and prediction of the physics of failure (PoF), optimisation of maintenance regimes and planning of maintenance to avoid unexpected failure. The leading institution researching PoF is the Center for Advanced Life Cycle Engineering (CALCE) at Maryland University where over 100 researchers are employed on PoF and research into prognostics to improve the level of understanding of failures and to produce methods and models to help in design.

In terms of maintenance, military equipment is especially challenging, much more than in the civilian sector. Military equipment is used in many different and extreme environments, sometimes outside of its original design specifications. Ministry of Defence (MoD) equipment is also not used in a uniform manner; there are heavy periods of operational or training use and there are often long periods of standby or storage. It is, therefore, questionable if modelling techniques from civilian applications that assume continuous and consistent patterns of usage are applicable. Many modelling techniques for reliability assume that equipment is used, if not continuously, then at least consistently—an assumption which should not be made about MoD equipment. At least three generic modes of usage can be identified for military equipment.

- Peacetime usage. Typical usage during peacetime is likely to see the equipment used in a range of training activities. Some of these could be relatively gentle in nature; others could be considerably more robust. An armoured unit undergoing live-fire training in the dust of the British Army Training Unit in Suffield (BATUS) Canada for example, will be deliberately pushed to extremes. This reflects the fact that all military training is designed, ultimately, to prepare service personnel for potential operations. During peacetime usage this heavy usage will be followed by established maintenance activity and then possibly storage. It would be wrong to assume, however, that all equipment in a particular fleet experience similar peacetime usage. The existence of fleets within fleets in which different equipment are modified to suit training requirements or particular roles or theatres is common.
- Storage Periods. The MoD has procured enough equipment for its operational capabilities and has now instituted Whole Fleet Management (WFM) for vehicles in order to ensure more equitable usage for all equipments within some of its fleets. A part of the fleet will be used for training and operations while the rest is kept in store and then the active fleet is rotated through the storage facility. These storage facilities will in future hold most of the stored vehicle fleet in Controlled Humidity Environment facilities which are known to reduce degradation.
- Operational Usage. In theory, operational usage is usually synonymous with war-fighting which is characterised by short periods of intensive, and to an extent, unpredictable activity. On such operations, the command imperative is absolute and the commander will adapt usage, tactics and implementation of procedures to counter the threat that prevails at the time whilst ensuring the delivery of operational success. Commanders may override any established or approved maintenance regimes and demand that equipment are used in novel and unforeseen ways; they would then call upon the improvisation skills of maintainers in the field to return equipment to full availability as soon as possible. What may be considered equipment abuse in peacetime may well be vital and necessary on operations. Modelling failures for war-fighting operations can, therefore, be extremely difficult, if not impossible. Other operational usage such as fishery protection or maritime survey will be "operational" and equally demanding but without the same pressures that war will generate.

These three quite generalised usage patterns are likely to have applicability across most MoD equipment but must be used with care. Not all equipments are used on operations and some will experience quite predictable usage patterns. Fishery protection vessels, for example, are unlikely to be used in any other role and thus their usage patterns can be relatively easily predicted.

The military environment is by its very nature demanding and hostile for equipment yet perhaps no more so than in the mining or construction industries. Yet military equipment seems to be less reliable than its commercial equivalents. This is often the result of being used outside its design specification, an issue that needs to be addressed by the MoD. Unfortunately the reality is that the defence environment and doctrine are constantly changing and the expected use and specification will inevitably be out of date by the time the equipment enters service. If specifications were originated to cater for all possible eventualities, an already expensive solution would be un-affordable. However, a contributory factor to the unreliability is the range of specialist equipment needed and environments that military equipment must operate in, both of which inevitably increase the complexity and the weight from the original design specification; adding more armour because the threat increases is a simple example here. Of course there is frequently a justified need to use the equipment in different ways and environments than it was designed for. Additionally another contributory factor to the unreliability is often the frequent modification of the equipment beyond its original design intent; Warrior and CVR(T) vehicles are two of many possible examples. Another influence on reliability is the level of equipment care provided by the user. The military environment again poses special challenges and military operators are well trained but naturally perceive military equipment as inherently robust and expect it to be tough enough to withstand severe handling in demanding environments. Yet often the required infrastructure for maintenance is not available on deployed operations and this can adversely affect reliability. The challenge for Whole Fleet Management (WFM) within the Army now is to overcome the reduction in a sense of ownership that WFM brings with the shared training fleets that result. It is not until users are issued with equipment in which they might have to go to war that the sense of ownership and total equipment care returns. Poor levels of husbandry care result in low availability and the Army tries very hard to promote higher standards of equipment care through its monthly Kit! Magazine (British Army 2010). It is nevertheless an on-going challenge.

14.3 Reliability Centred Maintenance (RCM)

RCM is an approach intended to improve maintenance policies and practices. It was first introduced in the USA through a study initiated by United Airlines who were concerned at the huge amount of scheduled servicing that would be required with the introduction of the Boeing 747 compared to their existing fleet of 707 s. The process was called MSG after the Maintenance Steering Group that formulated the process and in the airline industry is still used today as MSG-3 (Air Transport Association 1993). RCM and its industrial derivative process RCM2, developed by Moubray (2005) is very similar to MSG-3 and seeks, through a structured process, to develop the optimum maintenance requirements for the equipment or plant. Today's RCM is defined in the USA by the SAE JA1001 (Society of Automotive Engineers 2006) standard and in the UK military by AP 100C-22 (Ministry of Defence 1999) for the RAF and aircraft and by Defence Standard 00-45 (Ministry of Defence 2006) for the Army and RN. It is a powerful tool if applied correctly and ensures that the most appropriate schedule is developed for equipment coming into service and that experience of using the equipment in the real world can be used to modify the schedule once equipment have been in service for a time.

The RCM methodology is in fact a systematic approach which can be used to create cost-effective maintenance strategies to address dominant causes of equipment failure. RCM emphasises predictive maintenance in addition to preventive maintenance and determines the maintenance requirements by asking 7 questions:

- What are the functions and associated performance standards of the asset in its present operating context?
- In what ways does it fail to fulfil its functions?
- What causes each functional failure?
- What happens when each failure occurs?
- In what way does each failure matter—what is the impact?
- What can be done to predict or prevent each failure?
- What should be done if a suitable proactive task cannot be found?

RCM therefore determines the appropriate maintenance action for a given component and also lays the foundation for PHM. The RCM and MSG-3 processes also recognise that there are three major risks posed by equipment failure: threats to safety, threats to operations and threats to budget.

MoD requires RCM to be considered throughout the acquisition cycle. RCM analysis should first be conducted to determine a preventive maintenance programme for new equipment. Subsequently, after a period of real operational usage, it should be repeated to ensure that necessary changes to the schedule, which take account of actual usage and operating conditions, are incorporated. Further reviews should be carried out periodically to refine the schedule in the light of experience and changing circumstances. To achieve successful RCM application, it is necessary to prepare and implement a RCM plan. Defence Standard 00-45 (Ministry of Defence 2006) "Using Reliability Centred Maintenance to Manage Engineering Failures" is the top-level tri-Service standard that was first issued in 2006 (superseding Def Stan 02-45, developed for RN applications only). It states that RCM is accepted by MoD as the process to be employed to define maintenance programmes for new capabilities and is managed as part of the Integrated Logistic Support (ILS) process (Ministry of Defence 2006). The development of RCM as a tool to assist in prognostics and predicting what maintenance actions may be required, is yet to be really exploited due to the immaturity of prognostics.

It is believed by the RAF that RCM has saved countless millions of pounds, in particular through the rigorous review of maintenance schedules once an aircraft has been in service for a time. Particular maintenance actions might be extended or re-scheduled based on the review and the knowledge and experience of usage and environment. RCM in the RN has also made great savings and is being gradually applied to both existing and future ships. RCM is not mandatory however, for all Army vehicles and yet significant savings have been identified for several vehicle fleets which have undergone RCM. Implementation has often been slow because of a lack of resources and budget to make and publish the necessary changes to the schedules. The benefits of RCM are affected by fleet size and usage, the latter being influenced by the composition of force elements deployed on operations, the amount and type of training being undertaken and by the impact of WFM. It is probable that RCM could lead to significant potential savings across many more Army platforms but a major challenge would be securing the engagement of the MoD's Project Teams which are always operating under significant resource pressures. Additionally, the design authority (DA) bears a major responsibility for dealing with reliability issues, an aspect which may be addressed under availability contracting, where the DA and the service provider may be the same entity and will thus have a more direct stake in achieving improved reliability.

RCM has beyond any doubt proved its practical value both in industry and in the military although the lack of a consistent approach across the three Services suggests further significant savings could be achieved in the Army in particular. RCM cannot only be used to define maintenance regimes and provide financial and operational savings, but can also be very effective if supported by HUMS, CM, CBM and prognostics. The RCM process can potentially be used to identify areas where the introduction of HUMS, CM, CBM and prognostics could prove to be beneficial. It needs an acceptance that these technologies and processes will be cost-effective as solutions despite any initial costs for the technology insertion that is then necessary. This may be especially true for legacy systems where the potential is greatest but the technology insertion opportunities may be more limited. In the light of difficulties with justifying the operational benefits of HUMS, it is perhaps the cost benefits of HUMS that should be emphasised and the fact that it will inevitably lead to more operational effect. However, these aspects still require research and justification to make the cost benefit case. It continues to be difficult to engage the operational staffs on the advantages and the increased availability that RCM, HUMS, CM, CBM and prognostics can provide. Certainly one of the clear benefits of RCM, based on RAF experience, is the extended intervals between periodic maintenance which result from regular reviews of maintenance schedules using real usage and experience of equipments in operational conditions and use.

14.4 Health and Usage Monitoring Systems (HUMS)

Health and Usage Monitoring Systems have been used increasingly in modern equipment. However, CM is a similar technique which at its basic level might be a visual inspection but is more generally a technique such as oil health monitoring. In contrast to CM, HUMS is regarded as some form of a *system* that collects data and then provides some analysis on the collected data. We are accustomed to such systems in our modern cars and commercial aircraft. The Boeing 777 and the Airbus family routinely use HUMS (known variously as Aircraft Integrity Management System—AIMS) for example. The main reason was the advance in information technology that permitted cheap and reliable computers installed on vehicles and machinery. Additionally, developments in sensors and measuring technologies led to availability of a wide spectrum of relatively cheap sensors. These two technologies, combined together, allowed for sophisticated monitoring systems that could produce and store vast amounts of usage data that could later be analysed. However, Formula 1 motor sport provides the ultimate in what is capable with real-time monitoring of every aspect of the car and its use around the circuit. It shows what is possible, albeit with a huge investment, but does demonstrate and give a valuable insight into the real exploitation of this technology and the real-time monitoring of health and usage. Potentially HUMS opens up a wide range of opportunities such as allowing predictive maintenance, reducing the logistic footprint and reducing through life costs. Some practical problems with HUMS are that they may initially introduce extra costs, they may require changes in maintenance culture and the quality of collected data may not be sufficient for prognostic purposes. There are others. Currently the challenge within MoD is to prove the usefulness of HUMS and its operational and cost benefits. Experience thus far has been patchy and in particular is not coherent across the Services.

One of the limitations of HUMS and indeed CM, is the fact that instantaneous failure modes (random failures) cannot be prevented by monitoring systems. Some studies suggest that this type of failure is dominant for aircraft components and electronics. This may suggest that HUMS ought to be limited to certain application areas for which failures can be predicted. However, the value of HUMS does not only come from the fact that the data collected can be used for failure prediction. The usage monitoring itself can be very useful. For the components that fail unpredictably HUMS can *possibly* provide estimates of failure probability distributions (although this is currently not usually available). For components where we know the probability curve, HUMS can provide realistic estimates based on actual data. This can have immense value for defining maintenance schedules and RCM studies would benefit greatly from reliable failure data which are related to actual usage. The typical perception of HUMS is that it automatically implies prognostics. It is commonly believed that data can be easily used for prediction models. In fact currently collected data are often of insufficient quality for the predictive models, or the systems are too complex for creating practical models for prediction purposes. These models need to capture not only how the system works, but how it is being used in different environments and how it degrades over time.

The most active developments in the MoD in the past have been in the air domain, particularly for helicopters, although there is little commonality of HUMS equipments and technologies; consequently how it has been implemented and used across platforms is different and has varied significantly from the original aspirations. As a result a study is now underway to try to rationalise HUMS for all military helicopters. Typical problems include data download and recording which tend to be neglected by maintenance personnel. Part of the problem is that they see no benefits for themselves or for the squadron in collecting usage data. Implementation of HUMS has happened in a piecemeal way so causing the present situation and perception; some systems have limited on-board data storage capability and in reality there may be limited opportunities to download data before the on-board storage capacity is exceeded; this results in much lost data.

James Fisher Ltd. has worked with the Royal Navy since 1988 to install the MIMIC condition monitoring system, HUMS, as a part of the UK MoD's policy of RCM in the RN. Some 83 MIMIC products are installed on ships, submarines and shore establishments. However, the system is not considered to be operating to its full potential. The main problem is that whilst the data is being collected and stored, they are not being analysed. A number of factors contribute to this, amongst them the fact that initially the system was over-sensitive and triggered too many false alarms leading to a reduction in user confidence. Although its sensitivity levels were adjusted, this was not fully communicated to users and their confidence in the credibility of the alarms, therefore, has remained low.

MAN trucks have a comprehensive HUMS installed on their commercial trucks. The same system is now installed on the General Service Vehicle (GSV) a range of trucks delivered to MoD and MAN is very keen to be able to monitor them in real-time as part of the MoD support contract. The issue of security however, arises if the HUMS are downloading data as this may alert an enemy to the position of the truck; the willingness of MoD to allow a civilian commercial organisation to track their assets in real-time in an operational environment is also a security concern. Both parties continue to work to implement a workable solution but it is unlikely to be adopted for vehicles deployed on operations unless data download can be achieved simply and with transmission to UK enabled easily and securely at periodicities which allow for the security concerns and operational reality. One option could be to give MoD the ability to disable, at will, the vehicle tracking functionality in the HUMS fit when operational security dictates. MAN Trucks are currently monitoring MoD vehicles being used for training alone and this has allowed for detection of where driver training has not been fully effective or exceptional cases of misuse. This latter category has to be handled carefully by MoD commanders, using it to make drivers aware that the vehicle is under surveillance, but only in the interests of raising standards of equipment husbandry, usage and safety, rather than to catch them out. One example was the incorrect usage of the secondary braking system: the instructional staff were able to take corrective action, re-emphasising its proper use and thus rectifying the problem

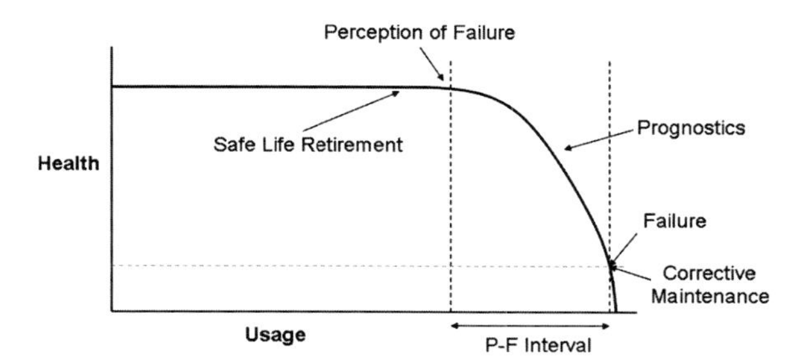

Fig. 14.1 The P–F Curve

quickly, and before it caused consequential unreliability. Without the HUMS it might have proved difficult to establish the cause.

HUMS addresses one of the key problems with estimating usage. It is known that the usage is not an indicator of the state of the system, but rather the intensity of usage is the key data required. Properly designed HUMS can collect data on actual usage and the loads and strains that have been put on equipment during that usage. This in itself provides a huge leap in terms of knowledge available to maintainers and the information for possible predictive models. An example is the helicopter undercarriage; flight hours are not a measure or a good indicator of the state of the undercarriage. It may happen that a helicopter experiences multiple landings within a short operational time, some of them being more intense than others, or it may be that it experiences long flights with few well controlled, softer, landings. HUMS can relate exact usage to other parameters and allow better estimation of the health of individual components and thereby the whole platform.

A key aspect of assessing health is the need to define healthy as an identifiable state. A useful way of doing this is to consider the P–F Curve (Fig. 14.1). By considering health, or an item's serviceability against its usage or time of operation, there is a point where the first opportunity is presented, point P, to predict an eventual failure, point F. The interval between these two points is known as the P–F interval. By knowing something about the life characteristics of the item and its usage, some decisions can be made about how rapid its decline will be to failure; with this information inspection intervals can be set so that the inspection interval is shorter than the P–F interval. Analysis of P–F intervals is a fundamental part of the RCM process to determine inspection tasks and intervals.

One of the key problems with HUMS is the need for expert analysis and interpretation of the data or an ability to turn the data into useful information. This may be done automatically on-board or off-board, but may also require some human analysis and interpretation. A common tendency and error is also to present the costs of HUMS as the cost of installation or embodiment and of the initial personnel training. Most of the cost analysis in presentations ignores the through life cost and through life benefits of HUMS. Other types of cost, might be the cost in terms of electrical power consumption and some small additional weight to the platform.

Even though HUMS are commonly associated with health monitoring their usage monitoring functionality is the easier function to deliver as it provides

Table 14.1 Benefits and costs of HUMS

Benefits	Costs
Increased life and availability	Cost of purchasing of new equipment
Reduced secondary damage and fewer maintenance-induced problems	New maintenance skills to learn
Reduced unplanned stoppages	On-going support costs
Focused and better planned maintenance effort with resources prepared and positioned for planned maintenance	Data storage
Reduced Logistic footprint	Data transfer processes
Pre-mission screening of unhealthy equipment	Data analysis
Increased crew awareness and overall safety, especially for airborne platforms	Power consumption and weight
Data for duty cycles, mission profiles, environments for design improvements	

information which is as valuable, or even more valuable, than that provided by the health monitoring functionality. First of all, usage monitoring is typically much easier to achieve than health monitoring. Secondly, automatic usage monitoring for systems that experience irregular use can bring useful information about system usage and any damage or untoward shocks or vibrations it has experienced. This could possibly be used to provide initial estimates of system condition (Table 14.1).

Usage monitoring may be especially useful for military vehicles, where patterns of usage are highly irregular, sometimes including long periods of inactivity between high usage in extreme conditions. Moreover, usage patterns differ between individual vehicles; some vehicles may spend most of their lives in storage or unused, while others may be used intensively during training or deployed on operations. Automated usage monitoring may also be an invaluable tool to assess the potential availability of the fleet, especially for equipment subject to WFM. Usage monitoring can challenge common beliefs on patterns of usage. For example, research conducted by the MoD's Material Integrity Group suggested that flight patterns for Chinook were significantly different from what was believed (Driver et al. 2007). This has an important implication and value in itself; it provides objective data on actual system usage, thereby enabling better informed assumptions on future usage to be made. Finally, any changes in usage trends can be instantly identified by an automated HUMS. Usage monitoring itself (without health monitoring) is also a quick win and can be used to promote HUMS. To illustrate the point, RAF pilots' flying hour recording was inaccurate by 40%. Once automatic flight recording was introduced, accuracy improved to the extent that reports were only 3% inaccurate. However, it should be noted that this result was mostly attributed to a certain number of individuals who had a disproportionate influence on the statistics (Condition Based Maintenance in the Air Domain Workshop 2008).

HUMS would seem to be necessary for successful availability contracting. It would be extremely difficult to monitor a contract for availability without usage monitoring, a fact not pressed hard enough by suppliers and not well understood by the customer. Usage monitoring is an enabling technology that allows the contractor to monitor usage of equipment to ensure that it is used in accordance with the terms of the contract and is satisfying the availability requirement. A good example is the Merlin helicopter. In the earliest design options HUMS was available, but it was only during 2008 that the contractor realised that they needed HUMS to meet contract requirements fully. They began to see HUMS as a critical technology for their business.

This also brings up the very sensitive issue of who owns the data collected by HUMS. There are two dimensions, legal and technical. Current HUMS produce vast amounts of data and their storage may require significant technology and storage on the platform and infrastructure off the platform, particularly for large fleets. A common perception is that the Joint Asset Management and Engineering System (JAMES) project, currently just used for Land vehicles, would serve this purpose for all three Services in the fullness of time, but this is not correct. JAMES does not have funding nor is it designed to act as a common data repository for all HUMS data. Typically, contractors are very interested in owning and getting all HUMS data, as this data potentially can be used to reduce maintenance costs and improve support. However, sensitivity of some of this data is an important and difficult issue and MoD in general is not being proactive in solving the issues of ownership, data security and giving oversight of fleet usage and abusage; the problem is probably not widely understood. The capability to exploit the HUMS data exists on several MoD vehicles but currently MoD has not endorsed any approved solution or issued directives on how to deal with these outstanding issues of data ownership and security. Without JAMES, there exists what has been termed the *Air Gap* between what is on the vehicle and the ability of the fleet manager, whether he be the operational or the contractual manager, to obtain, manage, exploit and use the data in a sensible timeframe. There is no recommended or authorised medium, memory stick, CD or electronic recording device that serves as the data transfer device. This is a major problem for the advancement of prognostic and predictive health management in the three Services.

Cook (2007) emphasises that it is the continuous nature of HUMS data that delivers the most important benefits for prognostics and consequently allows the streamlining of supply chains and ultimately leads to improved availability. For the MoD the real challenge is the acquisition of HUMS data from the platform and its storage in a dedicated data repository followed by timely analysis. Currently there is no infrastructure prepared to handle HUMS data in any general form. A typical practice is to collect data using memory sticks and post them using the normal mail system. Needless to say, such practice does not allow for real-time or even close to real-time analysis and the data can easily be lost.

There is no plan to address this problem within MoD. The challenges to be overcome include:

- No generic HUMS data standard (and nothing expected in the near future).
- The current communication infrastructure (the BOWMAN radio) was never intended to be able to handle HUMS data, especially because of the large volumes which might be involved and which would need excessive bandwidth.
- Reluctance to use wireless communication in operational situations.
- There are no proper data repositories identified to handle HUMS data.
- There is no capability to analyse the data and therefore no results being provided to show the operational benefits; hence, there is no perceived need by those at the front line to provide data to improve their operational effectiveness.
- There has been collation of cost benefit data or evidence to show the operational, maintenance and through life benefits.
- There is too big a gap between industry who might be able to improve operational effectiveness if they had the data and those at the front line who do have the data but are not able to extract it.

The conclusion is that any contractor, who intends to use HUMS data from MoD, should ensure that they provide an appropriate data repository and should expect that this data will not be accessible in real time.

HUMS will collect data from a great many sources around the vehicle depending on what has been decided as important data to collect. It may be data from accelerometers, sensors and data available from the CANbus.[1] Whatever sensors are fitted, they only provide data which needs to be turned into information, interpreted and used appropriately. Information is necessary in order to make decisions; however, information must have relevance, effectiveness and timeliness within the decision making process. In other words it must have operational value for it to be worth collecting. Nevertheless without data being available in the first place, no value will result. Moreover, decisions must depend on the accuracy and correctness of the information supplied or collected so the information needs to be clear, based on consistency, coherence and representing the complete picture of all the available sources; in addition it must be based on clear and explicit definitions and understanding. To realise the benefits successfully, the key requirement, which is to use information produced by HUMS to change maintenance regimes for improved safety, extended system life and reduced maintenance costs, it must involve bridging the *air gap*. Industry cannot do this alone. Although systems and contractual arrangements can be made to work in peacetime, it is during operations and extended deployments that the MoD must exploit the operational benefit for improved availability that is within their grasp.

HUMS in military applications do not mean prognostic models are available—yet! At present they only produce data that in time can be potentially used for prognostic models and further improvements in mission planning and operational availability. Some of the HUMS currently fitted and being developed have the

[1] CANbus is the multiplexed wiring system used to connect intelligent devices such as the engine management systems and other electronic control units on vehicles, allowing data to be transferred in a low-cost and reliable manner.

ability to include prognostic models, but too many do not. So we must now look at prognostics and how it can be developed with HUMS for military applications to improve operational availability.

14.5 Condition Based Maintenance (CBM)

From the 1980s new technologies such as CM, computational analysis, together with systematic approaches to understanding the nature of failures and their consequences, allowed for improved maintenance processes and more preventive maintenance. The key to the change was the realisation that failures are rarely due to the age of the components and that failure modes are typically much more complex processes. Furthermore, system designs were becoming more complex and delivering ever higher levels of reliability was becoming even more challenging.

Maintenance can be divided into corrective and preventive maintenance. Corrective maintenance is reactive, allowing failures in the system to occur. This approach has the disadvantage that it allows the system to suffer unplanned downtime, which in a military setting can mean serious consequences if the failure occurs during war-fighting. Unplanned failures in peacetime are also inconvenient at best and can place users in difficult situations. Preventive maintenance tries to address this problem by identifying components that might fail and, therefore, replace them, or components which might need some restorative maintenance and undertake their overhaul. The problem with this approach is that it is very difficult to predict failures, particularly how far into the future any failure might occur. The challenge is to determine when the failure may occur. CBM is a type of preventive maintenance that utilises any appropriate process to determine the condition of an item and to decide on whether any maintenance is needed. CBM comprises three stages: data acquisition, processing function, and decision on maintenance.

The data acquisition is typically carried out by HUMS which might be embedded sensors or portable equipment external to the system; the simplest technique though is a visual inspection. The second stage is related to processing the collected data. Other than for visual inspection it would be typically completed using some mathematical or statistical models that capture the nature of failures. Typically these models are implemented as computer-based analytical tools. Since the key element of success of CBM is the knowledge of the state of the item, it is crucial that these models are accurate in their assumptions and that they provide reliable and robust results. Additionally data invariably have to be processed, analysed and turned into information before it can be readily used. This is a common feature of CBM and its enabling technology of HUMS. Finally, the third stage, the decision on maintenance can then be made. This should take into account multiple factors, such as reliability of the prediction models used, operational requirements on the system and confidence in any subjective decision making which may occur such as for visual inspection for instance.

The MoD's Materials Integrity Group based at DE&S Fleetlands, Gosport prepared and published in September 2007 a MoD Joint Service Publication (Ministry of Defence 2007) which outlines the MoD policy for CBM and CM and covers the Air, Land and Maritime domains. Separate working groups have been established in all three domains with the intention of driving the process forward. The JSP 817 defines CBM as:

> Maintenance initiated as a result of knowledge of the condition of an item of equipment gained from routine or continuous monitoring.

The air domain (Ewbank 2008), however, feel a more wordy definition is required similar to the following:

> CBM is the forecasting of maintenance, based on the analysis of data captured by Condition Monitoring for maintenance significant items, to deliver the benefits of: effective logistic support, decreased costs, improved availability and mission achievement, maximised safety and airworthiness and maintenance alleviation (to improve mission success and fleet planning).

The US Army, for their air vehicles, have a definition which captures the essence of the task without trying to define how it is achieved, which may or may not be using automated technology (Ewbank 2008).

> CBM is the ability to translate aircraft condition data and usage into proactive maintenance actions that enable unit maintenance personnel to achieve and maintain higher aircraft operational availability. It also gives commanders a mission planning tool. Through the use of prognostics, the aircraft predicts its remaining mission availability and/or time to failure, providing valuable information for the commander to determine which aircraft are ready for battle and aircraft that require maintenance.

In the US, the Defense Acquisition Guidebook (Department of Defense 2004) reflects US Department of Defense (DoD) policy for CBM and it states that CBM is to be implemented to improve maintenance agility, increase operational availability and to reduce through-life operational costs. DoD Instruction 5000.2 (Department of Defense 2008) mandates the use of CBM. Early success of CBM pathfinder projects led to an enhanced and more ambitious *CBM Plus* initiative in 2004 one of whose targets was a review of the Joint Strike Fighter programme. It is a marked contrast that the US DoD mandates CBM and UK MoD does not.

In the air domain in particular, HUMS are more prevalent on helicopters and full aircraft integrity management and monitoring is something the RAF expects to get on the next generation of aircraft. Experience and lack of confidence with Tornado Built in Test (BIT) and Built-in-Test Equipment (BITE) and their contribution as a HUMS and CBM enabler has been mixed. The Typhoon fighter aircraft which has just entered service and the A400M transport aircraft, whose entry to service is expected in 2011, will provide the next increase in capability; real usage experience with Typhoon is yet to be collated however.

The experience of CBM and CM in the maritime domain is similar to that in the air domain. CM has been shown to provide benefits for the monitoring of heavy machinery where oil sampling has been shown to be effective at monitoring contaminates and wear debris. Nevertheless, the oil analysis needs sophisticated equipment in order to pinpoint accurately any problems and this is usually only available at a few locations so the expected benefits have not always been delivered. The new Royal Navy Type-45 destroyer will provide the next increase in both CM and CBM application and together with more automated monitoring and prognostic ability from within the automated systems, the benefits should be able to be exploited. In the Army, oil health monitoring has been applied to good effect on some vehicles such as the Challenger II tank. However, the problem of distance between the operational area and the analysis centre has meant that the failure has often occurred before the indicative results are received by the operator. Oil sampling is perhaps one of the basic levels of HUMS and CBM. Very little general progress has been made on adopting more general HUMS to existing or legacy platforms due to the expected cost of retro-fitment and the inevitable *air gap* in getting the data transferred to maintenance and operational managers. Those new vehicles entering service with HUMS fitted are enabled for CBM but have often suffered from contractual difficulties of getting the data downloaded and sent back to the Original Equipment Manufacturer (OEM). Data are often collected on a military classified network (such is the vehicle's network) and this can lead to problems when data have to be transferred from the vehicle and passed to a civilian organisation for analysis. In addition to the security issues of transmitting data from the battlefield an aggregation of seemingly non-sensitive data can suddenly become classified for no other reason than its bulk. There have also been issues with ownership of the data which have unfortunately not been solved in the initial contracts. The OEM needs the data for warranty reasons, improvements to support solutions and future designs. Solving the problem of data ownership is crucial for the Army before any benefits will be seen in whatever support solution is provided. Without the data being available both on and off the vehicle at nearly the same time, no support solution will be effective. Examples of projects in the US involving CBM and HUMS show a cost benefit and one such example quoted is the US Army CBM experience for the AH-64 Apache helicopter. It was able to show a 4–5% reduction in operating costs achieved, with the main factor being attributed to significantly better usage monitoring (Bayoumi et al. 2008).

Performance of any maintenance activity cannot be considered separately without taking into account logistics that deliver spares, tools and resources needed for the maintenance activity. CBM is not only more challenging in terms of maintenance, but requires more agile logistic support as the normal and predictable, fixed time maintenance activities will not take place; instead logisticians will have to manage more irregular requests for spares and consumables from maintenance personnel. Where their response is slow or inaccurate, for example providing the wrong spares, the perceived benefits of CBM will be quickly eroded. It is therefore important to include logistics in the process of modelling maintenance regimes, as logistics responsiveness is an important factor in determining optimal maintenance activities in CBM. Such modelling should take into account that the logistic system is not a constant factor and at least two reasonably distinct states of operations can be identified, covering peacetime and operational conditions.

It is clear that there is a strong relationship between CBM and HUMS. In a nutshell, HUMS provide the technology required for CBM and thus enables CBM. But CBM is not only about HUMS; it is also about prognostic models, policy changes and effective inspections that are appropriate to the situation among many more things. In addition, one cannot have CBM without first having an effective RCM process. RCM is required to understand the failure modes and decide the best maintenance activity.

The potential benefits of CBM include:

- Increased equipment availability through improved maintenance management.
- Increased logistic chain efficiency by exploiting data on spares requirements.
- Reduced logistic footprint by reducing spares inventory.
- Reduced maintenance time and cost.

The CBM challenges identified include:

- Leadership. Without strong and committed leadership at the top management levels of both the user and the provider, the required support will not be available to make the necessary changes.
- Policy. Without a clear policy in support of CBM, the implementation of new maintenance techniques is difficult.
- Maturity of technologies. It is not yet clear what data collection and predictive models are practical and suitable. Many technologies are in the early stages of development. Good understanding of the strengths and weaknesses is critical to avoid risks related to implementing CBM.
- Difficulty identifying benefits. The benefits of CBM are not always clear and they need to be identified and quantified. Good examples are needed to build confidence that CBM can offer financial savings and improved operational availability.
- Logistics support. Implementation of CBM will require changes in logistics support to meet new requirements for more agile and effective logistic support. This will require the introduction of new business processes, changes to logistics planning and improvements to the supply chain.
- Data repository. HUMS and CBM produce large quantities of data and information requiring the appropriate infrastructure for its transfer and storage. Development of specialised Military Information Systems is expensive and challenging and there are already too many systems with no coherent solution for operational HUMS data. The BOWMAN system, for example, was not designed for carrying this kind and quantity of data.

If a modelling technique is intended to identify optimal maintenance regimes for CBM, the modelling process should not just include the reliability of components. In real life, the cost of maintenance both in monetary and delay terms depends on many factors unrelated to reliability. Delays introduced by logistics, availability of the system for maintenance and repair, shortages of manpower or other resources, legal requirements on maintenance and repair activities and many other things, contribute to the total costs of downtime and are too important to be ignored in any realistic model.

14.6 Prognostics and PHM Prognostics and PHM

Prognostics is a process and enabling technology for CBM. Prognostic models should be developed in parallel with efforts to introduce CBM in practice. This is why since 2003, the US DoD has required prognostics to be included in all new military systems (Department of Defense 2007). In a nutshell, CBM requires prognostics to provide information about the remaining life of a component, subsystem or system. This estimate of remaining life should not only be reliable, but it should be accompanied with information regarding its accuracy. In practice it is not only a matter of providing an estimate of remaining life, but informing the user how much confidence he or she should have in the obtained result. It is perhaps this concept that has created the term PHM which embodies the notion of predicting the remaining life and thereby providing information as to the current health of the system to allow management decisions to be made.

There is a great variety of scientific methods that can be used to provide prognostic analysis of data. A general overview of the state of the art of prognostic modelling is discussed by Jardine et al. (2006). The paper offers an extremely broad literature review of 271 publications. This review is mostly focused on data-driven methods and practical aspects of the problem situation are not given much consideration by the authors. However, this is understandable given that the combination of a monitoring technique together with the nature of the monitored system and its operating environment, almost always creates a unique set of conditions; hence trying to generalise any practical prognostic applications would be extremely difficult, if not impossible.

An interesting practical problem with prognostics, noted by Cook (2007) was that prognostics itself may not be sufficient without proper diagnostics. The fact that a component deviates from an expected state does not imply that it will lead to failure. Again this underpins the notion of the health management part of PHM. In general unnecessary maintenance activities are costly and, therefore, if a prognostic model is incorrect it can result in increased maintenance costs. The modelling must be sufficiently rigorous and based on expert knowledge as well so that it deals with the chain of consequence that starts with an initiating event, such as a fault or error and its point of detection, followed by a failure with the resulting damaging consequences. It should be understood that there is a significant gap between monitoring to detect faults and fault prediction through a prognostic capability. The CM technique of oil health monitoring, is a good example; it allows for cost-effective and accurate monitoring and helps to diagnose failures and assess the state of deterioration. However, predictions based on the collected oil data can often prove to be unreliable and, therefore, not as valuable for assessing how much remaining life there might be. The idea and concept of PHM is that it seeks to encompass CM techniques, HUMS and CBM with prognostics in order to provide more effective maintenance and support.

14.7 Conclusions

The MoD is increasingly expecting operational effectiveness to be delivered by contracts for availability and HUMS, CBM, CM and PHM are vital enablers for their successful delivery. Yet the progress to achieve embodiment of these enablers is very slow. Whilst it is a challenging aspiration, it is achievable given the will and a willingness to spend to save. It could be achieved more realistically first on legacy systems with small fleets, or on subsystems of large systems, rather than for complex systems. It will be much more difficult on large fleets of diverse legacy vehicles where there are fleets within fleets due to the problems of configuration control and the variation and complexity of logistic support required. Newer vehicle acquisition projects will be designed with a CANbus so will be configured for HUMS and CM which will enable wider application of CBM; whether PHM will be possible in the near future will depend on more research.

Another dimension is the likelihood of the system to be deployed for operations; the complexity of understanding the possible different operational contexts and risks associated with them, will add challenges for contracts that seek to deliver better operational availability and support solutions using HUMS and CM. Yet both would enable greater operational availability and planning certainty if embodied and the available data used and managed in the operational theatre.

RCM is another enabler that is not achieving its full potential; it can invariably reduce the time spent on maintenance if the maintenance schedules are updated using experience of actual operational usage. It is often not given the credibility it needs though to enable changes to maintenance activities and periodicities. RCM must not only be used to prepare the initial maintenance schedules but must also be mandated once equipment has experienced actual operational usage. Changes to maintenance periodicities based on experience and actual usage invariably mean that the initial conservatism can be replaced by practical realism with the resultant increase in operational availability and reduction in down-time as well.

Contracting for availability and successful support solutions, almost certainly require HUMS and the data they generate. The significant shared benefits need to be recognised by both user and industry as early as possible so that successful contracts for availability and their underlying support contracts can be delivered. HUMS implementation should be an essential attribute and requirement on all future platforms if the user and contractor are serious about providing better equipment availability. The air gap problem for the download and transmission of HUMS data needs to be solved with a consistent and effective solution that gets the data and information to those that need it in real or near to real time; this needs to be enabled though for both peacetime and operations if the benefits for both user and supplier are to be realised.

Formula 1 motor sport, commercial vehicles and aircraft demonstrate that military applications and expectations for HUMS and prognostics lag way behind the commercial sector. Boeing and Airbus have considerable knowledge and experience that shows what is possible if the delivery of availability is critical to the success of the business or operation.

The UK Armed Forces could follow the US lead in mandating CBM and also mandate HUMS, CM, and PHM for future platforms in order to deliver improved operational availability. Whilst there will be some costs, the improvement in system effectiveness and operational availability is achievable and the delivery of predictive maintenance a realistic possibility with the development of PHM.

Further work in this area is now essential to highlight the cost benefits for operational availability and through life support, although it is always difficult to put a monetary cost on operational benefits. Nevertheless, candidate systems and projects are needed which can be used to demonstrate the benefits of applying these techniques to specific systems and projects.

14.8 Chapter Summary Questions

The chapter has identified that HUMS, RCM, CBM and PHM are enablers for increasing maintenance management, planning effectiveness and operational availability but are not being exploited in the military environment to their full potential. Each has a contribution to innovative and effective support solutions.

- How can we demonstrate more effectively the support benefits and savings from regular RCM reviews of in-service equipment?
- How can we better demonstrate the operational, support and cost benefits of HUMS applications to military equipment?
- Is it possible to overcome the operational concerns for management of HUMS and prognostic data in real time so as to benefit from predictive maintenance instead of reactive maintenance and support?
- How can we demonstrate more effectively the benefits of both CBM and PHM in delivery of both operational and support effectiveness?

References

Air Transport Association, ATA operator/manufacturer scheduled maintenance development (MSG-3) Revision 2. (ATA Publications, 1993)

A. Bayoumi, N. Goodman, R. Shah, L. Eisner, L. Grant, J. Keller, Conditioned-based maintenance at USC—Part IV: Examination and cost-benefit analysis of the CBM Process. The american helicopter society specialists' meeting on condition based maintenance, Feb 2008 (2008)

British Army, Kit! Magazine [Online] Available from: http://www2.armynet.mod.uk/armysafety/kit_mag.htm. Accessed 19 Mar 2010

Condition Based Maintenance in the Air Domain Workshop, Discussion during the workshop at RAF Wyton, 12 Mar 2008

J. Cook, Reducing military helicopter maintenance through prognostics, IEEE Aerospace Conference 2007

Department of Defense, DoD Instruction 4151.22. Condition based maintenance plus (CBM +) for materiel maintenance, DTD 2, Dec 2007 (2007)

Department of Defense, Defense acquisition guidebook. [Online] Available at: https://acc.dau.mil/CommunityBrowser.aspx?id=106830. Accessed 30 Nov 2009 (2004)

Department of Defense, DoD Instruction 5000.02. [Online] Available at: http://www.dtic.mil/whs/directives/corres/pdf/500002p.pdf. Accessed 30 Nov 2009 (2008)

S. Driver, M. Robinson E. Moses, H. Azzam, J. Cook, P. Knight, The UK MOD EUCAMS strategy and the FUMS developments. IEEE Aerospace Conference 2007

T. Ewbank, CBM in the air domain. [presentation] health and usage monitoring, condition-based maintenance and prognostics symposium. Shrivenham, April (2008)

A. Jardine, L. Daming, D. Banjevic, A review on machinery diagnostics and prognostics implementing condition based maintenance. Mech Sys Signal Process **20**, 1483–1510 (2006)

Ministry of Defence (MoD), Air Publication AP 100C-22—Procedures for developing preventive maintenance, 3rd edn, Apr (1999)

Ministry of Defence (MoD), Defence standard 00-45 using reliability centred maintenance to manage engineering failures. Part 3, Guidance on the Application of Reliability Centred Maintenance, Issue 1, Apr (2006)

Ministry of Defence (MoD), Joint service publication JSP 817—condition monitoring condition based maintenance policy (2007)

J. Moubray, *Reliability-centered maintenance*, 2nd edition edn. (Industrial Press, New York, 2005)

Society of Automotive Engineers, SAE standard JA1001: Software Failure Modes, Effects, and Criticality Analysis (SFMECA) Guide (2006)

Chapter 15
Modelling Techniques to Support the Adoption of Predictive Maintenance

Ken R. McNaught and Adam T. Zagorecki

Abstract Contracting for availability and contracting for capability are becoming increasingly common practices in the defence world. With these new service-oriented contracts, the responsibility for through-life support, including maintenance, has been shifted from the user to the service provider. In this new environment, innovative approaches to improving maintenance and reliability are necessary and create new, unique opportunities for value co-creation between stakeholders. This chapter focuses on investigating the applicability and implementation of an approach to predictive maintenance which combines prognostic modelling with Condition Based Maintenance (CBM) and its role in providing improved service provision for the repair and maintenance of complex systems. The role of prognostic modelling and Health and Usage Monitoring Systems as the emerging technologies that enable a value-oriented approach to maintenance are discussed. Bayesian networks are discussed as a modelling framework that is appropriate to capture uncertainties related to predictive maintenance. Special focus is placed on reviewing practical challenges and proposing solutions to them. The discussion is summarised in the form of a practitioner's guide to implementing prognostic modelling and CBM.

K. R. McNaught (✉) · A. T. Zagorecki
Department of Engineering Systems and Management, Defence Academy of the UK, Cranfield University, Shrivenham, SN6 8LA, UK
e-mail: k.r.mcnaught@cranfield.ac.uk

A. T. Zagorecki
e-mail: a.zagorecki@cranfield.ac.uk

I. Ng et al. (eds.), *Complex Engineering Service Systems*, Decision Engineering,
DOI: 10.1007/978-0-85729-189-9_15,

15.1 Introduction

In the defence sector, there is a growing tendency to extend the working life of complex engineering systems. With longer life spans, the challenge of providing adequate and cost-effective support, maintenance and management of the ageing systems is becoming of paramount importance. This trend has been recognised both by defence suppliers and customers. The new forms of contracting, contracting for availability and contracting for capability, are becoming the reality. The focus of the defence business is rapidly shifting from manufacturing to providing services to support complex systems. However, the new business environment has not been well understood. Service support for maintenance and repair of complex engineering systems requires all three transformations described within the Core Integrative Framework (Ng et al. Introduction chapter). We argue that to improve performance of maintenance and repair service through CBM one should not only focus on the technological aspect, but the organisational and wider business context should also be addressed and these can be achieved only through value constellation (Normann and Ramirez 1993). Those businesses which manage to adopt all three transformations swiftly will be in the most favourable position to win future defence contracts.

This chapter focuses on investigating the implementation of an approach to predictive maintenance which combines prognostic modelling with CBM and its potential role in providing improved service provision. While traditional maintenance activities have emphasised equipment transformation, it is information transformation that drives predictive maintenance. While recognising that people transformation will also be required to enable some aspects of the information transformation (e.g., training in condition monitoring technology), our focus in this chapter is firmly on information transformation.

In recent years, significant technological and scientific advances in areas such as information technology, intelligent systems, machine learning, sensors, communications, data mining and data fusion have taken place. As a result, Health and Usage Monitoring Systems are installed on new platforms and are capable of producing immense volumes of data. Some of these advances have also contributed new methods for monitoring, diagnostics and prognostics. However, the practical value of new approaches to repair and maintenance is yet to be determined and it is becoming apparent that their value is heavily dependent on the context.

We discuss the challenges and propose a set of practitioners' guidelines for implementing prognostic modelling and CBM. In particular, we consider the applicability of probabilistic graphical models as a modelling framework that supports information transformation, allowing for flexibly combining various sources of data and human expertise—a practical necessity for comprehensive prognostic models.

15.2 Prognostics and Condition Based Maintenance

Prognostics is concerned with predicting the residual life of an item or the time until the next failure is expected. The item under consideration can be a component or a complete system. Ma (2009) notes that Prognostics and Health Management (PHM) has been defined by the IEEE Reliability Society as "a system engineering discipline focusing on detection, prediction and management of the health and status of complex engineered systems". However, although the term 'prognostics', particularly when referred to in the context of PHM, is sometimes used in a more wide-ranging sense to also include fault detection and diagnosis, its use here is restricted to the narrower definition of prediction of future health.

In the defence sector, the F-35 Lightning II, the result of the Joint Strike Fighter (JSF) programme, provides the leading example of PHM application today (Brown et al. 2007), involving a mixture of on-board and off-board systems. The objectives of PHM on the JSF are to improve mission reliability, availability and safety, while reducing the overall costs associated with maintenance. CBM is an integral enabler for achieving these objectives, replacing fixed-interval maintenance actions, wherever possible, with maintenance only when required, as indicated by the monitored condition of the equipment.

A number of usefulness criteria which can be used to help define the minimum acceptable warning time for maintenance action are offered by Line and Clements (2006) and include items such as prevention of mission abortion, advanced maintenance planning, spares ordering and opportunistic maintenance.

CBM is appealing as theoretically it should lead to significant benefits including reduced maintenance effort, improved reliability, and reduced logistics costs due to better predictability of maintenance. If the remaining life of an item could be predicted with any degree of accuracy, this would allow necessary maintenance activities to be timed in an optimal manner as well as eliminating some unnecessary maintenance altogether. This in turn would provide many benefits, including reduced maintenance costs, reduced downtime and improved equipment availability. By reducing uncertainty associated with equipment failures, theoretically some reduction of the logistics footprint could be achieved.

In summary, the theoretical benefits of CBM are:

- Reduced maintenance costs achieved by eliminating unnecessary maintenance activities based on better knowledge of the health of the item.
- Improved reliability of the system through better knowledge of the state of the system and maintenance activities being conducted *just in time*.
- Improved reliability should allow for increased availability of the system due to reduction of unexpected failures and unnecessary maintenance activities.
- Reduced logistics footprint by reducing failures and reducing uncertainties related to failures.
- Improved reliability through re-engineering of the system by gaining a better understanding of the failure modes within the system, based on hard data collected from monitoring the system in use.

In reality, however, claims regarding the value of CBM are more in the realm of beliefs and expectations than proven facts. CBM is more of a broad concept rather than a well-defined set of techniques and applicable solutions to the problem of improving maintenance. There is a vast number of different maintenance problems and similarly there should be expected to be a large number of predictive maintenance techniques applicable to particular problems.

This does not mean that the lessons learned in one area of application should be ignored and not applied to other problems. But it should be done with considerable caution. For example, there may be two systems (such as land vehicles) that are very similar from a technical point of view, but with very different fleet sizes, e.g., 200 vehicles of one type and 2000 of the other. The difference in the number of vehicles may affect the total cost of implementing CBM to these two fleets. The cost of introducing CBM, which at this stage of development is considerable, should be taken into account.

Given the problem-specific nature of CBM, one should not realistically expect a general solution to it. The research on CBM should be driven by a small number of pilot projects that are carefully selected based on factors such as:

- Expected gains from application of CBM should outweigh development costs:
  - Current maintenance costs should be high due to factors such as fleet size, or large penalty costs related to failure.
  - Critical systems can be good candidates, their reliability can be valued more highly to justify the cost of the maintenance.
- Technical and organisational feasibility of reliable on-board data collection, its transfer and storage in off-platform repositories.
- Feasibility of implementing CBM in practice (for example, in the air domain, safety regulations can prevent implementation of otherwise beneficial changes in maintenance).
- Feasibility of defining predictive models based on available data (for example, a prognostic model may be able to detect problems, but it may do that in insufficient time to improve the maintenance regime). The problem relates to sudden failures versus slow progressive faults.
- Willingness on the organisational side to implement CBM and to introduce new approaches to maintenance in a general sense (this aspect highlights the importance of the three transformations identified by the CIF).

CBM relies on two technologies that are still relatively immature and are subject to active research. These technologies include:

- Health and Usage Monitoring Systems (HUMS)—even though HUMS are commonly installed on modern platforms, the usefulness of data collected by these systems is not well understood. The problem lies in the fact that HUMS are often driven by the availability of particular types of monitoring techniques rather than the overall value of the information gathered (which often is simply not known due to lack of experience). This value is really determined by the way this information is exploited to reduce maintenance costs, increase availability, etc., Additionally, the cost of maintenance and support of the HUMS system itself is still not well understood.

- Prognostic modelling—unlike diagnostic modelling, prognostics is a relatively new domain with few practical, fielded applications available. It should be expected that the field of prognostics will become a very active research area in the coming years. In particular, technologies like HUMS, advances in information technology, data mining, knowledge engineering and various modelling approaches have become sufficiently mature to enable rapid advances in the research on prognostics.

The ability to predict the remaining useful life of a system or a component also requires a reliable forecast of future equipment use. Although often mentioned in the same breath, prognostics and diagnostics are fundamentally different. Fault diagnosis is a post-event activity, aiming to assist recovery from a failure while prognostics is a pre-event activity, aiming to avoid failure. The fact that the future usage and the environment to which the item will be exposed are unknown makes prognostics a much more challenging problem than diagnostics.

Prognostics aims at predicting when the component of interest will fail, and in order to achieve this it would be desirable to identify and model the relevant deterioration processes. It is important to emphasise that in most cases time is not the only factor that determines remaining life. Usage patterns and shocks experienced are critical factors for remaining life prediction for mechanical components and electronics. A prognostic model should not only capture the degradation processes within the modelled system but also account for uncertainties related to predicted usage of the system under consideration.

Uncertainties arising from the future usage of the system are strongly dependent on the environment and intensity of system usage. These uncertainties for industrial machinery may not be challenging, as typical manufacturing equipment operates non-stop or in well-defined time periods under fairly constant environmental conditions typical to the factory floor. However, they may pose significant challenges for military systems, which are expected to be exposed to extreme usage patterns: ranging from months and even years of storage periods (typical for land systems) to intensive operational periods in conflict zones, where the systems are expected to operate outside of their design limits. Because the life of equipment is dependent on its usage patterns, prognostic models, in particular for military systems, should account for such variations in usage and environmental factors. This constitutes one of the key challenges for prognostic modelling and the development of CBM in general.

Uncertainty plays a fundamental role in prognostic modelling and there are many sources of uncertainties which affect our ability to predict the health of an asset. The point of failure is typically defined as some threshold level of accumulated wear or damage and this threshold itself might be variable or uncertain. The true current condition of the item will also be subject to some uncertainty, even if it has just been monitored, given that monitoring is imperfect. The future usage of the item may be uncertain but even if it can be reasonably known, the exact rate at which further damage accumulates will always be uncertain. Although a point estimate of the remaining useful life can be calculated, if the probability density function of the time until failure follows a symmetric distribution (such as the normal distribution), there is a 50% chance that the item will fail before the point estimate is reached. Consequently, we have to ask what probability of failure is acceptable for the item. For items where this is very low, some maintenance action will be needed well before the damage level gets anywhere near the critical threshold. For other items, however, it might be more cost-effective to wait until a higher level of damage has accumulated before taking action. Although this runs a higher risk of incurring a failure, it reduces the number of unnecessary and wasteful predictive maintenance actions. Identification of the acceptable failure probability is crucial for determining when action needs to be taken. The prognostic model should be able to predict the point in time when this critical probability is reached. Subtraction of the necessary lead time for maintenance and logistical activities then identifies the point where such actions need to be triggered.

Additional uncertainties relate to the item's failure modes and how well understood they are. If there might be additional failure modes, not covered by the prognostics, then the actual chance of failure may be higher than that suggested by the modelling and consequently the intervention time will have to be brought forward to compensate for this. The importance of Failure Modes, Effects and Criticality Analysis (FMECA) can not be underestimated in this regard (Department of Defense 1980). In fact, it has been suggested by Kacprzynski et al. (2001) that the role for FMECAs should be expanded to support prognostic modelling further. The task for diagnosis is to identify components that are acting in an unexpected, faulty manner. To achieve this, almost by definition a system of components is under consideration and the task is to identify a faulty sub-system. In prognostics, even if the task is to determine the remaining useful life of a system, each sub-system should be analysed to derive the remaining life of the whole system. This implies that prognostics should always be done on low-level components and when this step is achieved then the aggregation should take place. Therefore, prognostic models should be developed for the component level first, and later, aggregation models can be derived. This has a significant implication in practical terms. First, applications of prognostic models and CBM should be expected for systems that are of low complexity, or are independent sub-systems of larger equipments. And this seems to agree with fielded applications such as battery monitoring systems: in fact, it is a single component (battery) with well-understood remaining life models and well-monitored environmental variables (load on a system and temperatures).

Until now, PHM has been most widely applied to mechanical systems. A recent review relating to rotating machinery is provided by Heng et al. (2009a). Application to electronic systems has lagged behind, partly because of the perception that electronic systems fail in a more random fashion and that consequently little can be done in the way of condition monitoring. However, many such failures can be attributed to mechanical causes such as broken solder and recently, PoF models

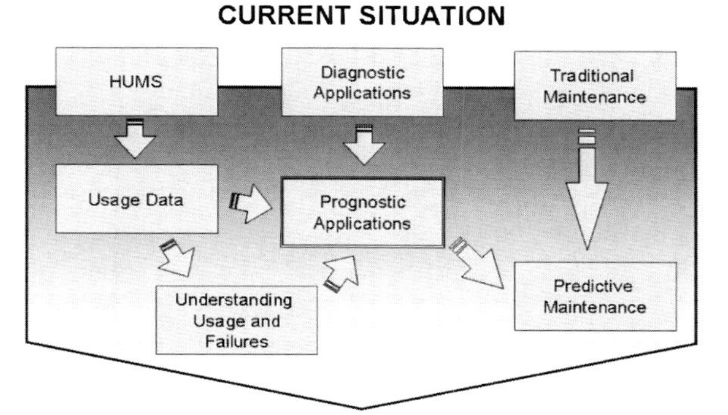

Fig. 15.1 Conceptual framework for Predictive Maintenance

have been developing rapidly in this area (Gu and Pecht 2007). The Center for Advanced Lifecycle Engineering (CALCE) at the University of Maryland is a leading developer of such models. An example of accelerated failure testing applied to a GPS receiver in order to identify features useful for PHM is provided by Orsagh et al. (2006).

In conclusion, the prognostic models of the future should be built based on a combination of domain-specific knowledge and available data. HUMS will most likely become the primary source of information for prognostic systems. Most of all, HUMS will be able to provide objective data on usage and reliability of systems, a new sea of knowledge that has not been available in the past. By itself, HUMS is an enabling technology for prognostic modelling and CBM. However, relying only on HUMS data may not be sufficient for constructing successful prognostic models. Other sources of knowledge should be exploited as well.

Figure 15.1 shows conceptual dependencies between enabling technologies that lead to Predictive Maintenance. In the top layer, there are technologies that are presently available and well-established. The further down, the more uncertain and less developed the methodologies are. We would like to indicate that the HUMS systems are an enabling technology, but availability of usage data should not be taken for granted just because HUMS is installed on the system. Getting the collected data to an off-platform data warehouse, ensuring quality and completeness of collected data, are real-life challenges, in particular in the defence sector where data security is a serious issue. Additionally, for availability and future capability contracts, issues around data ownership and access to it by stakeholders can pose significant challenges. It should be understood that it is not only the presence of HUMS itself that is important, but the availability of the data collected by HUMS is essential to enable prognostic modelling.

15.3 Prognostic Modelling

Prognostic models in the literature are divided between data-driven methods, model-based methods, statistical models and hybrid methods.

The data-driven methods rely explicitly on data collected on the system operations, maintenance, etc. These methods rely on data analysis to derive estimates of the remaining useful life without an explicit physical model. This approach is suitable in situations where (1) sufficient reliable data are available (2) a model-based approach is not feasible because of lack of knowledge regarding the system. Artificial neural networks appear to be the most popular data-driven approach to prognostics. Recurrent neural networks are particularly popular given their suitability for use with time series (Wen and Zhang 2004; Heimes 2008; Peel 2008; Heng et al. 2009b). The main drawback common to all neural network approaches, however, is the lack of transparency provided by the models.

Model-based approaches require an explicit model for capturing remaining useful life. They generally utilise expert knowledge about failure mechanisms and the physics of failure, and require less data as a result. Li et al. (2000) introduced a defect propagation model for rolling element bearings. A model-based approach to crack growth estimation employing particle filtering is described by Cadini et al. (2009), leading to estimates of the component's remaining lifetime given observations at a number of inspections. Model-based approaches are also increasingly important for electronic components (Gu and Pecht 2007).

Statistical approaches rely on data and information collected for populations of entities under consideration. Survival analysis is an example of such an approach applied to the estimation of remaining life for humans; it clearly addresses problems which are very close to those faced by reliability and PHM engineers. One aspect of the problem which it has been particularly developed to account for is censoring of observations. This could be especially useful in PHM as most of the time, actual failures will not be observed. However, despite some interest from the reliability community (Kumar and Klefsjo 1994; Kobbacy et al. 1997) the number of studies making use of survival analysis for machine prognostics seems much lower than might be expected. Consequently, we believe that this approach deserves further investigation. Another example is logistic regression, used by Yan et al. (2004) in conjunction with a time-series model to predict machine failure given the values of condition variables. Wang and Christer (2000) employ stochastic filtering theory to estimate the remaining life distribution of rolling element bearings conditional on the observed past history of the items. Markov chains are popular in reliability modelling of repairable systems and can be particularly useful for representing degradation mechanisms (Saranga and Knezevic 2000). Hidden Markov Models (HMM) have become increasingly popular for a wide range of applications, including medical diagnosis and speech recognition. The evolution of a system through time is represented as a Markov chain over a hidden or unobservable state space. At a number of discrete points in time, observations can be made which provide an imperfect indication of the true system state at those times. An example of their application to diagnosis and prognosis of machine cutting tools is given by Baruah and Chinnam (2005).

The hybrid models try to exploit advantages of two or more approaches. In practice, only hybrid models are expected to provide practical value. This is because HUMS data are readily available and can be exploited using data-driven methods. However, HUMS data by themselves are rarely sufficient. There are additional invaluable sources of knowledge that can be exploited by prognostic models that are not included in HUMS data. Examples of this can be: system design, models for physical processes, knowledge on certain usage patterns, and reliability data from repair records.

Since their introduction in the mid-1980s (Pearl 1988), Bayesian networks (BNs) have become increasingly popular as a framework for reasoning under uncertainty. While much of the initial growth stemmed from their popularity within the Artificial Intelligence community, the last decade or so has brought the attentions of the wider world of science and engineering, including the field of reliability or, more generally, dependability modelling. A recent review of BNs in reliability modelling is provided by Langseth and Portinale (2007).

Much of their appeal can be attributed to the flexibility which the modelling framework supports (Pourret 2008). For example, the same BN might be used to predict the probability of failure of some equipment and then to diagnose its cause once a failure has occurred. Furthermore, hard statistical data can be combined with softer expert opinion, perhaps regarding environmental factors or design considerations. Extensions of the basic BN further enhance the capability of the approach to not just match but exceed the power of traditional reliability modelling techniques such as fault trees, reliability block diagrams and state-space methods. Demonstrating the power of probabilistic representations, BNs have also been shown to offer numerous advantages in fault diagnosis and detection over alternative Artificial Intelligence approaches such as rule-based systems, neural networks and case-based reasoning.

15.3.1 Bayesian Networks

Bayesian networks (BN) belong to the family of probabilistic graphical models (Cowell 1999). Graphical models exploit graphs to express dependencies between variables in a modelled domain. The graphs are easily readable and intuitive for humans and at the same time are convenient tools from an algorithmic perspective. In graphical models, variables are represented by nodes in the graphical part and dependencies between variables are represented by links.

The ability of BNs to provide a flexible and powerful probabilistic modelling framework makes them suitable for applications across the field of reliability and maintenance. The earliest and still the most popular application is to fault diagnosis. Sanseverino and Cascio (1997) describe one such application to automotive electronic sub-systems which has been implemented in hundreds of Fiat repair centres across Italy. More recently, Romessis and Mathioudakis (2006) developed a BN for fault diagnosis of gas turbine performance in the jet engine domain. Chan and McNaught (2008) describe how BNs play a key role in a decision support system designed to provide advice on fault diagnosis and correction during the final system testing of mobile telephone base stations at Motorola. Another

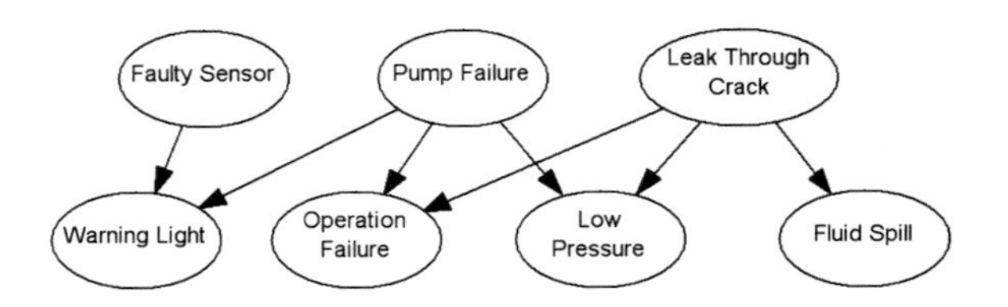

Fig. 15.2 Example of Diagnostic BN

application in the mobile telephone industry involves the use of a BN to help resolve problems in Nokia networks (Barco et al. 2005).

Indeed, many of the flagship applications of BNs are diagnostic applications, both medical (Shwe et al. 1991; Ogunyemi et al. 2000) and hardware (Przytula and Thompson 2001; Lerner et al. 2000; Forman et al. 1998). Among the different approaches to applying BNs to diagnostic problems (Lerner et al. 2000; Forman et al. 1998), one class of BN seems to be particularly successful. This class is named BN2O—a two-layer network in which arcs can only connect nodes from the top layer with nodes in the bottom layer, and nodes in the bottom layer are assumed to have a particular dependence on their parents called the noisy-OR interaction (a limitation on the probability distribution) (Pearl 1988). For the diagnostic BN2O networks, the nodes in the top layer represent failure modes (*faults*) while the nodes in the bottom layer represent *observations*, which can include error messages, symptoms, diagnostic tests, etc. The main development task is to define the fault variables in the model and for each fault, to identify relevant observations. This task is typically performed by experts who have little problem with naming all of the typical symptoms for each failure mode. There are two types of parameters that are required to populate these models: prior probabilities of faults and conditional probabilities of observations given faults are present. For hardware diagnostic domains, the prior probabilities of faults are typically easily approximated by using repair records or spares consumption records. More challenging are conditional probabilities of symptoms assuming the fault is present. However, in practice, for good performance of the models it is usually sufficient to use approximate values. One of the approaches is to elicit verbal descriptions (almost always, rarely, very rarely, etc.) from domain experts and convert them into numerical probabilities.

A simple example of a diagnostic BN is presented in Fig. 15.2. The example assumes three faults: *Faulty Sensor*, *Pump Failure*, and *Leak Through Crack*. The observable variables are: *Warning Light*, *Operation Failure*, *Low Pressure*, and *Fluid Spill*. In real models, observations are typically grouped in different categories. For example: mandatory observations, initial observations, and tests. Mandatory observations are assumed to be known for sure at the time when the diagnostic process starts. Typical examples would be fault and/or usage data from an onboard computer, that includes error messages that appeared or not during operation.

There are several factors that contribute to the success of this methodology: (1) the models capture how the system fails, not how it operates. In real applications, failures are driven by reliability of components and there are modes of failure that are difficult to predict in the design process—these can not be determined from design and functional blueprints. (2) This modelling technique is suitable for large, modular systems where failure modes and interactions between components are well defined, thus making it suitable for engineered systems. (3) The model is driven by probabilities, hence accounts for uncertainty and is relatively robust to partial or false information. (4) The statistical inference is objective, therefore it avoids known fallacies of human diagnosticians (such as confirmation bias, disregarding certain evidence, etc.).

To allow modelling of maintenance decisions, influence diagrams or decision networks are a natural extension of BNs, permitting additional types of nodes to represent decisions and capture costs, benefits and preferences via utilities. Early applications of these to reliability and maintenance modelling include Agogino and Ramamurthi (1990) describing the use of influence diagrams to control the real-time monitoring and control of milling machines, and Oliver and Yang (1990) describing their use within the setting of nuclear power safety analysis. Another application is troubleshooting where the costs of observation, testing and downtime can also be accounted for in deciding what intervention to take next, e.g., replace or repair a component, or perform a diagnostic test. An example of this approach applied to computer printers is provided by Heckerman et al. (1995).

Practical diagnostic models can easily reach several thousand nodes. Authoring tools such as GeNIeRate (Kraaijeveld and Druzdzel 2005) have been developed to support domain experts without requiring them to be familiar with BNs. A useful feature of BNs is that once a model is developed, it can still 'learn' from new cases. One possibility is to employ Bayesian updating of probability parameters Heckerman (1997). A simple example concerning application of a BN to machine tool maintenance is presented by Gilabert and Arnaiz (2006). In this they demonstrate a method called 'fading' which also supports adaptation of the network via updating of probability parameters based on observation of new cases.

There are many areas in engineering where it is beneficial to be able to account for 'soft' factors. In the process of designing and producing a new product, for example, it is clear that factors such as 'experience with similar products' and 'level of quality testing' will have an impact on final product reliability. In the past, such factors were largely ignored in reliability prediction given the difficulty of incorporating them in traditional tools. Neil et al. (2006) describe how this approach was implemented in the TRACS project, a system based on a BN which has been developed to assist the UK Ministry of Defence in the estimation of military vehicle reliability.

15.3.2 Dynamic Bayesian Networks

The traditional BN is a static model, representing variables at a snapshot in time. However, the approach has been extended to represent temporal aspects (Dean and Kanazawa 1989), in the form of dynamic BNs (DBNs). This permits traditional

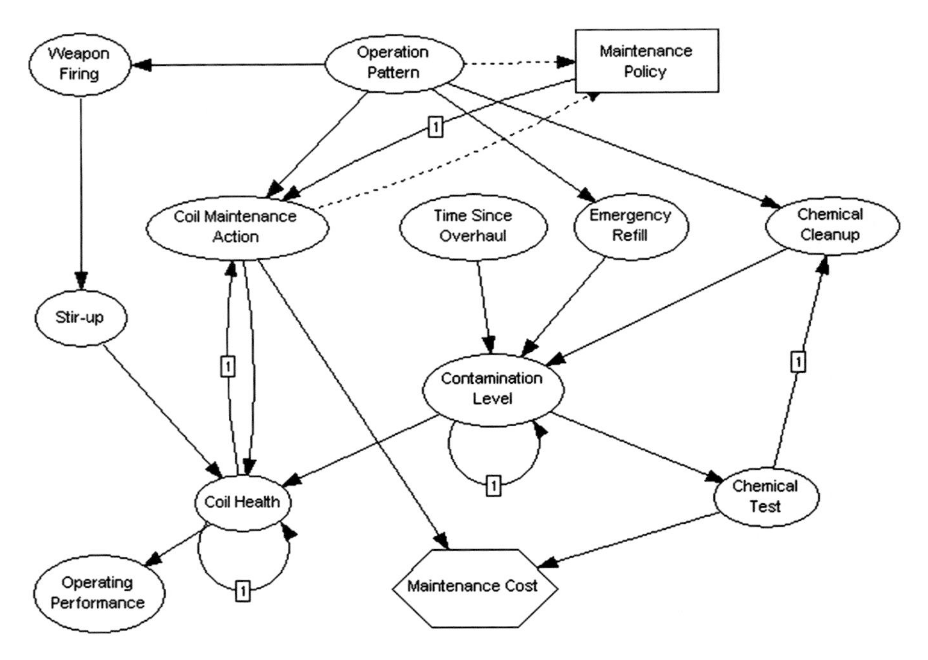

Fig. 15.3 Example of DBN for a Prognostic Problem

Markov Chain modelling and more, while retaining the flexibility of the BN framework. Essentially, to be able to represent the evolution of a system over time, variables are replicated in each time slice. While arcs between variables in the same time slice have the same meaning as in a normal BN, arcs between variables in different time slices correspond to temporal dependencies. Muller et al. (2008) describe a prognostic maintenance model employing a DBN with a manufacturing example involving metal bobbins. Dynamic variables are used to track a number of degradation mechanisms and the impact of different maintenance policies can then be evaluated.

An example of a DBN for the prognostic domain is presented in Fig. 15.3. The model presented here is an altered version of a model intended for modelling the value of maintenance policies for fan coils which are a part of the chilled water system of the Type 45 destroyer.

The arcs in the network represent dependencies between variables at given time. Arcs labelled with numbers (called *temporal arcs*) indicate dependencies between variables that occur over time and the numbers indicate the order of the time lag. For predictive calculations the model is expanded as a chain of models similar to the one presented here, but linked by temporal arcs. Strictly speaking, the model is an influence diagram, as it includes decision and utility nodes (denoted by rectangles and hexagons, respectively). These allow encoding of utilities (such as maintenance costs) and the calculation of optimal decisions (preferred maintenance policies).

Past information on maintenance, and other relevant facts (such as the operational pattern of the ship and any gun firings, which are known to affect the chilled water system) are entered into the model as evidence. Probabilities related to future usage can be adjusted as well, based on current knowledge about the platform. This information is then used to calculate probabilities and expected utilities for different maintenance regimes using inference algorithms.

15.4 Practitioner's Guide to Implementing CBM and Prognostic Modelling

The first step in model development is problem definition. However obvious this step seems to be, it should not be taken lightly. Prognostics and CBM are broad terms and the problem to be solved needs to be precisely defined. From lessons learned during this research, it is obvious that prognostics and CBM are understood differently by different individuals. This should not be a surprise, as the terms are broad and leave much room for interpretation.

For example, even such basic concept as remaining useful life is not really well defined. It requires clarification of what *useful* means and it should be expected that the predictions of remaining useful life will not be precisely defined to a particular date and time. Instead, definition of suitable confidence intervals or acceptable thresholds should be agreed.

The problem definition statement should identify which components or systems will be considered by the model. This step is a critical one in terms of future benefits that the model can bring. Some naïve futuristic visions often portray future vehicles as capable of monitoring every component of the vehicle and predicting failures accurately to the very moment of their occurrence. The reality is very different. Monitoring and prognostics makes sense from financial, safety and reliability perspectives only for selected components and systems. To identify such systems is the key to success of applying prognostics and CBM. A successful application of prognostics and CBM is one where the benefits outweigh the costs of development and implementation of the models and the new maintenance regime.

The costs of implementation of prognostic models and CBM can include:

- Development, installation and maintenance of the monitoring systems (HUMS).
- Cost and effort for data collection and storage (including required infrastructure and implementation of new practices).
- Cost of initial model development for prognostic purposes (including research time and tool development). Risk due to uncertainty of results may be significant for this element.
- Cost of model maintenance, updates due to system re-designs and new failure modes that may need to be included in the model.
- In case of CBM, the cost of introducing new processes, adapting logistic arrangements and organisational changes.

Factors that can significantly boost the benefits of introducing prognostics and CBM for particular components and systems are:

- Significant cost of unpredicted failures, such as cost of downtime, cost of shipping the asset to a workshop, cost of failure propagation through the system.
- Safety implications—can be related to safety implications of failure modes.
- Large number of similar components/systems in service.
- Failure rate of components—prognostics and CBM are more suitable for assets that fail more often and less suitable for high-reliability assets.
- Availability of monitoring techniques—either those already installed on the platform or the installation of additional low-cost HUMS.
- Affordable HUMS data collection and storage.

The second step is to identify the goals of the model. A model for defining optimal maintenance schedules based on constant condition monitoring is the ultimate goal. But such aims may not be realistic to achieve. Even though CBM is the ultimate goal, some intermediate solutions may be of great value. There are three groups of possible benefits that can be targeted.

- Prognostic models—this type of model is typically specific to the component level. They are concerned with prediction of remaining useful life of the component and they are focused on particular cases.
- Diagnostic model with elements of prognosis—this type of model is specific to more complex systems that consist of multiple components. Basically, these models extend diagnostic models by incorporating elements of prognostic modelling. The diagnostic part is necessary to identify elements responsible for failures or particular anomalies in HUMS data.
- Defining maintenance schedules—these models are focused on the broader problem of modelling the value of maintenance schedules. They differ from the previous types in that they include information on costs of operation, maintenance, failures, etc. To address the problem, they should be looking at the issue from the fleet perspective rather than at individual cases.

There are particular problems for which certain goals are more suitable than others. For example, for a component or a simple system, the prognostic model may be sufficient—information on the status of the asset can be sufficient for the maintenance personnel to decide on the most suitable maintenance activities without the need for an elaborate cost-benefit analysis.

15.4.1 Identification of Available Data

One of the main challenges in applying prognostic models (PGMs) to real-life domains is the model creation stage. The model itself is only as good as the data and expertise used to create it. Therefore, our main focus should be placed on

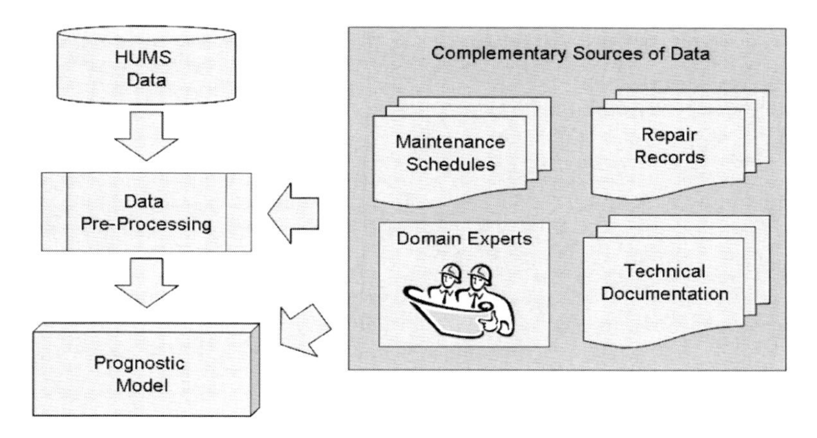

Fig. 15.4 Conceptual flow of data and information for a PGM

obtaining relevant, high-quality data and gaining a good understanding of the system to be modelled.

Even though modern HUMS provide large amounts of data, these data are often not directly applicable for prognostic purposes. This can be for many reasons. For example, HUMS are typically designed as monitoring systems and their primary goal is to provide real-time data for the system operators rather than for maintenance personnel. The operators may not be particularly concerned with the predictive modelling aspect of maintenance. Another factor is lack of understanding of the physics of failure—in some systems it is not clear which measurements are relevant to estimation of the residual life in the system. Therefore, it should be expected that many measurements which may be critical for the purpose of residual life estimation might simply not be collected.

In practice, some form of pre-processing of HUMS data will be necessary to make these data directly usable for the PGM. We expect that this pre-processing will be heavily problem-domain specific. For this purpose, we envisage using standard data-exploration techniques (for example, principle component analysis) and other data-mining techniques.

The HUMS data offer only partial information on the system. Modes of failure, dependencies between the system's components and their reliabilities should be obtained from other sources as HUMS simply does not cover this type of information. The basic interdependencies between these aspects can be deduced from simplified blueprints and easily provided by the domain experts. Similarly, reliability data should be provided based on the system's maintenance and fault reports. It is desirable to consider all readily available information sources. The conceptual flow of data and information needed to define a PGM is presented in Fig. 15.4.

15.4.2 Model Development

Model development involves the integration of available data and knowledge into a single framework which formalises them using a common language—for example, probabilistic graphical models. This process is very much problem specific and only a set of guidelines can be postulated.

The first step in model development should be the data exploratory phase. In this phase, familiarisation with available data, clarification of the meaning of variables and their relevance to the system should take place. At this stage, the data should be subject to initial statistical and data mining pre-processing. The goal is identification of variables that would be most relevant to the development of the domain-specific PGM and, in general, gaining insight into the quality of the data and its relevance to the problem.

The first version of the model should be constructed by a modelling engineer who is experienced in PGM—we will call this person a "knowledge engineer" who will develop the model based on initial meetings with domain experts and insights gained during the data exploration phase. At this point, the model is defined based on the understanding of the problem by the knowledge engineer. Once this is done, the initial model should be presented to the domain experts for discussion. Preferably, this should be done by means of a formal presentation which should cover the data used, assumptions made and initial results. A prototype model should also be presented by using some model authoring software. All variables, interactions between variables and parameters should be shown and discussed. Based on our previous experience, given the intuitive nature of BNs, it is not necessary for domain experts to have any familiarity with this modelling framework to be able to understand and discuss the model.

The interaction between the knowledge engineer, domain experts and prospective model users is one of the key elements for the success of the project. This is because it encourages knowledge sharing and facilitates better understanding of the problem. It relates not only to the knowledge engineer but also to the domain experts, who gain a deeper understanding of their domain by exercising different views of the problem.

Once the model is declared as finished from the perspective of including data and knowledge sources, a phase of model refinements may be necessary. This is especially true for models intended to capture the value of applying CBM, where modelling of the costs is also required. Sensitivity analysis allows testing models to identify variables and parameters that are most influential in producing results. Since the models are built from uncertain data and knowledge elicited in less than a sterile manner, it should be expected that, at least some of the information used to define the model will be more 'rough estimates' than precise measurements. Sensitivity analysis allows for testing the robustness of the results with respect to such factors.

15.4.3 Model Updates

The system introduced to service will undergo changes over time, some of which will be long planned for and some of which will arise at short notice for various reasons. Different mission profiles and operating conditions may emerge and costs and gains may also change over time. Fresh data and knowledge about the system and its failure modes will also keep accumulating. Consequently, the models for prognostics should be expected to require modification over time to reflect the changing system and its environment and improved knowledge. One of the possible benefits of prognostic models is to gain an understanding of the system and its failure profiles that can be exploited for informed system upgrades and improvement of its maintenance.

15.5 Summary and Conclusions

We discussed in detail the challenges of implementing prognostics, particularly in regard to the informational transformation required: lack of good understanding of physics of failure processes, uncertainties related to future use of the system, and relative immaturity of scientific approaches. We concluded that recent advances in monitoring techniques and their popularity on new platforms provide a basis for the development of prognostic models.

We discussed the suitability of probabilistic graphical models as a tool for prognostic models. As probabilistic graphical models have been a very successful tool for practical, intelligent diagnostics in the automotive and aircraft industries, it is worth considering if they have the same potential for prognostic applications. We presented an analysis of the sources of success of diagnostic BNs and discussed requirements for prognostic BNs.

It is clear that a mixture of methods will be required for prognostics and the development of predictive maintenance including physics-of-failure models, data-driven, statistical and knowledge-based approaches. We believe that several classes of model show significant promise for future development, including BNs, proportional hazard models, recurrent and ensemble neural networks, physics-of-failure methods and data filtering approaches. While Hess et al. (2006) suggest that some form of integrative framework will be required to handle various information sources, they do not say what that should be. In our opinion, BNs have much to offer in this regard. They assist with the management of uncertainty and the combination of expert opinion with imperfect observations, and are also extendable to permit the incorporation of costs, thus assisting with trade-off analysis.

Process modelling and functional decomposition are combined with the BN approach to provide a system level prognostic modelling tool in Muller et al. (2007). This is in contrast to most approaches in the literature which concentrate only on components and reaffirms our belief that the BN framework offers considerable promise for PHM and should be investigated and developed further. Future work may also seek to integrate some of the data-driven approaches discussed above within the overall BN framework.

15.6 Chapter Summary Questions

Although this chapter has focused on tools to support information transformation, the implementation of these tools along with HUMS and condition monitoring will require process changes which in turn will require people transformation, particularly in the customer organisation.

- What transformations within the customer organisation would be required to ensure proper implementation and adaptation of CBM?
- How can we identify subsystems where CBM can lead to cost-effective support and lead to value co-creation?
- How can we account for and quantify all of the benefits and costs related to multiple transformations?

References

A.M. Agogino, K. Ramamurthi, Real-time influence diagrams for monitoring and controlling mechanical systems, in *Influence diagrams, belief nets and decision analysis*, ed. by R.M. Oliver, J.Q. Smith (Wiley, Chichester, 1990), pp. 199–228

R. Barco, V. Wille, L. Diez, System for automated diagnosis in cellular networks based on performance indicators. Eur. Trans. Telecommun. **16**, 399–409 (2005)

P. Baruah, R.B. Chinnam, HMMs for diagnostics and prognostics in machining processes. Int. J. Prod. Res. **43**, 1275–1293 (2005)

E.R. Brown, N.N. McCollom, E.E. Moore, A. Hess, Prognostics and health management: A data-driven approach to supporting the F-35 Lightning II. IEEE Aerospace Conference 2007 (2007)

F. Cadini, E. Zio, D. Avram, Monte Carlo-based filtering for fatigue crack growth estimation. Probab. Eng. Mech. **24**, 367–373 (2009)

A. Chan, K.R. McNaught, Using Bayesian networks to improve fault diagnosis during manufacturing tests of mobile telephone infrastructure. J. Oper. Res. Soc. **59**, 423–430 (2008)

R.G. Cowell, A.P. Dawid, S.L. Lauritzen, D.J. Spiegelhalter, *Probabilistic Networks and Expert Systems* (Springer, Berlin 1999)

T. Dean, K. Kanazawa, A model for reasoning about persistence and causation. Comput. Intell. **5**, 142–150 (1989)

Department of Defense, MIL-STD-1629A: Procedures for performing a failure mode effects and criticality analysis (1980)

G. Forman, M. Jain, M. Mansouri-Samani, J. Martinka, A. Snoeren, Automated end-to-end system diagnosis of networked printing services using model-based reasoning. Software Technology Laboratory, HPL-98-41 (R.1) (1998)

E. Gilabert, A. Arnaiz, Intelligent automation systems for predictive maintenance: A case study. Robot. Comput. Integr. Manuf. **22**, 543–549 (2006)

J. Gu, M. Pecht, New methods to predict reliability of electronics. Proceedings of 7th international conference on reliability, maintainability and safety (2007), pp. 440–451

D. Heckerman, Bayesian networks for data mining. Data Min. Knowl. Discov. **1**, 79–119 (1997)

D. Heckerman, J.S. Breese, K. Rommelse, Decision-theoretic troubleshooting. Commun. ACM **38**, 49–57 (1995)

F.O. Heimes, Recurrent neural networks for remaining useful life estimation. International conference on prognostics and health management, PHM 2008

A. Heng, S. Zhang, A.C.C. Tan, J. Mathew, Rotating machinery prognostics: State of the art, challenges and opportunities. Mech. Syst. Signal Process **23**, 724–739 (2009a)

A. Heng, A.C.C. Tan, J. Mathew, N. Montgomery, D. Banjevic, A.K.S. Jardine, Intelligent condition-based prediction of machinery reliability. Mech. Syst. Signal Process **23**, 1600–1614 (2009b)

A. Hess, G. Calvello, P. Frith, S.J. Engel, D. Hoitsma, Challenges, issues and lessons learned chasing the Big P: real predictive prognostics part 2. (IEEE Aerospace Conference 2006)

G.J. Kacprzynski, M. Roemer, A.J. Hess, K.R. Bladen, Extending FMECA—health management design optimization for aerospace applications. IEEE Aerospace Conference (2001)

K.A.H. Kobbacy, B.B. Fawzy, D.F. Percy, A full history of proportional hazards model for preventive maintenance scheduling. Qual. Reliab. Eng. Int. **13**, 187–198 (1997)

P. Kraaijeveld, M.J. Druzdzel, GeNIeRate: An interactive generator of diagnostic Bayesian network models. Working notes of the 16th international workshop on principles of diagnosis (DX-05): (2005), pp. 175–180

D. Kumar, B. Klefsjo, Proportional hazards model: A review. Reliab. Eng. Syst. Saf. **44**, 177–188 (1994)

H. Langseth, L. Portinale, Bayesian networks in reliability. Reliab. Eng. Syst. Saf. **92**, 92–108 (2007)

U. Lerner, R. Parr, D. Koller, G. Biswas, Bayesian fault detection and diagnosis in dynamic systems. Proceedings of the 17th national conference on artificial intelligence (AAAI) (2000), pp. 531–537

Y. Li, T.R. Kurfess, S.Y. Liang, Stochastic prognostics for rolling element bearings. Mech. Syst. Signal Process **14**, 747–762 (2000)

J.K. Line, N.S. Clements, Prognostics usefulness criteria. IEEE aerospace conference 2006 (2006)

Z. Ma, A new life system approach to the prognostic and health management (PHM) with survival analysis, dynamic hybrid fault models, evolutionary game theory, and three-layer survivability analysis. IEEE aerospace conference 2009 (2009)

A. Muller, M.C. Suhner, B. Iung, Maintenance alternative integration to prognosis process engineering. J. Qual. Maint. Eng. **13**, 198–211 (2007)

A. Muller, M.C. Suhner, B. Iung, Formalisation of a new prognosis model for supporting proactive maintenance implementation on industrial system. Reliab. Eng. Syst. Saf **93**, 234–253 (2008)

M. Neil, M. Tailor, D. Marquez, N. Fenton, P. Hearty, Modelling dependable systems using hybrid Bayesian networks. Proceedings of 1st international conference on availability, reliability and security (2006), pp. 817–821

R. Normann, R. Ramirez, From value chain to value constellation: designing interactive strategy. Harv. Bus. Rev. (July–August): (1993), pp. 65–77

O. Ogunyemi, J.R. Clarke, B. Webber, Using Bayesian networks for diagnostic reasoning in penetrating injury assessment. Proceedings of the 13th IEEE symposium on computer-based medical systems (CBMS'00): (2000), pp. 115–120

R.M. Oliver, H.J. Yang, Updating of event tree parameters to predict high risk incidents, in *Influence diagrams, belief nets and decision analysis*, ed. by R.M. Oliver, J.Q. Smith (Wiley, Chichester, 1990), pp. 277–296

R.F. Orsagh, D.W. Brown, P.W. Kalgren, C.S. Byington, A.J. Hess, T. Dabney, Prognostic health management for avionic systems. IEEE Aerospace Conference 2006 (2006)

J. Pearl, *Probabilistic reasoning in intelligent systems: Networks of plausible inference* (Morgan Kaufmann, San Mateo, 1988)

L. Peel, Data driven prognostics using a Kalman filter ensemble of neural network models. International conference on prognostics and health management (2008)

O. Pourret, P. Naïm, B. Marcot, *Bayesian networks: a practical guide to applications* (Wiley, New York 2008)

K.W. Przytula, D. Thompson. in *Proceedings of SPIE Component and Systems Diagnostics, Prognosis, and Health Management*, ed. by P.K. Willett, T. Kirubarajan. Development of Bayesian diagnostic models using troubleshooting flow diagrams. 4389: 110–120 (2001)

C. Romessis, K. Mathioudakis, Bayesian network approach for gas path fault diagnosis. J. Eng. Gas. Turbines Power Trans. ASME **128**, 64–72 (2006)

M. Sanseverino, F. Cascio, Model-based diagnosis for automotive repair. IEEE Expert. **12**, 33–37 (1997)

H. Saranga, J. Knezevic, Reliability analysis using multiple relevant condition parameters. J. Qual. Maint. Eng. **6**, 165–176 (2000)

M.A. Shwe, B. Middleton, D.E. Heckerman, M. Henrion, E.J. Horvitz, H.P. Lehmann, G.F. Cooper, Probabilistic diagnosis using a reformulation of the INTERNIST–1/QMR knowledge base: I. The probabilistic model and inference algorithms. Methods Inform. Med. **30**, 241–255 (1991)

W. Wang, A.H. Christer, Towards a general condition based maintenance model for a stochastic dynamic system. J. Oper. Res. Soc. **51**, 145–155 (2000)

G. Wen, X. Zhang, Prediction method of machinery condition based on recurrent neural network models. J. Appl. Sci. **4**, 675–679 (2004)

J. Yan, M. Koc, J. Lee, A prognostic algorithm for machine performance assessment and its application. Prod. Plan Control **15**, 796–801 (2004)

Chapter 16
Component Level Replacements: Estimating Remaining Useful Life

W. Wang and M. J. Carr

Abstract Condition based maintenance modelling can be used to maximise the availability of key operational components that are subject to condition monitoring processes, such as vibration or oil based monitoring and thermography. This chapter addresses the operational need for a component replacement decision analysis problem utilising available condition monitoring data and incorporating various prognostic modelling options for the estimation of remaining useful life which is essential in a prognostics model. Guidelines are presented in the chapter which enable the selection of an appropriate prognostic model for a given application based on the characteristics of the scenario and the availability of historical data to train the model. Consideration is then given to scenarios where historical data are scarce or unavailable and new modelling developments are presented to cater for this contingency. The objective of the modeling process is to maximise availability whilst avoiding the occurrence of costly component failures. The developed model has been programmed into a prototype software package for facilitating the implementation of the methodology.

16.1 Introduction

When developing a contract for service provision, it is important to define the necessary capability and reliability requirements for the equipment in question. In this chapter, we investigate the tactical planning issues and short term decision

W. Wang (✉)
Salford Business School, University of Salford, Salford, M5 4WT, UK
and
School of Economics and Management, University of Science and Technology, Beijing, China
e-mail: w.wang@salford.ac.uk

M. J. Carr
School of Medicine, The University of Manchester, Manchester, UK

I. Ng et al. (eds.), *Complex Engineering Service Systems*, Decision Engineering,
DOI: 10.1007/978-0-85729-189-9_16,

making associated with high value, high risk operational components, parts and sub-systems. The components in question are those which are subject to condition monitoring (CM) processes and require adaptive dynamic solutions for utilisation in condition based maintenance (CBM) models. The use of CM techniques is rapidly increasing for key operational components and has lead to increased availability and a reduction in the occurrence of failures and the costs associated with maintenance. The availability of CM information data provides a decision maker an additional information to assess the condition of the plant and make appropriate maintenance decisions. One of the obvious decisions is whether to replace the component monitored now or later based on observed CM history. However, how to best use the available CM information is still a question not answered properly. This leads to the focus of this chapter, which is on the utilisation of monitoring techniques in maintenance and replacement decision models. The objective is to provide informed decision support for the application of CBM/replacement models in a limited or incomplete information environment. The type of CM information that we consider can be 'indirect' or 'direct'. Direct CM processes are those where the measurement or observation relates exactly to the condition or state of the component, e.g., component wear or degradation, and the objective is to predict the evolution of the degradation. In the case of indirect CM processes, the observed information is assumed to be stochastically indicative of the underlying condition. As such, we are concerned with estimating the state and then predicting how it will evolve over time.

Indirect CM techniques include vibration based monitoring, oil analysis, infrared thermography, acoustic emission analysis and motor current analysis. The overall vibration level (or total energy of the vibration signal) combined with a detailed spectral analysis of the signal is commonly used in the fault diagnosis of rotating machinery. Vibration signals often correspond to a two stage process over the life of a component with the first stage being 'normal' operation and the second being 'defective' but operational; see Wang (2002). Most machine types can potentially benefit from vibration monitoring except those that operate at very low speeds. Techniques for sampling and analysing engine, transmission and hydraulic oils are used to monitor wear and contamination and can be used to identify impending failures before they occur. There is also the potential for reducing the frequency of oil changes. Available oil-based monitoring techniques include spectrometric oil analysis, electron microscopy, X-ray analysis and ferrous debris quantification. See Collacott (1997) for a detailed review of CM techniques.

The literature on diagnostic techniques that utilise CM information to determine the current state of a component or system is vast; see Jardine et al. (2005). The literature on prognostic modelling techniques that use the indicatory condition information for extrapolation purposes, in order to predict future states such as component/system failures, is far less comprehensive. In many situations, condition monitoring can be ineffective if it is not combined with appropriate prognostic modelling. CBM techniques are used to minimise the occurrence of failures and increase the availability of the component for operational purposes in a cost effective manner. CBM is superior to breakdown maintenance or frequent as possible policies as action is typically only undertaken when deemed to be necessary. Available prognostic techniques include stochastic filtering (Wang and Christer 2000; Lin and Makis 2003; Wang 2002, 2006), proportional hazards modelling (Kumar and Westberg 1997), Makis and Jardine (1991) and Jardine et al. 1998), accelerated life models (Kalbfleisch and Prentice 1980), and Hidden Markov models (Bunks and McCarthy 2000). The typical objective of a CBM application is to develop an automated on-line maintenance decision system incorporating CM processes, prediction abilities, scheduling routines and updating capabilities.

CBM techniques are often misused or implemented ineffectively in operational contexts. Complete past datasets of CM information with associated failure times are rarely available to establish and initialise the maintenance model and the costs associated with collecting suitable data can be high. With the research documented in this chapter, we are attempting to address these practical issues in order to develop a prognostic model that is as robust as possible for future CM and CBM prognostic applications.

This chapter is organised as follows. Section 16.2 presents a condition based replacement model. Section 16.3 discusses the guidelines for the selection of prognostic models for remaining life prediction. Section 16.4 introduces the prognostic models outlined in Sect. 16.3. Section 16.5 highlights the new developments made in the case of direct monitoring, while some part of modelling is provided in an appendix. Section 16.6 presents the numerical examples based on one of the new developments in Sect. 16.5 and Sect. 16.7 concludes the chapter.

16.2 Replacement Decision Modelling

The prognostic models proposed in this chapter involve the development of a conditional probability density function (pdf) for the time until failure of an individual component using observed CM information. A key advantage in establishing a conditional pdf over the evaluation of a single point estimate of the failure time is the availability of the conditional cumulative density function (cdf). At each CM point throughout the life of a component, an optimal replacement time can be scheduled using renewal-reward theory and the long run 'expected cost per unit time' (Ross 1996). The conditional pdf and the resulting cdf are used in the development of the replacement decision model that incorporates the probability of failure before a particular instant conditioned on the CM history to date.

Figure 16.1 illustrates a CM process over time for an individual component where, t_i is the time of the ith CM point and y_i is the observed CM information. The failure time of the component is represented by T.

For the modelling of an individual component, we employ the notation;

- T_f—random variable representing failure time
- m—number of CM points before failure

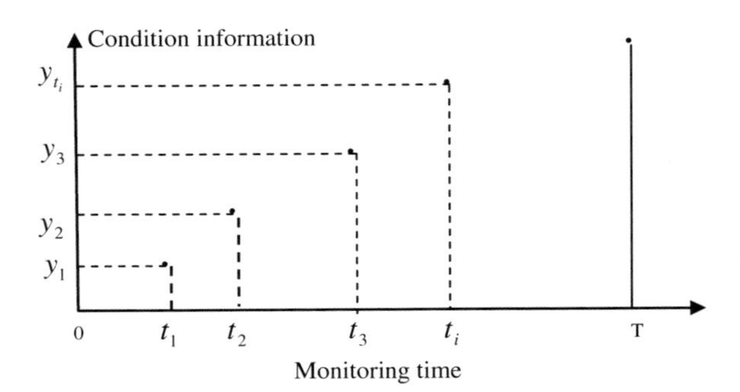

Fig. 16.1 The CM process for an individual component

- t_i—time of ith CM point (irregularly spaced)
- $Y(t)$—random variable representing CM information at time t
- $y(t)$—observed realization of $Y(t)$
- $\mathbf{Y}(t_i)$—history of CM observations at ith CM point; $\mathbf{Y}(t_i) = \{y(t_1), y(t_2),\ldots, y(t_i)\}$
- T_R—planned replacement time (to be optimised)
- c_p—preventive replacement cost
- c_f—failure replacement cost
- $C_i(T_R)$—expected cost per unit time given T_R at t_i

We seek to determine, at a given CM point, whether or not a preventive replacement is required before the next scheduled CM point. At the ith CM point at time t_i, the expected cost per unit time is given by

$$C_i(T_R) = \frac{\text{Expected Cost Given } T_R}{\text{Expected Life Given } T_R} \tag{16.1}$$

which is minimised to obtain the optimal replacement time T_R^*. As such, the replacement decision at the ith CM point is obtained via the minimisation of

$$C_i(T_R) = \frac{c_p + (c_f - c_p)\Pr(T_f \le T_R|\mathbf{Y}(t_i))}{t_i + (T_R - t_i)\{1 - \Pr(T_f \le T_R|\mathbf{Y}(t_i))\} + \int_{t_i}^{T_R} u\,p(T_f = u|\mathbf{Y}(t_i))du} \tag{16.2}$$

for $T_R \ge t_i$ where,

- Pr $(T_f \le T_R|\mathbf{Y}(t_i))$ is the conditional probability of failure before the planned replacement time given the history of CM observations.
- $p(T_f = u|\mathbf{Y}(t_i))$ is the probability density function of failure at time u.

Clearly, $p(T_f = u|\mathbf{Y}(t_i))$ is the key in Eq. 16.2, which is the subject in the following sections of this chapter. Illustrations of the potential replacement decisions are given in Fig. 16.2. The solid curve represents a situation where replacement is scheduled at optimal replacement time T_R^* before the subsequent CM point. In the case of the dashed curve, the decision produced by the model is to

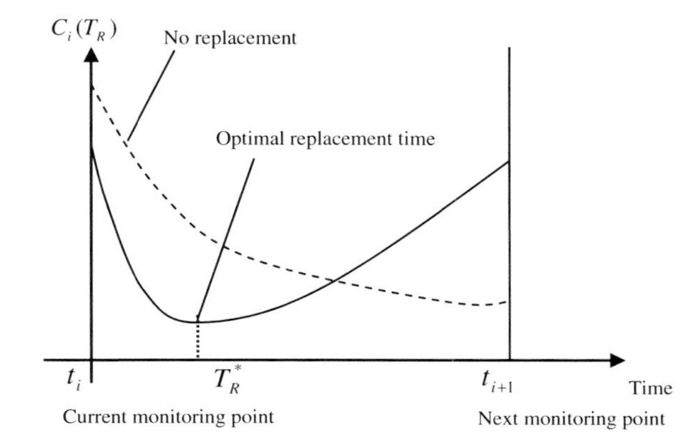

Fig. 16.2 Illustrating the replacement decision at the *i*th CM point

leave the component in operation until the next CM point and then re-assess the situation.

A simple alternative role for decision making would be to compare the product of the probability of failure and the consequence of such a failure for each key component and then those components with higher ranking should be scheduled at the next preventive maintenance window or as soon as possible depending on the maintenance capacity. This would provide a list for a maintenance dash board where appropriate components can be selected for maintenance execution.

16.3 Review of Prognostic Models

This section reviews the prognostic models discussed in this chapter.

16.3.1 Non-Linear Stochastic Filtering

Non-linear stochastic filtering is a prognostic technique that has been tailored for CBM applications; see Wang and Christer (2000), Wang (2002, 2006) and Carr and Wang (2008). The model has been employed in a number of previous case studies including applications involving rolling element bearings, aircraft engines and naval diesel engines. A principal difference of the non-linear filtering approach over alternative prognostic techniques, such as proportional hazards modelling, is that the observed CM data are assumed to be a function of the component's state, which is generally accepted in practice. The filter based approach also has an ability to incorporate the entire history of monitored information when making predictions about the time until failure of a component.

Figures 16.3, 16.4, 16.5 illustrate some typical output from the stochastic filtering model. At a given CM point at time t, the stochastic filter utilises the history

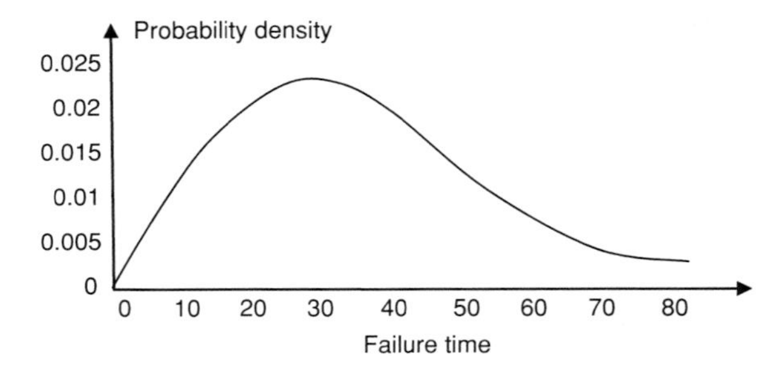

Fig. 16.3 The initial failure time distribution

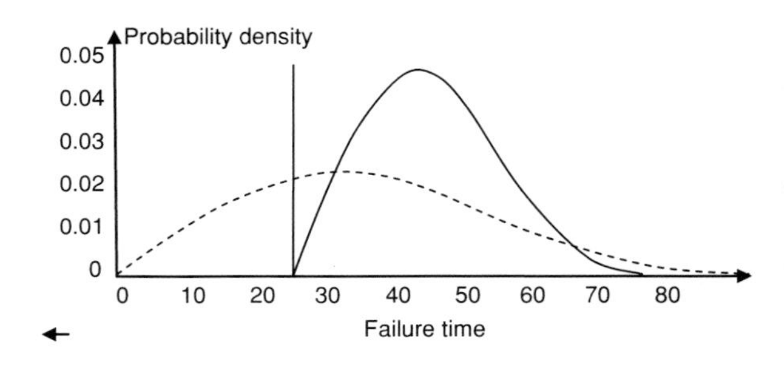

Fig. 16.4 The conditional failure time distribution at the 1st monitoring point

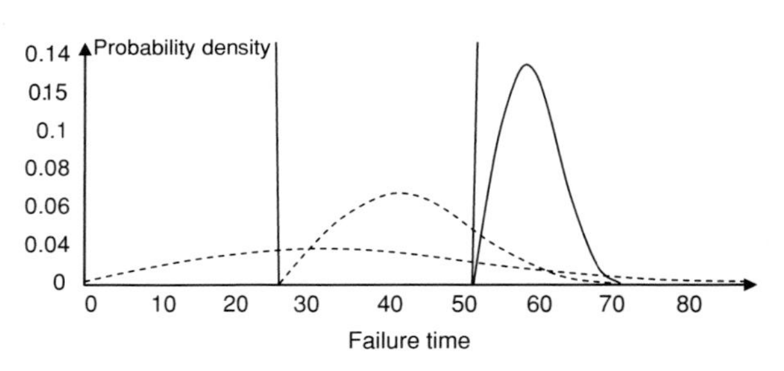

Fig. 16.5 The conditional failure time distribution at the 2nd monitoring point

of CM information to construct a conditional distribution for the failure time of the component. Initially, a prior component failure time distribution is specified and initialised using historical failure time information, as shown in Fig. 16.3.

Bayesian theory is used to iteratively update the failure time distribution using the age of the component and the available CM information. As more CM information is obtained, the error in the distribution about the actual upcoming failure time should decrease, assuming that a relationship exists between the observed CM information and the underlying condition of the component. The increasing accuracy of the failure time distribution is illustrated in Figs. 16.4 and 16.5 for monitoring points 1 and 2 respectively.

The prognostic failure time distribution is then incorporated into the replacement decision model in order to evaluate the optimal replacement time from the current CM point. Figure 16.6 illustrates some typical output from the replacement model with a stochastic filter as the prognostic input.

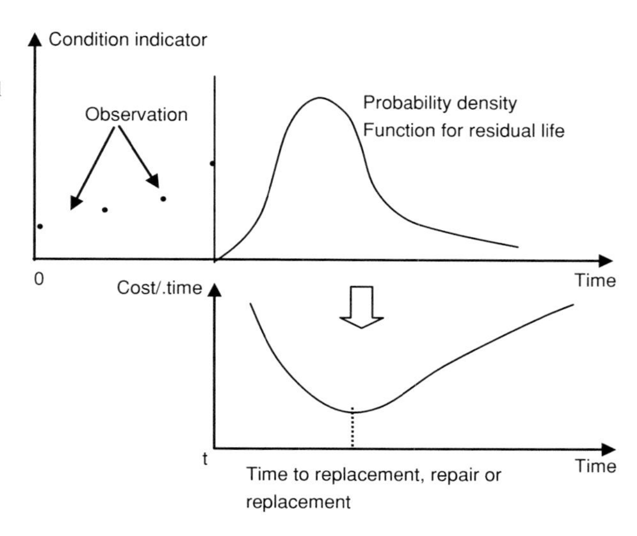

Fig. 16.6 Illustrating some typical output from the residual life estimation model

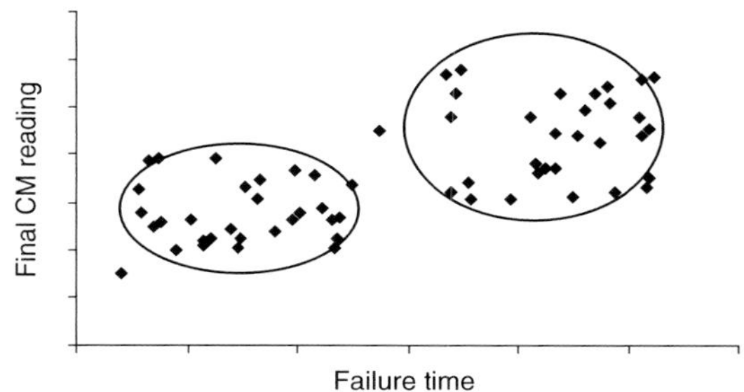

Fig. 16.7 Clustering of failure times under the influence of two different failure modes

16.3.2 Failure Mode Filtering

Many monitoring scenarios provide evidence that the operational components involved may potentially be subject to a number of individual distinct failure modes, rather than a single dominant failure mode as modelled previously. Figure 16.7 provides a typical example where the failure times of the components and the final CM reading before failure are observed to be clustered in the historical data.

The modelling procedure proposed to handle this scenario is based on the assumption that an individual monitored component will fail according to one of a number of predefined failure modes. Individual non-linear stochastic filters are constructed to facilitate the failure time prediction under the influence of each potential failure mode. The output from each filter is then weighted according to the probability that the particular failure mode is the true underlying (unknown) failure mode. The probabilities associated with each failure mode are recursively derived using a Bayesian model and the CM information obtained to date. The failure mode analysis filter is described in detail in Carr and Wang (2010a).

16.3.3 Extended Kalman Filtering

Many established prognostic modelling techniques for CBM can be computationally expensive and the input from multiple simultaneous sources of CM cannot be incorporated without resorting to data reduction algorithms or other approximate techniques, where much of the multivariate information is lost. The use of different locations for vibration sensors or multiple oil contamination readings is now very common and many established techniques are not satisfactory on a theoretical or performance based level. To overcome these significant shortcomings, we have developed an approximate methodology using extended Kalman filtering and the historical CM information to recursively establish the conditional failure time distribution. The one we developed is different from other extended Kalman filters since we use a Log-nomal distribution for the state variable. The filter can be used to construct the distribution in a computationally efficient manner for a large number of components and CM variables simultaneously. The model has been documented by Carr and Wang (2010b).

16.4 Guidelines for Prognostic Model Selection

For a given application, the choice of which prognostic model to use for the development of the conditional failure time distribution (used in the replacement decision model) is dependent on the nature of the CM information and the availability of historical data for initialisation purposes. Figure 16.8 presents a decision tree that can be used for prognostic model selection based on the type of CM employed.

In the next section, we discuss the new developments (highlighted in Fig. 16.8) during the course of the project.

16.5 New Prognostic Modelling Developments

As discussed previously, new developments are needed to model operational CBM process requirements. The approach presented here is a new methodology for failure time prediction and attempts to address some of the issues associated with operational scenarios and applications. The models developed are a type of direct monitoring. The approach has a number of advantages over the techniques including an ability to initialise the model using expert engineering opinion in the absence of failure time information. The model also incorporates updating capabilities when operational data become available. The predicted time to failure is based on the history of CM and is established using an initial threshold level and the characterisation of a failure zone for the condition information. This is a novel

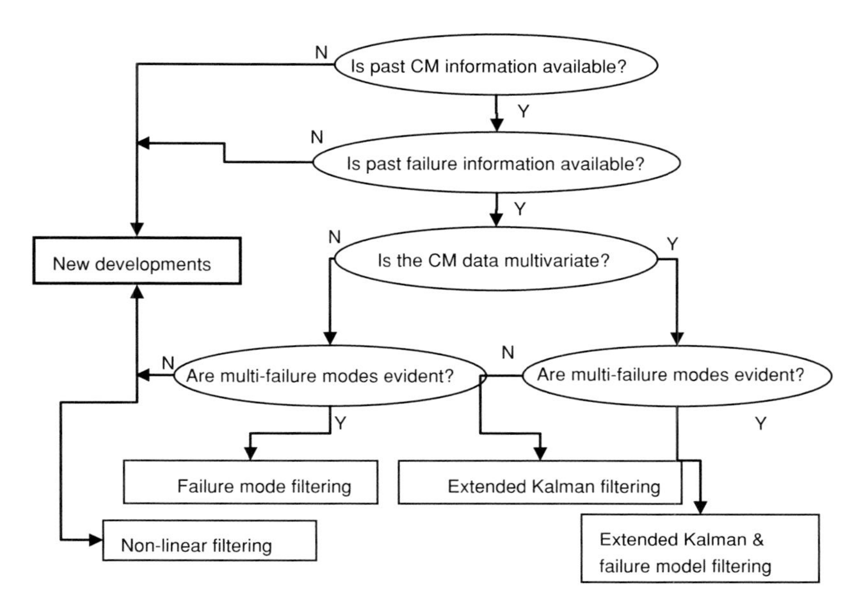

Fig. 16.8 Prognostic model selection decision tree

approach as most of the available research on threshold modelling involves the employment of a constant failure threshold level, whereas, here we use a distribution on the assumption that no such convenient threshold level exists and that each component should be treated individually. This implies that even when the CM reading has exceeded the initial threshold level, the component may only fail according to a probability distribution. We call this distribution a failure zone distribution. The model is practical, robust and applicable to any type of component or monitored information.

Figure 16.9 illustrates the modelling principles. When the CM information $Y(t)$ reaches a failure zone bounded by an open interval $[Y_L, \infty)$ and characterised by a random variable y_f with a pdf $f(y_f)$, where y_f is the level of y_t when the failure occurs, a failure is deemed to be imminent and the system needs to be shutdown for immediate attention. Both Y_L and $f(y_f)$ may not be fixed and can be initialised using subjective expert information then updated as objective data becomes available. The future CM path $Y(t)$ is described as a function of time with a random scaling coefficient and is modelled using two different variants and updated using Bayesian theory and the observed CM information. The first variant is a Kalman filtering algorithm and the second is based on adaptive Gamma process theory. The output from the model is consistent with that shown in Figs. 16.4, 16.5, 16.6 for the non-linear stochastic filter. Naturally, as more CM information is obtained, the models predictive capability is enhanced.

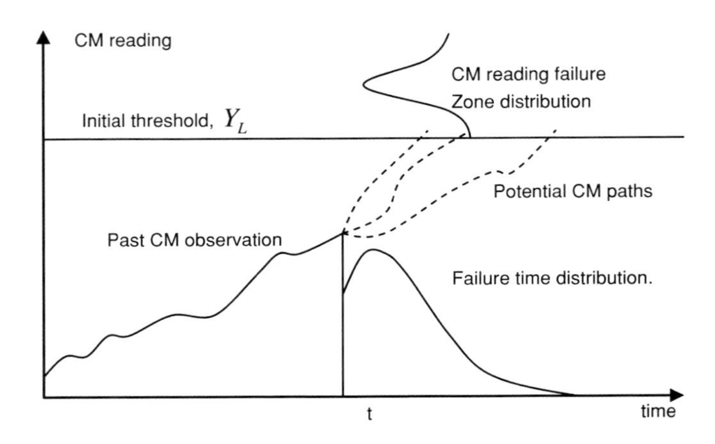

Fig. 16.9 Illustrating the modelling principles

There are two key issues for consideration when modelling the remaining useful life in an operational context:

1. modelling the CM reading failure zone
2. modelling the deterioration process of $Y(t)$

16.5.1 Modelling the Failure Zone

Here we assume that YL is the lower boundary of a random variable which characterises the failure zone in terms of observed monitoring signals. When the observation exceeds the threshold YL the probability of failure is described by another distribution. Both YL and the failure zone distribution are updated when new information becomes available. We use a Bayesian approach for the development of the failure zone. An initial distribution for the failure time CM reading is specified using historical threshold levels, failure time CM readings (when available) or expert engineering opinion. The initial distribution is then updated when additional information becomes available from components monitored after the initialisation of the model. For a detailed description of $f(y_f)$, see Wang and Carr (2010).

If at time t_i, $f_i(y_f)$ is available, then

$$\Pr(T_f \leq t|\mathbf{Y}(t_i)) = \int_{\max(y(t_i),Y_L)}^{\infty} \Pr(T_f \leq t|\mathbf{Y}(t_i), y_f)f_i(y_f)dy_f \qquad (16.3)$$

Now we discuss the models for $P_r(T_f \leq t|Y(t_i), y_f)$, that is, the probability of the residual life is less than t given the history of the observed CM data and a threshold level of y_f.

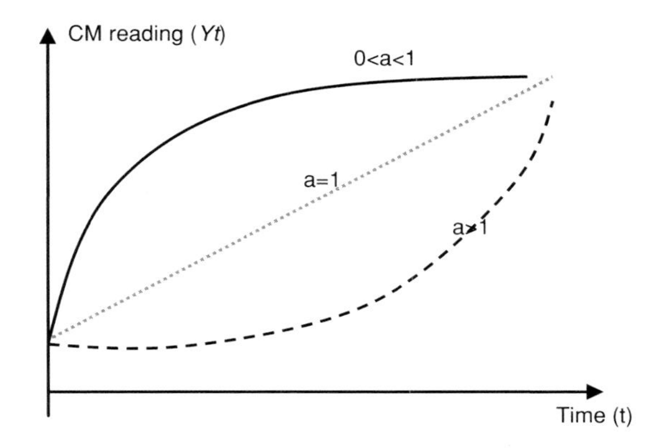

Fig. 16.10 Illustrating the different types of trend catered for by the deterioration model

16.5.2 Modelling the Condition Monitoring Process

Now we discuss the modelling of a CM process over time and consider the potential for updating our knowledge of the process when observations become available. The objective is to accurately predict the path or trend of the CM observations. Initially, we consider the modelling of $Y(t)$ without updating. If we know the starting reading for the CM process, $y(0)$, the evolution of the monitored variable over time can be described using a deterioration model. The CM reading at time t is modelled as:

$$Y(t) = y(0) + bg(t) + \theta(t)$$

where, b is a scaling parameter, $g(\theta t)$ is some function of time and $\theta(t)$ is the random error in the prediction over the interval $(0, t)$. A number of different functional trending forms for $g(t)$ have been considered including (i) $g(t) = t$, (ii) $g(t) = t^a$, (iii) $g(t) = e^{at}$, and (iv) $g(t) = e^{-a/t.}$

For option (i), the scaling parameter, b, in the equation for $Y(t)$ becomes the linear gradient of the process. Naturally, this model is only appropriate for cases where the CM variable changes linearly over time. After some initial numerical studies, the modelling form that we have decided to adopt is option (ii). As such, the CM variable is modelled over time as:

$$Y(t) = y(0) + bt^a + \theta(t)$$

This functional form provides a lot of flexibility in the different trend behaviours it can model, as shown in Fig. 16.10.

As illustrated in Fig. 16.10, when the parameter a is set to 1, the model reverts to the linear case (option 1 above). When we have $0 < a < 1$, the CM variable can be modelled as increasing steadily and then levelling off asymptotically as the component approaches failure. This type of behaviour is often observed when oil based indicators such as metallic contamination levels are used as the CM input. When $a > 1$, the value of the CM parameter increases steadily at first and then begins to increase more rapidly as the component approaches failure. This type of behaviour is consistent with vibration monitoring applications where the initiation of a hidden defect in the component typically triggers a rapid increase in the resulting vibration level.

We now consider the iterative modelling of the CM process $Y(t)$ over time. If a CM observation, $y(t_i)$, is available at the ith CM point, we are then interested in modelling the behaviour of $Y(t)$ from this point onwards. At time t_i, we have

$$y(t_i) = y(0) + bt_i^a + \theta(t_i)$$

and for $t \geq t_i$, we model the difference in the CM variable over time, since t_i, as

$$Y(t) = y(t_i) + b(t^a - t_i^a) + (\theta(t) - \theta(t_i))$$

This formulation uses the current monitoring value when making predictions about the future trend of the CM variable over time but it does not utilise the history of monitoring information. To incorporate the CM history, we have developed a couple of updating procedures for the scaling parameter b.

Finally, we present the deterioration model that iteratively describes the trending behaviour of the CM information over time. At the ith CM point we model the future behaviour of the CM variable as

$$Y(t) = y(t_i) + b(t_i)(t^a - t_i^a) + \theta(t_i, t) \tag{16.4}$$

where, $\theta(t_i, t)$ represents the random error in the prediction over the interval (t_i, t) and, for $t_1 < t_2 < \cdots < t_i \leq t$, we have $b(t_i)$ as a function of $\{y(t_1), y(t_2), \ldots, y(t_i)\}$. The modelling of $Y(t)$ and the updating of b can be undertaken using one of two new models proposed in this chapter. These are a Brownian motion model and a Gamma process based model. It is via the modelling of $Y(t)$ that, at any given monitoring point, the distribution of the time to failure is inferred. The failure time distribution is then utilised in the replacement decision model discussed earlier to evaluate the optimal replacement decision. For the Brownian motion based model, see Wang and Carr (2010). The Gamma based model is presented in the appendix.

16.6 An Example of the Gamma Process-based Model

We now present a case example using a simplified linear version of the proposed Gamma process based model (with parameter $a = 1$) to obtain the failure time distributions used in the replacement decision model in order to evaluate the optimal decision at each CM point. The case we are considering is a limited data case. The monitored information consists of the overall vibration level recorded at irregular time points during the operational life of 6 components and the associated failure-time information. The data were obtained in a fatigue experiment. The data from 4 of the components are defined as historical data used to fit the model and the other 2 are treated as new data to test and demonstrate the application of the

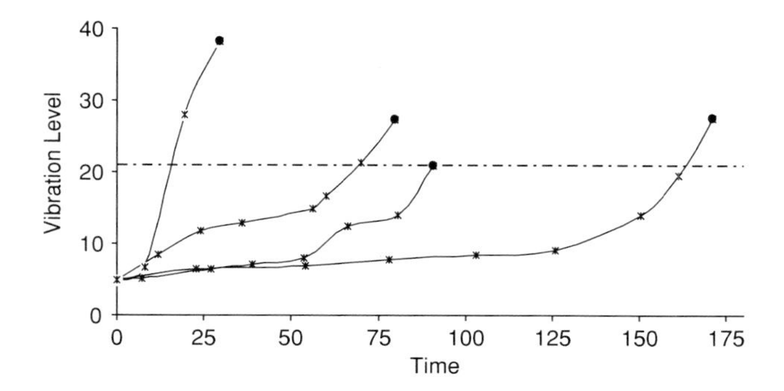

Fig. 16.11 The vibration monitoring histories used to fit the model

Table 16.1 The estimated parameters

Parameter	Estimate
σ	1.1711
α_0	1.2318
β_0	2.9051
α	0.2061
β	0.0179
Y_L	20.9
λ	0.1311

model. Figures 16.11 and 16.12 illustrate the vibration histories for the 4 components used to fit the model and the two component datasets used to demonstrate the application of the model respectively. The black dots represent failures. The costs associated with failures and preventive component replacements are £6,000 and £2,000 respectively. Using the maximum likelihood method, the estimated parameters are given in Table 16.1.

Figures 16.13 and 16.14 illustrate the tracking of the residual life using the Gamma process model at successive monitoring points for the first and second new component, respectively. The solid line represents the actual remaining life and the dotted line represents the mean residual life.

It is evident that as more information becomes available over time, the residual life estimation process is substantially improved. This demonstrates the potential advantages of utilising CM information. Figures 16.15 and 16.16 illustrate the fit of the distribution about the actual failure at successive monitoring points. The dots represent the actual failure time. As more information is received, the fit improves giving increased confidence in the replacement decisions.

Figures 16.17 and 16.18 illustrate the evaluation of the optimal replacement decisions (represented by boxes) and the time remaining before failure (represented by dots) for the two demonstration components.

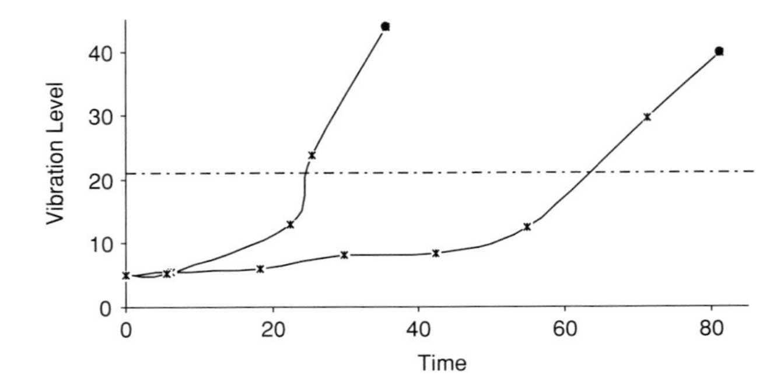

Fig. 16.12 The vibration monitoring histories used to test and demonstrate the model

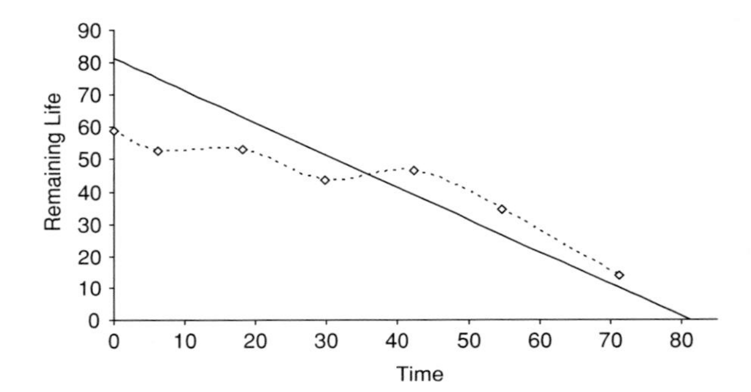

Fig. 16.13 Tracking the remaining life of the 1st demonstration component

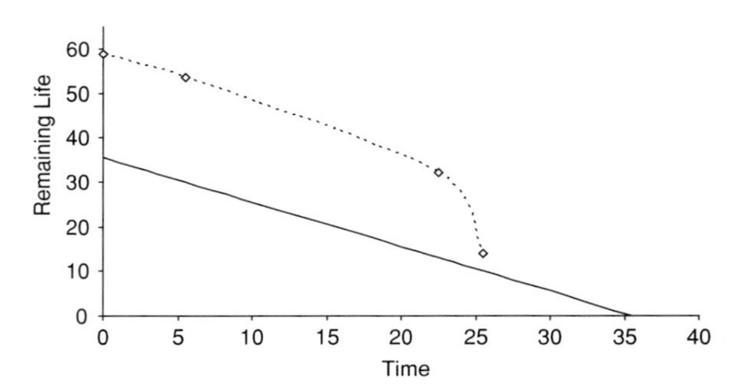

Fig. 16.14 Tracking the remaining life of the 2nd demonstration component

At all CM points in Figs. 16.17 and 16.18 except the final one before failure, the replacement is scheduled to occur after the subsequent CM point, at which time the replacement decision is re-evaluated. As such, both components are kept in operation until the final CM point and a failure is still averted. This illustrates how the proposed prognostic technique and the replacement decision model can maximise component availability, without compromising reliability, and in a cost effective manner.

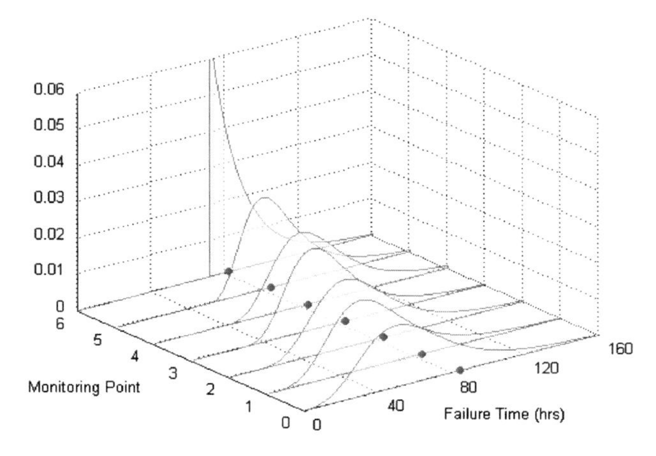

Fig. 16.15 Illustrating the fit of the failure time distributions about the actual failure time for the 1st demonstration component

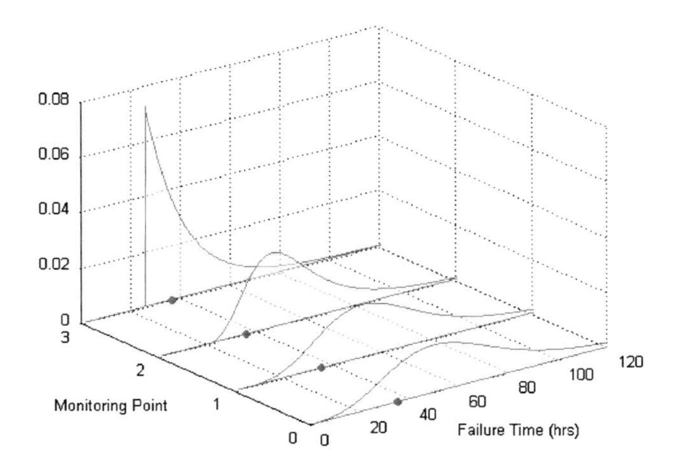

Fig. 16.16 Illustrating the fit of the failure time distributions about the actual failure time for the 2nd demonstration component

16.7 Summary of Developments

The models presented in this chapter tackle some of the issues associated with scheduling replacements for operational components used in operational applications. The models can be used to increase the availability and reliability for key components and the costs of planned maintenance can be reduced by eliminating unnecessary interventions. The newly developed model is a practical and robust procedure for operational scenarios and incorporates subjective model initialisation techniques in the

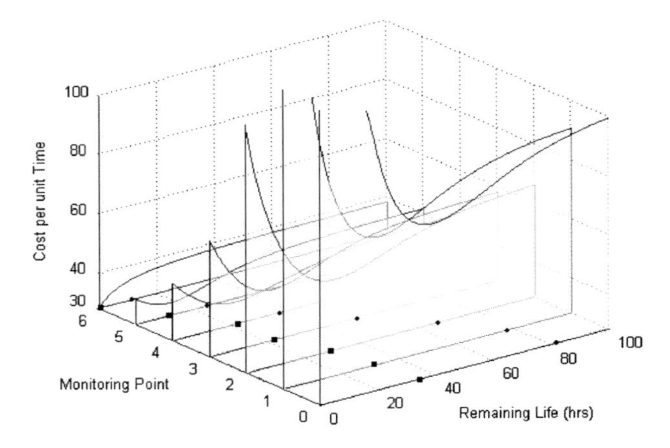

Fig. 16.17 Illustrating the optimal replacement decision at each CM point for the 1st demonstration component

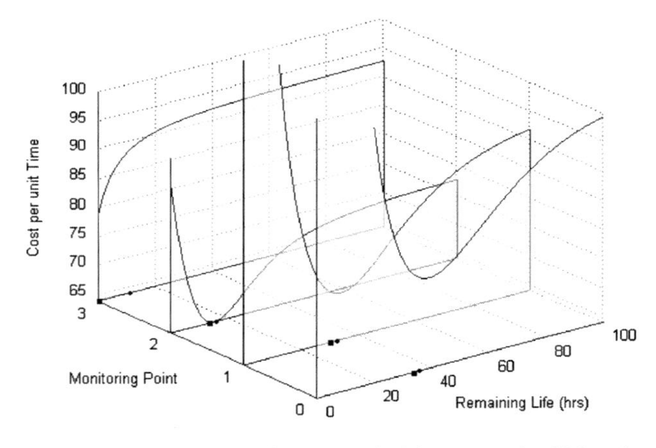

Fig. 16.18 Illustrating the optimal replacement decision at each CM point for the 2nd demonstration component

absence of past life histories. In addition, an optimal replacement decision model has been constructed which utilises the failure time prediction models and learning algorithms have been developed for updating purposes.

16.8 Chapter Summary Questions

The first question from this chapter is why residual life prediction is important. This leads to another question as what can we gain by predicting such as residual life from observed condition information. The last question is do we need failure data to enable the model introduced in this chapter?

- Why do we need to predict the residual life of operating equipment?
- What benefits can we obtain by predicting the residual life?
- What key information is needed for the new developments introduced in the chapter?

Appendix

From Eq. 16.4, we describe the updating of $b(t_i)$ in the Gamma process model using the observed $\mathbf{Y}(t_i) = \{y(t_1), y(t_2), \ldots, y(t_i)\}$. The updating process for $b(t_i)$ is difficult in this case as an analytical form is not available. As such, we make a few assumptions and attempt to obtain a closed form updating equation for $b(t_i)$ using the observed monitoring information. Firstly, at time t_i,, before considering the observed $y(t_i)$, we assume that $b(t_i) = b(t_{i-1})$ and we define $v(t_i) = \{y(t_i) - y(t_{i-1})\}^{-1}$ and $\mathbf{V}(t_i) = \{v(t_1), v(t_2), \ldots, v(t_i)\}$. Secondly, we assume that the predicted pdf $f(b(t_i)|\mathbf{V}(t_{i-1}))$ is available using $b(t_i) = b(t_{i-1})$. With these assumptions, we have

$$f(b(t_i)|\mathbf{V}(t_i)) = \frac{f(v(t_i)|b(t_i))f(b(t_i)|\mathbf{V}(t_{i-1}))}{\int_0^\infty f(v(t_i)|b(t_i))f(b(t_i)|\mathbf{V}(t_{i-1}))db(t_i)} \quad \text{(A16.1)}$$

Assuming $f\ (b(t_i)|\mathbf{V}(t_{i-1}))$ is Gamma and $f\ (v(t_i)|b(t_i)) = \mathrm{Ga}\{v(t_i);\ \alpha,\ \beta b(t_i)(t_i^a - t_{i-1}^a)\}$, it follows that $E[v(t_i)|b(t_i))] = \alpha/\beta b(t_i)(t_i^a - t_{i-1}^a)$ and $\mathrm{var}[v(t_i)|b(t_i))] = \alpha/(\beta b(t_i)(t_i^a - t_{i-1}^a))^2$. Letting $f(b(t_i)|\mathbf{V}(t_{i-1})) = \mathrm{Ga}\{b(\mathrm{ti});\ \beta_{i-1},\alpha_{i-1}\}$ where the parameters contain the history $\mathbf{Y}(t_{i-1})$, then after some manipulation we obtain

$$f(b(t_i)|V(t_i)) = \mathrm{Ga}\{b(t_i); \alpha + \alpha_{i-1}, \beta(t_i^a - t_{i-1}^a)v(t_i) + \beta_{i-1}\}$$

with shape and scale parameters $\alpha_i = \alpha + \alpha_{i-1}$ and $\beta\ (t_i^a - t_{i-1}^a)v(t_i) + \beta_{i-1}$ respectively. When the hyper-parameters for $f(b(t_0)|\mathbf{V}(t_0)) \equiv f(b(t_0))$ are α_0 and β_0, we have

$$f(b_{t_i}|\mathbf{V}_{t_i}) = \mathrm{Ga}\left(b_{t_i}; \alpha_0 + i\alpha, \beta_0 + \beta\sum_{k=1}^{i}\left(t_k^a - t_{k-1}^a\right)v_{t_k}\right) \quad \text{(A16.2)}$$

and a closed form updating expression is available as

$$\hat{b}_{t_i}|\mathbf{V}_{t_i} = \frac{\alpha_0 + i\alpha}{\beta_0 + \beta\sum_{k=1}^{i}\left(t_k^a - t_{k-1}^a\right)v_{t_k}} = \frac{\alpha_0 + i\alpha}{\beta_0 + \beta\sum_{k=1}^{i}\frac{\left(t_k^a - t_{k-1}^a\right)}{(y_k - y_{k-1})}} \quad \text{(A16.3)}$$

References

C. Bunks, D. Mccarthy, Condition-based maintenance of machines using hidden Markov models. Mech. Syst. Signal Process **14**(4), 597–612 (2000)

M.J. Carr, W. Wang, A case comparison of a proportional hazards model and a stochastic filter for condition-based maintenance applications using oil-based condition monitoring information. Proc. Inst. Mech. Eng. Part O J. Risk Reliab. **222**, 47–55 (2008)

M.J. Carr, W. Wang, *Failure mode analysis and residual life prediction for condition based maintenance applications*. IEEE Trans. Reliab. (forthcoming) (2010a)

M.J. Carr, W Wang, *An approximate algorithm for CBM applications*. Under review (2010b)

R.A. Collacott, *Mechanical fault diagnosis and condition monitoring* (Chapman & Hall, London, 1997)

A.K.S. Jardine, V. Makis, D. Banjevic, D. Braticevic, M. Ennis, A decision optimisation model for condition-based maintenance. J. Qual. Main. Eng. **4**(2), 115–121 (1998)

A.K.S. Jardine, D. Lin, D. Banjevic, A review on machinery diagnostics and prognostics implementing condition-based maintenance. Mech. Syst. Signal. Process. **20**(7), 1483–1510 (2005)

J.D. Kalbfleisch, R.L. Prentice, *The statistical analysis of failure time data* (Wiley, New York, 1980)

D. Kumar, U. Westberg, Maintenance scheduling under age replacement policy using proportional hazard modelling and total-time-on-testing plotting. Eur. J. Oper. Res. **99**, 507–515 (1997)

D. Lin, V. Makis, Recursive filters for a partially observable system subject to random failure. Adv. Appl. Probab. **35**, 207–227 (2003)

V. Makis, A.K.S. Jardine, Optimal replacement in the proportional hazards model. INFOR **30**, 172–183 (1991)

S. Ross, *Stochastic processes* (Wiley, New York, 1996)

W. Wang, A model to predict the residual life of rolling element bearings given monitored condition information to date. IMA J. Manag. Math. **13**, 3–16 (2002)

W. Wang, Modelling the probability assessment of the system state using available condition information. IMA. J. Manag. Math. **17**(3), 225–234 (2006)

W. Wang, M.J. Carr, An adaptive Brownian model for plant residual life prediction. Proceedings of the 2010PHM Conference, Macau (2010)

W. Wang, A.H. Christer, Towards a general condition based maintenance model for a stochastic dynamic system. J. Oper. Res. Soc. **51**, 145–155 (2000)

Chapter 17
Scheduling Asset Maintenance and Technology Insertions

W. Wang and M. J. Carr

Abstract This chapter presents a methodology that has been developed to assist in the planning, scheduling and outsourcing associated with maintenance and service contracting of complex engineering systems. The methodology is based on the delay-time modelling concept and can be used to construct and specify a model of the failure and inspection maintenance processes for a given system or asset. The resulting process model can be used for optimising the maintenance service interval with respect to a number of different criteria including system cost, downtime, availability or reliability. It also caters for a number of analytical scheduling options. Most notably, the model can be used to determine the optimal inspection interval for multiple types of maintenance and service, and can also be used for validatory purposes or to improve the actual maintenance processes. This chapter also discusses the application of the methodology in the context of fixed service life contracts and the potential for incorporating technology insertion processes over time in the modelling process. The model developed can be programmed into a software package to facilitate the actual use of the methodology. The chapter concludes with some simple examples to demonstrate the concepts.

W. Wang (✉)
Salford Business School, University of Salford, Salford, M5 4WT, UK
e-mail: w.wang@salford.ac.uk

W. Wang
School of Economics and Management, University of Science and Technology, Beijing, China

M. J. Carr
School of Medicine, The University of Manchester, Manchester, UK

I. Ng et al. (eds.), *Complex Engineering Service Systems*, Decision Engineering,
DOI: 10.1007/978-0-85729-189-9_17,

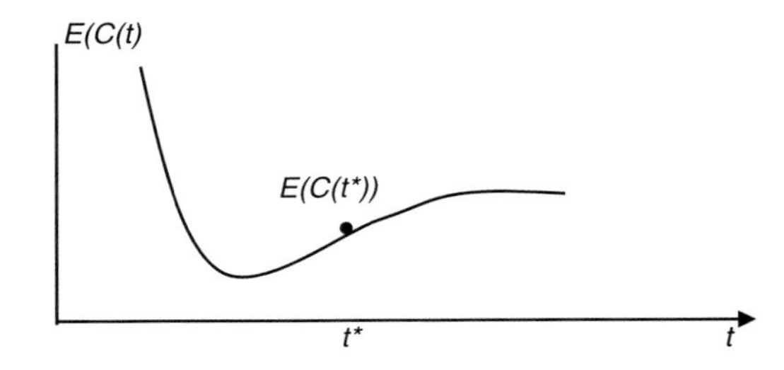

Fig. 17.1 The observed and expected cost model v service interval

17.1 Introduction

Due to advances in technology and production, longer product life cycles and increased asset availability are increasingly common requirements in many sectors and represent important considerations for the development of service plans and contracts. This is particularly so when both the equipment providers and the maintainers are under the pressure of performance-based contractual arrangements. Improving the provision of contracts to incorporate servicing throughout the proposed life of an asset should theoretically result in a reduction in the acquisition of new engineering systems and equipment and ultimately a behavioural transformation within and between the relevant parties. Longer lifecycles mean that the likelihood of technological obsolescence is substantially increased and the necessity for improved maintenance, repairs and spare parts planning becomes more apparent if the goal of complex system capability enhancement is to be attained.

An important aspect of service contract provision is the development of a cohesive cost-effective maintenance framework with the provision for technological insertion planning and obsolescence management. The transformation from a traditional product supply arrangement to a product supply and service provision contract will involve the realisation of multiple support processes into a single coordinated multi-tier plan. The maintenance plan should support the contract and demonstrate the impact of the transition in terms of cost, downtime, risk, availability or any other criterion.

To clarify the objective of the type of the cost benefit analysis we are concerned here with, considering a plant item with a maintenance practice, or concept, of servicing every period t hours, says, weeks or months, with repair of failures undertaken as they arise. The service consists of a check list of activities to be undertaken, and a general inspection of the operational state of the plant. Any defect identified leads to immediate repair and the objective of the maintenance concept is to minimise the operational cost. Other objectives could be considered to include downtime, availability, and output, but for now we consider cost reduction.

Conceptually, there is a relationship between the expected cost per unit time $E(C(t))$, and the service period t, see Fig. 17.1. If t was small, the cost per unit time would be large because the plant would frequently be unavailable due to servicing, and if t were sufficiently large, the cost per unit time would essentially be under a breakdown maintenance policy. If the chosen service period is t^*, all that can expect to be known of E($C(t)$) is the observed value E($C(t^*)$), that is the current cost measure. One wishes to reduce E($C(t^*)$), and if a model such as the curve in Fig. 17.1 was available, there would be little difficulty in identifying a good operational period for t, which could be infinite, that is do not inspect. Unfortunately, in the absence of modelling, all that is generally available is the data of the dot in Fig. 17.1.

To obtain the curve in Fig. 17.1 requires maintenance modelling, and the curve in Fig. 17.1 is a graphical representation of a model with only one type of the service interval.

Plant inspection with only one type of the service interval has been well addressed in the literature, Wang (2008a). However it is noted that multi-level service provision is increasingly common, particularly in the defence sector. The different types of maintenance can be lines with different activities, differing levels of depth, or different sub-systems. In many applications, service provision is grouped into two different levels. The first level consists of on-base processes for maintenance and upgrading activities. The second level consists of on and off-base support providing; labour and management resources, material and repair services, technical information and capability development services. A number of authors have presented models for maintenance service contracts. Lugfigheid et al. (2007, 2008) discussed finite horizon problems under a maintenance contract. Lisnianski et al. (2008) and Jackson and Pascual (2008) model increasing failure intensities over the planned service life of an asset. However, the effects of planned inspection maintenance and multiple types of maintenance intervention are neglected in all of these studies.

This chapter presents a methodology developed for determining the optimal multiple inspection services and maintenance intervals. The methodology can be utilised for a given application to integrate and schedule maintenance and service activities over different lines or at different levels of depth. The maintenance plan is constructed in an optimal manner and the scheduling is undertaken over the lifetime of an asset in order to provide through-life support while balancing the competing objectives of capability and affordability. The methodology has been developed with consideration given to the limitations associated with operational scenarios and can be tailored to build a model of the failure and planned maintenance processes for any given application. Upgrading and technology insertion activities are regarded as part of the maintenance planning process and capability enhancement trade-off decisions are readily incorporated.

The chapter is organised as follows. Section 17.2 introduces the background of the modelling technique used. Section 17.3 presents the methodology and the model. Section 17.4 develops the model based on a fixed system life, while Sect. 17.5 considers the impact of technology insertions. Section 17.6 illustrates numerical examples and Sect. 17.7 concludes the chapter.

17.2 Background

Two maintenance policy extremes are typical in industry. These are (i) 'breakdown' maintenance policies where the system is only attended to upon the

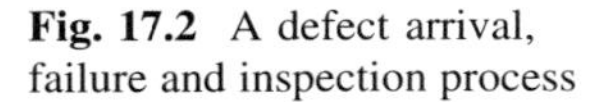

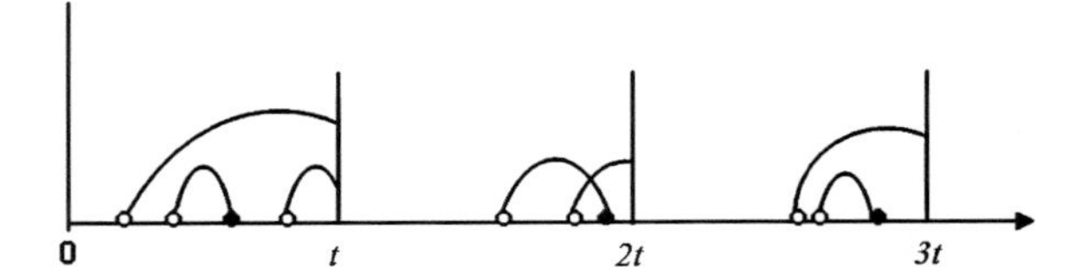

Fig. 17.2 A defect arrival, failure and inspection process

occurrence of a failure, and (ii) 'frequent-as-possible' maintenance scheduling. However, it is rare for either of these policies to produce satisfactory results. Figure 17.2 illustrates a typical defect arrival, failure and planned inspection process for inspections undertaken on a regular interval *t*, where multiple defects can be present in the system at the same time and defects that are present are identified and removed or corrected as part of the inspection process. The circles represent the initiation of random defects and the dots represent failures caused by the defects if no maintenance intervention takes place. The arc linking a circle and a dot is the delay time of the defect. Due to the impact of planned inspection interventions, some defects in Fig. 17.2 are identified at inspection and rectified. This reduces the number of system failures from 7 to 3. Clearly, the interval of such an inspection is important since more frequent inspections will identify and remove more defects and consequently reduce the cost associated with failures but will also result in an increased total inspection penalty. We introduce a modelling technique called delay-time modelling which can be used for such a cost benefit analysis to balance the trade-off between these two sources of costs.

Delay-time modelling can be used to optimise the inspection process and avoid the obvious pitfalls associated with 'breakdown' or 'frequent as possible' maintenance policies. The methodology has proven to be very useful for complex system maintenance scheduling since its introduction in Christer and Waller (1984). Useful summaries of the key delay-time modelling developments are provided in Christer (1999) and Wang (2008a). Typical case studies using the delay-time modelling can be found in, Baker and Wang (1991) for medical equipment, Christer and Waller (1984) for high speed canning lines, Akbarov et al. (2008) for a baking production line, Christer et al. (1995) for an extrusion press, Christer et al. (1997) for a hot steel mill, Jones et al. (2009) for a manufacturing plant and Pillay et al. (2001) for fishing vessel to name a few. Virtually all companies which use some kind of time-based inspection on their assets can benefit from the delay-time model based analysis. Delay-time modelling provides a framework that is applicable to any complex asset maintenance scheduling problem. The delay-time concept defines the failure process of an asset as a two-stage process. The first stage is the normal operating stage from new to the origination of an identifiable defect in the system. The length of this stage depends on the type and nature of the inspection process, the available technology and the associated maintenance team. The second stage is defined as the delay-time of the defect until failure. During the course of the delay-time, preventive maintenance actions can be undertaken to remove the defect and prevent the impending failure.

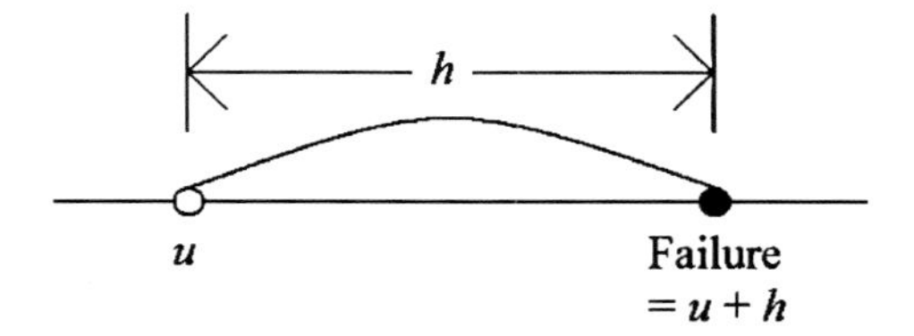

Fig. 17.3 The delay-time of a defect from conception to failure

By modelling the process in this manner, an optimal inspection interval can be established based upon downtime, cost, availability, risk or any other criterion. The delay-time concept is similar but has a number of advantages over the much used P–F interval concept in reliability centred maintenance (RCM). However, the P–F interval is only part of the plant life, and the time to P since new is more important than the P–F interval, which was not considered in RCM at all. Moreover, RCM only provided such a P–F interval concept, but the delay-time modelling proposed a comprehensive way to estimate the time to P and the P–F interval. No RCM literature has reported any way for the quantification of such a P–F interval.

Initially, we are concerned with complex assets and inspections taking place in every t units of time, where the aim of the interventions is to minimise the cost associated with system failures. Figure 17.3 shows that the origin of a particular defect is defined as time u, while h denotes the delay-time until failure. A maintenance intervention in the period $(u,\ u + h)$ would prevent the failure from occurring.

If the averaged costs of failure, inspection and defect removal are available, we wish to establish the curve depicted in Fig. 17.1 using the delay-time concept. The expected cost per unit time is the sum of the cost of failures, inspection and defect removals occurred within the interval divided by the interval. To present this in a model we have

$$\mathrm{E}(C(t)) = \{c_f \mathrm{E}(N_f(t)) + c_s + c_d \mathrm{E}(N_d(t))\}/t \tag{17.1}$$

where $C(t)$ denotes the total cost per unit time over t, c_f denotes the average cost per failure, $\mathrm{E}(N_f(t))$denotes the expected number of failures in $(0,\ t)$, c_s denotes the average cost of an inspection service, c_d denotes the average cost of a defect removal, and $\mathrm{E}(N_d(t))$ denotes the expected number of defects identified at the service point. The cost parameters may be available from the clients, but to get $\mathrm{E}(N_f(t))$ and $\mathrm{E}(N_d(t))$ we need to know the average rate of defect arrival over time and a probability distribution for the delay-time $f(h)$. For the derivation of $\mathrm{E}(N_f(t))$ and $\mathrm{E}(N_d(t))$ see Wang (2008a). Clearly Eq. 17.1 is a function of t through $\mathrm{E}(N_f(t))$ and $\mathrm{E}(N_d(t))$, and once they and the cost parameters are available, Eq. 17.1 can be readily calculated to assess which t is the best. Equation 17.1 is the fundamental objective function we shall use throughout this chapter while all others are variants of it.

17.3 Multiple Types of Maintenance

17.3.1 Methodology

Until recently, the delay-time modelling approach has only been used to optimise the maintenance process for a single line or type of service provision. For complex asset systems with many sub-systems and components, the planned inspection and maintenance will normally be undertaken at different intervals, levels, and depths. Typical practice involves a routine inspection on a more frequent basis to some sub-systems or components which may be subject to more frequent failures and then a longer interval applied to the system as a whole. An example of such maintenance service practice can be seen in aircraft maintenance where up to four different maintenance intervals may be in place (Sriram and Hagham 2003). These different intervals are usually nested, as up-level maintenance with a longer interval will include the content of lower levels of maintenance with shorter intervals (Wang 2000, 2008b). The following summarises the methodology devised for modelling multiple types of maintenance and the associated assessment activities:

1. Collection, storage and pre-processing of events based data.
2. Drawing up a bill of material/product tree to show the structure of the system.
3. For each defined sub-system that will be serviced by a line or type of maintenance service:
   - estimate the defect arrival rate,
   - specify and parameterise the delay-time distribution,
   - estimate the downtime or cost associated with failures and scheduled maintenance servicing.
4. Establish the interactions between the different lines or types of maintenance.
5. Options for analysis include:
   - optimal scheduling algorithms,
   - candidate policy comparisons,
   - reliability and availability evaluation,
   - validation of existing scheduling activities,
   - evaluate efficacy of maintenance activities,
   - technology insertion sensitivity analyses.
6. Assess the performance by comparing the observed and expected output over time.
7. Assess the availability and capability of the system using an expected downtime and reliability evaluation. System reliability functions can be used to assess the system from a risk perspective.

The methodology requires specification and adaptation for a given scenario and is dependent on; the nature of the application, limitations on the availability of data

Table 17.1 The modelling notation

Notation	Definition
k	Defect type, where $k = 1$ or 2
λ_k	Rate of type k defect arrival
h	Delay-time of a defect
$f_k(h)$	Type k defect delay-time distribution
c_{kf}	Average cost measure for a type k failure
c_{ks}	Average cost measure for a type k inspection
c_{kd}	Average cost measure for a type k defect removal at a type k inspection
$\mathrm{E}[N_{kf}(t)]$	Expected number of type k failures over an interval $(0, t]$
$\mathrm{E}[N_{kd}(t)]$	Expected number of type k defect removals at a type k inspection after an interval of duration t
$\mathrm{E}[C(\bullet)]$	Expected cost per unit time with $\bullet$ as the decision variables

and access to design/engineering expertise. However, regardless of the application, the same building blocks are used to model the defect arrival, failure and preventive maintenance processes. It is an extendable and broadly applicable methodology that can be applied to any system. As noted in point 6, for a defined test case we are able to compare the expected cost against the observed operating cost for a specified maintenance service policy when initiated. As such, the comparison can be used to validate the delay-time model of the complex system.

17.3.2 Two Types of Maintenance

In this section, we address a simple case when two types of inspection process are in operation. The model can be readily extended to more than two types of inspections. We consider the categorisation of minor and major defects and failures and the scheduling of the corresponding minor and major inspection-based interventions with different cost measures. The different defect arrival rates are estimated independently as λ_1 and λ_2 for minor and major defects, respectively. Similarly, the individual delay-time distributions are specified as $f_1(h)$ and $f_2(h)$ and the average cost associated with failures, c_f, inspections, c_s and preventive defect removal, c_d, are estimated independently for each defect type. Table 17.1 summarises the modelling notation. We specify the following additional assumptions:

- There are two broad types of defects and therefore failures, namely minor and major.
- Minor defects can be identified and rectified by a minor inspection service, while major defects can only be identified and removed by a major inspection service. A major inspection service includes the content of a minor inspection service and occurs at one of the minor inspection service interventions.

17.3.3 Common Inspection Model

Firstly, we consider the scenario where minor and major defects and failures are identifiable but the minor and major inspections are undertaken on the same interval. As major inspections incorporate all the activities of minor inspections, we only incorporate the cost associated with major inspections. Extending the concept used in Eq. 17.1, the expected cost per unit time is

$$\mathrm{E}[C(t)] = \left\{\mathrm{E}[N_{1f}(t)]\, c_{1f} + \mathrm{E}[N_{1d}(t)]\, c_{1d} + \mathrm{E}[N_{2f}(t)]\, c_{2f} + \mathrm{E}[N_{2d}(t)]\, c_{2d} + c_{2s}\right\}/t \tag{17.2}$$

for a common inspection interval t.

17.3.4 Multiple Inspection Model

Now we consider the multiple inspection case where minor and major inspections are scheduled. As discussed in the introduction to this section, major inspections contain all the activities of minor inspections. As such, major inspections are scheduled to occur at integer multiples of the minor inspection interval and the cost associated with minor inspections is assumed to be incorporated in the major inspection penalty. With minor and major inspection intervals of duration t_1 and $t_2 = mt_1$, respectively, we have

$$\begin{aligned}\mathrm{E}[C(t_1, m)] = &\{m(\mathrm{E}[N_{1f}(t_1)]\, c_{1f} + \mathrm{E}[N_{1d}(t_1)]\, c_{1d}) + (m-1)c_{1s} \\ &+ \mathrm{E}[N_{2f}(mt_1)]\, c_{2f} + \mathrm{E}[N_{2d}(mt_1)]\, c_{2d} + c_{2s}\}/mt_1\end{aligned} \tag{17.3}$$

Naturally, when $m = 1$ the model is equivalent to the common inspection model.

17.4 Fixed Service Life

A key assumption used in the construction of the models in the previous section is that of an infinite planning horizon which allows the use of renewal reward theory in the evaluation of the expected cost per unit time. However, many contracting applications involve the scheduling of maintenance and service activities over a proposed service life. As such, further modelling developments are required. Initially some further notation is introduced:

- T is the planned service life, and
- $\mathrm{E}[C(\bullet)]$ is the expected cost over T with $\bullet$ as the decision variables

Note that for fixed service life applications, we are interested in the total expected cost over the planned service life, E[$C(\bullet)$], rather than the expected cost per unit time, E[$C(\bullet)$], used previously. Continuing the multiple inspection interval case with minor and major inspection types, the total expected cost over the service life of the asset is the summation of the cost associated with minor and major failures and minor and major defects identified and removed at inspections. Again, we assume that minor defects can be rectified by minor and major inspections but major defects can only be rectified at a major inspection of the system.

17.4.1 Common Inspection Model

If a common inspection on interval t is scheduled for minor and major inspections, then the total number of inspection intervals during the service life is $\lfloor T/t \rfloor$ where, $\lfloor x \rfloor$ is the largest integer less than or equal to x. As there will be no inspection at the end of the asset service life, it follows that the total expected cost over the planned service life is

$$\begin{aligned} \mathrm{E}[C(t)] \approx \sum_{k=1}^{2} \big((|T/t| - 1)\{\mathrm{E}[N_{kf}(t)]c_{kf} + \mathrm{E}[N_{kd}(t)]c_{kd}\} \\ + \mathrm{E}[N_{kf}(T - (|T/t| - 1)t)]c_{kf} \big) + (T/t - 1)c_{2s} \end{aligned}$$

However, as we are interested in large values of T and comparatively small values of t, the expected cost can be approximated as

$$\mathrm{E}[C(t)] \approx \sum_{k=1}^{2} ((T/t - 1)\{\mathrm{E}[N_{kf}(t)]c_{kf} + \mathrm{E}[N_{kd}(t)]c_{kd}\} + \mathrm{E}[N_{kf}(t)]c_{kf}) + (T/t - 1)c_{2s} \quad (17.4)$$

17.4.2 Multiple Inspection Model

In the multiple inspection case, we define t_1 and t_2 as the minor and major inspection intervals respectively. Assuming that $t_1 \cdot t_2$ (i.e., the minor and major inspections do not coincide), the expected cost over the contracted service life is

$$\mathrm{E}[C(t_1, t_2)] \approx \sum_{k=1}^{2} \big((T/t_k - 1)\{\mathrm{E}[N_{kf}(t_k)]c_{kf} + u_{ks} + \mathrm{E}[N_{kd}(t_k)]c_{kd}\} + \mathrm{E}[N_{kf}(t_k)]c_{kf} \big) \quad (17.5)$$

In the nested case, where major inspections are undertaken on an interval $t_2 = mt_1$ which is an integer multiple of the minor inspection interval t_1, the expected cost over the service life becomes

$$\mathrm{E}[C(t_1,m)] \approx (T/t_1 - 1)\{\mathrm{E}[N_{1f}(t_1)]c_{1f} + u_{1s} + \mathrm{E}[N_{1d}(t_1)]c_{1d}\} + \mathrm{E}[N_{1f}(t_1)]c_{1f} + ((T/mt_1) - 1)\{\mathrm{E}[N_{2f}(mt_1)]c_{2f} + (c_{2s} - c_{1s}) + \mathrm{E}[N_{2d}(mt_1)c_{2d}\} + \mathrm{E}[N_{2f}(mt_1)]c_{2f} \quad (17.6)$$

It is straightforward to show that if technology insertions are incorporated, which reduce the major defect arrival rate over time, the expected cost will be smaller than in the no insertions case. This can be used partially to justify the need for technological insertions if the cost associated with the actual insertion process is not excessive. However, it should be noted that, if the major defect arrival rate is not constant, then a variable inspection interval policy should be employed. This is shown in the next section.

17.5 Incorporating Technology Insertions

During the lifecycle of major platforms, the implementation of technology insertions is often carried out to increase the capacity of the platforms. However, technology insertions should only be considered on the assumption that they increase the capability and availability of the system, reduce the cost of planned and unplanned maintenance interventions or prevent technological redundancy over time. Technology insertions can be incorporated into maintenance and service scheduling in a number of different ways and the objective may always not be able to improve the operational capability of equipment. For the example in this chapter, we consider a technological 'upgrading' mechanism that is manifested in step-wise decreasing defect arrival rates as more reliable components are assumed to be inserted over time. In many cases, technology insertion may be used to combat natural degradation and obsolescence. This can be regarded as 'maintaining' capability for consistent requirements. Alternatively, technology insertion may be used to modify the asset for changing/evolving requirements over time. In these situations, the delay-time modelling scheduling process should be adapted to produce a plan designed to maintain levels of reliability.

Technological insertions are considered in many managerial papers in a qualitative manor; see Kerr et al. (2008a), Strong (2004), and Dowling (2004). These studies mainly address the framework, concepts, components and dimensions of such insertions. However, regardless of the particular application, justification for the technological insertions will be required in the form of a cost benefit analysis that should demonstrate the expected contribution with regard to the minimisation of associated risks and costs. To incorporate an effective technology insertion programme, it is necessary that system architectures are designed on an 'open' and 'modular' basis, Kerr et al. (2008a). Sensitivity analyses and a comparison of the reduced maintenance costs attributable to failures and inspections and the cost of the insertion activities can then be undertaken. Technological insertions must have an impact upon the failure behaviour of the asset and therefore on the associated inspection policy as well; Dowling (2004).

To incorporate the impact of technological insertions over time for fixed horizon inspection models, we introduce the following additional assumptions;

- New technology that may lead to component upgrading occurs during major inspections at discrete time points and is assumed to impact upon the arrival rate of major defects.
- The arrival rate for major defects is defined as a function of the time since the last major inspection.
- The interval between minor inspections is constant, but the interval between major inspections, which is influenced by aging and upgrading, varies.

Minor inspections are undertaken on an interval of duration t_1 and the time of the ith major inspection is denoted by $t_{2,i}$. To model the impact of technological insertions over time, $\lambda_2(t_{2,i})$ is defined as the major defect arrival rate after the ith major inspection. If the major defect arrival rate is influenced by technological insertions over time, we must have a variable inspection scheme for the major inspections. Assuming that there are n major inspections over the course of the service life T, where n is a decision variable, we have the total expected cost associated with maintenance as

$$\begin{aligned}\mathrm{E}[C(t_1, n, \{t_{2i}; i1, 2, \ldots, n\})] &= (T/t_1 - 1)\{c_{1f}\mathrm{E}[N_{1f}(t_1)] + c_{1s} + c_{1d}\mathrm{E}[N_{1d}(t_1)]\} \\ &+ c_{1f}\mathrm{E}[N_{1f}(t_1)] - nc_{1s} + \sum_{i=1}^{n-1}\{c_{2f}\mathrm{E}[N_{2f}(t_{2,i} - t_{2,i-1})] + c_{2s} \\ &+ c_{2d}\mathrm{E}[N_{2d}(t_{2,i} - t_{2,i-1})]\} + c_{2f}\mathrm{E}[N_{2f}(t_{2,n} - t_{2,n-1})]\end{aligned} \tag{17.7}$$

For the expected number of minor failures over a minor inspection interval, $\mathrm{E}[N_{1f}(t_1)]$ and the expected number of minor defects removed at a planned minor inspection, $\mathrm{E}[N_{2d}(t_1)]$, see Wang (2008a). Similarly, the expected number of major failures over an interval between major inspections, $\mathrm{E}[\mathrm{N}_{2f}(t_{2,i} - t_{2,i-1})]$, and the expected number of defect removals at the ith major inspection, $\mathrm{E}[\mathrm{N}_{2d}(t_{2,i} - t_{2,i-1})]$, are also given in Wang (2008a). The expected cost can be compared with proposed budgetary constraints on the maintenance and service activities. This measure can also be used for contract negotiation if the maintenance service is outsourced.

Assuming that major inspections occur at integer multiples of the planned minor inspection interval t_1, the objective function becomes

$$\begin{aligned}\mathrm{E}[C(t_1, n, \{t_{2,i}; i = 1, 2, \ldots, n\})] &= (T/t_1 - 1)\{c_{1f}\mathrm{E}[N_{1f}(t_1)] + c_{1s} + c_{1d}\mathrm{E}[N_{1d}(t_1)]\} \\ &+ c_{1f}\mathrm{E}[N_{1f}(t_1)] - nc_{1s} + \sum_{i=1}^{n-1}\{c_{2f}\mathrm{E}[N_{2f}(m_i t_1)] + c_{2s} + c_{2d}\mathrm{E}[N_{2d}(m_i t_1)]\} \\ &+ c_{2f}E[N_{2f}(m_n t_1)]\end{aligned} \tag{17.8}$$

where $m_i \geq 1$ are the integer values and the major defect arrival rate after the ith major inspection is

Table 17.2 The historical failure and inspection data

Type	Total no. of failures	Total failure cost	Total inspection cost	Total defect repairs	Total defect repair cost
1	$TN_{1f} = 300$	$TC_{1f} = 1{,}200$	–	$TN_{1d} = 900$	$TC_{1d} = 900$
2	$TN_{2f} = 40$	$TC_{2f} = 800$	–	$TN_{2d} = 150$	$TC_{2d} = 750$
Both	$TN_f = 340$	$TC_f = 2{,}000$	$TC_s = 200$	$TN_d = 1{,}050$	$TC_d = 1{,}650$

$$\lambda_2(t_{2,i}) = \lambda_2\left(\sum_{j=1}^{i} m_j t_1\right).$$

An optimisation algorithm is required to jointly optimise the objective function given by Eq. 17.8 with respect to t_1, n and m_i for $i = 1,2,\ldots, n$.

17.6 Illustrative Examples of the model developed

17.6.1 Example 1: Scheduling multiple inspection service intervals

This example illustrates the potential applications of the proposed methodology for scheduling multiple types of maintenance and the benefits that can be realised for the development of service contracts.

17.6.1.1 Historical Data

We assume that we are presented with $r = 50$ life cycles of historical failure and inspection maintenance data. A constant cycle length of duration $t = 7$ days has been used and the cumulative data are illustrated in Table 17.2 for two different failure types. Although the different defect/failure types are identifiable, a single inspection policy has been employed historically.

17.6.1.2 Single Inspection Interval

In the single inspection interval case, the data is analysed collectively, i.e., no distinction is made between different types of defect or failure, as described in Sect. 17.2. As such, the process parameters are estimated from the data corresponding to both failure types as given in the row ‘both’ in Table 17.2. We assume a single defect arrival rate λ and employ an exponential form for the delay-time distribution $f(h)$ with parameter α. The cumulative delay-time distribution is therefore $F(h) = 1 - \exp(-\alpha h)$. Using the relevant values from Table 17.2 and

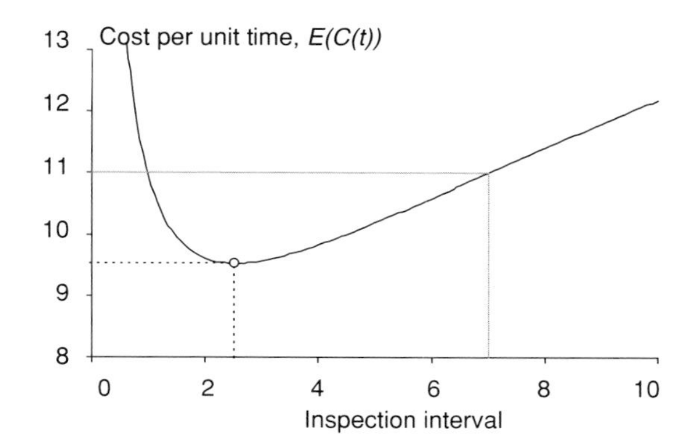

Fig. 17.4 The expected cost per unit time, E[$C(t)$], as a function of the inspection interval, t

applying a standard statistical procedure we obtain the estimated parameters $\lambda = 9.267$ defects per day and $\alpha = 0.197$. The total number of events and the total associated costs from Table 17.2 are used to estimate the average costs. The average failure cost is estimated as $c_f = 2{,}000/340 = 5.882$, the average inspection cost is $c_s = 200/50 = 4$ and the average cost for an inspection-based repair is $c_d = 1{,}650/1{,}050 = 1.571$.

Using the process parameters and the estimated costs, the single inspection type model of Eq. 17.1 can be used to analyse the effects of varying the maintenance interval on the expected future performance of the asset. Using Eq. 17.1, Fig. 17.4 illustrates the expected cost per unit time, E[$C(t)$], against the inspection interval t. The optimal inspection interval is found to be 2.526 days rather than 7 days employed historically. As shown in Fig. 17.4, changing the single inspection interval from the original 7 days to the optimal interval of around 2.5 days is expected to reduce the expected cost per unit time from 11 to 9.524 which is a 13.4% reduction in the costs associated with maintenance. This illustrates how modelling can be used to develop a tool for decision analysis that is used to inform and assess the benefits of potential maintenance policy changes.

17.6.1.3 Two Types of Maintenance

Now we consider the modelling of two different types of maintenance, as described in Sect. 17.3. We consider an infinite planning horizon and seek to minimise the expected cost per unit time. The parameters are estimated for each defect/failure type independently using the same process described in the previous sub-section and the relevant events data from Table 17.2. The estimated parameters are; the rate of type 1 defect arrival $\lambda_1 = 3.429$, $\alpha_1 = 0.087$ (for exponentially distributed type 1 delay-times) and the rate of type 2 defect arrival $\lambda_2 = 0.543$, and $\alpha_2 = 0.07$ (for exponential type 2 defects). We introduce an assumption that 15% of the inspection activities noted in Table 17.2 are attributable to the investigation of type 1 defects. As such, we have the cost associated

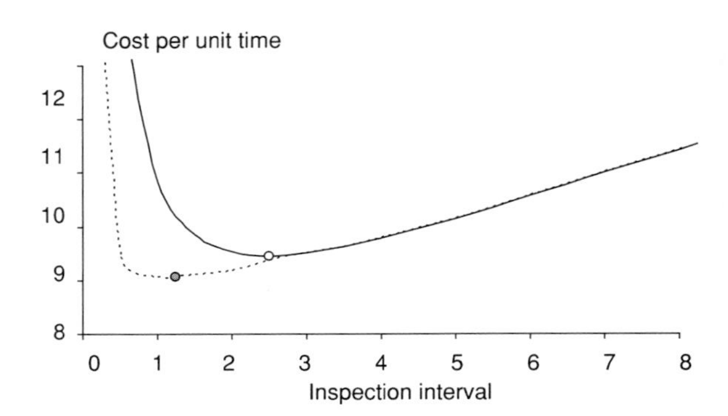

Fig. 17.5 The expected cost per unit time, for the common and multiple inspection models, as a function of the inspection interval (type 1 interval in the multiple case)

with type 2 inspections as $c_{s2} = c_s$ (from the previous section) and the cost of a type 1 inspection as $c_{s1} = 0.15 \times c_{s2}$. The average costs are;

- Type 1: $c_{1f} = 1{,}200/300 = 4$, $c_{1s} = 0.15 \times 200/50 = 0.6$ and $c_{1d} = 900/900 = 1$
- Type 2: $c_{2f} = 800/40 = 20$, $c_{2s} = 200/r = 4$ and $c_{2d} = 750/150 = 1$

Attention is now turned to comparing the common inspection interval and multiple inspection intervals models, as proposed in Sect. 17.3.3 and Sect. 17.3.4 and given by Eqs. 17.2 and 17.3, respectively. Figure 17.5 illustrates the expected cost per unit time against the inspection interval, where the multiple inspection intervals model is represented by the dashed line. Note that, in the multiple interval case, the representation is for the type 1 inspection interval, t_1, with the integer value m maximised for each potential value of t_1.

It is evident from Fig. 17.5 that a multiple inspection interval policy is superior for this case. In the common interval case, we obtain the optimal interval $t^* = 2.498$ days, giving the expected cost per unit time $E[C(t^*)] = 9.456$. In the multiple interval case, we obtain the optimal type 1 interval $t_1{}^* = 1.234$ days and $m^* = 3$ giving the optimal type 2 interval $t_2 = m^* \times t_1{}^* = 3.702$ and the expected cost per unit time $E[C(t_1{}^*, m^*)] = 9.053$.

17.6.2 Example 2: Fixed Service Life and Technology Insertions

A hypothetical system with associated maintenance policy is proposed over a specified service life. The potential impact of technology insertion is then assessed by comparing the expected costs over time with and without upgrading activities. The example illustrates how the potential benefits of modelling and incorporating technology insertion can be demonstrated. Alternative candidate maintenance policies and a sensitivity analysis on the impact of upgrading activities can be carried out in a similar manner for a given application. Figure 17.6 illustrates the major aspects of the hypothetical complex system.

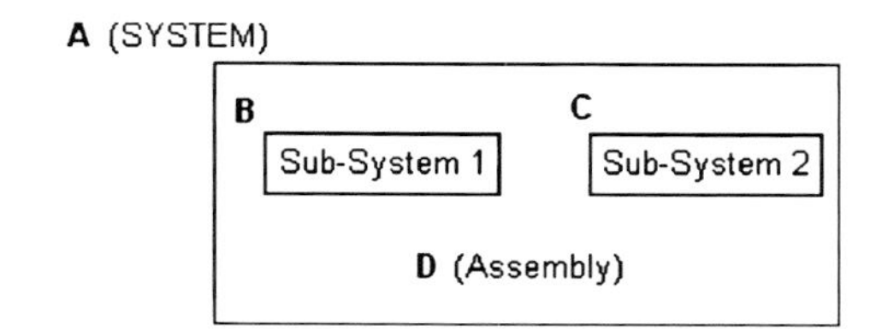

Fig. 17.6 The example system

Table 17.3 The activities and intervals for each type of maintenance/service

Type	Maintenance activities	Interval duration
1	Assembly (D) inspection and repairs	$\Delta t = 1$ month
2	Sub-System 1 (B) inspection and repairs	$L_2\Delta t$
3	Sub-System 2 (C) inspection and repairs	$L_3\Delta t$
4	System (A) overhaul and technology insertion	$L_4\Delta t$

The system is composed of an assembly supporting two sub-systems and we have A = B + C + D. The maintenance/service types, the associated activities and the interval between interventions are given in Table 17.3 where, L_2, L_3 and L_4 are positive integers. The proposed service life is $T = 120$ months.

We assume that an overhaul of the system (type 4 intervention) is comprehensive in that it includes all of the activities undertaken in type 2 and 3 interventions and more. As such, when an overhaul is scheduled to occur at the same time as type 2 or 3 interventions, no type 2 or 3 takes place. Technology insertion takes place at the same time as a system overhaul and its impact is manifested in reducing rates of defect arrival for sub-systems 1 and 2 (B and C). Defects are assumed to arise at an average rate of $\lambda_1 = 2$ a month in the assembly (D). For the purpose of exploring the impact and analysing the cost benefit, we consider the same maintenance schedule with and without technology insertion. Without insertion activities, we adopt average defect arrival rates of 8 and 10 a month in sub-systems 1 and 2, respectively. We now consider the effect on the arrival rates when insertion activities are incorporated. If $t_{2,i}$ is the time of the ith overhaul, the average defect arrival rate in sub-system 1 at time s when we have $t_{2,i} \leq s < t_{2,i+1}$ is assumed to be $\lambda_2(s) = 8 - 7t_{2,i}/120$.

Similarly, in the case of sub-system 2, we assumed the rate to be $\lambda_3(s) = 10 - 8t_{2,i}/100$. The mean delay-time from the initiation of an individual defect until the resulting failure is taken to be 1 month for all the different aspects of the system and we assume that the delay-times are exponentially distributed. The various process, system and cost parameters are summarised in Table 17.4.

A total expected cost model of the activities relating to the hypothetical system over a proposed operational life is established using the methodology proposed in Sect. 17.5 but extended to the 4 inspection type case. From the description of the proposed maintenance schedule, we define $E[C(T)]$ as the expected cost over the proposed asset service life T and $t_{2,i} = iL_4\Delta t$ is defined as the time of the ith overhaul. The parameters c_f and c_s represent the costs associated with failures and service interventions respectively and λ and $F(h)$ are the defect arrival rate and the

Table 17.4 The activities and intervals for each type of maintenance/service

Parameter	Value	Parameter	Average	Parameter	Estimate
T	120	c_{1f}	250	α_1	1
Δt	1	c_{2f}	300	α_2	1
L_2	4	c_{3f}	250	α_3	1
L_3	3	c_{1s}	60	λ_1	2
L_4	12	c_{2s}	75	$\lambda_2(s)$ (with TI)	$8 - 7\ s/120$
		c_{3s}	50	$\lambda_3(s)$ (with TI)	$10 - 8\ s/100$
		c_{4s} (with TI)	10,000	$\lambda_2(s)$ (no TI)	8
		c_{4s} (no TI)	125	$\lambda_3(s)$ (no TI)	10

cumulative delay-time distribution for the relevant aspect of the system. As the delay-time distributions are assumed to be exponential, the parameters are the reciprocals of the mean delay-time. We derive $\mathrm{E}[C(T)]$ as the summation of the expected total cost for each aspect of the system and the associated types of maintenance activity as follows;

Type 1—Expected Cost

$$\mathrm{E}[C^{(1)}(T)] = (T/\Delta t)c_{1f}\mathrm{E}[N_{1f}(\Delta t)] + \{(T/\Delta t) - 1\}c_{1s}$$

Type 2—Expected Cost

$$\mathrm{E}[C^{(2)}(T)] = \sum_{i=1}^{T/(L_4\Delta t)} \{(L_4/L_2)c_{2f}\mathrm{E}[N_{2f,i-1}(L_2\Delta t)] + \{(L_4/L_2) - 1\}c_{2s}\}$$

Type 3—Expected Cost

$$\mathrm{E}[C^{(3)}(T)] = \sum_{i=1}^{T/(L_4\Delta t)} \{(L_4/L_3)c_{3f}\mathrm{E}[N_{3f,i-1}(L_3\Delta t)] + \{(L_4/L_3) - 1\}c_{3s}\}$$

Type 4—Expected Cost

$$\mathrm{E}[C^{(4)}(T)] = \{(T/L_4\Delta t) - 1\}c_{4s}$$

The total expected cost over the proposed service life is simply

$$\mathrm{E}[C(T)] = \sum_{\mathrm{Type}=1}^{4} \mathrm{E}[C^{(\mathrm{Type})}(T)]$$

The specific schedule under consideration is a policy as specified in Table 17.4 with $L_2 = 4$, $L_3 = 3$ and $L_4 = 12$ and the question is whether or not to incorporate technology insertion activities over a contracted service life of 10 years. The average costs associated with failures of the assembly, sub-system 1 and sub-system 2 are taken to be 250, 300 and 250, respectively. The average costs associated with interventions of type 1, 2 and 3 are 60, 75 and 50, respectively. A type 4 intervention costs an average of 125 without technology insertion. The cost of a type 4 intervention with upgrading activities (this includes the predicted costs associated with the inserted technologies) is 10,000.

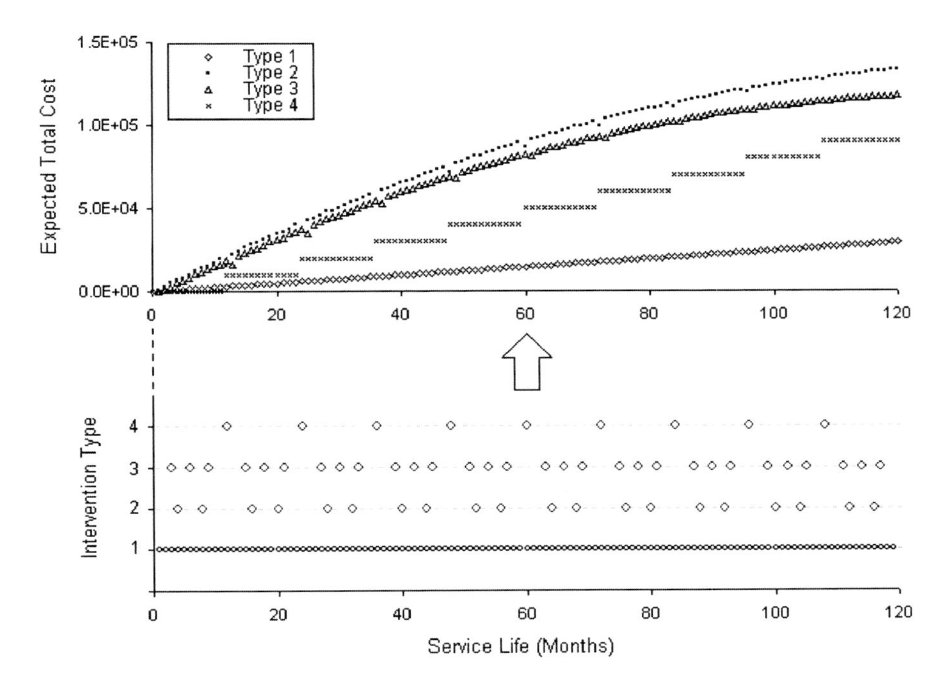

Fig. 17.7 The events and costs associated with the different lines of maintenance when technology insertion is incorporated

Figure 17.7 illustrates the events and expected costs incurred over time for the four different types of maintenance intervention when technology insertion activities are incorporated. Using the cost model, Fig. 17.8 illustrates the total expected cost incurred over time for the maintenance/service policy with and without technology insertion. From Fig. 17.8, it is evident that half way through the service life, the benefits of the technology insertion programme have not yet been realised. However, by the end of the proposed 10 years service life, the insertion version of the policy is expected to produce a substantial saving; the total expected cost associated with the insertion policy is 370310 compared with 455635 without.

We have illustrated how the methodology can be used to demonstrate the potential benefits gained in the development of an integrated maintenance plan incorporating the various sub-systems and a technology insertion programme for a complex asset. The example demonstrates that any complex system can be modelled if the defect arrival rates and the average costs associated with failures and maintenance interventions can be estimated for each type of inspection or sub-system. The failure process must be characterised and the defect arrival rates estimated using historical events data and/or expert information. The example also illustrates how the potential impact of technology insertion can be assessed and the manner in which a sensitivity analysis could be undertaken for a given case.

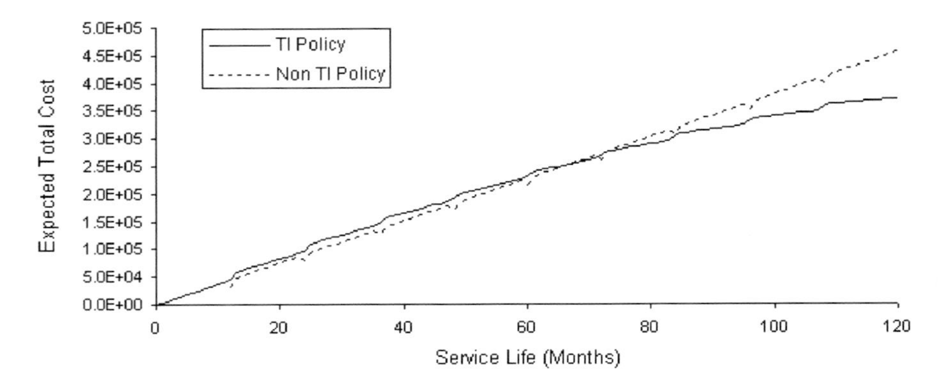

Fig. 17.8 The total expected cost over the asset service life with and without technology insertion

17.7 Conclusion and Summary

The methodology presented in this chapter can be used to build a process model for any asset, system or maintenance set-up. In this chapter, the methodology has been used to investigate a two-type maintenance scenario. However, the approach is not limited to the modelling of two types but can be tailored for any number of types of maintenance, with associated interactions, if the process parameters can be estimated using historical data (or subjectively using expert opinion). We have also discussed the potential for scheduling activities over a fixed service life and presented some ideas for the incorporation of technology insertion activities. For the developed models to be useful in practice, software packages based on the models reported in this chapter must be developed. We have developed demonstration prototype software which can be partly used for this purpose. The package can be part of an existing computerised maintenance management system or as a module in an ERP system.

17.8 Chapter Summary Questions

The discussion in this chapter raises several questions:

- Whether the delay-time concept used to model the inspection service interval exists or not in practice?
- Why is such a delay-time concept be useful for optimising the service interval?
- Can other measurements such as downtime, availability or capability be optimised instead of the cost measures used in the chapter?

References

A. Akbarov, W. Wang, A.H. Christer, Problem identification in the frame of maintenance modeling: A case study. Int. J. Prod. Res. **46**(4), 1031–1046 (2008)

R.D. Baker, W. Wang, Determining the delay-time distribution of faults in repairable machinery from failure data. IMA J. Math. Appl. Bus. Ind. **3**, 259–282 (1991)

A.H. Christer, Developments in delay-time analysis for modelling plant maintenance. J. Oper. Res. Soc. **50**, 1120–1137 (1999)

A.H. Christer, W.M. Waller, Reducing production downtime using delay-time analysis. J. Oper. Res. Soc. **35**, 499–512 (1984)

A.H. Christer, W. Wang, R.D. Baker, J.M. Sharp, Modelling maintenance practice of production plant using the delay-time concept. IMA J. Manag. Math. **6**, 67–83 (1995)

A.H. Christer, W. Wang, J.M. Sharp, R.D. Baker, A stochastic modeling problem of high-tech steel production plant, in *Stochastic modelling in innovative manufacturing. Lecture notes in economics and mathematical systems*, ed. by A.H. Christer, S. Osaki, L.C. Thomas (Springer, Berlin, 1997), pp. 196–214

T. Dowling, Technology insertion and obsolescence. J. Def. Sci. **9**(3), 151–155 (2004)

C. Jackson, R. Pascual, Optimal maintenance service contract negotiation with aging equipment. Eur. J. Oper. Res. **189**, 387–398 (2008)

B. Jones, I. Jenkinson, J. Wang, Methodology of using delay-time analysis for a manufacturing industry. Reliab. Eng. Syst. Saf. **94**, 111–124 (2009)

C.I.V. Kerr, R. Phaal, D.R. Probert, Technology insertion in the defence industry: A primer. Proc. Inst. Mech. Eng. Part B J. Eng. Manuf. **222**, 1009–1023 (2008)

A. Lisnianski, L. Frenkel, L. Khvatskin, Y. Ding, Maintenance contract assessment for aging systems. Qual. Reliab. Eng. Int. **24**, 519–531 (2008)

D. Lugfigheid, A.K.S. Jardine, X. Jiang, Optimizing the performance of a repairable system under a maintenance and repair contract. Qual. Reliab. Eng. Int. **23**, 943–960 (2007)

D. Lugfigheid, X. Jiang, A.K.S. Jardine, A finite horizon model for repairable system with repair restrictions. J. Oper. Res. Soc. **59**, 1321–1331 (2008)

A. Pillay, J. Wang, A.D. Wall, T. Ruxton, A maintenance study of fishing vessel equipment using delay-time analysis. J. Qual. Maint. Eng. **7**(2), 118–128 (2001)

C. Sriram, A. Hagham, An optimization model for aircraft maintenance scheduling and re-assignment. Transp. Res. Part A Policy Pract. **37**(1), 29–48 (2003)

G. Strong, Technology insertion—a worldwide perspective. J. Def. Sci. 9 (3): 114–121 in press doi: 10.1016/j.ress.2008.06.010 (2004)

W. Wang, A model of multiple nested inspections at different intervals. Comput. Oper. Res. **27**(6), 539–558 (2000)

W. Wang, Delay-time modelling, in *Complex system maintenance handbook*, ed. by K. Kobbacy, D. Prahakbar Murth (Springer, London, 2008a)

W. Wang, Condition based maintenance modelling, in *Complex system maintenance handbook*, ed. by K. Kobbacy, D. Prahakbar Murth (Springer, London, 2008b)

Chapter 18
Simulation Based Process Design Methods for Maintenance Planning

Joe Butterfield and William McEwan

Abstract The primary objective of this work is to use simulation methods for the development of optimal service support procedures using an integrated Product, Process and Resource (PPR) structure. Mid life fatigue modifications required on a sub-system within an existing aerospace platform that have been used to demonstrate the utility of this approach in supporting the transition of an original equipment manufacturer (OEM) to availability contracting. Other objectives of the work were to show how significant cost drivers can be identified and quantified through the virtual development of work breakdown structures (WBS) for service support processes and to show how the mechanisms used to develop optimal processes, can produce animated instructional materials which enhance organisational learning as processes evolve. This supports the effective communication of methods and work breakdown structures between the technical author and service personnel. Although this work has been completed using an aerospace sub-system as a demonstrator, the outcomes are equally applicable to other platforms which utilise complex engineering systems. The results of this work have system level significance in reducing risk in service provision through the support of value co-creation. The simulation outputs also provide data for higher level cost modelling which in turn, informs the strategy required for supply chain engagement through incentivisation.

18.1 Introduction

The last 10–15 years have seen momentous changes in global socio-economic conditions as well as a significant shift in military strategy as world orders changed after the fall of communism. According to Spreen, this resulted in a drop in

J. Butterfield (✉) · W. McEwan
School of Mechanical and Aerospace Engineering, Queen's University Belfast, Ashby Building, Stranmillis Rd, Belfast, BT9 5AH, UK
e-mail: j.butterfield@qub.ac.uk

I. Ng et al. (eds.), *Complex Engineering Service Systems*, Decision Engineering,
DOI: 10.1007/978-0-85729-189-9_18,

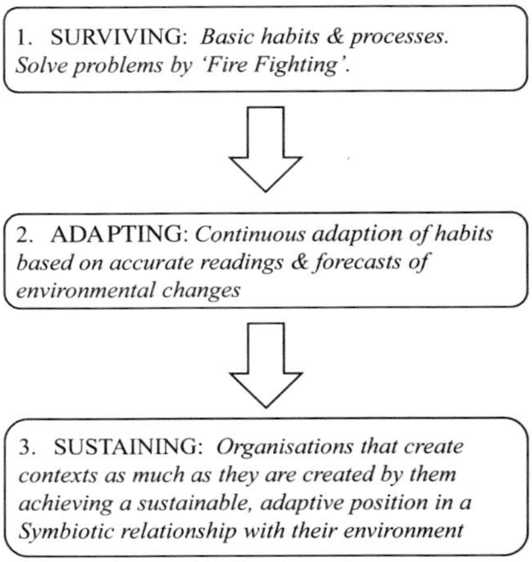

1. SURVIVING: *Basic habits & processes. Solve problems by 'Fire Fighting'.*

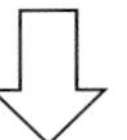

2. ADAPTING: *Continuous adaption of habits based on accurate readings & forecasts of environmental changes*

3. SUSTAINING: *Organisations that create contexts as much as they are created by them achieving a sustainable, adaptive position in a Symbiotic relationship with their environment*

Fig. 18.1 Evolutionary stages of a learning company (Pedler et al. 1996)

demand for military aircraft particularly in Europe during the 1990s and although the emergence of new terrorist related threats prompted a recovery in the demand for defence related equipment in the US after 2001, the probability that the number of new military aircraft programs will recover to cold war levels is exceptionally low. These circumstances have resulted in a contraction in the demand for new product development and an increase in the pressure to keep legacy systems in service for longer. OEMs are therefore, increasingly evolving from product suppliers to service providers. Challenging commercial conditions have also resulted in end users re-structuring their own core businesses. A faltering world economy has placed enormous pressures on government finances and despite clear operational needs, defence spending has been highly constrained in recent years. The net result of these circumstances and changing behaviours has been the need for the OEM to *integrate and manage* support service activities *in partnership* with the customer to deliver the availability of a given platform or product. This improves the likelihood of operational sustainability for the OEM through shared commercial risks while reducing the cost of ownership for the customer with equivalent or improved platform availability.

In order to complete the transition to '*availability contracting*', the OEM must move beyond core business activities based on pure manufacture and supply to become involved in the broader product lifecycle. This fundamental shift in commercial practice must take place in partnership with the customer and it represents a significant challenge at all organisational and technical levels within *both* organisations. A successful, sustainable outcome will depend on the extent to which both parties become a '*Learning Organisation*'. This will require the effective amalgamation of key business elements of the previously separate enterprises and both the OEM and the customer will have to pass through the evolutionary stages proposed by Pedler et al. (1996) as shown in Fig. 18.1, from simply surviving as separate entities, through adaptation to joint sustainability.

Table 18.1 Characteristics of the Learning Company

Learning approach to strategy	Pilots and experiments used to inform direction
Participative policy making	All stakeholders contribute to policy making
Informating	Use of I.T. to automate and distribute information thereby empowering operatives to act on their own initiative
Formative accounting and control	One aspect of informating opening up budgeting, reporting and accounting to broader enterprise
Formative accounting and control	One aspect of informating opening up budgeting, reporting and accounting to broader enterprise
Internal exchange	All internal units and departments are effectively customers and suppliers in a supply chain leading to the customer. Contracting with and learning from one another
Reward flexibility	Greater participation creates the need for flexible and creative monetary/non- monetary rewards
Enabling structures	Roles, departments, organisational charts, procedures and processes can be viewed as structures that can be used to meet job, user or innovation requirements
Boundary (interface) controls	Used for 'environmental scanning' by those who have contact with external users, customers, suppliers, clients, business partners etc. Processes used for data handling passing into (and out) of the enterprise
Inter-company learning	Through joint ventures and other learning alliances, organisations learn from other companies. This is facilitated by joint meetings for mutually beneficial information exchange
Learning climate	Management facilitating experimentation and learning from experience through questioning, feedback and support. This is subsequently exported to the broader commercial and technical activity and any business partner

A learning organisation is one that can change from within itself. Easterby-Smith (1999) states that implicitly, its underlying principle is an adaptive one where adaptation to environmental conditions or pressures leads to organisational learning and changes. Learning and changing faster than others can help to ensure the maintenance of any competitive edge or at the very least, increase the organisation's chances of survival. According to Pedler et al. (1996) the characteristics of the learning company are listed in Table 18.1.

This work has come about as a result of the recognition that evolution or transition to availability contracting will become a significant part of the operational activities of BAE Systems moving forward. BAE Systems is already evolving into a 'learning organisation' and the characteristics listed in Table 18.1 are all engendered in the EPSRC funded program entitled Support Service Solutions: Strategy and Transition (S4T). The material presented in this chapter was carried out as part of Work Package 3 within this program. It presents a strategy for the virtual simulation of maintenance planning processes and their execution for enhanced service provision. The main outputs required were WBS, the identification and quantification of sub-system level cost drivers (for subsequent use in

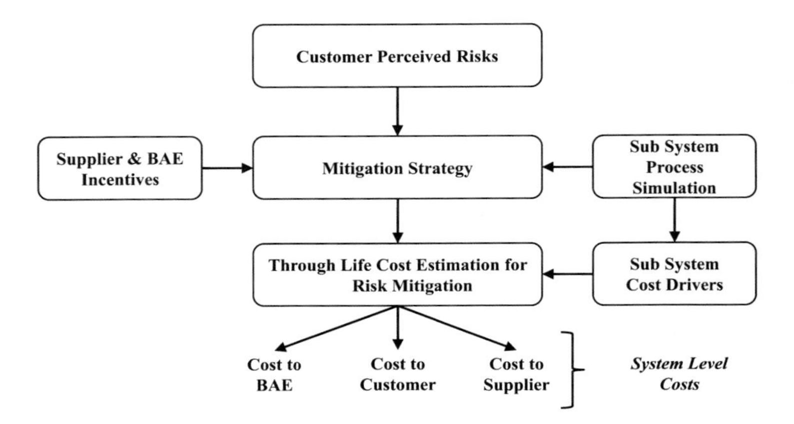

Fig. 18.2 Position of sub-system simulation activity within work package 3

system level cost models) and the provision of digital instructional materials. Digital manufacturing tools were used to deliver these requirements. The data flow diagram showing how simulation fits with the other research streams within Work Package 3 can be seen in Fig. 18.2.

Butterfield et al. (2007) have stated that the need to minimise the risk of commercial failure by developing leaner products has lead to the development of manufacturing simulation software systems which use simultaneous or concurrent engineering concepts for the design and engineering of new products across a range of industries. The way in which any complex assembly is constructed or de-constructed for maintenance, repair or overhaul (MRO) is an important consideration if a whole life approach is taken to product design and process development. Traditionally process design has not been supported by predictive technologies to the same degree as other areas of product design and development. To deal with this problem concurrent engineering concepts are now available in computer aided design (CAD) and computer aided engineering (CAE) tools and they are used to identify optimal *process* designs. Although to date, these tools have been used in a *product* design and manufacturing context, they are equally applicable to *process* design for the maintenance repair and overhaul (MRO) activities that are at the core of availability contracting. The simulation techniques that these systems offer allow process planners to define, validate, manage and deliver fully optimised manufacturing or MRO process specifications.

Although software vendors can provide 'off the shelf' solutions for the analysis of typical manufacturing processes, very often OEMs have company specific activities and requirements which do not fit into standard software versions. The CATIA[1] (Computer Aided Three-dimensional Interactive

[1] CATIA is a 3D CAD modelling environment used for the design and specification of engineering components.

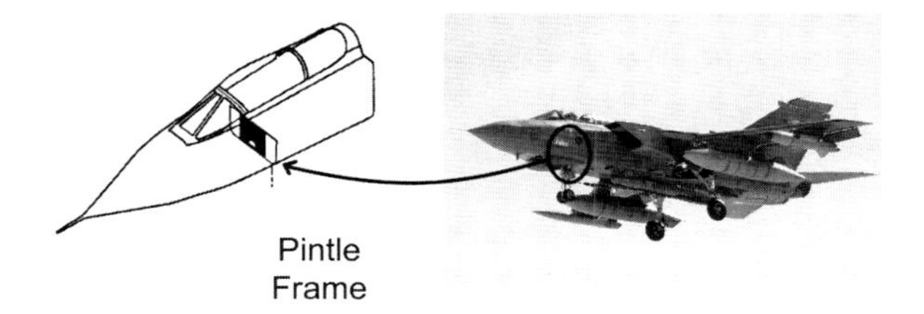

Fig. 18.3 Pintle frame location within the front section of Panavia Tornado

Application) and DELMIA[2] (Digital Enterprise Lean Manufacturing Interactive Application) software modules used for this work were enhanced and customised where required, to suit the specific requirements of service support process development. An exemplar is presented which provides an evaluation of simulation methods on a 'proof of concept' basis. The exemplar is based on a mid life fatigue modification required on a sub-system within the twin-engined Panavia Tornado combat aircraft. The work shows how digital methods, originally developed for purely manufacturing applications, can be applied to a low volume/high complexity task based on the replacement of a pintle frame section. The pintle is a significant piece of internal structure in the front end of the aircraft, see Fig. 18.3. The application in this instance is based on the simulation of the removal and replacement of key structural components in order to demonstrate how the entire process can be designed and captured in a virtual environment. This, in turn, demonstrates how the methods used can be applied to the service of elements within a complex asset which were never designed to be replaced. This is an important example of the increasing complexity of defence-based service provision.

The work focuses on the development and validation of an optimal WBS and the main outputs are the process definition including task completion times (to facilitate cost prediction), the development of digital animations for process design purposes as well as instructional delivery. Butterfield et al. (2007) have demonstrated the utility of the method in a manufacturing context for improving operator learning through the use of animated work instructions. Butterfield et al. (2008) have shown improved management learning using concurrent digital working practices and Jin et al. (2009) for simulating manufacturing assembly times.

The work detailed in this chapter uses a CAD-based collaborative environment to develop this approach to bring simulation and optimisation strategies in line with the planning required for aerospace-based availability contracting at BAE Systems. The structure uses, defines and retains process data for a single sub-system in this case, but the hierarchical nature of the structure means that complete systems can be modeled. The end result is an integrated, collaborative environment containing the WBS which is developed from the manufacturing bill of

[2] DELMIA is an integrated, 3D CAD based environment used for engineering process design and optimisation.

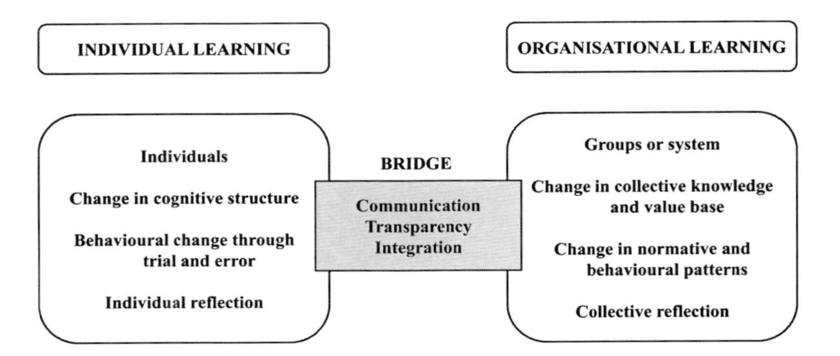

Fig. 18.4 Transformational bridge between individual and organisational learning, Probst

materials (MBOM). It defines the processes associated with service support activities and includes time and cost data for product assembly and dis-assembly as well as the animated process definitions. According to Probst and Buchel (1997) this then places the digital tools used for this work, in the role of a 'bridge' between the behaviours associated with the types of individual learning discussed by Butterfield et al. (2008) (managerial, operator etc.) and those required for the system level organisational learning which will aid the transition to availability contracting, see Fig. 18.4.

18.2 Sub-System Level Simulation

18.2.1 Background

The recent extension to the service life for the Tornado has given rise to the Mid Life Fatigue Programme (MLFP) within BAE Systems. A physical test programme has identified several critical airframe components which are approaching the end of their fatigue life. The original service expectation on the aircraft meant that these components were designed to operate up to the temporal point now reached by the aircraft. The extension to service means that these components must now be replaced.

The original design constraints and demands on these components did not include repair or replacement as they were expected to perform within design tolerances for the specific life span originally dictated. Many of these components are embedded deep within the airframe superstructure. Issues of access and method of change remain the dominant feature of the MLFP programme. Limitations of available data, as well as test case aircraft to assess concepts represents a strong case for digital simulation of MLFP activities in order to arrive at optimum, cost-effective solutions.

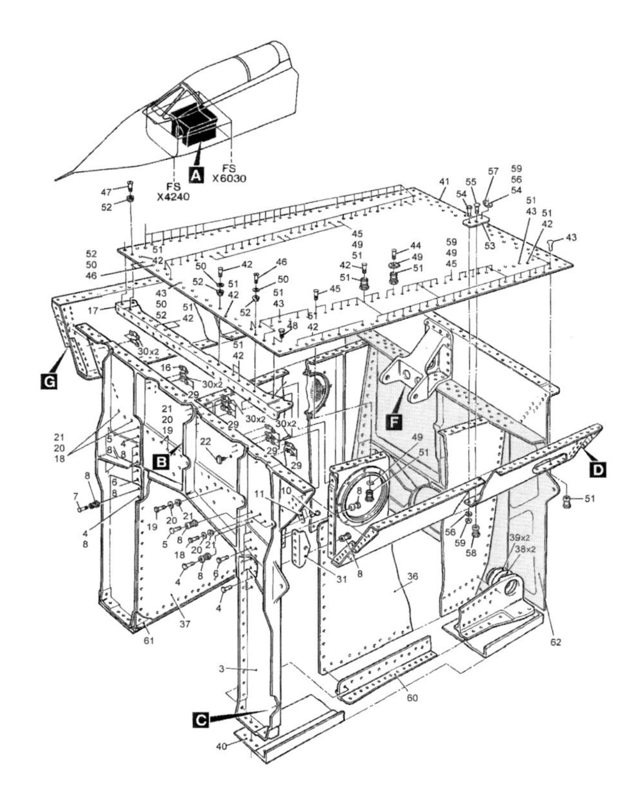

Fig. 18.5 Pintle frame location

For the purposes of this work, Queen's University Belfast (QUB), in conjunction with BAE Systems Warton, have digitally simulated the replacement process of a major airframe structural component, the pintle frame, with a view to establish an optimum solution in both technique and cost. Utilising an integrated PPR structure, it was possible to model the relevant airframe components, associated fixtures and spatial constraints using DELMIA V5. The working methodology through the BOM within the DPM (Digital Process for Manufacture) environment and the associated work flow of the physical methodology. One specific proposal was then animated and ergonomics and clash detection utilised to establish the WBS.

18.2.2 Test Case 'The Sub-System'

Figure 18.5, shows the major portion of the airframe investigated. The panel to the rear of the seatwell box assembly, shown in exploded format, is the pintle frame.

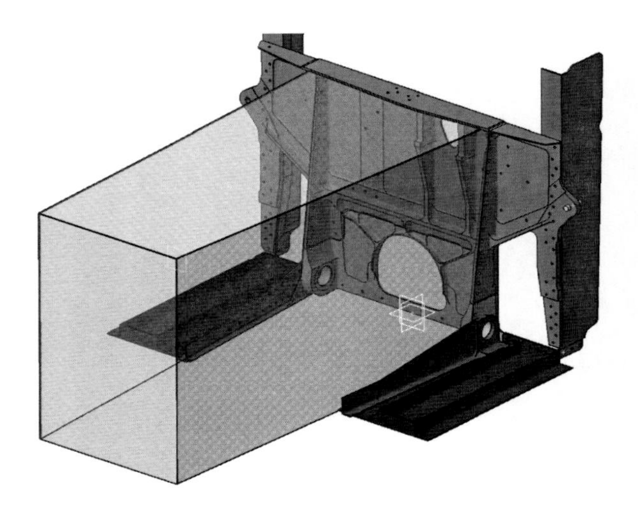

Fig. 18.6 Simulation of work space environment surrounding pintle frame

18.2.2.1 Establish Maintenance Activities as Part of Standard BAE Systems Protocol

Generic fitting activities are well established under the normal operating routines at BAE Systems. Pintle frame replacement constituted a significant challenge even with experienced fitting operatives. To gain access to the panel prior to replacement, it was first necessary to dis-assemble and remove the undercarriage. With access to the seatwell established, the hardware bolted to the interior surfaces of the surrounding structures was also removed and/or loosened from position to gain access to the pintle frame and create space within the seatwell, for the technician's working convenience.

From this point onwards, all activities are modelled within the digital environment. All movements, associated tooling and fixtures were input as time dependant variables representing the sub-system level activities contained in the pintle replacement work package.

18.2.2.2 Assembly Model of Environment

Beginning with the major structural components, provided by BAE Systems in CATIA V4 format, all CAD data was translated into the V5 environment. A set of idealised shear walls were added to simulate the surrounding structural elements (see Fig. 18.6) for which only 2D data in the form of lofted drawings was available. This was the standard method for providing engineering data for a given part at the time when the Tornado was designed.

With the interior working volume established, the exterior surfaces were modelled to accurately reproduce the working environment. Again a simple idealised surface model of the exterior of the forward fuselage skin was used to provide relevant information pertaining to the movement of technicians, tooling

Fig. 18.7 Human analysis simulation in pintle frame area

and fixtures around the work area. Information related to the distance between the workshop floor and the fuselage model was required to establish the limitations on technician movement around the work area.

18.2.2.3 Human Considerations

The shear wall surfaces were idealised to provide the physical boundaries or ergonomic constraints for the human analysis model. Figure 18.7 shows how a mannequin was used make an ergonomic assessment of the work area in front of the pintle frame. Here the mannequin can be seen in a position of disadvantageous medial rotation. This resulted from the lateral limitations of the work space in conjunction with the forced positioning with respect to the work piece height from the hanger floor. This example is an ideal illustration of how the effect of physical limitations can be analysed to express the optimum solution for operator motion. In this case the analysis of posture indicates a limitation on lateral and transverse application of force available to the operator, and hence a clear design choice to focus on longitudinal motions to execute the task. In response, the associated tooling design to meet this particular objective was designed with longitudinal motion as the key ergonomic requirement.

18.2.2.4 Identification of Parts and Fasteners

Fastener type and count were recorded based on the original manufacturing BOM. It must be noted here that CAD definitions of the fasteners does not exist and is not common practice in aerospace design. Fastener type, number and associated time for removal and replacement, however was defined for the purpose of time generation and their locations were determined from the 2D data provided by BAE Systems. To improve the efficiency of this task, a Visual Basic script was also devised based on an automated search which recovered hole sizes, quantities and depths directly from the feature tree in the CAD model see Fig. 18.8. Visual Basic

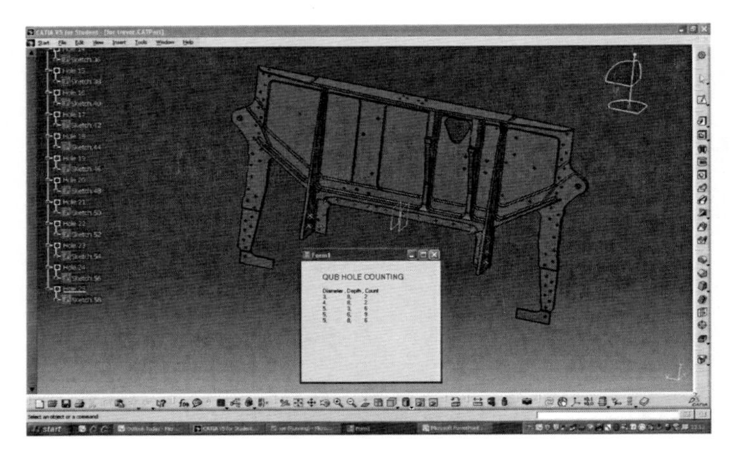

Fig. 18.8 Automated extraction of part data from feature tree in CATIA

is a programming language that enables the development of graphical user interface applications as well as access to databases using Data Access Objects, Remote Data Objects, or ActiveX Data Objects. In simple terms, this was a computer program designed to extract hole properties (sizes, quantities etc.) from the CAD model. By associating fastener types with given hole sizes, fastener counts could be established. The hole size and depth was equated directly with drilling time.

18.2.2.5 Develop Simulation for Pintle Replacement

Initial product assessments showed that the major dis-assembly of the airframe that was required to free the pintle frame for replacement was neither practicable nor cost effective. From this, the decision to re-work the pintle in situ was made. From this position, the first design constraints were established for any proposed engineering solutions.

A preliminary trade-off study was performed to establish the optimum method for the pintle replacement from a high level engineering perspective. The three solutions considered were:

- *Full Automation*—Computer controlled robotic machine installations
- *Semi Automation*—Hand held machine tools and guide templates
- *Manual*—The use of manually guided hand tools only

Each of the potential methods was rated according to six different metrics:

- *Time*—Required to perform tasks including set up, measurement or metrology and inspection; all value added activities
- *Re-use*—A measure of the potential utilisation of any hardware for future tasks not associated with the pintle replacement

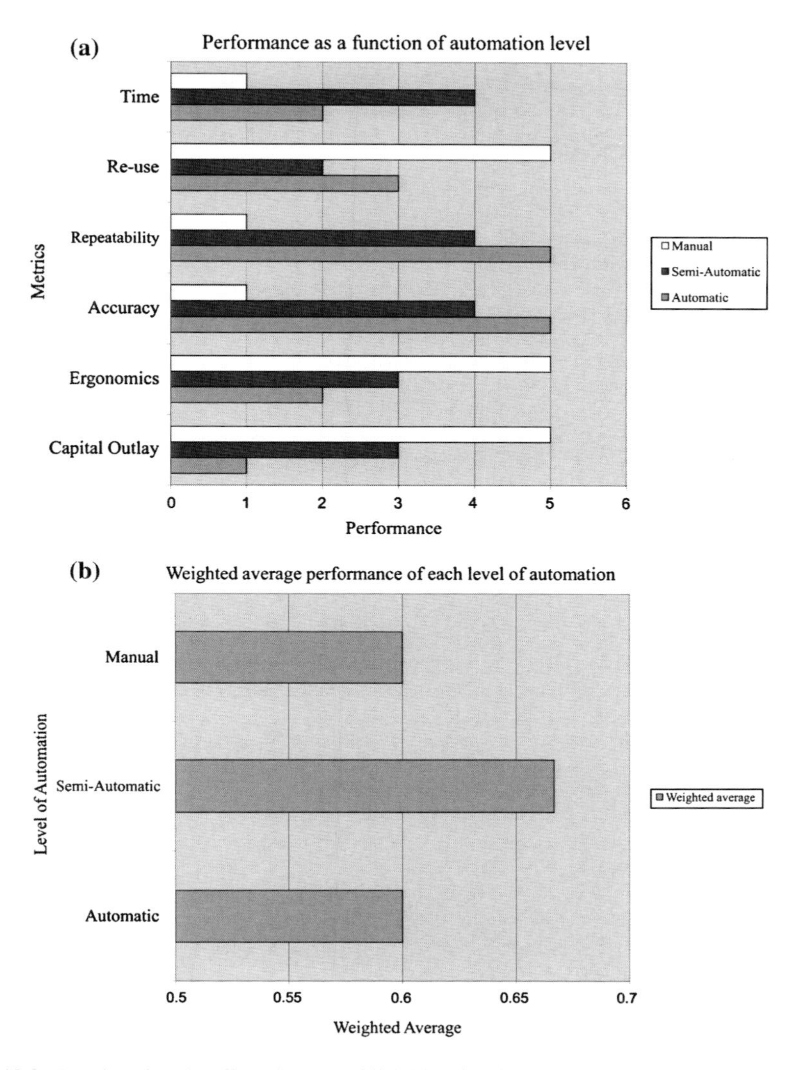

Fig. 18.9 Results of trade off study to establish high level methodology

- *Repeatability*—The recurrent holding of tolerances in all worked parts
- *Accuracy*—The capacity to meet geometric requirements with regard to the position of all worked surfaces with respect to existing geometry
- *Ergonomics*—Ease of operator movement (Includes measure of fatigue)
- *Capital outlay*—Cost of any associated hardware or consumables

The graphs shown in Fig. 18.9 illustrate the overall result of this decision making process. Graph (a) shows how each of the three methodologies performed with regard to the six metrics listed previously, as a mark out of a possible maximum of five where: 1 = worst/low rating, 5 = best/high rating. Graph (b) in Fig. 18.9 shows the weighted average of these values as a non-dimensional expression between 0 and 1. This graph shows that, for the conditions used in this trade study, the semi-automatic methodology had the best combination of performance properties.

Having established the benefits of a semi-automated solution, a secondary set of design criteria were established. The need to escape the uncertainty in both repeatability and accuracy with the fully manual methodology, as well as the intrinsic high temporal cost, predicated a maximum utilisation of suitable machine tools. The limitations on such tooling, while including capital cost, were driven primarily by physical size. A survey of available hardware, performed for the trade off study, identified a number of miniaturised milling machines capable of carrying the appropriate tool bits and performing with the necessary power to cut the maximum skin thickness identified from the pintle geometry.

18.2.2.6 Fixtures and Tooling

The crux of the engineering problem presented here involves the fact that the replacement pintle pieces are newly manufactured. These have to be fitted to pre-existing, well used airframes of variable, and most importantly un-quantified, dimensional stability. With this in mind, a decision was made to use a machining pattern or fixture that would facilitate accurate and repeatable removal of the redundant pintle material.

With the idealised machine tool modelled in the CAD environment, to provide the required dimensions for the interface with the new piece, it was possible to test the positions and motions of the tool to establish the optimum path. This was carried out as follows:

- *Tool path*—derived from the final geometry of the portion of the pintle which was to remain in situ, and the replacement piece to be introduced and fitted. The minimum number of movements and set ups required by the operator in execution of cutting. Clash detection with all surrounding components with respect to the mill dimensions
- *Tool choice, cutting speed and feed rate*—again derived from the final geometry, variation in material thickness, minimum tool movement during operation and best possible proximity of the mill to the work piece without clash

Choice of tooling implemented in the final design solution was established iteratively through the simulated environment including design, assembly and human analysis methods. For example, the mini-mill chosen for the machining operations was one of a small number of machines initially selected for its size characteristics. All primary tool choices had to be small enough to fit within the confined space of the pilot seatwell with the operator. From within the design environment, the tool path was established for the removal of material from the original pintle. From the assembly environment, it became apparent that the proximity of the shear webs and pintle bush lugs to the tool path negated the use of some of the primary tooling choices due to a physical clash between the tools and the surrounding structure. From the human analysis environment, it could be seen that the physical limitations on the operator posture resulted in a severe reduction in the operator's capacity to apply any meaningful degree of lateral force when handling the tool. From this an engineering decision to select a tool which could be handled in a predominantly longitudinal manner was made.

One major problem associated with the overall task was the establishment of reliable datums on which to base all hardware. The replacement sections of the pintle frame will be manufactured off site using a fixed set of dimensional constraints, but the surrounding airframe components and their position with respect to one another cannot be assumed to be as per original design or replicant from one aircraft to another. With these conditions in mind, it was decided to select the main undercarriage bearing positions on the longerons as the most suitable datum. Since the longerons are major structural components responsible for the positioning of much of the rest of the airframe, it can be assumed that they are prone to the least digression from original specification, both in terms of intrinsic dimensions, and also with regard to position in respect to one another. By extension it is logical to assume that the digression of the pintle frame position with respect to the longerons is also limited. Finally, since the pintle frame shares the same bearing locations as the longerons it is sensible to fix the position of the new pintle section from these points and base the cutting geometry of the original pintle with respect to the insert. This ensures that when the new section is inserted and located correctly with respect to the bearing locations, the machined surfaces on the cut out of the original pintle are aligned correctly with the mating surfaces of the insert, regardless of any distortion present in the original pintle.

The design solution arrived at consists of a flat guide plate which locates on the bearing positions and alters between a horizontal position for the early cutting stages and vertical orientation for the latter stages. This plate has guide paths on its surface which correspond to pins on the mill allowing it to follow set paths which mimic the required cut out profile on the pintle frame. The translation between horizontal and vertical positions is to allow the change in orientation necessary to make the set of cuts that exist in different planes. This plate, like the mill, is available for repeated use on each aircraft without the need for any measurement or alteration. This particular feature represents a large time saving in the overall time required to perform the task. In addition to this plate, a drilling template for the positioning of holes in the outer webs of the pintle is also utilised.

Major structural components, idealised bounding surfaces and where data is available, the sub-components fastened to the pintle frame were assembled within the CAD environment. Parts, fasteners and the positioning of all relevant components were fixed. The BOM, held within the PPR hub, was then available for interrogation by the user. The final choice in milling machine was the *Ricci Corp Mini-mill.*

For the given material properties of the components and the maximum material thickness to be encountered, engineering rules were applied and a size up regime

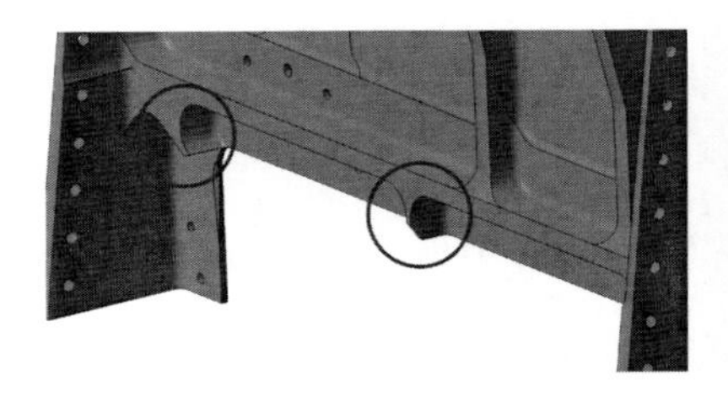

Fig. 18.10 Pintle frame with web material still in place

for the drilling operations was established as 2 mm to 3.5 mm to 5 mm. A long taper flute drill is best suited to this material, however, the time deriving software for drilling operations that was used in this project, SEER Design for Manufacture,[3] does not support such bits, hence, normal twist drills were used in this model. Since drilling templates were used where new holes were to be established without a pre-existing guide hole in a mating component, there was no requirement for spot drilling to ensure accurate hole placement. This represents a classic engineering decision within this model, to create a process which requires a minimum of specialised skill on the part of the operator, and hence mitigating against problems associated with the availability of highly skilled operators.

As mentioned previously, drilling operation times were established via SEER where the times associated for the drilling of any specific hole was established as a factor of drill size, hole depth, material properties and rotational speed of the drill. In a similar manner, the cutting speed of the mill was established through SEER also. The first milling operation involves the removal of the major portion of extraneous material from the pre-existing pintle. A 2 mm diameter plunge and cut bit of high speed steel with 4 flutes was chosen for this task. SEER has the capacity to model milling operations according to tool specifications (as listed above), the manner of cutting and the length and depth of the cutting path, or alternatively the volume of material to be removed. As the depth of the material, through any given cutting path was not consistent, each cut was modelled as a separate operation, the number of which was equal to the number of step changes in material depth through the entire cutting path. This provided optimal accuracy in the output time from the SEER model.

With the first cut complete and the waste material removed, the mill was fitted with a 10 mm diameter edge cutting bit of high speed steel with 8 flutes. With this tooling the internal edge of the pintle was finished to final dimensions. The final milling operation involves the removal of web material from the internal face of the pintle. The web material is highlighted in Fig. 18.10.

A 28 mm diameter face and edge cut tool of high speed steel and 16 flutes was selected. The final face profile is shown in Fig. 18.11. A three stage cutting plan was followed which avoided the possibility of machining in-accuracies due to removing too much material in one run.

---

[3] SEER is a registered trade mark for cost management software marketed by Galorath inc.

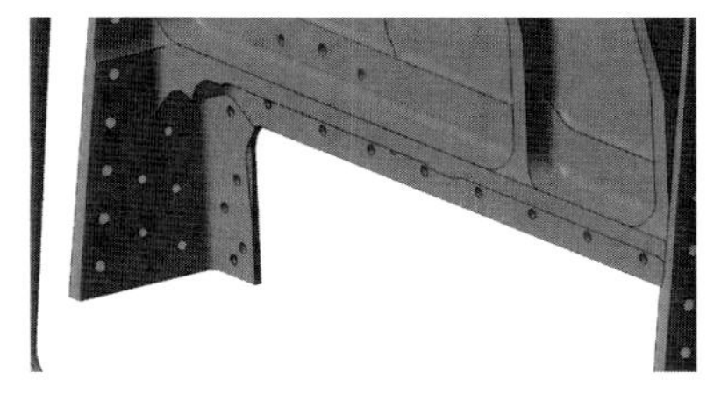

Fig. 18.11 Pintle frame in final state

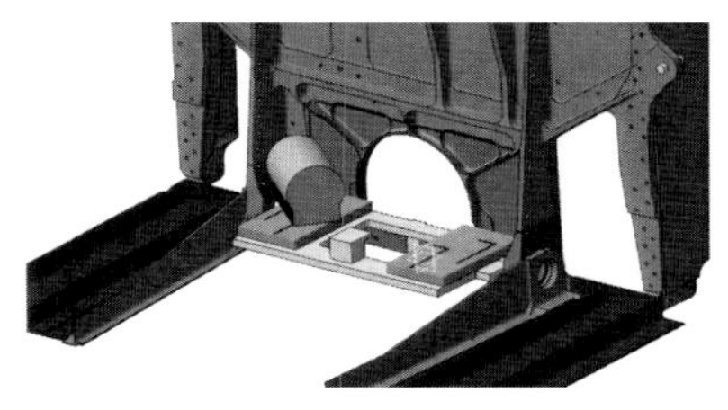

Fig. 18.12 Placing of mill and tool lateral cuts

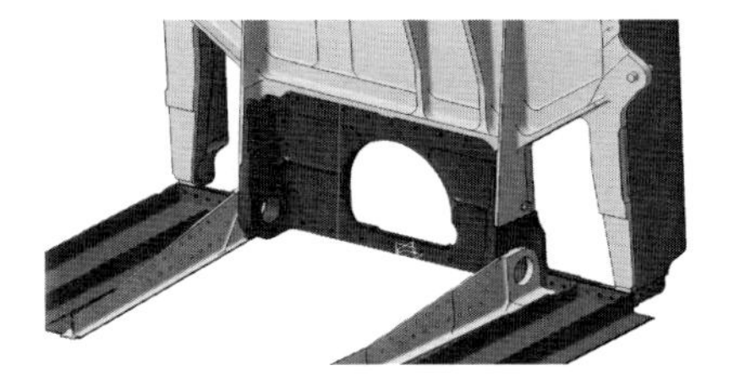

Fig. 18.13 Introduction of new pintle part

Each discrete drilling operation was modelled within DPM where the tooling can be seen moving around the work piece as per the process order. This in turn is also seen in the output work instructions.

18.2.3 Process Definition

The dis-assembly activities and all associated data were established using the simulation methods detailed above. A preliminary discussion with BAE Systems led to the following broad work stream order: Fastener strip, disassembly and removal of sub-components, clean down of exposed pintle frame, introduction of fixture(s) for cutting, machining and dressing, cut and dressing process for pintle frame, removal of now extraneous material from pintle frame, clean down, assembly of new pintle sub-part, fastening, re-assembly of sub-components, fastening and inspection. Each of these activities was broken down into sub-activities with different requirements in tooling, labour and hence time. The sub-activities relating to the work focussed directly on the pintle frame are derived directly from the simulation as optimised solutions.

Figures 18.12 and 18.13, show 2 stages of the process required to replace the pintle frame.

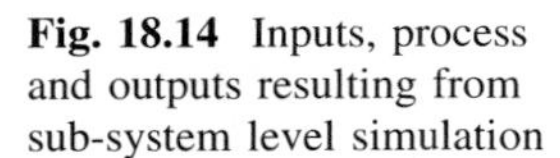

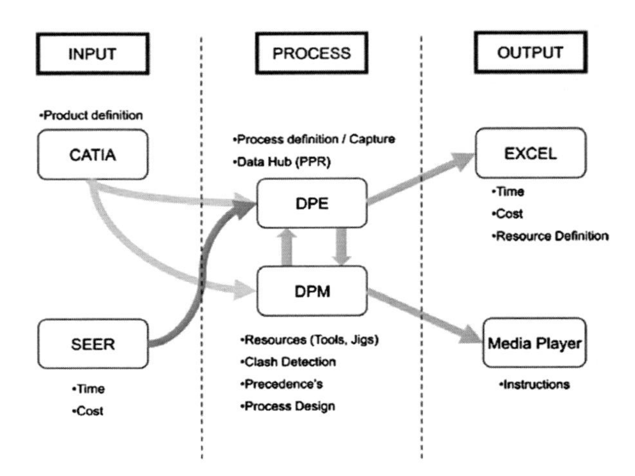

Fig. 18.14 Inputs, process and outputs resulting from sub-system level simulation

Each discrete component movement was animated within the CAD environment as they were removed or replaced. Clash detection studies were carried out to establish optimum pathways for each component movement. In conjunction, mannequins were utilised to perform an ergonomic assessment of the simulated activities.

With a working order established to the physical activities, each was recorded within Delmia Process Engineer (DPE)[4] as discrete actions with their associated time and tooling requirement.

18.2.4 Simulation Outputs

A number of outputs were derived directly from the digital simulations. These were: Optimised work breakdown structure, identification of cost drivers, delivery of representative times for activities and the generation of instructional materials.

Figure 18.14 shows the overall process structure for the use of simulation methods to yield these outputs. Section 2.3 details the WBS which was defined in the CAD environment and subsequently recorded in DPE. The WBS was supplemented with process times and part counts.

The second output from this work package component was the identification of cost drivers. The drivers arising from the parameters within the sub-system simulation were found to include: Replacement parts, tooling, fixturing, fasteners and operator time (for dis-assembly/assembly and inspection/quality control).

The timing data generated from the simulations is in the form of the 'hands on' time associated with specific activities. When this is used for process costing at a

[4] DPE is the process modelling arm of the Dassault Systemes suit where discrete processes can be modelled, in order, and with associated attributes such as time.

system level, these times must be supplemented with data associated with labour type and source, i.e., BAE Systems staff or customer personnel. A utilisation factor must also be applied to allow for 'non hands on' activities which are an integral part of an operator's working day. Consideration of learning curve impact is also critically dependant on labour type. Both utilisation and learning curve allowances will vary depending on the labour source and should be an important part of the system level costing strategy.

The final deliverable is represented by the animated work instructions. Clash detection, and ergonomic considerations, generated the most time conservative solution to the task of pintle frame repair. From the same digital simulation, instructional materials were generated to assist the execution of the pintle replacement activities. Butterfield et al. (2008) have shown that the use of animated instructional materials can reduce the learning curve resulting in lower manufacturing costs. It is expected that a similar result can be achieved with regard to maintenance and repair activities if operator learning is driven by animated work instructions. Here the process order and the specific details of each process have been shown as virtual movement of the components complete with fixtures and tools, as would occur in the real world scenario. Accompanying these animations are written instructions which appear as dialogue boxes at the points most appropriate to the specific processes. These instructions contain the details that are not immediately apparent from the animation. This includes such details as specific tool size or positioning of fixturing.

18.3 Discussion

The transition from a conventional supplier/customer relationship to an availability contract arrangement, constitutes a significant challenge for both parties at all levels within the enterprise—commercial, technical and human. According to Butterfield et al. (2009), the nature of the commercial relationship changes as previously separate business entities merge and self interests are replaced by the need for optimal 'co-performance'. Operational sustainability remains a challenge for any business entity under current political and economic conditions. Companies must become learning organisations which can react efficiently to changing market needs to retain any operational advantage or in extreme cases, simply to survive. The fact that BAE Systems have recognised the need for and initiated the S4T program means that it is evolving as a learning organisation. The characteristics listed in Table 18.1 are engendered in the S4T program as a whole and the simulation methods used for this work can be linked directly or indirectly to most of the same characteristics.

The aerospace exemplar presented here shows how simulation and the principles of digital manufacturing can be used to design service solutions for complex platform elements which were never intended for replacement. The functionality of the simulation framework allows complete system definition through the hierarchical summation of process definitions and parameters within individual sub-systems. The development, optimisation and retention of process knowledge in the virtual environment, increases the likelihood of a positive outcome for both the OEM and the customer by providing a framework by which the attributes of value co-creation discussed by Ng and Nudurupati (2010) and Ng et al. (2010a), can be facilitated using a controlled, measureable methodology. This enables the transformation of process structures (*information*), product definitions (*materials and equipment*) and people (*instructional materials*) in line with what will be required for the provision of platform availability. The simulations deliver process details from an engineering perspective, which include WBS for maintenance operations, delivery of representative times for 'hands on' activities (which can be used for resource optimisation and cost analysis), identification of cost drivers such as labour, tooling, fixturing and inventory and the generation of instructional materials. The outcomes can be used in turn, for system level cost modelling. The approach to the delivery of instructional materials in this case has been to use animated maintenance processes. The ability to carry out significant modifications in small batches is an important factor in service provision. This emphasises the need for an efficient means of transferring knowledge and instructing operators who may not be under the direct control of the OEM. In a service context they may have to carry out multiple tasks on a given system or across several platforms. The availability of animated instructional materials facilitates the operational agility and breadth of knowledge that an operator must have to deliver these efficiently.

The digital methods presented here are best suited for complex 'hard' engineering problems such as the pintle frame replacement. For other test and inspection activities the added value of this approach is questionable but the digital systems used for this work can still play a part in developing and delivering process definitions. Test and inspection activities can still be supported using a digital approach but the instructions would be based on static CAD images.

Digital manufacturing provides an integrated framework for the generation, management and delivery of engineering process definitions. Within a single enterprise, this facilitates concurrent activities between the engineering disciplines. The transparent and collaborative environment can be enhanced to include maintenance activities, as the process node in the manufacturing hub is extended to include the broader product lifecycle. The definition and optimisation of maintenance processes in a virtual environment using virtual assets reduces the uncertainty associated with availability contracting. Simulated process scenarios allow the process designer and implementer to reduce or eliminate exposure to risk and uncertainty by generating realistic data for important cost components.

This work has shown how digital methods can be used in support of the transition to availability based contracting. The benefits of using process simulation methods in support of this business model has been demonstrated on a proof of concept basis. The next step would be the implementation of the methods developed here starting with a pilot program covering a broader platform application working from sub-system to system level simulations. In order to exploit the utility of the methods presented here the question of how the simulation platform is managed has to be addressed. Process data related to MRO activities is generated at source by the platform manufacturer. When held as a central data repository within the PPR hub, process data can be accessed by individuals and disciplines across the enterprise. Levels of access and user privileges can be controlled by setting the relevant permissions. With current software versions, this can be achieved across existing enterprise wide networks but access can be a problem for outside parties due to security set ups etc. Data should also be available to both OEM and the customer and this has been recognised as an issue. Future versions of the software will be web mounted making access between enterprises possible.

18.4 Conclusions

A sub-system level exemplar based on the partial replacement of an aerospace structural panel has shown how simulation can deliver virtual process definitions for service support activities including work breakdown structure, key cost drivers and instructional materials. The utility of manufacturing simulation techniques in maintenance process design has been demonstrated. This requires the extension of a hierarchical PPR environment into the extended platform lifecycle where the assembly and dis-assembly processes associated with MRO activities can be modelled. The resulting process structure enables the determination of process sequences, labour hours and part counts. Instructional materials are developed concurrently with process design thereby eliminating the need for separate instructional authoring. This has been demonstrated at a sub-system level and the outcomes can feed into system level cost models. The PPR structure facilitates the process of knowledge generation, capture and retention thereby facilitating the transition of an OEM to availability contracting as a learning organisation.

The outcomes integrated with the other WP3 themes with digital methods supporting the Attributes of Value Co-Creation as developed by Ng et al. (2010a). The identification and quantification of sub-system cost drivers provided inputs for the cost modelling activity carried out by Roy et al. (2009), which in turn, informed the incentivisation work carried out by the University of Bath. Outside WP3, the QUB simulation outcomes provide the 'What?' and the 'How?' in terms of service support process definition and this information can be integrated with the University of Salford, WP4 activities related to the question of 'When?' in availability contracting.

18.5 Chapter Summary Questions

Having demonstrated the utility of digital methods for service provision on complex assets, any decision to move forward with this technology should be based on the outcomes presented here as well as the following:

1. To what extent will existing customer assets be expected to perform beyond their original design life?
2. How compatible are original platform designs with the types of service required to extend their design life?
3. If provision for 1 and 2 above is required how will this be addressed in emerging services?

References

J. Butterfield, S. Crosby, R. Curran, S. Raghunathan, D. McAleenan, Network analysis of aircraft assembly processes using digital methods. ASME J. Comp. Inf. Sci. Eng. **7**(3), 269–275 (2007)

J. Butterfield, R. Curran, G. Watson, C. Craig, S. Raghunathan, R. Collins, T. Edgar, C. Higgins, R. Burke, P. Kelly, C. Gibson, Use of digital manufacturing to improve operator learning in aerospace assembly. 7th AIAA Aviation Technology, Integration and Operations Conference (ATIO), 2nd Centre of Excellence for Integrated Aircraft Technologies (CEIAT) International conference on innovation and integration in aerospace sciences (Hastings Europa Hotel, Belfast, Northern Ireland, 2007)

J. Butterfield, A. McClean, Y. Yin, R. Curran, R. Burke, B. Welch, C. Devenny, Use of digital manufacturing to improve management learning in aerospace assembly. 26th ICAS Congress including the 8th AIAA ATIO Conference, Anchorage, Alaska, 2008

J. Butterfield, I. Ng, R. Roy, W. McEwan, Enabling value co-production in the provision of support service engineering solutions using digital manufacturing methods. Winter Simulation Conference, Austin, 2009

M. Easterby-Smith, L. Araujo, J. Burgoyne, *Organizational learning and the learning organization: Developments in theory and practice*, 1st edn. (Sage, London, 1999), pp. 130–156

Y. Jin, R. Curran, J. Butterfield, R. Burke, B. Welch, Intelligent assembly time analysis using a digital knowledge-based approach. J. Aerosp. Comput. Inf. Commun. **6**(8), 1542–9423 (2009)

I.C.L. Ng, S. Nudurupati, Outcome-based service contracts in the defence industry—Mitigating the challenges. J. Serv. Manag. **21**(5), 656–674 (2010)

I.C.L. Ng, S. Nudurupati, P. Tasker, Value co-creation in the delivery of outcome-based contracts for business-to-business service. AIM working paper series, WP No 77—May 2010 http://www.aimresearch.org/index.php?page=wp-no-77. Under review at J. Serv. Res. (2010a)

M. Pedler, J. Burgoyne, T. Boydell, *The learning company: A strategy for sustainable development*, 2nd edn. (McGraw-Hill Professional, NJ, 1996), pp. 3–10

G. Probst, B. Buchel, *Organizational learning: The competitive advantage of the future* (Prentice Hall, London, 1997), pp. 15–31

R. Roy, P. Datta, F. Romero, J. Erkoyuncu, Cost of industrial product-service systems (IPS2). CIRP International Conference on Life Cycle Engineering (LCE), Cairo, Egypt, 2009

Chapter 19
Integrated Approach to Maintenance and Capability Enhancement

Proposal of a Maintenance Dashboard Framework

Emma Kelly and Svetan Ratchev

Abstract In the current climate, the industrial sector is increasingly characterised by longer product life cycles and asset availability demands. There is likely to be a reduction in the number of major acquisition projects in the future. This, combined with organisational changes and the fact that both governmental and commercial sectors are steering towards contracting for capability, has led to an internal shift in manufacturing centric companies. These traditional companies are now providing service offerings for their products, thereby reducing customer oriented risk. The services aspect includes the use of new technologies and methods for managing technical products over their life cycle and ensuring that customers' required capability and availability demands are met. This imposes new challenges on subsequent maintenance, repair and capability enhancement procedures. A framework for the development of a Maintenance Dashboard is proposed. The underlying purpose being to establish an approach that supports the decision-making process on whether to maintain, repair, upgrade or update a given asset. Through incorporating maintenance and capability enhancement, both facets are considered in their entirety as opposed to in isolation. The proposed framework aims to direct key asset status information at the various stakeholders involved while an asset is deemed 'active', thus aiding consistent decision-making which may ultimately be related to KPI management.

19.1 Introduction

The time between major platform acquisition projects in both the commercial and defence sector is increasing. Thus, there is a need to prepare for a potential 'ramp up' in the pace of technology insertion to enhance asset capability, as well

E. Kelly (✉) · S. Ratchev
Precision Manufacturing Centre, Faculty of Engineering, University of Nottingham, Nottingham, NG7 2RD, UK
e-mail: Emma.Kelly@nottingham.ac.uk

I. Ng et al. (eds.), *Complex Engineering Service Systems*, Decision Engineering,
DOI: 10.1007/978-0-85729-189-9_19, © Springer-Verlag London Limited 2011

considering robust maintenance management procedures as a means to sustain a working asset. This can be achieved through taking into consideration technological advances, allowing a response to both 'evolution and revolution in capability'. There is a requirement for open architectures which permit the incremental insertion of technology in a 'plug and play' manner. In order to enable this, platforms and systems need to be designed with adaptability in mind. New roles for existing systems are also defined in response to changing market requirements. Targeting those systems that need to be modular in nature requires specific identification of technologies which evolve rapidly. Alongside capability enhancement is the need to reliably predict system status. Prognostics is key to aiding both operational and support planning. Effective maintenance planning allows for improved fleet management and reduced support costs. The consideration of these factors sets the scene for the development of an integrated approach to maintenance and capability enhancement.

19.1.1 Retrospective Viewpoint

'We have now reached a crossroads. We are seeing a shift away from platform oriented programmes towards a capability-based approach, with corresponding implications for the demand required of the traditional defence base', A1.4 (MoD 2005). While this quote is taken from a defence perspective, it is equally applicable to the commercial world where companies such as Rolls-Royce have coined the phrase *Power By The Hour*® for their performance-based contracts.

The term capability is commonly used in a variety of contexts, and thus it is necessary to define the term in relation to the context in which it is being referenced. The UK Ministry of Defence (MoD) Acquisition Operating Framework (MoD 2009) describes Capability as the 'enduring ability to generate a desired outcome or effect, and is relative to the threat, physical environment and contributions of coalition partners'. The Oxford English Language Dictionary defines Capability to be 'the power or ability to do something'. Capability can thus be defined on a number of levels ranging from high-level operational viewpoints through to specific requirements, such as providing heating at a set temperature. For the purposes of future discussion, capability is considered from a platform/system perspective; 'the ability of a platform or system to deliver a specific requirement in support of an overall goal'.

Capability enhancement through the insertion of technology, either for the purposes of upgrade or update (discussed in Sect. 19.2), falls within capability management. At a simplified level, this would occur in response to an influencer. An influencer may be either an internal or external factor which must be taken into account if a given platform or system is to deliver the required effect. Influencers, at a strategic level, may comprise a number of factors including changing market requirements, threats, opportunities, environmental factors and/or internal policy changes. A further notable relationship occurs between capability management and technology management. Technology management (Shanks 2008) is key to the successful implementation of capability management since it involves developing an awareness of available and upcoming technologies, and can thus inform the decision-making process in any future integrated approach of the critical technologies available to meet a noted capability requirement.

Specific areas which require addressing in implementing an effective integrated approach, which includes capability-based planning, cover; requirements management, integration on two levels—technologies into systems and systems into platforms, and robust decision processes for determining potential solutions to previously identified capability requirements.

Capability management employs a top-down approach to its delivery. Key to the integrated approach towards maintenance and capability enhancement is a rigorous planning and requirements specification phase. Maintenance involves maintaining and securing systems in, or restoring them to, a state in which they can perform their required function or functions. One of the challenges for maintenance planning is to identify the actions for preventive maintenance and to ensure that necessary resources are available (Rosqvist et al. 2009). The role of maintenance has changed from simply being a repair solution to having an intrinsic role in through life management. To this end, models for predicting the remaining useful life of components or systems and prognostic methods for determining future system defects can be utilised (Jardine et al. 2006; Wang 2008b).

From the viewpoint of examining maintenance within through life management there are a number of issues which require consideration (Takata et al. 2004) and are key to any proposed integrated approach:

- Adaptation to changes in platform capability requirements during the life cycle.
- Adaptation to platform changes due to technology insertion.
- Integration of past and future maintenance information.

One of the fundamental measures of success for a system is the degree with which it meets its intended purpose. The eliciting of capability requirements (Nuseibeh and Easterbrook 2000) is vital to the successful implementation of a maintenance and capability enhancement decision process. Requirements Engineering has developed into a key stage in the overall systems engineering process (Stevens et al. 1998) with respect to interpreting and understanding user requirements and their successful transformation and implementation. Consequently, parallels may be drawn between knowledge enriched approaches (Ratchev et al. 2003) and the specific requirements of an integrated maintenance and capability enhancement approach.

Industry examples (Pagotto and Walker 2004; Rotor & Wing 2005; Aerosystems International 2003), indicate that maintenance management and capability enhancement programmes have been managed in isolation. Both governmental and commercial organisations across the world are now declaring that they are more likely to contract for capability rather than purchase specific products (MoD 2005). The priorities within acquisition projects are shifting towards procuring the capability to carry out an operation. From a customer perspective this offers reduced risk and support costs. This chapter proposes a framework which will allow maintenance and capability enhancement to be viewed in an integrated manner. These two aspects are embodied within the move towards servitization, especially where the capability being delivered is underpinned by data collection and information processing/analysis techniques. The framework forms the high-level architecture for a Maintenance Dashboard.

19.1.2 New Approach to Maintenance and Capability Enhancement

The reasoning behind the integrated approach stems from the fact that both preventive maintenance and capability enhancement techniques are based on the same fundamental comparison of capabilities. At a basic level, in the case of preventive maintenance, the performance of a capability enabling system fluctuates between design operating and actual values. When considering capability enhancement, the current system capabilities are compared with future and/or desired capabilities. Preventive maintenance and capability enhancement are thus two facets within the lifecycle of a system which can be addressed in an integrated approach and are considered to embody the decision-making process behind the proposed framework.

The majority of traditional manufacturing companies' customers face continuing budget pressure and as a consequence are implementing service methods which are better value for money while maintaining equipment availability levels. There is likely to be a reduction in the number of large acquisition projects in the future. This, combined with organisational and market changes, has led to an internal shift within traditional manufacturing companies from a manufacturing centric organisation to a service centric organisation concerned with establishing a service-based capability. Focus in this case is required in Enabling Through Life Capability and a systems engineering approach. Customer organisational changes have ultimately led to the manufacturer managing the risk. As such, customer emphasis has been refocused towards minimising the cost of ownership while maintaining high levels of equipment operability and functionality. Under the umbrella of 'refocused customer emphasis', practical linkages can be drawn between the potential outputs of the integrated approach in terms of asset status and asset performance indicators. An analysis of maintenance data in conjunction with capability enhancement information would enable clear reporting of asset status under the guidance of noted key performance indicators and measures.

The systems engineering approach (Stevens et al. 1998; MoD 2005) can be adapted from a support services perspective (Fig. 19.1). The top tier represents the customer, companies such as BAE Systems often perform the mid-tier integrator roles and, in conjunction with lower tier partners, produce a support service solution. Systems engineering provides an inter-disciplinary approach to problem

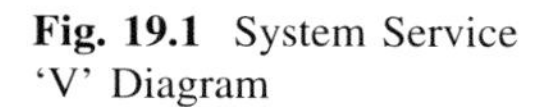

Fig. 19.1 System Service 'V' Diagram

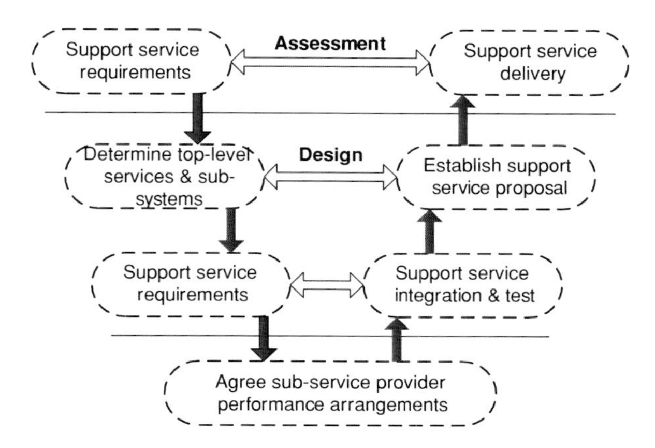

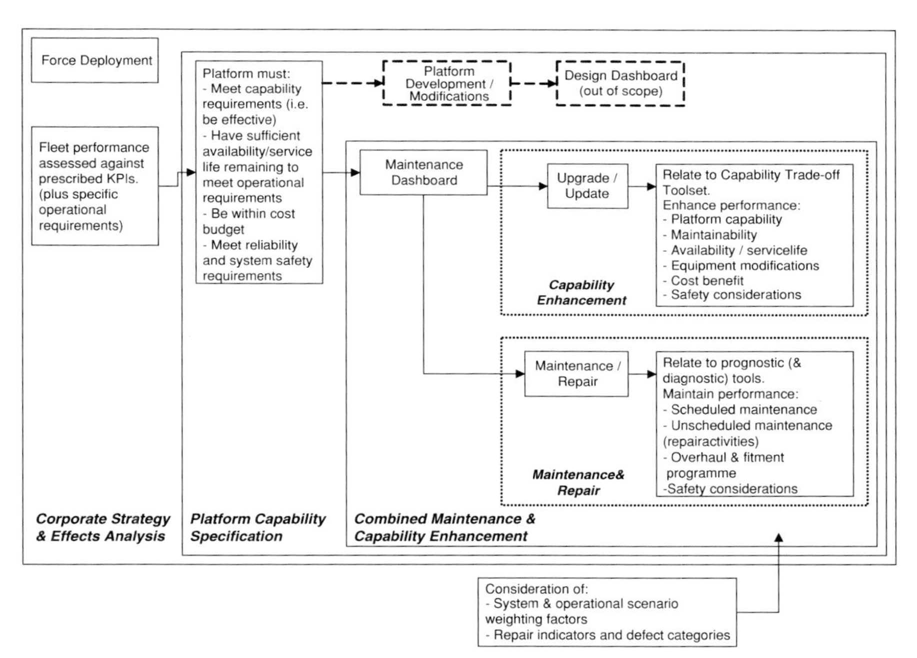

Fig. 19.2 Service requirements for an integrated approach towards maintenance and capability enhancement

solving and is the backbone to delivering the end aim. The end aim being to create a support service structure which satisfies the defined customer requirements while remaining within cost and schedule constraints. An industrial consultation conducted by the Engineering and Physical Sciences Research Council (EPSRC 2009) noted product life-cycle as one of the key manufacturing research challenges, with emphasis placed on providing a whole systems approach towards servitization.

The proposal of a Maintenance Dashboard is directly aimed at extending product life cycles and contributes towards the creation of a support service structure. Extensions to product life cycles are achievable through lengthening the 'useful phase' of the product, where the 'useful phase' is defined as the period during which the product has a functional value. It is proposed that this may be realised through a balanced integrated strategy towards preventive maintenance and product modifications. In this case, modifications cover system adaptations due to changing capability requirements, thereby conducted via capability enhancement. In order to increase product life, the application and addition of new technologies is imperative as a means to ensure the possibility of permanent upgrading. Figure 19.2 illustrates the specific service requirements of importance in developing an integrated approach within the context of the System Service 'V' Diagram.

19.2 Development and Integration of Preventive Maintenance and Capability Enhancement

The concept behind the Maintenance Dashboard is discussed in the following section. The main principle being a *decision process which involves assessing platform capability requirements alongside information retrieved from both preventive maintenance and capability enhancement trade-off programmes*; the aim being to provide an indication of platform status. In this manner it is possible to specify, for example, whether a particular sub-system requires maintenance due to a predicted defect or, if due to capability trade-off information, upgrade should take place at the next maintenance opportunity. *The decision process thus acts to integrate preventive maintenance and capability enhancement and may be viewed as a maintenance management assistance tool.* Maintenance planning, with regard to the ordering of parts and organisation of resources, can be managed in conjunction with optimising maintenance activities associated with scheduled maintenance and repair tasks or system upgrade.

In Sect. 19.1 it was noted that the Maintenance Dashboard proposal could be related to the design of sustainable product life cycles. Sustainable manufacturing for the next generation is also focused on enhancing use-productivity in the total product life cycle (Selinger et al. 2008). Sustainability, from a technological point of view, can thus be viewed to be associated with ensuring the 'useful phase' of a product's life is extended. Figure 19.3 has been adapted (Nieman et al. 2009) to represent the design of sustainable product life cycles from the perspective of creating a support service structure.

The three central boxes illustrate that the success of the proposed Maintenance Dashboard is reliant on service requirements, information and the customer. From a holistic point of view the Maintenance Dashboard supports life cycle management; within the framework the asset status is modelled and evaluated according to available data. The solution proposed by the dashboard with respect to maintain,

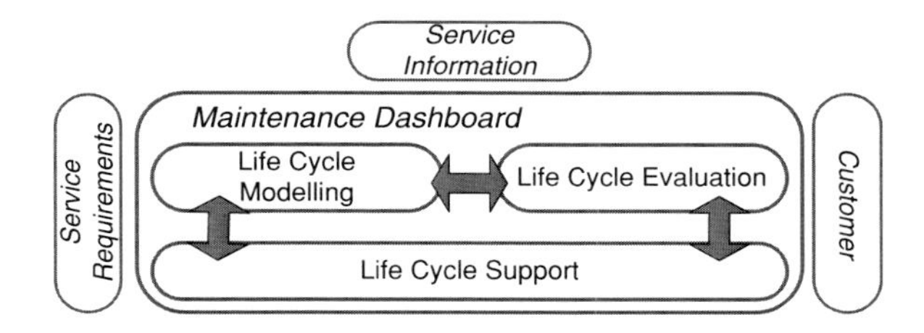

Fig. 19.3 Sustainable product life cycle—support service perspective

repair, upgrade and update is arrived at through a decision process that optimises the required asset capabilities against a known scenario.

19.2.1 Methodological Design Progression

The proposed framework for the Maintenance Dashboard is adapted from the precedent set by the reference architecture for the CommonKADS system (Schreiber et al. 1994; 2000; Kingston 1998). The CommonKADS methodology is a collection of structured methods for building knowledge-based systems (KBS); permitting a structured, detailed analysis of knowledge-intensive tasks and processes.

In brief, CommonKADS is a methodology for KBS development which proposes the creation of different models between which implicit links are identified. The models are thus both related to each other and depend on each other. The methodology comprises six key models of which the design model is the main element of regard in relation to the development of a framework for the Maintenance Dashboard. The model proposed in this chapter comprises the technical design process of the Dashboard.

The technical design consists of three main stages. Generic descriptions of the stages are summarised for reference. Section 19.2.3 describes the stages with specific reference to the Maintenance Dashboard in greater detail.

1. System Architecture Design: The general architecture of the Maintenance Dashboard is specified.
2. Identification of Implementation Platform: The constraints with respect to the implementation platform are identified i.e., environment in which the Maintenance Dashboard is set.
3. Specification of Architecture Components and Application: The individual architectural components of the dashboard are defined in greater detail, in particular the interfaces between components. All knowledge-based information is then mapped onto the system architecture. This includes tasks to be performed (i.e., aim of dashboard), knowledge bases, associated inferences and decision process mechanisms. The application specific sections within the architecture are also specified.

Table 19.1 Maintenance dashboard solution definitions

Decision	Function
Maintain	Conduct scheduled maintenance according to prescribed maintenance plan
Repair	Conduct unscheduled maintenance (and indirect capability enhancement if upgraded technology spares used) through the replacement of faulty system components
Upgrade	Conduct capability enhancement through the replacement of systems or components containing newer technologies capable of increased functionality (also taking into account component replacement as parts near end of life)
Update	Conduct scheduled maintenance to maintain system capability through the replacement of obsolete components

19.2.2 Decision Process

The principal aim of the decision process is to provide a means for comparing current and future required capabilities in order to determine if the platform (or constituent systems) requires maintaining, repairing, upgrading or updating. The decision process thus presents a solution that indicates whether the maintenance and/or capability enhancement route should be followed. This gives rise to one of four solutions; maintain, repair, upgrade or update. The definitions assumed for each of these terms are detailed in Table 19.1.

Taking into account factors raised in Requirements Engineering, the decision process adopts a three-phase evolutionary approach:

1. Identification: Identification of specific capabilities (either individually or collectively) to be taken into account. This phase also involves defining the system/platform capability requirements (i.e., requirements elicitation).
2. Analysis: The majority of the analysis phase involves developing, refining and evaluating all possible solutions to the given capability requirements from (1) above. The decisions which govern whether a system should be maintained or upgraded depend on, in brief; (i) the obsolescence attributes of the technology in question that provide the capability, (ii) the 'utility' (e.g., performance) realised to the system or platform by changing the technology and (iii) associated capability priorities.
3. Solution: Determination of the best solution for meeting the required capability(s) identified in (1).

19.2.3 Development of a Maintenance Dashboard Framework

The first stage, as noted in the design process, involves defining the structural framework (architecture) of the Maintenance Dashboard. There are three main components identified in the reference architecture. If the framework for the

Maintenance Dashboard were to be viewed as a system, then these components would comprise the principal sub-systems. The three sub-systems are termed Controller, Views and Application Model (Gamma et al. 1995). The framework also structures the information flow from requirements elicitation through the three major components, thereby delivering a solution.

The *Controller* represents an integral "command and control" centre which handles external information (i.e., User Input) in order to activate application functions. The input requirements model is composed from two data streams obtained through requirements elicitation: (i) current system capability levels, and (ii) required system capability levels. The integration of preventive maintenance techniques and capability ranking data provide the system status with regard to (i) maintenance schedule, (ii) urgent actions to be addressed e.g., component failure, (iii) failure profiles and (iv) technology insertion programmes.

The *Application Model* specifies the functions and data that together deliver the functionality of the dashboard application. Additionally it contains the reasoning functions, including information and knowledge structures, which give rise to the decision approach. It primarily contains the elements that realise a solution from the functions and data specified during analysis.

Within the dashboard application, the *View* allows static and dynamic information from the application to be available. Utilising information from the decision process, the View delivers the output. This comprises one of four options; (i) Maintain system (according to maintenance plan), (ii) Initiate repair strategy, (iii) Update or (iv) Upgrade.

Figure 19.4 illustrates the formation of the structural framework for the Maintenance Dashboard based on the reference architecture (Schreiber et al. 2000). Detailed component functional information specific to the Maintenance Dashboard framework is expanded under Sect. 19.2.3.1, where the architecture components are specified as per Stage 3 of the design process.

The platform specification phase (design process Stage 2) involved in developing the Maintenance Dashboard further is twofold; (i) specification of the asset (e.g., aircraft) and (ii) specification of the computational infrastructure which will support the Maintenance Dashboard. From a theoretical and concept generating perspective, the design process behind the development of the framework for the Maintenance Dashboard can be conducted independently of the computational implementation platform.

19.2.3.1 Specification of Architecture Components and Application

The final stage in the design process involves defining the three major components (Controller, Application Model, Views (Gamma et al. 1995)) in the proposed Maintenance Dashboard framework. In addition, the interfaces between the components and application specific facets of the framework are also identified. The main purpose of this stage is to decompose the knowledge base into 'chunks of information' which may be then used to determine a solution to the system status query. In the case of the Maintenance Dashboard, functional decomposition

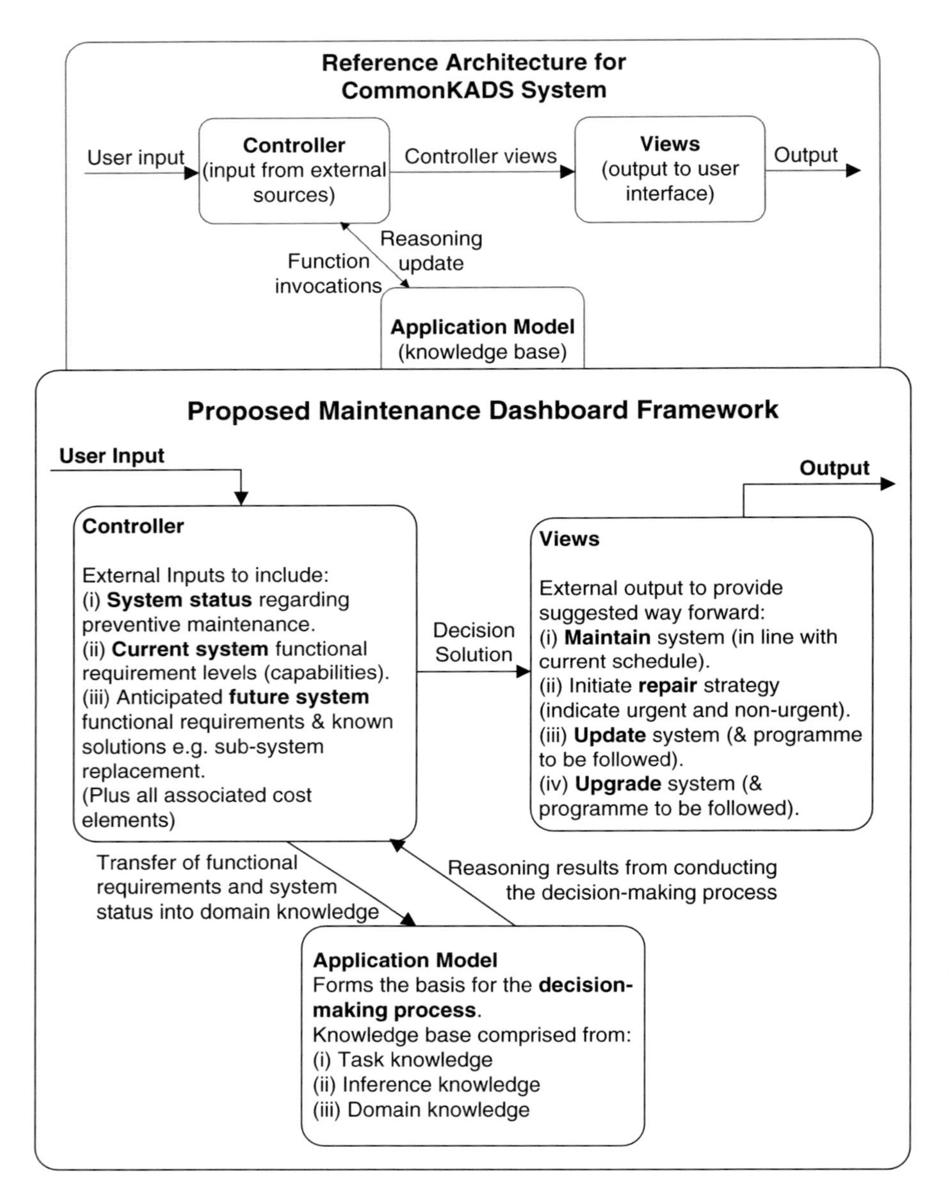

Fig. 19.4 Proposed maintenance dashboard framework

is employed; inferences between knowledge elements are preserved according to their functionality. The framework component of most importance, with regard to the functionality of the Maintenance Dashboard, is the Application Model. The Application Model embodies the decision process which ultimately determines the solution with reference to maintain, repair, upgrade or update.

Under the third stage of the design process the application specific sections of the framework are also defined. The proposed Maintenance Dashboard framework is not platform dependent and thus fully adaptable since the decision process is performed based on the data held within the knowledge base. The processes contained within the framework are however scenario dependent. Therefore, the Maintenance Dashboard is only applicable within the remit of determining platform status (maintain, repair, upgrade or update) based on the current and required platform (and/or system) capability levels. The processes and scenario-related information are changeable, and thus the Maintenance Dashboard can be adjusted to cover a variation in situation if the 'maintain-repair-upgrade-update' scenario-related decision approach is inappropriate (i.e., to cover a change in requirements to the desired output for assisting decision support). The Maintenance Dashboard framework components are summarised:

The *Controller* represents an integral "command and control" centre which handles both external and internal information (e.g., user input) in order to activate application functions. Within the Maintenance Dashboard framework the controller represents the central information hub and its function is threefold:

1. The principal purpose of the Maintenance Dashboard is to provide the platform status with regard to maintain, repair, upgrade or update. Related information (e.g., operational defect data, system concessions, alterations and additions system operational priority ratings) is input into the Controller component to aid this task. This data is then transferred to the Application Model for analytical purposes and comprises data streams on (i) current system capability levels, (ii) required system capability levels, (iii) preventive maintenance results (prognostics, failure profiles), (iv) urgent actions to be addressed (e.g., failure/imminent failure) and (v) capability trade-off ranking results taking technology insertion into consideration (capability enhancement route).
2. The Controller component informs the Application Model when to conduct analysis. The Controller thus initiates the first task that is necessary in order to determine the status of the asset (and associated systems).
3. The results from the Application Model are relayed to the Controller. The Controller 'handles' this information and transfers the solution of the analysis conducted by the Application Model to the View Component. The Controller also retrieves all data from the Application Model with regard to the analysis reasoning process. The added handling of this data provides traceability of information and permits clear reporting routes, if applicable.

Within the dashboard application, the *View* allows static and dynamic information from the application function to be available. In brief, the View realises the presentation of the Maintenance Dashboard purpose to the users. All data presented by the View is transferred from the central information hub (Controller). The View simply acts as an interface between the decision process and the end user. The solution output comprises one of four options; (i) Maintain system, (ii) Initiate repair strategy, (iii) Update system and (iv) Upgrade system.

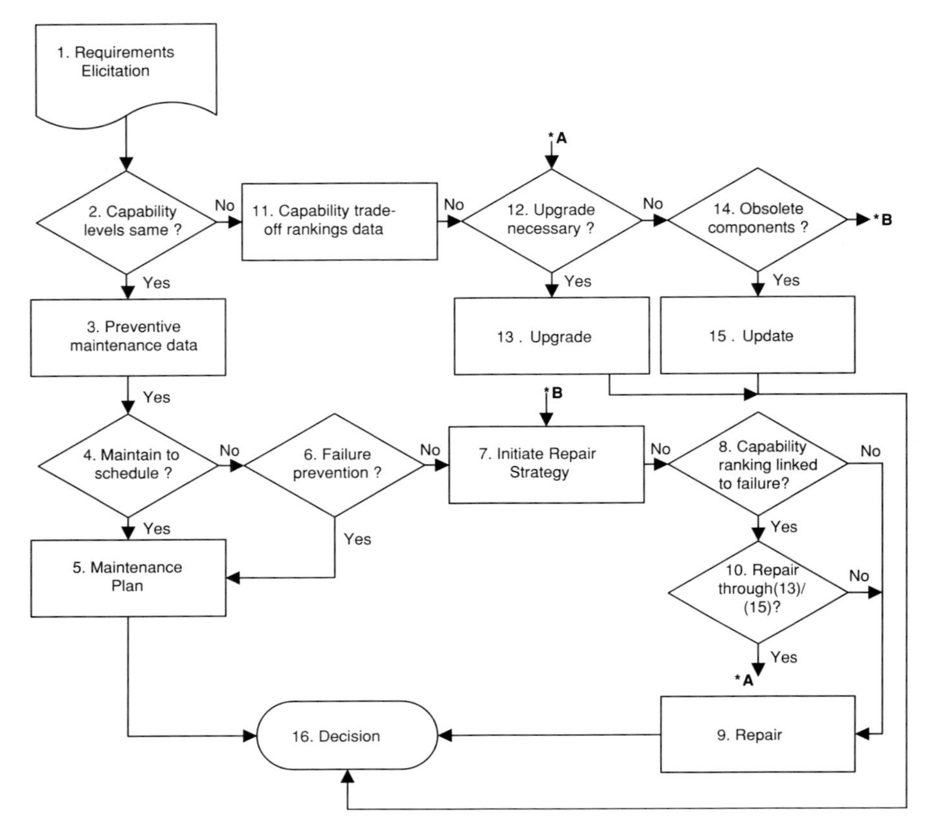

Fig. 19.5 Representation of application model component decision algorithm

The *Application Model* specifies the functions and data that together deliver the functionality of the Maintenance Dashboard. The Application Model knowledge base is formed from the data streams input into the Controller component as well as the reasoning functions behind the decision approach. The Application Model therefore contains the elements necessary for realising a solution. Within the Maintenance Dashboard framework the Application Model provides the user with decision support and assistance in determining the asset status relative to the 'maintain-repair-upgrade-update' scenario-related decision algorithm. The task initiated by the Controller component results in the commencement of the decision process. This task defines a single operation; conduct a comparison of current capability levels with required capability levels. The execution of the task invokes the decision algorithm (simplified illustration presented in Fig. 19.5), which in turn results in the initiation of specific decision tasks (represented by $\diamond$).

In performing the decision process, data is sought from the knowledge base. This includes the current system status obtained from preventive maintenance techniques, system specific technology insertion programme data retrieved from

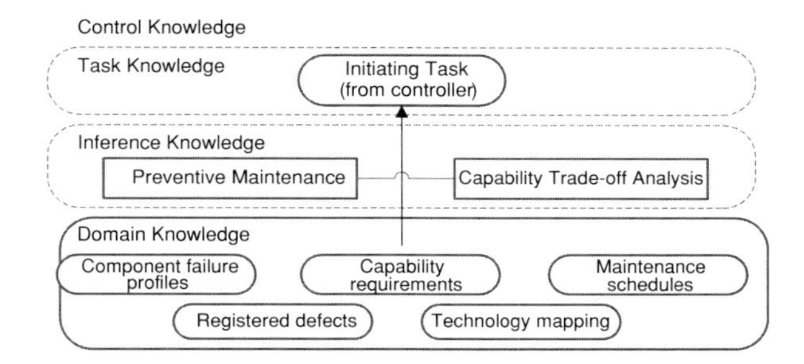

Fig. 19.6 Knowledge base 'snap-shot' (requirements mapping illustration)

capability ranking procedures and associated inferences between the variables contained in the data bases and the decision tasks. A 'snap-shot' of an example knowledge base structure is illustrated in Fig. 19.6.

Control Knowledge defines both the content and structure of the task and inference specific data: (i) *Task Knowledge* is defined by a key goal and describes the decomposition process involved in the decision algorithm (decision tasks referenced in Application Model). (ii) *Inference Knowledge* describes the inference steps that are to be followed in completing the key goal (task) through utilising information obtained from capability ranking procedures for example.

Domain Knowledge contains the concepts, relationships and facts that are required in order to reason a given application domain. For example, in cases where preventive maintenance techniques are used to describe inference steps, system failure profiles are contained within domain knowledge.

19.2.4 Maintenance Dashboard Application: Asset Status

The framework developed to integrate preventive maintenance and capability enhancement is realised through the representation of a Maintenance Dashboard. The Maintenance Dashboard illustrates the functionality of the decision approach and is structured to relay asset status information to the service provider at system, platform and fleet levels (Fig. 19.7). A generalised schematic representation of the Maintenance Dashboard, based on the structural framework illustrated in Fig. 19.4, is presented in Fig. 19.7. An example of application of the Maintenance Dashboard is illustrated in Fig. 19.8. In the given example, the dashboard is linked to an overall support service offering by relating it to the defined Ministry of Defence Defence Lines of Development.

At the highest level (Fleet/Class) the Dashboard offers a single overview representation of all of the assets and their availability for a particular scenario and duration. This availability is determined based on the status of the key systems required for that particular scenario. The platform level (Asset/Warship) illustrates the status of the asset systems (i.e., fully operational or issue identified) and

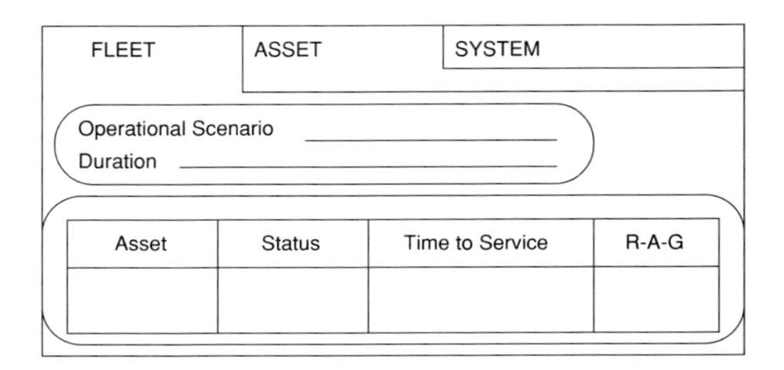

Fig. 19.7 Maintenance dashboard—generalised schematic representation

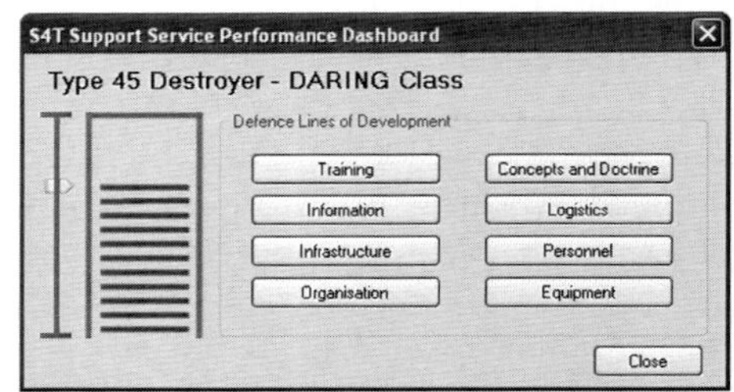

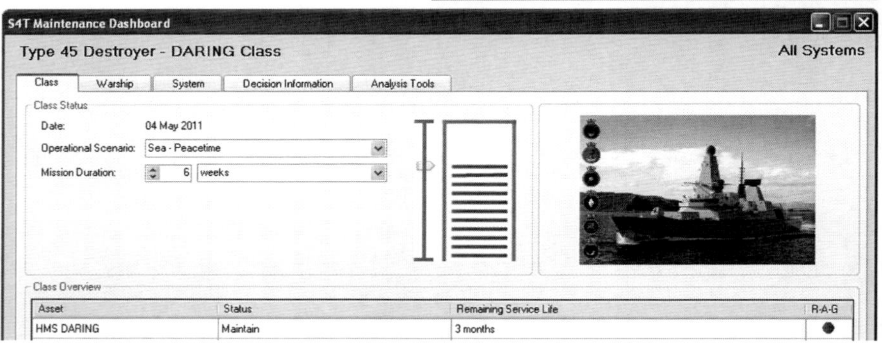

Fig. 19.8 Maintenance dashboard—an example application

associated priority for reparative action. Relevant capability enhancement programmes are noted and information relating to enhancement options, cycle time and fitment opportunities are detailed.

The decision support process (application model) summates this data into a single operational capability level representation for instant service information acquisition. The system level of the Dashboard provides similar analysis utilising a greater degree of detail with regard to scheduled maintenance, unscheduled maintenance, current and required system capabilities. The outputs from the Maintenance Dashboard feed directly into maintenance planning decision support. This is a significant component in the delivery of a service support offering and directly links to the implementation of service methods as noted in Sect. 19.1.2.

19.3 Maintenance Dashboard and Services Transition

Servitization is noted to involve the innovation of an organisation's capabilities and processes so that it can better create mutual value. This is specific to the relationship between customer and supplier, through a shift from selling products to selling associated systems (Ng et al. 2010b). Consequently, service information and related decision support tools, such as the proposed Maintenance Dashboard, are instrumental to the successful delivery of a support service offering.

Popular advice to manufacturing based companies has been that 'in order to remain competitive they should move up the value chain and focus on delivering knowledge-intensive products and services' (Baines et al. 2007; Hewitt 2002). Occurring in conjunction is that fact that governments are now declaring that they will contract for capability (Neely 2009) and, as such, the support service aspect is outsourced to the supplier. From a supplier perspective, servitization may be viewed as a way in which sales revenue can be increased, whilst from a customer perspective, servitization offers reduced risk and improves the way in which costs and budgets may be set.

Servitization involves the use of the new technologies as well as methods for managing technical products (i.e., asset) over their life cycle. Service science is interdisciplinary and focuses on service as a system of interacting parts that include people, technology and business (Demirkan et al. 2008). Within this environment, product life can be evaluated. The Maintenance Dashboard proposal is aimed at extending the 'useful phase' of a product's life. This 'useful phase' being defined as the period during which a product performs a particular function (in the context of this chapter, meets a given capability). This extension is achievable through conducting preventive maintenance and modifications, where modifications involve capability enhancement. To enable product longevity, the option for permanent upgrading and updating must remain viable. The Maintenance Dashboard provides an option for initiating a balanced strategy with regard to maintenance and capability enhancement and thus better serves to increase the 'added value' of the product during its 'useful phase' in the life cycle. The solution retrieved from the Maintenance Dashboard is of specific relevance to the different management levels within a business/organisation. Figure 19.9 illustrates the information flow.

1. Corporate with regard to business strategy; this may relate to budgetary control for example.
2. Operational and tactical assessment of working scenarios/missions, required fleet capabilities and determining the 'most sensible time' to conduct asset maintenance, repair, upgrade or update.
3. Equipment management from the perspectives of those involved in maintenance management and logistics planning (resources, spares, and work-space).

In the context of the Core Integrative Framework (CIF) (Ng et al. Introduction chapter) for complex engineering service systems, the Maintenance

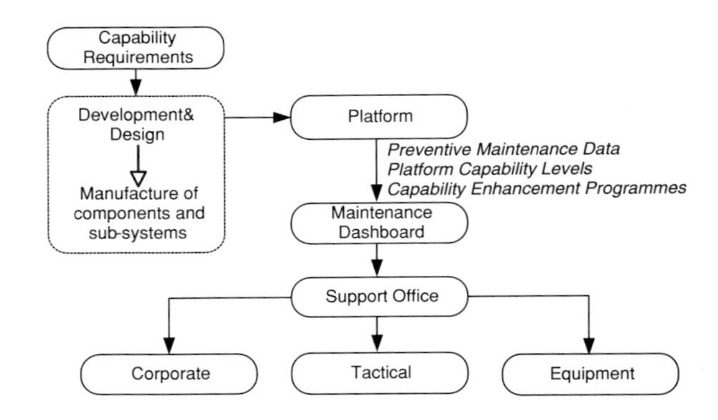

Fig. 19.9 Information flow from maintenance dashboard

Dashboard fits within transformations (a) and (b). These transformations cover transforming materials and equipment and transforming information. Ng et al. document that if a customer were to consider contracts from an outcome based perspective then value is likely to be added through the adoption of all three transformations noted in the CIF. The 'mutual value' referenced in the definition of servitization is thus delivered through the interactions between the three transformations of information, materials and equipment and people. The Maintenance Dashboard proposal does not dovetail with 'transforming people', however the decision support provided by the dashboard may be used to, in effect, persuade the adoption of different mindsets. The proposed Maintenance Dashboard is able to provide support for day-to-day operations conducted under the service paradigm plus address longer-term strategic perspectives from a technical point of view. This serves to combine technical asset considerations and interdependencies with the strategy challenges that may be faced through the analysis of information. A simplified representation of the information interactions between transformations (a) and (b), with reference to the Maintenance Dashboard, is depicted in Fig. 19.2.

19.4 Concluding Remarks

'The quality and shelf-life of current technical products is no longer determined by wear and attrition but by being technically out of date' (Niemann et al. 2009). Current, and future, strategies for achieving maximum product utilisation are required to consider longer-term planning for product life-cycles. The concept behind life-cycle management aims to optimise product performance. This covers the three main phases within a product life-cycle; (i) manufacture, including design and development, (ii) usage and (iii) disposal/recycling. The Maintenance Dashboard framework is directly aimed at extending product life cycles and aiding the associated planning process. Extensions to product life cycles are achievable through lengthening the 'useful phase' of the product. Within the Maintenance Dashboard this is realised through a balanced integrated strategy towards preventive maintenance and capability enhancement. Through incorporating maintenance and capability enhancement data, both facets are considered in their entirety as opposed to in isolation. As a result, the proposed framework takes into account the various stakeholders involved during the 'useful phase' of the product's life-cycle as illustrated by Fig. 19.9. This entails directing asset status information at the required level to the various stakeholders. For example, an engineering director is likely to place greater emphasis on knowing the high-level status of the fleet under his control, i.e., understand *x* out of *y* assets are active and their respective status with regard to downtime relative to maintenance programmes and planned modifications. An engineering manager however will be interested in the details behind planned maintenance programmes and their effect on items such as concessions.

The framework supports the analysis involved in implementing an integrated maintenance and capability enhancement approach. The structural framework architecture provides a concise representation of the processes involved in determining the status of an asset with regard to maintain, repair, upgrade or update; while the knowledge base for the decision process is suitably customisable dependent on the platform and related systems under consideration.

It has been reported (Neely 2009) that new business models for manufacturers have implications for operations management frameworks and philosophies. This is of particular relevance to situations where the delivery of an operational capability is underpinned by the data collection and information analysis techniques. The decision process associated with the Maintenance Dashboard is initiated by a single task; comparison of current and required capabilities (assumed from a platform/system perspective). The Dashboard is dependent on the results obtained from asset maintenance information and capability trade-off analyses. These, in turn, are dependent on the data that can be retrieved from the asset and planners.

The provision of service support is gaining in importance as more organisations move towards capability and performance-based contracts. There are an increasing number of reasons for traditional manufacturing companies of high value complex systems to include services into their product offerings. This list is by no means exhaustive, however reasons include facilitating the sale of high value products, further strengthening customer relationships and addressing business growth and demand requirements. With the introduction of service contracts customers are far more likely to expect integrated service solutions i.e., goods and services integrated into customer specific packages (Davies 2003; Brax 2005). A service provision should offer value to both the customer and manufacturer. It is postulated that the Maintenance Dashboard may form part of a higher-level Integrated Support Service Dashboard and that the creation of a dashboard hierarchy would enable increased planning of maintenance activities alongside other functions.

19.5 Chapter Summary Questions

This chapter has brought to the forefront the consideration that service information and related decision support tools can be instrumental in the successful delivery of a support service offering. The services aspect of a support offering can be said to include the use of new technologies and methods for managing products over their life cycle, while also ensuring that the customer's required capability and availability demands are met. This drives the following questions:

- What process can be followed to ensure the service provider is able to consider maintenance and capability enhancement in their entirety as opposed to in isolation?
- How can these facets be structured into a framework for an integrated approach towards a decision-making mechanism?
- Within the transition to services, do the outputs from the integrated approach encourage the appropriate distribution and analysis of decision support results? Are these results suitably directed at the three levels depicted within the framework (i.e., fleet, asset, system)?

References

Aerosystems International, Sapphire. Available at: www.aeroint.com/products/techpubs/sapphire.pdf (2003)

T. Baines, H. Lightfoot, S. Evans et al., State-of-the-Art in Product Service Systems, in Proceedings. IMechE Part B: J. Eng. Manuf. **221**, 1543–1552 (2007)

S. Brax, A manufacturer becoming service provider: Challenges and a paradox. Manag. Serv. Qual. **15**, 142–155 (2005)

A. Davies, Are firms moving downstream into high value services?, in *Service innovation, series on technology management*, vol. 9, ed. by J. Tidd, F. Hull (Imperial College Press, London, 2003)

H. Demirkan, R. Kauffman, J. Vayghanb et al., Service oriented technology and management: Perspectives on research and practice for the coming decade. J. Electr. Commer. Res. Appl. **7**, 356–376 (2008)

EPSRC, Research challenges in manufacturing: An industry consultation. Available at: www.epsrc.ac.uk/CMSWeb/Downloads/Other/ManIndReport.pdf (2009)

E. Gamma, R. Helm, R. Johnson et al., *Design patterns: Elements of reusable object-oriented software* (Addison-Wesley, New Jersey, 1995)

P. Hewitt, The Government's Manufacturing Strategy (Department for Trade and Industry) http://www.berr.gov.uk/files/file25266.pdf (2002)

A. Jardine, L. Daming, D. Banjevic, A review on machinery diagnostics and prognostics implementing condition based maintenance. Mech. Syst. Signal. Process. **20**, 1483–1510 (2006)

J. Kingston, Designing knowledge based systems: The CommonKADS design model. Knowl. Based. Syst. **11**, 311–319 (1998)

Ministry of Defence (MoD), *Defence industrial strategy. Defence white paper (CM6697)* (Her Majesty's Stationery Office, London, 2005)

Ministry of Defence (MoD), Acquisition operating framework. Available at: www.ams.mod.uk/aofcontent/tactical/techman/content/trl_applying.htm (2009)

A. Neely, Exploring the financial consequences of the servitization of manufacturing. Oper. Manag. Res. **1**, 103–118 (2009)

I.C.L. Ng, R.S. Maull, L. Smith, in *2010 volume in Service Science: Research and innovations (SSRI) in the Service Economy Book Series*, ed. by H. Demirkan, H. Spohrer, V. Krishna. Embedding the new discipline of service science (Springer, New York, ISSN: 1865-4924, forthcoming)

J. Niemann, S. Tichkiewitch, E. Westkamper, *Design of sustainable product life cycles* (Springer-Verlag, Berlin, 2009)

B. Nuseibeh, S. Easterbrook, Requirements engineering: A roadmap. Proc. Conf. Future Softw. Eng. **1**, 35–46 (2000)

J. Pagotto, S. Walker, Capability engineering—Transforming defence acquisition in Canada. Proc. SPIE Def. Secur. Symp. **1**, 89–100 (2004)

S. Ratchev, E. Urwin, D. Muller et al., Knowledge based requirement engineering for one-of-a-kind complex systems. Knowl. Based Syst. **16**, 1–5 (2003)

T. Rosqvist, K. Laakso, M. Reunanen, Value-driven maintenance planning for a production plant. Reliab. Eng. Syst. Saf. **94**, 97–110 (2009)

Rotor and Wing, Making maintenance manageable. Available at: www.aviationtday.com/rw/ (2005)

G. Schreiber, B. Wielinga, R. de Hoog et al., CommonKADS: A comprehensive methodology for KBS development. IEEE Expert. **9**, 8–37 (1994)

G. Schreiber, H. Akkermans, A. Anjewierden et al., *Knowledge engineering and management: The CommonKADS methodology* (The MIT Press, Massachusetts, 2000)

G. Selinger, H.-J. Kim, S. Kernbaum et al., Approaches to sustainable manufacturing. Int. J. Manuf. **1**, 58–77 (2008)

A. Shanks, Technology management in warship acquisition. Proc. Int. Naval. Eng. Conf. **1**, 1–11 (2008)

R. Stevens, P. Brook, K. Jackson et al., *Systems engineering: Copying with complexity* (Prentice Hall, Hemel Hempstead, 1998)

S. Takata, F. Kimura, F. van Houten et al., Maintenance: Changing role in life cycle management. Ann. CIRP Manuf. Tech. **53**, 1–13 (2004)

W. Wang, Condition based maintenance modelling, in *Complex system maintenance handbook*, ed. by K. Kobbacy, D. Prahakbar Murth (Springer, London, 2008)

Chapter 20
Mapping Platform Transformations

Clive I. V. Kerr, Robert Phaal and David R. Probert

Abstract Technology insertion provides the means to proactively sustain and enhance the functionality and associated performance levels of legacy product platforms. It aims to deliver in-service technological innovations in response to the need for new capabilities that address emerging threats, obsolescence concerns and affordability issues. Platform modernisation via technology insertion is an interaction between the three principal stakeholders of end-user, acquisition authority and product-service system provider. To bring these three groups together for the vision setting and planning activities, a transformation mapping approach has been developed. It requires the participants to populate three visual templates that respectively map the future strategic context, the portfolio/fleet of complex product platforms and the key functional systems that generate utility. The adoption of this approach provides the ability to outline future capability requirements, determine product development options, and align these with the associated technology upgrade paths against the time dimension. To illustrate the implementation of the method, a case study from the defence industry is employed to depict the typical outputs that can be generated.

20.1 Introduction

Military product platforms have exceptionally long lifecycles, in the order of decades, and given the state of defence budgets there is the general trend of sustaining the operational capability of those legacy platforms for much greater

C. I. V. Kerr (✉) · R. Phaal · D. R. Probert
Centre for Technology Management, Institute for Manufacturing,
Department of Engineering, University of Cambridge, 17 Charles Babbage Road,
Cambridge, CB3 0FS, UK
e-mail: civk2@cam.ac.uk

I. Ng et al. (eds.), *Complex Engineering Service Systems*, Decision Engineering,
DOI: 10.1007/978-0-85729-189-9_20,

periods than their original design intent. In the context of through-life management for complex product-service systems, one of the key challenges is how to direct and co-ordinate the flow of technology for platform modernisation during the in-service phase of the product lifecycle. With a platform in active use by the end-user, inserting the latest technological innovations is achieved principally through technology insertion mechanisms. They provide the means to upgrade and update the capability of a platform in terms of its functional enhancement and associated performance attributes. The requirements for such enhancements is driven by the likely changes in future military operations, the pace of change in technology and the increasing pressures on lifecycle cost reduction.

To guide practitioners, from both the military and industrial communities, a mapping approach has been developed to support the front-end strategic dialogue in respect of the vision setting and planning activities. This approach allows these different stakeholders to visually represent and examine the most appropriate paths to support the operational sustainment of a product platform's capability. To demonstrate the application of the mapping approach a case study on the Royal Australian Navy's Surface Combatant Fleet will be used. This illustration can act as a key reference case. From the position of complex engineering systems, examples from the defence industry provide very rich real-world content and context for exploration. In terms of the defence industry itself, maritime platforms are "inherently technology sensitive and capital intensive. This creates an imperative—an imperative to manage change" (Smith 2001). Thus, they lend themselves to being applicable for exchanging learning outcomes across to the air and land domains. Additionally, in the maritime environment the emphasis on the surface combatant fleet as the main focal point is very pertinent as it represents the Navy's largest investment (Davies 2008). Finally, bounding the discussion and illustration of this case example to the combat system perspective ensures relevance to the current and future issues faced by the Navy and naval industrial-base as it represents the heart of a maritime product platform (SPC 2009).

20.2 Transforming Platforms Through Technology Insertion

Technology insertion is defined as "the utilisation of a new or improved technology in an existing product" (Kerr et al. 2008a). It is concerned with how to manage the flow of technology from the research and development stages into the fielded arena where the recipient platforms are actively on operational service with the end-users (Kerr et al. 2008b). However, it is critical to first address the issue of modernisation viability, i.e., is it actually worth considering the avenue of technology insertion to enhance the capabilities of a given platform? [Note: the use of the term 'capabilities' in relation to a platform signifies its functionality and associated performance levels].

In terms of managing the defence capital stock, there are the two fundamental choices (Dowling et al. 2007; Kosiak 2004) of either (i) continuing with the legacy platform or (ii) replacing the current platform with a new generation. Modernisation viability is a consideration of a legacy platform's capacity to still have a role to play in future military tasks weighed against the level of investment that would be needed to satisfy these operational needs. "In general, upgrading is less costly in the near term, but is only reasonable if the system that is being upgraded can perform well enough and last long enough" (Balaban and Greer 1999). "And even with upgrades, if the lifecycle period under consideration is long enough, eventually the upgraded system will need replacement, so acquisition is deferred, not avoided" (Balaban and Greer 1999). From underneath this legacy versus replacement decision, technology insertion can be described as the best method to utilise "limited resources to sustain weapon systems and grow fielded capability" (Milas and Vanderbok 2006). In that regard, technology insertion considers: (i) the use of better technology, and (ii) a product in need of improvement (Kerr et al. 2008a). Thus, technology insertion provides a mechanism to "enable faster and cheaper capability upgrade" with a "focus on the pull-through of new technology" (MoD 2005); in essence transforming a platform's capabilities, and even role, through selective enhancement packages.

The product platform dimension is obviously the mainstay for technology insertion activities. Recognising that "much of a platform's capability is delivered through its subsystems" (MoD 2005), the primary consideration is thus on product-centric (i.e., systems, equipment and component) changes as opposed to either process orientated improvements or the betterment of associated platform support arrangements. Therefore, in this chapter, the unit of analysis will be the platform (product). Product-centric changes are in response to only three fundamental reasons, as outlined in Kerr et al.'s (2008a) rationale model. They are:

- The threat of obsolescence
- The requirement for additional or new capability
- The challenge of affordability

Additionally in the paradigm of technology insertion (Kerr et al. 2008a), there are four classes of changes (Kerr et al. 2008b) that feed product-centric modifications, namely:

- Changes from the external environment
- Changes from user evaluation
- Changes from technology developments
- Changes from funding availability

Therefore, the promise of technology insertion is the repeated introduction of the latest technology into an in-service platform as both the threats and the technologies, themselves, evolve.

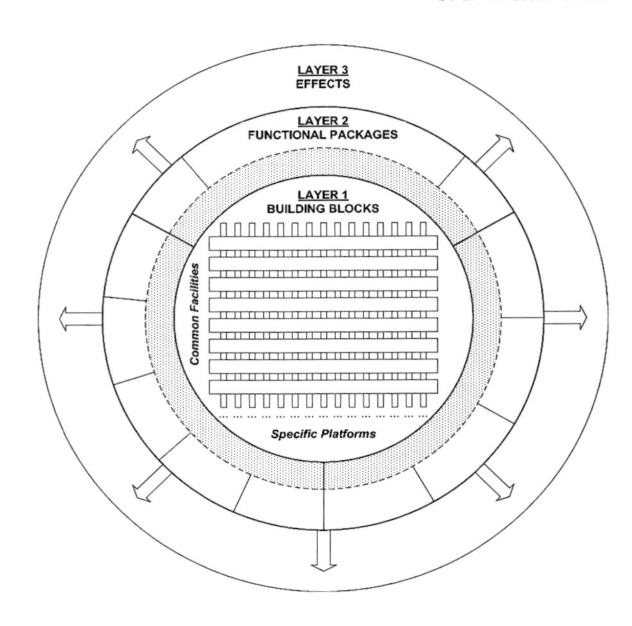

Fig. 20.1 Template for the military capability framework

20.3 A Mapping Approach for Visualising Platform Transformations

This chapter describes the mapping approach developed by the University of Cambridge's Centre for Technology Management. The approach is based on highly visual methods that can be used to clearly articulate and present the capability transformation of product platforms. It is based on the integration of Kerr et al.'s (2006, 2008c) visual framework for military capability and the visually-orientated roadmapping scheme (Phaal et al. 2007; Kerr et al. 2009). There are three steps in the general approach, namely:

- Map the strategic context
- Map the fleet
- Map the functional systems

First, there is the need to vision and articulate the future force for the specific service branch of the military. This is a top-down holistic view that provides the strategic context in which a platform will operate and from where the critical drivers for product-centric changes are derived. It is framed according to the layered structure of the military capability representation developed by Kerr et al. (2006, 2008c). The approach requires the stakeholders to populate the end-state template with their view of the future warfighting effects, functional force packages and building blocks. The template is depicted in Fig. 20.1. The building blocks represent the inner layer of capability consisting of the lines of common facilities, i.e., the support system, interwoven with the spectrum of strategic platform types. Bonded onto the capability building blocks are the functional

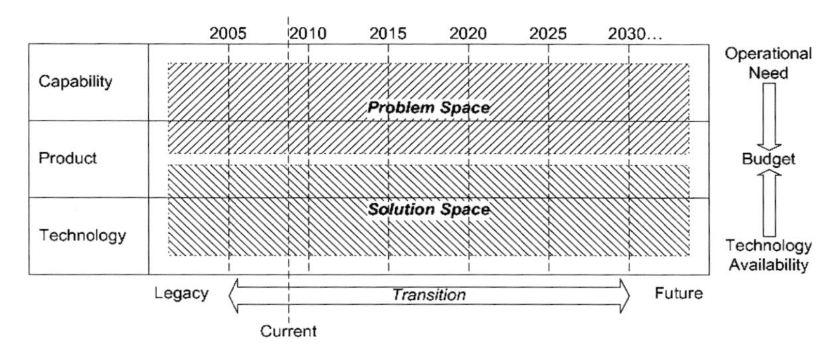

Fig. 20.2 Template for the roadmap

packages which represent the warfighting force functions. The final layer integrates the effects-based approach.

The second and third steps in the approach utilise the roadmapping method. Figure 20.2 shows the template used to structure the architecture of a roadmap. Although it is a tailored configuration for the issue of platforms transformations, the template's layout has been constructed from and conforms to the basic form and principles outlined by Phaal et al. (2004). The vertical axis of the template is composed of three principal layers, i.e., capability/product/technology that reflects the problem-structure for technology insertion. These three layers are broad categories that are to be unpacked to reveal higher levels of granularity. The horizontal axis in Fig. 20.2 consists of a chronological timeline. This axis represents the temporal shifts that take place as legacy capabilities/products/technologies are transitioned to future states. These future embodiments may be evolutions of the current generation, resulting in an incremental change, or a next generation that results in a step/disruptive change. Through the roadmap template the transition paths from the current to end-states can be plotted at both the fleet and functional systems levels. Since capabilities are path dependent (Birchall and Tovstiga 2005), the visual mechanism of roadmapping allows these paths to be clearly plotted that is, mapping "the timeline of product needs and requirements against technology advancement and obsolescence" (Milas and Vanderbok 2006). The power of roadmapping comes from having a recognition that there exists windows of opportunity for inserting technologies into future upgrade blocks, driven from the four classes of change that feed product-centric modifications (as outlined in the previous section), and matching these with the windows of availability from the technology development activities (Kerr et al. 2008b).

Taking a psychosocial stance on the work of Phaal et al. (2007), the adoption and practice of roadmaps provide a mechanism/vehicle to cogitate, articulate and communicate (Kerr et al. 2009). The generated roadmaps are effectively boundary objects because as entities they forge the links between the differing stakeholders and communicate their shared viewpoints (Kerr et al. 2009). They provide a locus for communication and co-ordination (Yakura 2002). Additionally, the resultant roadmaps "should be reviewed and shared on a regular interval to ensure that

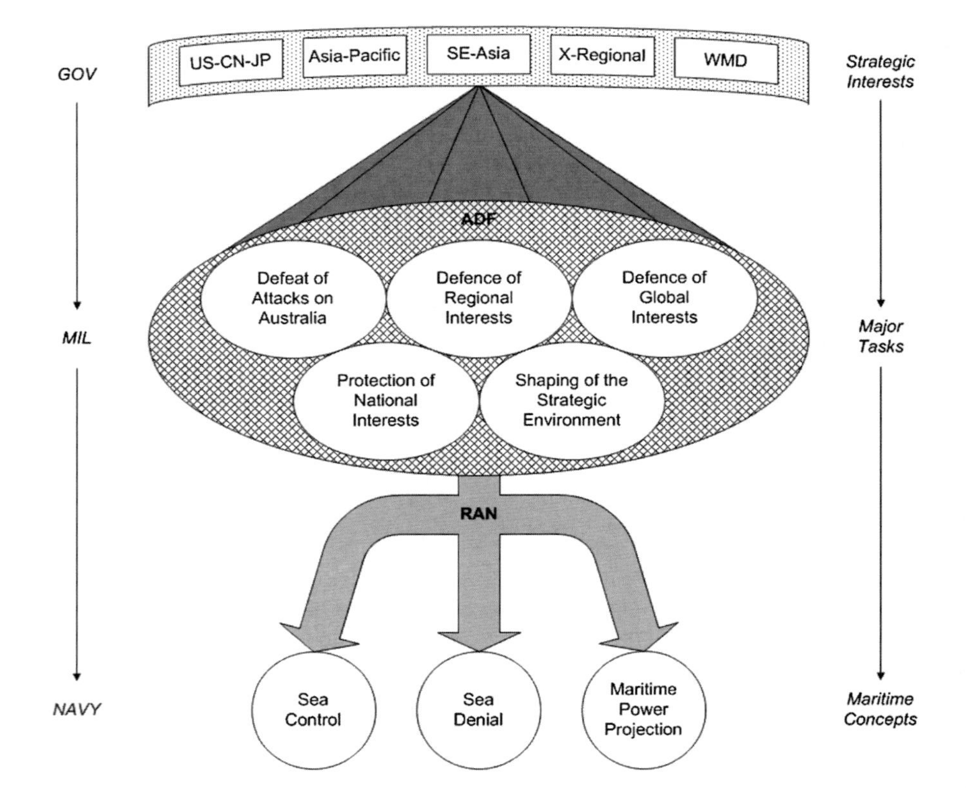

Fig. 20.3 Mapping national and maritime strategy

discontinuances, developments, requirements and changes are known and monitored" (Milas and Vanderbok 2006). It must be noted that the Australian Government's Department of Defence (DoD 2006) highlight the power of roadmapping in terms of its contribution in improving the quality of decision-making and improving the subsequent implementation of those decisions.

20.3.1 Mapping the Strategic Context

The mapping of the strategic context starts with a top-down analysis beginning at a country's strategic and national interests, moving down through its defence priorities to the concepts of operations for the relevant service branches of the military. For each service branch, in turn, the future warfighting elements and effects are determined. Finally, the complex product-service systems are surrounded by this high-level definition of the future strategic context in which they will operate. Figures 20.3 and 20.4 present the case of Australia from the viewpoint of the Royal Australian Navy.

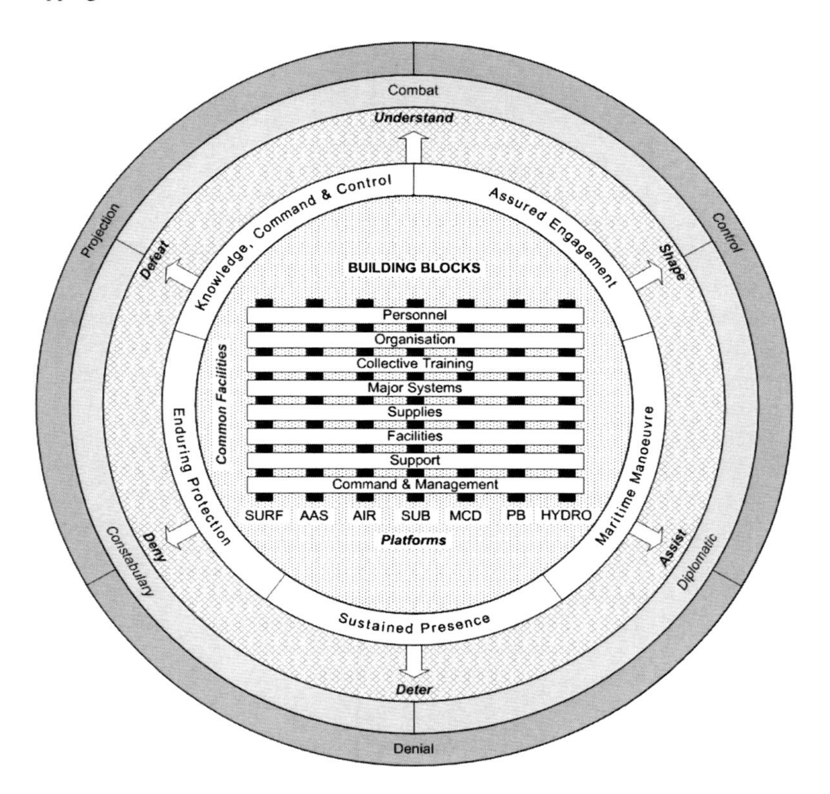

Fig. 20.4 Mapping maritime capability

The Australian Government has identified five enduring strategic interests (RAN 2000):

- "Avoidance of destabilising strategic competition developing between the United States, China and Japan as the power relationships between the three evolve and change.
- Prevention of the emergence within the Asia–Pacific region of a dominant power, or group of powers, whose strategic interests are hostile to those of Australia.
- Maintenance of a benign environment in South East Asia, particularly maritime South East Asia, which respects the territorial integrity of all states.
- Prevention of the positioning of extra-regional military forces in neighbouring countries which might be used contrary to Australia's strategic interests.
- Prevention of the proliferation of weapons of mass destruction."

Below these, in Fig. 20.3, are the associated five major tasks expected of the Australian Defence Force in order to fulfil the nation's security requirements, namely:

- Defeat of attacks on Australia
- Defence of regional interests
- Defence of global interests
- Protection of national interests
- Shaping of the strategic environment

From the major tasks, the Navy's contribution is composed of three maritime concepts namely those of control, denial and power projection. Sea control is defined as "that condition which exists when one has freedom of action to use an area for one's own purposes for a period of time" (JFADT 2004). This condition "includes the air space above and the water mass and seabed below as well as the electro-magnetic spectrum" (RAN 2000). On the other hand, sea denial aims to prevent the use of the sea by another force which implies a more passive posture where the emphasis is on defence (JFADT 2004). For instance, the maintenance of a blockade of enemy forces through the operation of exclusion zones or specific campaigns against an adversary's trade or logistic systems (RAN 2000). The concept of maritime power projection is the delivery of force from the sea. For instance, "the landing of amphibious or Special Forces or the delivery of seaborne land forces, or bombardment by guided or unguided weapons from seaborne platforms" (RAN 2000). These three elements encapsulate Australia's maritime strategy.

Figure 20.4 depicts how the complex product-service systems operated by Royal Australian Navy map into its maritime strategy. The outer ring illustrates the three concepts of sea control, denial and power projection providing the surrounding strategic context in which the product platforms must provide end-users with utility. Aligned to these concepts is the maritime mission space (Crane 2007) that reflects the span of maritime operations (Booth 1977). There are three broad categories:

- Combat
- Constabulary
- Diplomatic

Combat can be classified from two standpoints, i.e., operations at sea versus operations from the sea. Operations at sea include such functions as strike, interdiction, cover and containment (RAN 2000); whereas operations from the sea include the functions of amphibious assault and support to operations on land. Of course, the primary goal of Australia's maritime strategy is to control the air and sea approaches to its continent. Thus, the key requirement is to "maintain an assured capability to detect and attack any major surface ships, and to impose substantial constraints on hostile submarine operations, in the extended maritime approaches" (DoD 2000). The constabulary category reflects the range of policing duties. It includes such operations as anti-piracy, fisheries protection, embargos and sanction enforcement (RAN 2000; Crane 2007). Diplomatic operations embody "the use of maritime forces in support of foreign policy" (RAN 2000). These are either as: (i) a demonstration of capability used to reassure, impress and

Table 20.1 Maritime capability enablers (RAN 2006)

Package	Definition
Assured engagement	The capability of maritime forces to decisively engage target sets across the battlespace using networked systems to provide the required responsiveness, weight of fire, precision and assure success by employing lethal and non-lethal weapons
Maritime manoeuvre	The capability of maritime forces to move freely between the open ocean and the littoral environments and to project force through exerting local sea control to facilitate the delivery of support to the joint or combined mission
Sustained presence	The ability for a joint maritime force of significant combat weight to operate for an extended period at potentially long distances from Australia
Enduring protection	The ability of each maritime force element to successfully achieve designated missions and tasks through the combined capability of defensive, staying and fighting power
Knowledge, command and control	The exploitation of superior battlespace awareness and, through people, innovatively applying operational art and adaptive command to gain decision superiority over an adversary

warn, or (ii) a demonstration of readiness to deploy a degree of combat power (RAN 2000).

The next layer towards the centre of Fig. 20.4 is the effects layer. The Australian Defence Force describes it as the capacity or ability of the force to achieve a particular operational effect (DoD 2006). The Navy has five strategic capability-based effects to deliver, namely:

- "Understand the geopolitical and operational context and maintain appropriate situational awareness.
- Shape (and deter) the choices of potential adversaries seeking to directly attack Australia or its interests.
- Defeat any potential adversary seeking to launch attacks on Australia.
- Deny operational freedom to any potential adversary or security threat within the immediate neighbourhood.
- Quickly and decisively assist the civil authorities of Australia by providing military assistance" (Houston 2007).

Capability must also be considered from the perspective of combat functions (Kerr et al. 2008c). Therefore the functional packages layer, underneath the effects (Fig. 20.4), encapsulates the warfighting force structures of a future force. The template of military capability illustrated in Fig. 20.1 has been populated with the Royal Australian Navy's Future Maritime Operating Concept Year 2025. The definitions of the functional packages are given in Table 20.1.

Finally, the centre of Fig. 20.4 presents the key building blocks of military capability, i.e., the product-service system space. The supporting infrastructure consists of eight key lines of development (DoD 2006):

- Personnel
- Organisation
- Collective training
- Major systems
- Supplies
- Facilities
- Support
- Command and management

These lines are interwoven in a matrix fashion with the spectrum of strategic platform types. The Royal Australian Navy has seven product area groups (ANAO 2003):

- Surface combatants (SURF)
- Amphibious and afloat support (AAS)
- Aviation (AIR)
- Submarine (SUB)
- Mine clearance diving (MCD)
- Patrol boat (PB)
- Hydrographic, meteorological and oceanographic (HYDRO)

As a visual summary Fig. 20.4 represents Australian maritime capability for 2025. It is a high-level mapping of the future strategic context and presents the future end-state for the Navy to work towards. An analysis of this future military capability poses three challenges (DoD 2000):

- The adequacy of ships' defences against the more capable anti-ship missiles that are proliferating in the region.
- The requirement for a long-range air-defence capacity in the fleet.
- The future provision of support ships to increase the maritime capability by keeping ships at sea longer and at greater ranges from port.

The next step is to then map the future shape of the fleet given these priorities.

20.3.2 Mapping the Fleet

Roadmapping provides a canvas upon which to map the maritime fleet as it transitions from its current form to the future state required to fulfil the Navy's contribution and commitments to national strategy as outlined in Figs. 20.3 and 20.4. A fleet roadmap can be generated for each of the Navy's product area groups. The Surface Combatant (SURF) fleet is used to demonstrate the output that can be achieved by adopting the methods developed by Phaal et al. (2007). This is presented in Fig. 20.5. The architecture of the roadmap is based on a grid layout. The columns are formed from the timescale, in this case 2000–2025 and beyond, with the rows splitting the map between product platforms via classes and technology

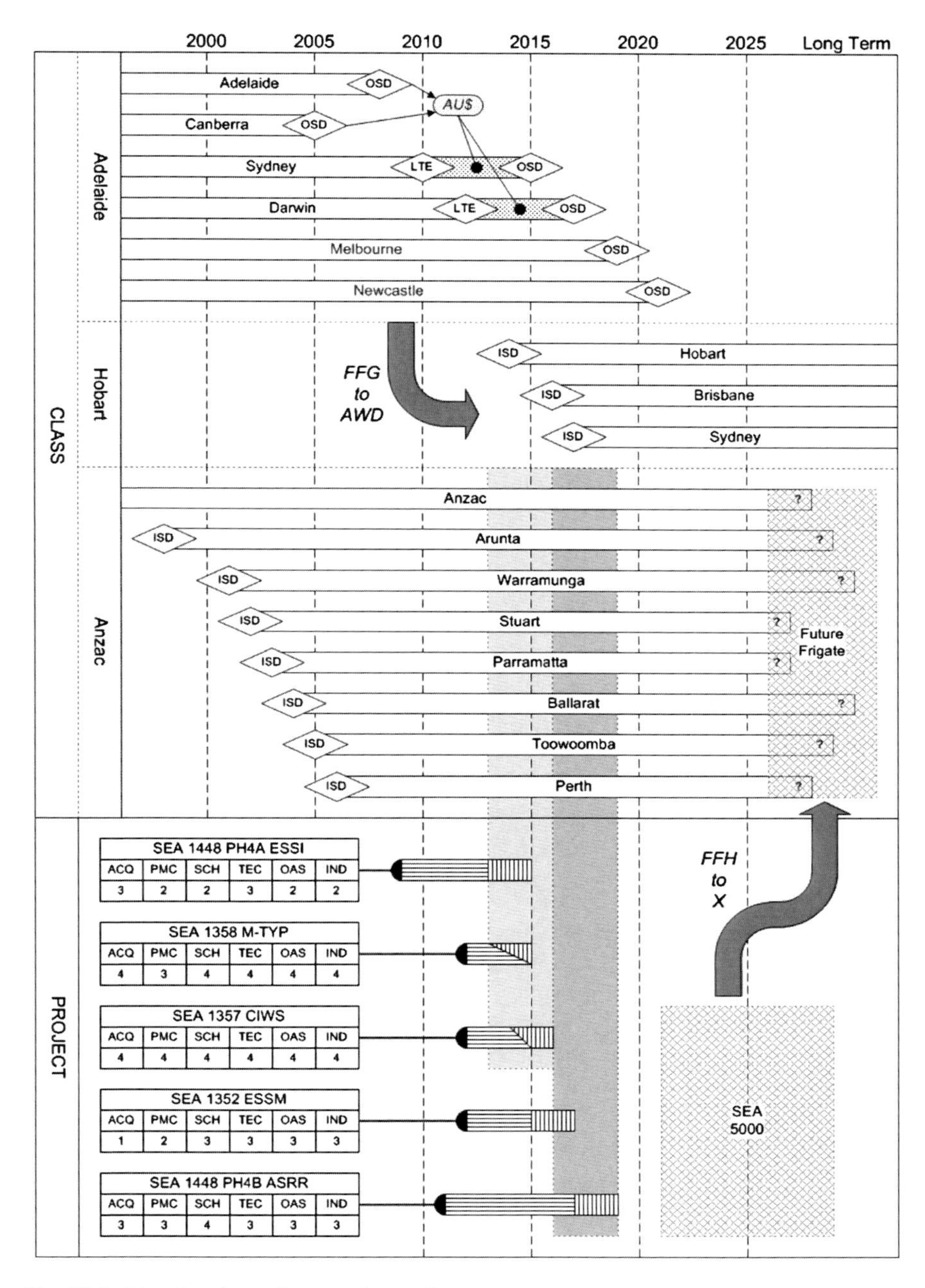

Fig. 20.5 Mapping the surface combatant fleet

insertion projects. In the top half of the fleet roadmap the in-service dates (ISD), out-of-service dates (OSD) and life-of-type extensions (LTE) of the various ships are plotted against the considered timeframe. The bottom half is populated with the key update and upgrade projects.

As can be seen from Fig. 20.5, the current surface fleet consists of two classes of frigate, i.e., Adelaide and Anzac. The Adelaide ships entered service between 1980 and 1993 (DoD 2000). They were based on the US Navy Oliver Hazard Perry design. The first four of the class were built in the United Sates, with specific modifications undertaken in Australia, whereas the last two ships (HMAS Melbourne and HMAS Newcastle) were wholly constructed in Australia. The Anzac's entered service in the early 2000s. They are based on the German MEKO 200 design, with all eight ships constructed in Australia. These two classes, Adelaide and Anzac, are the mainstay of the Royal Australian Navy's fighting ships currently in-service (Thomson 2008).

However, Fig. 20.5 also highlights that the surface combatant fleet is a force in transition (Davies 2008). The Adelaide class frigates will be replaced by a new Hobart class of air warfare destroyer (DoD 2000). This transition is mapped in Fig. 20.5 as the arrow labelled 'FFG to AWD'. The Hobart ships will cost nearly AU$8 billion (DoD 2007) and the first-in-class will enter service in 2014 (SPC 2009). The Hobart class is using the Spanish F100 for its baseline platform design. The three ships will be equipped with an Australianised version of the Aegis combat system. This will then be coupled with the SM-6 surface long-range anti-aircraft missile (Davies 2009; DoD 2009a). The other transition, labelled 'FFH to X', is the likely candidate that will replace the Anzac class in the long-term. The associated new product development programme is being budgeted under 'Project SEA 5000' (Davies 2008). It will potentially see the production of a fleet of eight new 'Future Frigates' (Davies 2009; DoD 2009a) to enter service sometime post-2025.

In the period preceding the two transitions in class replacements, more than AU$3 billion has been committed towards the upgrade of both the Anzac and Adelaide class frigates in order to ensure that they remain at the forefront of regional naval capability (DoD 2007). In respect of the Adelaide class, four of the original six ships will receive an upgrade. As can be seen in Fig. 20.5, the fleet was reduced from six to four vessels. HMAS Canberra was retired in 2005 and HMAS Adelaide was withdrawn from service in 2008 (DMO 2008). These two ships were decommissioned in order to offset the cost of upgrading the remaining four ships. It is worth noting that AU$3 million was allocated for gifting HMAS Adelaide to the New South Wales Government to prepare the ship for sinking as a dive wreck (Thomson 2008). Additionally, HMAS Sydney and HMAS Darwin are to receive life-of-type extensions (ANAO 2007) so that their service on active duty is prolonged by an addition 5 years as illustrated in Fig. 20.5. The details of the technology insertion programme for the Adelaide class will be outlined in Sect. 20.3.3 when the mapping of the upgrade paths for the functional systems is described.

For the Anzac class, the bottom half of the roadmap (Fig. 20.5) depicts the technology insertion projects that will be conducted in the future so as to ensure that these vessels embody operationally relevant functionality and performance. The aim of the upgrades is to enhance the air warfare capabilities and level of anti-ship missile defences against the threats of the next decade (DoD 2008). There are five key projects arranged into two phases. The upgrade paths are the shaded boxed areas in Fig. 20.5 that link the projects with the ships in the Anzac class. In the first bundle of upgrades, three technologies will be inserted into the fleet namely:

- SEA 1448 Phase 4A Electronic Support System Improvements—This reflects an improvement of the CENTAUR electronic system in order to maintain regional capability parity and ensure the sensors align with future threats (DoD 2009b).
- SEA 1358 Mini Typhoon—This is to provide short-range surface defence against asymmetric threats (DoD 2009b).
- SEA 1357 Close-In Weapons System—This is a block upgrade for the Phalanx weapon for improved ship self-defence against anti-ship missiles, helicopters and small craft (DoD 2009b).

Upon successful insertion of these technologies, a subsequent two further upgrades will be incorporated onto the vessels during the second phase between 2016 and 2019. These are:

- SEA 1352 Evolved Sea Sparrow Missile—This is the addition of the named missile system to defend against the evolving anti-ship cruise missile threats (DoD 2009b).
- SEA 1448 Phase 4B Air Search Radar Replacement—This is a replacement of the ageing AN/SPS-49 radar with a modern digital version which will complement the phased array radar system recently installed (DoD 2009b).

On the roadmap, each of the five upgrades has specific metadata contained underneath their named project title captions. The metadata consists of six attributes and associated scores. This representation is based on the ACAT (acquisition categorisation) method (DoD 2009b) as advocated in the Australian Department of Defence's AIS (acceptance into service) framework (DoD 2006). Table 20.2 describes the six attributes used to provide a measure of complexity for the different elements of the technology acquisition process and management activities. Each of the attributes is scored on the basis of a 1–4 rating with 1 representing the very high end and 4 being low.

From inspection of the insertion projects' metadata, it can be seen that there are two upgrades that stand out as being the riskiest. The Evolved Sea Sparrow Missile (SEA 1352) upgrade presents the greatest contributor to cost; it equates to over AU$1,500 million. Whereas, in terms of scoring across all of the attributes, the Electronic Support System Improvements (SEA 1448 PH4A) rates as high. The other three remaining upgrades rate as moderate (SEA 1448 PH4B) and low (SEA 1358 and SEA 1357).

To summarise, the fleet roadmap is a visual tool for visioning the future class transitions and also the key technology enhancements for the current in-service vessels. It provides a high-level depiction for use by planners and contains

Table 20.2 Project attributes (DoD 2009b)

Element	Definition
ACQ (acquisition cost)	This includes the cost of the materiel system (i.e., mission system plus support system), plus facilities costs. This does not include ongoing sustainment budgets
PMC (project management complexity)	This highlights complexity beyond that associated with traditional project management knowledge areas, which are characterised by a project execution environment which is novel and uncertain with very high-level political interactions
SCH (schedule)	This recognises the complexity brought about by schedule pressures on the project requiring the application of varying levels of sophistication in schedule management.
TEC (technical difficulty)	This reflects the inherent complexities which are associated with technical undertakings of design and development, assembly, integration, test and acceptance
OAS (operation and support)	This embodies the readiness of the organisation and environment into which the system will be operated and supported
IND (commercial)	This recognises the capability of industry to deliver and support the required system/equipment, the complexity of the commercial arrangements being managed including the number and level of interdependency of commercial arrangements

metadata orientated for programme managers. The next level down in granularity, which provides the detailed content that describes the actual technology insertion projects is contained in the mapping of a ship's functional systems, i.e., the ship roadmap.

20.3.3 Mapping the Functional Systems

The third and final step in the mapping approach is at the ship level. This is focused on the functional systems and mapping the contents of the bundles of technology upgrades. To illustrate the mapping of the functional systems for a ship, the historic example of the Adelaide class is used. The resultant roadmap is given in Fig. 20.6, in which the focus is limited to the combat system. This map visually outlines the upgrades to the four frigates that are to improve both their self-defence and offensive capabilities until the delivery of the new Hobart destroyers. The most significant enhancement was to address the threat from anti-ship missiles (DoD 2008); this was listed as one of the top priorities for the surface combatants in the Navy's maritime strategy as stated at the end of Sect. 20.3.1. In terms of threat, proliferated anti-ship cruise missiles are the principal access-denial weapon against surface ships (Barber and Gilmore 2001) and the availability of such weapons is increasing (Brooks et al. 2005). Due to the "proliferation of high-capability anti-ship missiles such as Harpoon, Exocet and their Russian equivalents" (DoD 2000). "Such weapons are relatively inexpensive to build or buy.

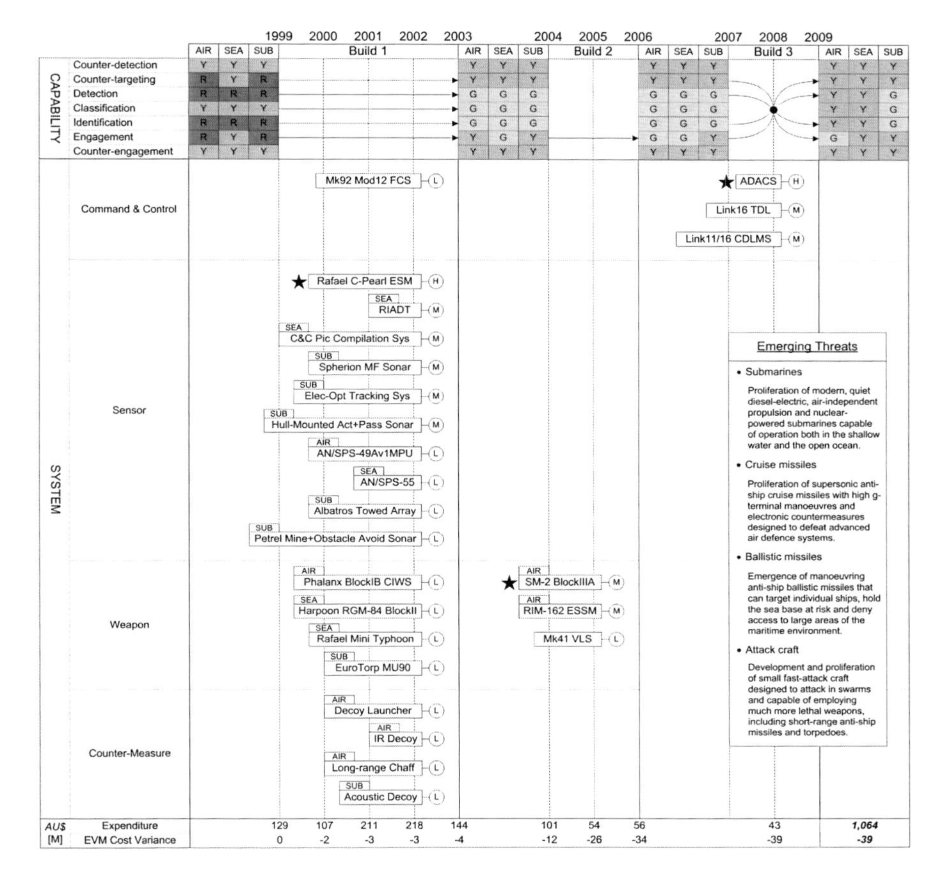

Fig. 20.6 Mapping the Adelaide class' functional systems

New technologies—global positioning receivers, compact gas turbine engines, composite aerostructures—are available to virtually any state or non-state actor wishing to fashion a precise and lethal cruise missile" (Thompson 2004). The roadmap also provides an overview summary of all of the emerging threats facing maritime forces (Mahon 2009); this is contained in the enclosed bullet list on the right-hand side.

At the top of the roadmap (Fig. 20.6) there is the provision of a capability assessment matrix using a traffic light grading scheme where red (R) signifies major deficiencies, yellow (Y) for minor deficiencies and green (G) for sufficient capability. The capability of the ship-level combat system is graded for each three battlespace domains (air, sea surface and sub-sea/underwater) against the key kill-chain functions. The functions are those of counter-detection, counter-targeting, detection, classification, identification, engagement and counter-engagement (RAN 2006). The matrix requires the combat capability to be reassessed after each phase of the technology insertion activities. For example, in Fig. 20.6 the matrix is completed for each of the three builds.

The main body of the ship roadmap outlines the individual upgrades that take place. They are categorised into vertical layers based on the classification of command and control, sensors, weapons and countermeasures systems (RAN 2006). The horizontal time dimension also ties the main columns to the respective builds. In this case there are three phases to the Adelaide's enhancement. The first build is orientated to upgraded fire control, early warning radars and underwater warfare systems (DMO 2008). Build 2 is a focused weapons upgrade which inserts both the Evolved Sea Sparrow and SM-2 missile systems. The third build is the replacement of the original command and control system with the Australian Distributed Architecture Combat System (ANAO 2007) and a new dedicated Link 16 Joint Tactical Information System (DMO 2008). Each of the three build phases has been populated with the relevant technology upgrades. Each project is tagged with their operational battlespace domain (air, sea surface and sub-sea/underwater) to which they contribute to and also a consolidated risk metric (high, medium and low). For instance, the EuroTorp MU90 upgrade, in Build 1 under the weapons layer, is for the underwater domain and has been assessed as a low risk project. The EuroTorp MU90 is an off-the-shelf anti-submarine torpedo. The area of anti-submarine warfare was shown to exhibit a capability shortfall (Davies 2008) and this is reflected in the 1999 capability assessment matrix. The EuroTorp MU90 is the replacement for the older Mk46 lightweight torpedo already installed on the Adelaide class (DID 2009). "The MU90 has greater performance and lethality than the Mk46 and requires less logistic support. It is 3 m long, weighs 300 kg, has a range of more than 10 km and is designed to track and attack quiet-running submarines at depths ranging from 25 m to more than 1,000 m" (Thomson 2008). Finally, at the bottom of the roadmap there is a row of financial data corresponding to the expenditure for each year and the associated cost variance. As can be seen from Fig. 20.6 the upgrades for the four Adelaide ships totalled AU$1,064 million, which is AU$ 39 million over its budgeted cost (ANAO 2007). In summary the ship-level roadmap, which is populated with the functional systems upgrades, provides a visually concise overview of the technology insertion projects that are to be implemented for a given class of product platform. Although the examples used in this chapter are those from the maritime sector, the approach can be applied across to the product platforms in the air and land environments.

20.4 Summary

This chapter introduced an approach for visually mapping the future transitions in complex product platforms. Using the case example of the Royal Australian Navy's Surface Combatant Fleet 2025, the steps in the approach were both described and illustrated. This essentially provides a guide through the method for practitioners and additionally acts as a key reference case that depicts the outputs which can be achieved by utilising the approach. The approach, itself, uses a visual framework for visioning military capability together with a dynamic systems orientated roadmapping template. Using a top-down process, the strategic context in which a product-service system will operate is first determined. This effectively provides a view of the future force for the respective service branch of the military. The defence priorities, concepts of operations, operational-based effects and warfighting elements are all determined. These then provide the surrounding context to the product platforms and the supporting service infrastructure. In order to satisfy this future end-state, the next step is to map the transitions between the current-future portfolio of platforms. It involves plotting the life-of-type extensions and out-of-service dates of the active fleet along with the in-service dates of the next generation replacements. Aligned to the timeframe are the key update and upgrade projects that must be conducted to resolve any capability gaps during fleet transitions. Measures for acquisition cost, complexity of project management, schedule constraints, technical difficulty, operation and support readiness, and commercial arrangements are provided for each technology insertion project. Finally, the phased introduction of the specific technologies within work packages is mapped to the functional systems of the platform together with a capability assessment of their utility with the end-user.

20.5 Chapter Summary Questions

- How do you maintain the technological edge in products that will remain on active service for decades?
- How do you facilitate the engagement of all stakeholders in determining future capability?
- How do you balance operational need, budget constraints and technology availability?
- How do you manage the flow of technology for platform modernisation?

References

Australian National Audit Office (ANAO), *Navy operational readiness. Report Number: 39* (Australian National Audit Office, Canberra, 2003)

Australian National Audit Office (ANAO), *Management of the FFG capability upgrade. Report Number: 11* (Australian National Audit Office, Canberra, 2007)

H.S. Balaban, W.L. Greer, *Model for evaluating the cost consequences of deferring new system acquisition through upgrades. Report Number: P-3424* (Institute for Defense Analyses, Alexandria, 1999)

A.H. Barber, D.L. Gilmore, Maritime access: Do defenders hold all the cards? Def. Horiz. **4**, 1–8 (2001)

D. Birchall, G. Tovstiga, *Capabilities for strategic advantage: Leading through technological innovation* (Palgrave Macmillan, Basingstoke, 2005)

K. Booth, *Navies and foreign policy* (Croom Helm, London, 1977)

T.A. Brooks, H. Jenkins, N. Polmar, R. Pirie, T.D. Ryan, J. Sommerer, W. Weldon, J. Wolbarsht, *Science and technology for naval warfare 2015–2020. Report Number: NRAC 05–3* (Naval Research Advisory Committee, Arlington, 2005)

R. Crane, ADF responses to meet future contingencies. Aust. Def. Force J. **173**, 68–72 (2007)

A. Davies, *ADF capability review: Royal Australian Navy. Policy Analysis* (Australian Strategic Policy Institute, Barton, 2008)

A. Davies, Australia's defence white paper 2009. Def. Syst. **12**, 34–37 (2009)

Defense Industry Daily (DID), Australia's hazard(ous) frigate upgrade. Available at: http://www.defenseindustrydaily.com/australias-hazardous-frigate-upgrade-04586/. Accessed 28 Oct 2009 (2009)

Defence Materiel Organisation (DMO), *SEA 1390 phase 2.1: FFG upgrade project* (Defence Materiel Organisation, Canberra, 2008)

Department of Defence (DoD), *Defence 2000: Our future defence force. Report Number: Defence White Paper* (Department of Defence, Canberra, 2000)

Department of Defence (DoD), *Defence capability development manual. Report Number: DCDM, Capability Systems Division* (Department of Defence, Canberra, 2006)

Department of Defence (DoD), *Australia's national security: A defence update 2007* (Department of Defence, Canberra, 2007)

Department of Defence (DoD), *Defence annual report: Volume 1* (Department of Defence, Canberra, 2008)

Department of Defence (DoD), *Defending Australia in the Asia Pacific century: Force 2030. Report Number: Defence White Paper* (Department of Defence, Canberra, 2009a)

Department of Defence (DoD), *Defence capability plan. Report Number: DCP* (Department of Defence, Canberra, 2009b)

T. Dowling, R. Hood, R. Hirst, D. Barker, E. Lomas, E. Phillips, A. Griffiths, B. King, D. Field, *Agile capability and adaptable systems. Report Number: QINETIQ/EMEA/TECS/CR0700304, TI MPA (Technology Insertion Major Programme Area)* (QinetiQ, Farnborough, 2007)

A. Houston, The ADF of the future. Aust. Def. Force J. **173**, 57–67 (2007)

Joint Standing Committee on Foreign Affairs, Defence, Trade (JFADT), *Australia's maritime strategy. Joint Standing Committee on Foreign Affairs, Defence and Trade* (Department of the House of Representatives, Canberra, 2004)

C. Kerr, R. Phaal, D. Probert, A framework for strategic military capabilities in defense transformation. The 11th International Command and Control Research and Technology Symposium (ICCRTS 2006)—Coalition command and control in the networked era, Cambridge, Sep 26–28

C.I.V. Kerr, R. Phaal, D.R. Probert, Technology insertion in the defence industry: A primer. Proc. Inst. Mech. Eng. Part B. J. Eng. Manuf. **222**, 1009–1023 (2008a)

C.I.V. Kerr, R. Phaal, D.R. Probert, Aligning R&D with changing product requirements in an evolutionary acquisition environment. The R&D management conference 2008, Ottawa, June 18–20 (2008b)

C. Kerr, R. Phaal, D. Probert, A strategic capabilities-based representation of the future British armed forces. Int. J. Intell. Def. Support. Syst. **1**, 27–42 (2008c)

C.I.V. Kerr, R. Phaal, D.R. Probert, Cogitate, articulate, communicate: The psychosocial reality of technology roadmapping and roadmaps. The R&D management conference 2009—the reality of R&D and its impact on innovation, Vienna, June 21–24 (2009)

S.M. Kosiak, *Matching resources with requirements: Options for modernizing the US Air Force* (Center for Strategic and Budgetary Assessments, Washington DC, 2004)

M. Mahon, US Navy surface warfare: Future requirements and capability. Def. Syst. **11**, 40–44 (2009)

M.J. Milas, R. Vanderbok, Beyond proactive DMSMS, what's next: Coordinated technology management. 9th Joint FAA/DoD/NASA conference on aging aircraft, Atlanta, March 6–9 (2006)

Ministry of Defence (MoD), *Defence industrial strategy. Defence white paper (CM6697)* (Her Majesty's Stationery Office, London, 2005)

R. Phaal, C. Farrukh, D. Probert, Customizing roadmapping. Res.Technol. Manag. **47**, 26–37 (2004)

R. Phaal, C.J.P. Farrukh, D.R. Probert, Strategic roadmapping: A workshop approach for identifying and exploring strategic issues and opportunities. Eng Manag J **19**, 3–12 (2007)

Royal Australian Navy (RAN), *Australian maritime doctrine. Report Number: RAN Doctrine 1, Sea Power Centre* (Royal Australian Navy, Canberra, 2000)

Royal Australian Navy (RAN), *Plan blue. Royal Australian Navy* (Department of Defence, Canberra, 2006)

G. Smith, Stating the problem: Facing the challenge. Maritime war in the 21st century: The small and medium navy perspective, Papers in Australian Maritime Affairs Number 8, ed by D. Wilson, Sea Power Centre, Royal Australian Navy, Canberra, Australia (2001)

Sea Power Centre (SPC), *The Navy's new Aegis. Report Number: Semaphore 07, Sea Power Centre* (Royal Australian Navy, Canberra, 2009)

L. Thompson, *Cruise missile defense: Connecting theater capabilities to homeland needs* (Lexington Institute, Arlington, 2004)

M. Thomson, *The cost of defence: ASPI defence budget brief 2008–2009* (Australian Strategic Policy Institute, Barton, 2008)

E.K. Yakura, Charting time: Timelines as temporal boundary objects. Acad. Manag. J. **45**, 956–970 (2002)

Part V
Integrating Perspectives for Complex Engineering Service Systems

Glenn Parry

Any discussion of complex engineering service system requires links to be made between of a number of different areas of academic study, including, but not limited to, management and engineering practice. The subject area draws upon a breadth of knowledge and requires integration of numerous disciplines. This final section seeks to provide more diverse perspectives on service and also seeks to provide integration. We will provide forms of meta-analysis or meta theory, descriptions and frameworks which we hope will help build a debate over the elements involved in Complex Engineering Service Systems.

The first chapter explores the design of service solutions and explores how this may be done in concert with the customer. The work explores the social context of service, identifying the dominant thinking style in social space as "service thinking", and counterpoints this to systems engineering. It identifies that the social dimension is key to service design and delivery, highlighting the need for research in 'discerning the mind of the customer'.

The second chapter proposes an Activity Based Framework for Services (ABFS), which has three aims: it seeks to situate and relate different disciplinary definitions of services; understand the foci of different service design approaches; and produce models of service systems. The ABFS seeks to move beyond the transformational model presented in the CIF and provide more detail about what is being transformed, the artefacts used and the activities undertaken by the system.

The final chapter reflects upon the contributions of the authors within the context of the broader subject area. It provides a further development of the CIF, attempting to provide a visualisation of the challenge of delivering service excellence against changing customer contexts that permeate processes and seek to push variety into a service system. The resultant evolution provides a framework which may be used to place current research and locate areas requiring future work.

Chapter 21
Service Thinking in Design of Complex Sustainment Solutions

L. A. Wood and P. H. Tasker

Abstract Delivering contracted performance levels for service based on the sustainment of complex engineering systems is a necessary but not sufficient condition for user satisfaction. Service is received in a context that is shaped by the state of mind of the customer—perceptions, biases, memories, intentions and patterns of thinking. Service teams need to understand the "mind of the customer", complementing the "voice of the customer" used in requirements development. The chapter considers how service solutions are designed and suggests that the state of mind of the customer needs greater consideration during solution development. The service team functions in the social dimension to understand the customer's mind and harmonises the service solution. The dominant thinking style in social space is characterised as "service thinking", complementing the system thinking style which dominates in the conceptual space of product-service systems.

21.1 Thinking Styles for Product-Service Systems

This chapter explores the need to introduce a new thinking style, labelled "service thinking", for the design of service solutions based on the sustainment of complex engineering systems. These systems can be characterised as one-off (or few-off)

L. A. Wood (✉)
The University of Adelaide, Adelaide, SA, Australia
e-mail: lincoln.wood@adelaide.edu.au

P. H. Tasker
The University of Cambridge, Cambridge, UK
e-mail: pht25@cam.ac.uk

P. H. Tasker
Cranfield University, Cranfield, UK

P. H. Tasker
University of Kent, Kent, UK

I. Ng et al. (eds.), *Complex Engineering Service Systems*, Decision Engineering,
DOI: 10.1007/978-0-85729-189-9_21,

designs with high engineering complexity, high performance, long life and likely to incorporate in-service upgrades or change of use. Examples include military equipment, large scale infrastructure and installations, major buildings and high value production facilities. The hypothesis is that manufacturers of complex systems who offer end user service based on the assured availability of equipment performance in operation (the product-service system) will be unable to achieve service excellence in the mind of the customer without a paradigm shift in thinking at corporate, team and individual levels. The need arises largely as a result of the maturity of the engineering "system thinking" method (system engineering) used for the development of large scale bespoke products. System engineering tools and thinking style ably address the conceptual space of the manufacturing design and integration process, ensuring that the customer's expectations for system performance are met by providing a structured approach to managing explicit requirements—the "voice of the customer". They are of little help in ensuring a connection with the "mind of the customer" necessary to complete the service solution design in the social space, nor in ensuring that the conceptual and social visualisations are harmonised into an integrated solution which satisfies both the tangible requirement for system performance and the intangible concept of service need in the customer's mind.

The proposition is that for a complex product-service system, service satisfaction resides in the mind of the customer so that the sustainment of fit-for-purpose goods is a necessary but not sufficient condition for customer satisfaction in the overall service. The end state of service is a mental concept that is intangible and its achievement is heavily shaped by the context in which it is received. The context includes the emotional state of mind, its perceptions and preconceptions, its patterns of memory and patterns of automated processing, and plans for the future. Service satisfaction therefore requires the service provider to have a highly developed concept of the mind of the customer; whether the customer is an individual or an organisation, the same principle applies. The desired end state of service provision is trust, wherein the customer habitually seeks more.

This language is unfamiliar in the product domain where achievement of functional performance requirements is regarded as paramount. It is often tacitly assumed that service is an inherent feature of the product, somehow embedded within it: from the perspective of the product developer, the user is an extension of the product. The user's viewpoint is quite the opposite: the product is an extension of the user, sometimes even anthropomorphising certain classes of products. Intel Corporation has established a User Experience Group staffed by anthropologists and ethnographers in order to better understand customer patterns of life that determine how its product offerings are used. This group seeks to understand, for example, how customers (situated users) accept certain products as "symbolic markers of cultural practices, gender and even religious identities".

The system engineering thinking style is deeply embedded in the development of complex systems, especially defence systems where it has become a universal language; systems of regulation and governance are built on it. Notwithstanding the development of soft systems and similar approaches, system engineering thinking is perceived to be inextricably rooted in the product paradigm ensuring a cultural rigidity inhibiting development of the method to accommodate design in the social space necessary for the attainment of service excellence. It is therefore important to develop complementary techniques for service design in the social space based on "service thinking" and the theory of mind, leading to new perspectives on leadership and the building of expert teams.

If service thinking is to be established as a separate and identifiable thinking style that is complementary to system thinking, what is it, and how is it to be implemented in the design of complex sustainment systems? This chapter explores these questions from the practitioner viewpoint by firstly examining features of the established system thinking style, then offering an argument to include the social space in service design suggesting cardinal features of a service thinking style, followed by suggestions as to how the two styles may be harmonised. This inevitably leads to rich research questions and some suggestions for new practice.

21.2 Design in the Conceptual Space: Product-Service Systems

Design of product and processes is a journey through *conceptual space*, while service is a journey through *social space*. They are parallel journeys. Each must take into account the other; effective business solutions seek to harmonise them.

21.2.1 Current Design Methodologies

Like the trend to servitization, product-service systems have evolved over time, with different industry sectors taking different paths. Servitization has been a route to diversification for many manufacturers (Ng et al., Introduction chapter; Neely 2008), adding services to create enhanced value to cope with intensified competition. For some industry sectors, such as automotive, truck and mobile equipment industries, service in the form of maintenance and repair (sustainment) has long been a part of the enterprise business model, often implemented through a franchise relationship with local businesses to enable broad geographic coverage. For other industry sectors, different models have applied. In the civil airframe sector, airline operators for many years have been the maintainers, while more recently third party enterprises have become established as maintenance and repair organisations.

In the defence industry sector, where complex, large scale and bespoke systems are typical, sustainment has often been delivered organically by the defence operators. In this situation, the product lifecycle is clearly divided into two very distinct phases: acquisition and sustainment. The historic focus of this industrial sector has been on acquisition, driven by the need for outstanding performance, sometimes at the cost of sustainability. Over time, greater emphasis has been placed on sustainability by applying "design for" constraints, such as *design for reliability*, *design for maintainability*, *design for availability* and many others. The separation of acquisition and sustainment has gradually broken down in recent years, especially in the context of new defence industry contracting policies as exampled by the UK MoD Ministry of Defence in its *Defence Industrial Strategy: Defence White Paper* (2005). Many western defence organisations have adopted similar approaches. An increasing proportion of military sustainment activity is now outsourced to commercial contractors, with outcome-based contracting (Ng et al. 2009) evolving as a preferred commercial approach. Contracting in non-defence sectors is also evidencing similar trends (Johnstone et al. 2008, 2009). Defence companies have long been engaged in sustainment functions for the military, but the outcome-based contracting approach has several important differences. The historic sustainment activity has been for discrete tasks or components within the larger customer organic solution (a traditional logistics support function is based on the deep technical knowledge of the products being supported). In comparison, the outcome-based contracting environment requires that the contractor provide a comprehensive integrated business solution wherein significant contractual risk is transferred to the contractor.

The design of complex product systems is enabled by a comprehensive system engineering methodology and tool set. Similarly, the traditional logistics support activity is enabled by analytical methods (logistics engineering analysis), with design and implementation supported by very detailed logistics support and management processes based on system engineering (Blanchard 1998). A similar situation applies for design and management of the supply chain, and also for maintenance activities, all core elements of a sustainment solution. Business process modelling also has a long history (Williams 1967), with a supporting enabling suite of processes and software tools. Business modelling capability has increased dramatically following the development and routine deployment of software tools, with UML (Unified Modeling Language) and IDEF0 (Integrated Definition function modeling) being among the dominant approaches.

The term "product-service system" is used in outcome-based contracts as they are necessarily service-like in their objectives and approach; the contractor is providing an intangible output to the customer within the user environment. In the defence sector the output is typically the availability of a component of military capability such as trained operators or assured availability of performance. Systematic design of service has been addressed by various authors (Shostack 1982; Ramaswamy 1996). The subject of service design has often been approached through the marketing and management disciplines, where service task elements are identified and connected with adjacent tasks into a network that seeks to produce a common goal, the service output.

Morelli (2006) has described product-service systems as "social constructions", especially due to the value co-production aspects of service design. An important feature of this work is the identification and understanding of the roles of the actors in the network associated with the socio-technical process. The social approach to design has been elaborated by Bijker et al. (1987) who analyse the sociology of technological construction. Bijker (1995) produced three comprehensive case studies of socio-technical constructions (bicycles, bakelite and fluorescent light bulbs). He points out that "society" and "technology" are both human constructs, and that they are inevitably linked. A social approach to design leads to fundamentally different thinking about the nature of design. This basic idea is mirrored to an extent in the product-service literature, where the primary drivers of academic research have been sustainable ecology and economically sustainable business.

Nevertheless, some authors have reported that technical product-service systems are still designed without the benefit of a systematic methodology (Aurich et al. 2006), observing that "service design is frequently performed detached from product design", with very little interaction between the two. This has been described as "intuitive design". Aurich et al. propose a systematic design approach and evaluate its application to a business that produces heavy road construction equipment. In this context, they identify three "dimensions" of technical service design: the product dimension, such as maintenance and spares provisioning; the process dimension, which describes the activities in the service operations; and the information dimension, which describes information gathering and exchange. They then compile an overall process methodology that responds to the top level requirements of the product-service system. Alonso-Rasgado et al. (2004), Alonso-Rasgado and Thompson (2006) have considered the design of systems of hardware combined with a service system, which they term "total care products". They pay particular attention to the customer–supplier relationship throughout the design process of the total care product which they envision as a series of iterative engagements.

One aspect of product-service systems that is challenging to incorporate into the design process is customer satisfaction. Although product and service design requirements attempt to capture customer needs, this is difficult to comprehensively achieve. Bullinger et al. (2003) have introduced "service engineering" as a means to systematise the development of services. Sakao and Shimomura (2007) describe service engineering as "a discipline to increase the value of artifacts and to decrease the load on the environment by reason of focusing service", in their view changing the focus of product development from functionality to customer satisfaction. Although engineering practice has arguably already moved to incorporate non-functional aspects, Sakao and Shimomura's approach is of interest in that it seeks to explicitly incorporate the customer state through "receiver state parameters" which may be dynamically assessed. The conventional engineering approach would be to introduce customer characteristics through static requirements that constrain the design. Receiver state parameters apply to a "Persona", an imaginary target user. Persona data is divided into two classes: demographic data; and psychological data (values and life style). The latter includes such parameters as sense of belonging, relationships, self-fulfilment, enjoyment, respect, and excitement. Although this is directed at traditional service industries, it is a significant attempt to incorporate intangible values into a service design methodology. A different approach, through the "design" of expert teams, is suggested later.

Earlier steps to include psychological functioning were made by Hollnagel and Woods (1983) (also Rasmussen et al. 1994). In the context of safety–critical nuclear power plant operations, they sought to go beyond the conventional approaches of "man–machine systems" (or human–machine interfaces). Using what they termed "Cognitive Systems Engineering" (CSE), they incorporated the psychological factors of operators in addition to the conventional physical and physiological factors. They use the term "cognitive" to incorporate a spectrum of psychological factors, including emotions, as well as attitudes, aspirations, motives, etc. The sense is clear—"a different interdisciplinary synthesis is required". Hollnagel and Woods continue:

> The central tenet of CSE is that a [man-machine system] needs to be conceived, designed, analysed and evaluated in terms of a cognitive system. Like the Gestalt principle in psychology, a [man-machine system] is not merely the sum of its parts, human and machine. The configuration or organisation of man and machine components is a critical determinant of the outcome or output of the system as a whole.

There are numerous instances of failure due to humans being forced to interact on the machine's terms. The CSE approach has attempted to avoid such problems through a design process that accounts for human thinking characteristics. While cognitive systems engineering was developed for the machine-human operator design case, it has some parallels with customer thinking in product-service systems, where the customer can be conceived in the place of the operator.

The range of functions in complex product-service is one the challenges in the architecting of solutions. There is a growing literature on the architecting of complex systems (for example, Rechtin 1991). Many of the concepts are borrowed from software engineering, which in turn draws extensively on system engineering. Further, software architecting is itself established within the software engineering discipline (Shaw and Garlan 1996). As a professional methodology, architecting draws heavily on professional practice rather than academic theory. Architectures are the "bones" of a design, a higher level of abstraction, upon which the design may be "fleshed out", but it is not the actual design. Architecting is often described as an "art", as opposed to a science (Maier and Rechtin 2002). It makes extensive use of heuristics derived from experience, underscoring the professional practice origins of architecting. Operating at a layer above design, architectural requirements are less precise and more interpretive, capturing concepts and principles rather than specifics. There are now various frameworks developed for architecting complex systems; widely used one is the US DoD Architectural Framework or DoDAF (now at Version 2.0, US Dept of Defense 2009). It is quite general in its potential applications and may be used in any enterprise, public or private, to develop enterprise architectures. Importantly, it incorporates a service perspective alongside systems, defining a "Systems and Services View" as one of its four views. DoDAF enables the visualisation and understanding of complex, large scale systems and underlying processes of almost any type, including product-service systems. The authors know of several instances where it has been used to devise sustainment solutions which have significant service features.

Service-oriented architecture is a particular IT industry approach to business architectures that incorporates customer, user and provider views (Allen et al. 2006). The approach recognises services as significant, not secondary to the technology. Nevertheless, the services considered are software services, not the broader total solution approach required for output-based contracts.

This brief overview of existing design approaches has highlighted a number of parallel streams of development, all developed with different objectives and from different origins:

- Conventional product design, where after-sales service is added
- Product-service systems, developed with environmental objectives in mind
- Service engineering, developed with customer satisfaction as the key focus
- Cognitive systems engineering, developed for safety critical systems with humans in the loop
- Complex systems architecting, derived from professional practice
- Software as a service

It is apparent that there are many methods, practices and tools already available that may be brought to bear on complex engineering service systems, including the design of service processes. Yet in many instances there is often not a designated "chief designer" for the business solution. As Maier and Rechtin (2002) point out, a notable feature of successful unprecedented systems is that they have a "clearly identifiable architect or a small architect team".

21.2.2 System Thinking and System Engineering

System: a complex whole; a set of connected things or parts; An organised body of material or immaterial things (The Concise Oxford English Dictionary, 8th Edition 1990)

System thinking has different meanings for different professional disciplines—it is heavily loaded with interpretations. In the design of technical products, system thinking is equated to "design thinking"; in other contexts, it is limited to "holistic thinking". Both are valid in context, but it is important to understand the context.

Engineering professionals would describe system thinking as the dominant thinking style in complex product systems. Its specific implementation is *system engineering*; it has become a ubiquitous methodology, a universal language, especially for defence systems. This language, not in itself complex and certainly easily learned, is nonetheless regarded as arcane and therefore not widely understood beyond the discipline of engineering. For practitioners, system engineering is *the* generic engineering method that enables them to manage complexity.

Like many useful words, *system* is a widely used term across many fields, often carrying qualifying words, such as health system, library system, or as a qualifier for another term, such as system biology or system ecology. Although retaining its core meaning throughout, each application of this very versatile word usually carries with it additional nuances that give it specific meaning and usefulness within that application. Some systems approaches, such as general system theory (Bertalanffy 1962) focus on holistic behaviours at system level. Service is fundamentally concerned with system level behaviours, at the level of the outputs to the customer. Because of its focus on system level behaviours, general system theory is sometimes cast as an antithesis of scientific reductionism.

The system engineering method is *both reductionist and holistic*. This is an important point. Product design requirements are reduced through several levels to component requirements, always taking into account the interfaces that influence behaviours at each level (and thus ultimately at the system level). The system design has meaning at every level, and is rigorously verified at every level. Furthermore, the physical instantiation of the design is integrated from its components and at each level the physical design is validated by test. The devolution of requirements is often not visible to observers external to the system engineering process (hence is poorly understood), but until this reductionist phase is complete, the highly visible physical integration and test phases cannot commence.

The system engineering process is often symbolised by the icon "V" that illustrates both reduction and integration. From the top left of the "V", system level requirements are reduced to fundamental units of design; they are then integrated through a verification and test process upwards to the top right. The physical instantiation of the design is integrated as the deliverable system. Understanding the properties and behaviours of subsystems is critical to managing risks to performance at the level of the delivered system. Product design is tightly choreographed by the system engineering process, and is closely coupled with the processes of business management (especially project management). The system engineering process enables comprehensive budgets and schedules to be developed and risks to be identified and mitigated; system engineering and project management are intertwined in execution.

When invoked in the business literature (including the services literature), system approaches usually focus on the properties and behaviours of the whole system. It is when we set out to describe product-service systems that systems terminology becomes challenged. For service systems supporting safety–critical technical products (such as a fleet of aircraft or power generation assets), confusion over language must be assiduously avoided. The system engineering language is the language of the technical regulatory environment and it is reflected in the contracting environment. It is the language of customer requirements and of project management. It is ubiquitous in the defence sector. The integrity of the language is critical to product safety as well as product performance, and understandably, there exists an inertia of deeply embedded meanings.

Although diverse and even intangible constraints may be incorporated through the requirements process, its default style is to consider constraints in product (machine) terms. An example is human factors engineering (Stanton et al. 2005), where product design is influenced by ergonomic requirements, both physical and cognitive. Product performance and other requirements are elicited from the customer before commencing design. Griffin and Hauser (1993) and Blanchard (1998) describe requirements as the "voice of the customer", enabling designers to

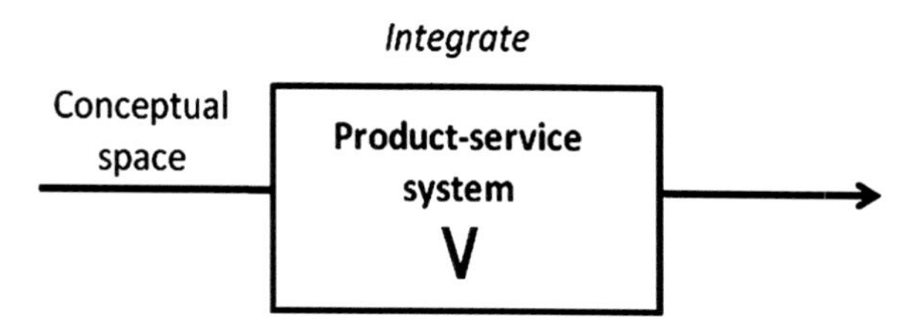

Fig. 21.1 Top level depiction of the design of a product-service system. Design occurs in conceptual space; its physical realisation is inferred rather than explicitly depicted here. "Integrate" is used as the primary verb describing the design activity. "V" is a symbol for the system engineering process. It is also symbolic of "Voice of the customer"

understand customer intentions. Mostly it is an early activity in the design process, enabling a requirements baseline to be struck. Once the baseline is set, the "voice of the customer" becomes an historical recording. This is very different to service, where there is a continuing dialogue between the customer and service personnel.

Here we focus on the design of the product-service system which comprises the product and the technical service system that supports it. The act of service delivery will be considered in a later section of this chapter. The system engineering approach is as valid for this product-service system as it is for the product system and is routinely applied. In Fig. 21.1 where the design of a product-service system is symbolically depicted, the same symbol "V" is used to describe the design process in conceptual space (in contrast, service is conceived in social space). "Integrate" is used as a verb to capture the idea of assembly of the product-service system from its many product and process components. "V" is also conveniently symbolic of "voice of the customer", the customer's requirements for the design.

There are numerous published variations on the generic system engineering process in product design (and undoubtedly even more unpublished variations), but the ideas of eliciting and evaluating customer requirements, developing a design concept, validating the design against the requirements and integrating the design are all core. The generic process uses a defined and structured language, supported by project dictionaries, to provide an appropriate degree of linguistic precision. In contrast to engineered products, natural systems such as biological systems (for example) are not "constructed" or "integrated" in the same sense as product. The engineering language of construction is not paralleled in the system language employed for natural systems. The systems biology language is predominantly focused on the need to preserve existing system level behaviours, rather than to construct them.

21.3 Case Example of a Complex Sustainment Solution

This case example is used as an illustration of a complex sustainment system. The Lead-in-Fighter In-Service Support Contract for the Royal Australian Air Force (RAAF) is a performance-based contract involving the sustainment of an aircraft fleet and requiring availability of mission-ready aircraft on a daily basis. It is similar in many respects to the UK MoD Tornado support contract (NAO 2007).

This contract is for the sustainment of the Hawk fast jet training fleet operated by the RAAF at bases on the east and west coasts of the continent. Under the terms of the contract, the contractor provides:

- Deeper maintenance for on- and off-aircraft equipment, at two bases
- Engineering services for management of the aircraft to military airworthiness standards
- Logistics supply support

There are several output measures of contract performance, the principal measure being the availability of a contracted number of mission-capable aircraft to meet the RAAF's daily flying requirements. The contract is relevant here because of the extent of the skilled engineering and maintenance operations, the nature of the availability measures and the nature of the relationships between the stakeholders: contractor, customer and the independent military airworthiness regulator. As an early performance-based contract, it was established in a "product thinking" environment within both the customer and contractor communities and has to consciously incorporate "customer-oriented thinking".

The role of the independent airworthiness regulator should be noted. While it is not party to the contract, it is a stakeholder in the technical systems aspects of the solution and places hard constraints on the solution space. It functions as an independent authority, setting engineering and maintenance standards that are applied to both parties in the contract. It is typical for safety–critical engineering sustainment systems to have an independent regulator as a stakeholder.

The original design for the Hawk training aircraft was conducted in the UK many years prior to its service in Australia, although like many aircraft the type has been subjected to continuing development over many years. The variant in-service with the RAAF was developed specifically to meet its requirements as a lead-in fighter trainer. Engineering support is provided by a small team of engineering personnel in Australia who have access to a larger engineering support organisation in the UK. All maintenance and logistics support is delivered on-site in Australia at the end of a global supply chain.

In this example of a complex sustainment solution, teams of highly skilled personnel function in tightly controlled and highly regulated environments to deliver the contracted outputs. Some of these personnel have "back office" roles with little direct contact with the military customer, while others ("front office" personnel) have frequent and crucial contact. Project managers, senior engineers, maintenance managers and integrated logistics personnel have direct customer contact. Front office personnel requires appropriate social skills to perceive the customer's intentions and to manage the complex relationships among the stakeholders. But back office or front office, all are part of the delivery team.

21.4 The Social Construction of Service

Personnel responsible for service delivery in the case example typically employ a social language of description for service activities. They may talk of "a journey with the customer", "an evolving relationship", "understanding the customer's thinking", or similar term. These terms are suggestive rather than precise and are in contrast to the structured and categorised descriptions necessarily employed in the design of the aircraft product-service system.

21.4.1 Importance of the Social Dimension

If product system design requires the "voice of the customer" to be heard, then service requires the "mind of the customer" to be understood. Many contractor "front office" personnel in the case example are in daily face-to-face contact with their customer counterparts. This may be to manage defined tasks and plans that have been routinised (such as flight training missions), or to negotiate variations to standard programs (such as fleet maintenance schedules), or to review status of major work tasks and plans. Arisings that occur from routine operations may require the attention of both contractor and customer personnel to devise work-arounds; these may have flow-on effects to other standardised tasks that will then also require attention. Especially in the defence sector, the customer may impose urgent (or surge) operational requirements (National Audit Office 2007) that necessitate significant changes to the design of the sustainment solution, possibly requiring suspension of an existing solution element and inclusion of a new element. To meet these urgent needs, there may be insufficient time to meticulously define the requirements in the usual manner of system engineering; in these circumstances it is essential that the contractor be able to "read between the lines" of the sparse urgent requirement. When the relationship is mature and validated, the contractor can rapidly infer the customer's intentions and act accordingly.

As social beings, we create representations of the minds of others based on schemas of our own mind (Rizzolatti et al. 2001; Frith 2002; Gallese et al. 2004). The brain is an organ of adaptation, and developing a mental model the mind of others is part of its work; this model is sometimes called "theory of mind". Neural measurements have provided clear evidence that our capacity to develop an image of the mind of others is related to our capacity to develop an image of our own mental processes (Decety and Chaminade 2003). The fact that a mind can develop a model of itself may be the basis of consciousness, and thereby the basis of social cognition (Damasio 2000); without a sense of self it is impossible to develop a sense of other. Building representations of the minds of others incorporates not merely their focus of attention, but also their desires and intentions, beliefs, attitudes, memories, emotions, patterns of thought and perceptions. Theory of mind is mindful of the future consequences to others of current actions. Insensitivity to consequences is an indicator of an undeveloped theory of mind, or an immature relationship.

Living organisms move with intention and by possessing a theory of mind can build maps of intention of others, enabling them to attune and adapt to their perceived internal states (Siegel 1999). Secure social relationships are an essential enabler of this construction and develop when a state of trust exists. Trust is a mutual state; it can be expressed as confidence in the intention of others, and, in the context of contracts, supported by sustained prior performance and behaviours. The social aspect of relationships acknowledges the influence of emotional processing together with abstract reasoning; cognition and emotion are not independent and each influences the other (Damasio 1996). In most instances of typified by the aircraft availability example, the execution of relationship tasks requires many-to-many—as opposed to one-to-one—relationships. Teams of people create the relationship structure (relationships between networks/groups), although of course it incorporates many one-to-one relationships. Extension from one to many is common in cognitive science. For example, Hutchins (1995) in a detailed study of team performance in maritime navigation tasks defined the boundary of the cognitive unit as the team, not the individual. The team was evaluated as a socially distributed system of cognition.

Milo et al. (2004) proposed the idea of superfamilies of networks (sociological, biological and technological) and explored patterns within the networks—"network motifs". In biological systems, recurring patterns were associated with local circuits to perform key information processing tasks. Similar pattern behaviours were found in all three types of networks, but it is this feature in the sociological network type that is relevant here. Greenfield (2008) observed certain aspects of functional scalability between individual neurons, individual brains (networks of neurons) and groups (networks of brains). Cozzolino (2006) has made similar observations of cascading networks in the context of psychotherapy, introducing the idea of the "social synapse" to describe communication between brains. Sensing across the social synapse enables the construction of a theory of mind. This functional scalability enables the understanding of human relationships at the level of the mind-to-mind dyad to be applied at the level of teams.

21.4.2 Design in Social Space: The Creation of Expert Teams

The delivery of service is achieved by integrated teams of skilled personnel. In the aircraft availability example, these teams comprise personnel from many disciplines, including engineering, maintenance, logistics, supply chain, contract management and project management. They employ the product-service system in combination with their domain expertise and social skills.

Effective teams exhibit socially distributed cognition. Hutchins (1995) conducted his extensive study of team expertise and performance in the framework of social cognition, and in the complex context of the workplace characterised by

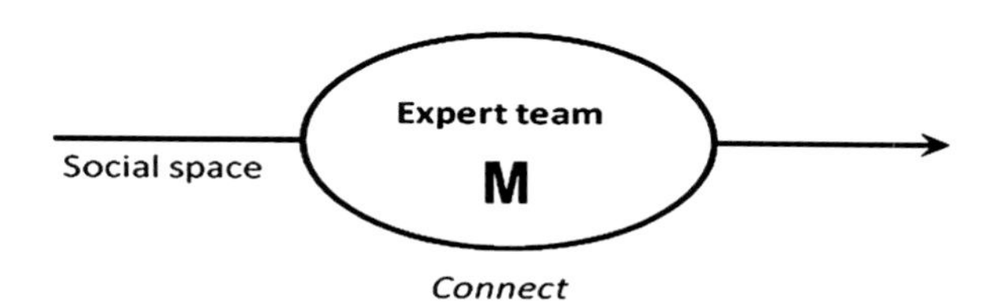

Fig. 21.2 A top level depiction of design in the social space. The symbol "M" represents the primary task to perceive the "Mind of the customer". "Connect" is the primary verb describing the team activity. A *rounded shape* distinguishes it from system design in the conceptual space (Fig. 21.1) which uses *squared corners*

changing and stressful circumstances. Not all teams composed of experts become expert teams (Hackman 1990), so clearly there is more to effective teams than individual expertise which is where the concept of socially distributed cognition becomes important. Neuroimaging has provided supporting evidence that social cognition results in altered neural connections at the level of binary (one-to-one) relationships; the progression to relationships among many is a natural extension. Salas et al. (2006) have summarised evidence from studies of teams in business, aviation, healthcare and military environments. They categorised the characteristics of expert teams as follows:

- Expert teams hold shared mental models and anticipate each other's needs
- They optimise resources by learning and adapting
- They have clear roles and responsibilities
- Expert teams have a clear, valued and shared vision
- They engage in a cycle or discipline of pre-brief, performance and de-brief
- Expert teams have strong team leadership
- They develop a strong sense of collective, trust, teamness and confidence
- They manage and optimise performance outcomes
- Expert teams cooperate and coordinate

This instructive list demonstrates a strong sense of reflection, assessment of the states of others and adaptation to those states. This summary does not specifically address the states of those outside the team (for example, an external customer team) but the ability to develop representations of other minds is generic. This is then coupled with the concept of social cognition that enables a shared understanding to be achieved. Arguably the most important skill set of the team comprises the social skills of attunement and adaptation to the intentions of others. This is supported by observations and anecdotal evidence from industry practice and the case example.

Thus, the primary hypothesis offered here is that, for complex engineering systems, design in the social space is about the creation of expert teams, and the primary skill set of expert teams is the ability to develop a "theory of mind" of others. For this reason, in Fig. 21.2 the symbol for design in the social space is chosen to be "M", for "Mind of the customer". This has some symmetry with the symbol "V" employed in system design which represents the "Voice of the customer". The team "connects" with the cognitive state of the customer, attunes and adapts its actions to the customer's perceived intentions.

21.5 Service Performance: Harmonising the Social Space With the Conceptual Space

Service performance via a heterogeneous socio-technical system combines ("harmonises") the intangible social cognition of the service team with the tangible product-service system to achieve the solution. A symbolic representation of harmonisation is depicted in Fig. 21.3. Some members of the service team in the case example speak of judgment and creativity as critical enabling skills but these skills often cannot be reliably replicated elsewhere as they are context-dependent. They must always be used in conjunction with highly developed technical skills.

This chapter offers the idea of connecting with and modelling the minds of others as central to service thinking for complex engineering service solutions. It enables teams to attribute desires and intentions to others and to evaluate and explain their likely actions. In a repeating cycle of attunement, evaluation and adaptation, teams learn to anticipate customer aspirations and can adapt in advance. When the cycle becomes habitual, backed up by technical performance (delivering value), the customer develops confidence in the values (future performance) of the team and a state of trust is achieved. In this sense, service satisfaction may be described as *value underpinned by values*.

Three phases can be synthesised in the construction of representations of another's mind for the purpose of deciding on future courses of action (after Siegel 2007):

- Attunement—this involves orienting the focus of attention on the internal world of another, free of judgmental bias, to perceive signals relating to their intentions. These intentions may then be mirrored internally. Interaction styles may be classified as: mature/regulated; disorganised; ambivalent; dismissive. A mature/regulated style is essential if signals are to be correctly sensed. Effective attunement provides motivation and energy for downstream action.
- Evaluation—evaluation is a deliberative conceptual activity that appraises the received perceptions to classify or categorise likely responses into schemas of action. Expert teams will have established patterns of practice that facilitate rapid and routine (but effective) responses, but the evaluation must consider whether the required response fits the pre-existing schemas. If not, then adaptation is required.
- Adaptation—when the need for change is identified, the team must then re-align its activities accordingly. Expert teams will adapt quickly to achieve resonance with the customer's intentions, although a degree of negotiation may be required.

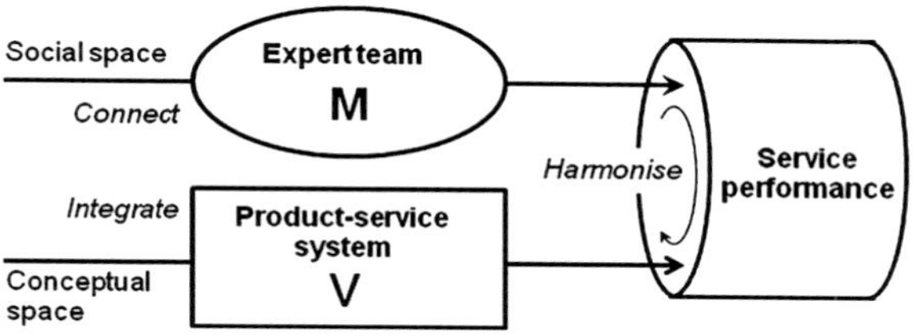

Fig. 21.3 Combining "mind of the customer" (M) with "voice of the customer" (V): a symbolic depiction of solution delivery as actions *Integrate/connect/harmonise*. A situated solution in a field setting would be a chain of recurring activities symbolised in the figure

Effective adaptation requires persistence beyond the initial connection so that harmonisation occurs in the long term.

The idea of Fig. 21.3 recurring on a regular (even daily) basis to generate a chain of attunement, evaluation and adaptation has some similarity with "appreciative systems" described by Checkland and Casar (1986) and attributed to Vickers (1965). Vickers referred to the interacting flux of events and ideas over time (a "two-stranded rope"). The social features of appreciative systems highlighted by Checkland (2000) are judgment and relationship management. Checkland separates judgment into two aspects: judgments about reality (the world in being), and judgments about values (distinguishing good from bad). Judgment in both senses is present in the harmonisation chain, but arguably theory of mind is a richer description than relationship management. Both are living, open systems.

21.6 The Need for Service Thinking as a Style Distinct From System Thinking

We have developed, at least an initial argument, that the success of service systems design and operation is as much dependant on consideration of the social dimension as the more conventional conceptual approach using system engineering methods. This is of critical importance to ensure that the expert team is configured to understand the "mind of the customer" and hence deliver on the expectations held in the customer's own mental model of service excellence. It is proposed that an extension of current systems thinking and system engineering cannot for the foreseeable future successfully address this need as the reductionist method is too deeply embedded in the development of complex systems. A new complementary thinking style needs to be developed to design expert teams able to achieve service excellence in the mind of the customer.

The president of Fuji Xerox Co. has been quoted in the daily business press as acknowledging the difficulties in trying to shift his business model to a service-oriented company: "it is a struggle to shift his back office experts from their customary focus on manufacturing QCD (quality, cost, delivery) and indeed to shift them out of the back office altogether. He wants his experts in the sales force, developing solutions for clients and customers, and understanding that their copier manufacturing company is becoming a service provider" (Alford 2009). There is anecdotal evidence that other companies making similar shifts are experiencing similar challenges.

It is difficult to change patterns of activity when existing schemas have served well over long periods. Such is the case with system thinking as the dominant cognitive style in the domain of complex product design implemented in system engineering: it is too deeply embedded to support cultural change through incremental development. The intention is not to reduce its effectiveness, but to complement it with a distinctively different style. Making yet another adjustment to the plethora of existing instantiations will probably mean that, in professional practice where it really counts, its value will be depreciated merely by association with the familiar. Variations and enhancements to such a dominant style, no matter how important, tend to be subsumed within entrenched patterns of practice. If we contend that service thinking is indeed critical to service delivery, then the lesson from this account is clear: it must be offered as a distinctive style that stands on its own, clearly distinguished from system thinking which, for system designers, is equated to system engineering. Education is easier to achieve than re-education.

For the purposes of contrasting the two styles, systems thinking, driven from the product paradigm, is characterised by:

- A product approach, with the customer at arm's length
- Creation of a technical design solution and is an episodic activity
- Segregation of solution design from delivery
- Understanding of how things connect, incorporating mainly objective requirements
- Incorporation reductionist method, supporting both holistic and elemental views

At a basic level, service thinking requires the service provider to account for customer thinking, and availability contracts will usually (even if indirectly) incentivise this behaviour. Managing the customer relationship becomes a critical success factor in these contracts and it is here that social features (beliefs, attitudes, memories, emotions, etc.), which cannot be encoded in product design, are highlighted. Consequently, and as a comparator against systems thinking, service thinking is characterised by:

- A social activity wherein the customer co-creates and co-delivers the solution
- Describing a business solution within which design is a continuing activity
- Integrating design and delivery which are closely coupled
- Understanding how minds connect and incorporates social content
- Not yet amenable to method, supporting only holistic views

These comparisons of thinking styles are heuristics. Product system practitioners are accustomed to using heuristics, which is probably why they find these contrasts to be useful. Heuristics are common in practical system architecting as noted earlier. Service thinking has strong social and business flavours. It recognises affective content in solution delivery, something that does not fit comfortably with system thinking which presents as an entirely dispassionate and objective process, yet, the authors have found that among experienced systems practitioners, when offered as a style distinct from system thinking, service thinking has achieved acceptance.

The concept of co-production of service drives the need for service thinking. In the context of complex engineering service systems, co-production is a continuing activity where the contractor team repeatedly engages with the customer to determine customer intentions. In contrast, requirements elicitation is an episodic activity at arm's length, rendering design requirements as historical records that may be occasionally updated. Compared with product design, service is forward-looking and gauges the intentions of the customer. It achieves this in the social space where the service team seeks to connect with the mind of the customer. This is a rich idea, involving both the person-to-person and team-to-team dyads, as well as recognising the non-deliberative functions of the mind, including emotion. Service design is about creating expert teams that generate shared awareness of customer intentions, and acting on that shared cognition. The generation of trust in service provision requires that the team deliver service performance, but it also requires that the customer has confidence in the future performance of the team, which in turn requires discernment of customer intentions.

Consideration of the case example offers some basic insights into the application of these thinking styles to practical service systems. The face-to-face protocols of the aircraft availability contract in the case example are centred on technical processes which utilise system thinking, but the relationships extend beyond that. A cultural appreciation is required for customer fleet management, maintenance and engineering operations, all of which must be conducted in strict compliance with the regulations of the military technical airworthiness management system. Contact between the contractor and customer occurs daily and the social demands in relationship management are important. The cultural and social demands of the contract are very real and critical to service satisfaction.

21.7 Chapter Summary and Questions

Based on professional experience and observations from current practise, this chapter has offered some practical insights regarding the nature of service system design for complex services. Some pragmatic suggestions can be made for the design of future complex service systems to more nearly achieve "service excellence" in the mind of the customer:

- A product-service system must have a clear "chief architect" who is attuned to working in the social and conceptual space, and who is equipped to understand the mind of the customer and contractor

- The service solution must be co-developed with the customer with openness and honesty so that both customer and contractor can understand each other's mind. The expert (delivery) team needs to include the customer as a joint enterprise and needs to be initiated when the service need is identified—it is difficult to see how successful service can be procured through competition
- The concept of measuring "service excellence" through system engineering methodology is illusory—the social dimension needs to be considered, possibly using team behaviours as an indicator

The key concepts of social construction of service, namely, service thinking, designing expert teams, and discerning the mind of the customer, offer a rich field for future research. The chapter suggests several research questions:

- Service thinking: what are the differentiators for service thinking which enable excellence in service delivery?
- Expert teams: how do expert teams harmonise the social and system dimensions in complex sustainment solutions, what are the characteristics that enable it, and how can they be embedded in practice?
- Theory of mind: how do teams capture and use the mental models of others, particularly with regard to intentions, to enable mutual trust and service satisfaction?

Further work needs to be done to review and re-analyse case study material to provide more specific evidence and further practitioner interviews are required. It is nevertheless proposed that service thinking is sufficiently different from system thinking (and important) to warrant a separate identity. Achieving its acceptance by practitioners will still require patient explanation and example. As former industrial practitioners, the authors recommend the pragmatism of this approach.

References

P. Alford, Copy giant sees sense in services. The Australian, 10 June 2009 p. 26

P. Allen, S. Higgins, P. McRae, H. Schlaman, *Service orientation* (Cambridge University Press, Cambridge, 2006)

T. Alonso-Rasgado, G. Thompson, A rapid design process for total care product creation. J. Eng. Des. **17**(6), 509–531 (2006)

T. Alonso-Rasgado, G. Thompson, B.-O. Elfstrom, The design of functional (total care) products. J. Eng. Des. **15**(6), 515–540 (2004)

J.C. Aurich, C. Fuchs, C. Wagenknecht, Life cycle oriented design of technical Product Service Systems. J. Clean Prod. **14**, 1480–1494 (2006)

L. Bertalanffy, *Modern theories of development: An introduction to theoretical biology* (Harper, New York, 1962). (originally published in 1933 in German)

W.E. Bijker, *Of bicycles, bakelites, and bulbs: Toward a theory of sociotechnical change* (MIT Press, Cambridge, 1995)

W.E. Bijker, P.T. Hughes, T.J. Pinch, *The social construction of technological systems* (MIT Press, Cambridge, 1987)

B.S. Blanchard, *Logistics engineering and management*, 5th edn. (Prentice Hall, Upper Saddle River, 1998)
H.-J. Bullinger, K.-P. Fahnrich, T. Meiren, Service engineering—Methodological development of new service products. Int. J. Prod. Econ. **85**, 275–287 (2003)
P. Checkland, Soft systems methodology: A thirty year retrospective. Syst. Res. Behav. Sci. **17**, S11–S58 (2000)
P. Checkland, A. Casar, Vicker's concept of an appreciative system: A systematic account. J. Appl. Syst. Anal. **13**, 3–17 (1986)
L. Cozzolino, *The neuroscience of human relationships* (WW Norton & Company, NY, 2006)
A. Damasio, The somatic marker hypothesis and the possible functions of the prefrontal cortex. Phil. Trans. Roy. Soc. Lond. Ser. B (Biol. Sci.) **151**, 1413–1420 (1996)
A. Damasio, *The feeling of what happens: Body, emotion and the making of consciousness* (Vintage Books, London, 2000)
J. Decety, T. Chaminade, When the self represents the other: A new cognitive neuroscience view on psychological identification. Conscious Cogn. Int. J. **12**, 577–596 (2003)
C. Frith, Attention to action and awareness of other minds. Conscious Cogn. Int. J. **11**, 481–487 (2002)
V. Gallese, C. Keysers, G. Rizzolati, A unifying view of the basis of social cognition. Trends Cogn. Sci. **8**(9), 396–403 (2004)
S. Greenfield, *The quest for identity in the 21st century* (Hodder & Stoughton, London, 2008)
A. Griffin, J.R. Hauser, The voice of the customer. Mark. Sci. **12**(1), 1–27 (1993)
J.R. Hackman (ed.), *Groups that work* (Jossey-Bass, San Francisco, 1990)
E. Hollnagel, D.D. Woods, Cognitive systems engineering: New wine in new bottles. Int. J. Man. Mach. Stud. **18**, 583–600 (1983)
E. Hutchins, *Cognition in the wild* (MIT Press, Cambridge MA, 1995)
S. Johnstone, A. Dainty, A. Wilkinson, In search of "product-service": Evidence from aerospace, construction and engineering. Serv. Ind. J. **28**(6), 861–875 (2008)
S. Johnstone, A. Dainty, A. Wilkinson, Integrating products and services through life: An aerospace experience. Int. J. Oper. Prod. Manag. **29**(5), 520–538 (2009)
M.W. Maier, E. Rechtin, *The art of systems architecting* (CRC Press, Boca Raton FL, 2002)
R. Milo, S. Itzkovitz, N. Kashtan, R. Levitt, S. Shen-Orr, I. Ayzenshtat, M. Sheffer, U. Alon, Superfamilies of evolved and designed networks. Science **303**, 1538–1542 (2004)
N. Morelli, Developing new product service systems (PSS): Methodologies and operational tools. J. Clean Prod. **14**, 1495–1501 (2006)
National Audit Office, Transforming logistics support for fast jets. Report by the Comptroller and Auditor General, UK. HC 825 Session 2006-2007 (2007)
A. Neely, Exploring the financial consequences of the servitization of manufacturing. Oper Manag Res. doi: 10.1007/s12063-009-0015-5 (2008)
I.C.L. Ng, R. Maull, N. Yip, Outcome-based contracts as a driver for systems thinking and service-dominant logic in service science: Evidence from the defence industry. Eur. Manag. J. **27**, 377–387 (2009)
R. Ramaswamy, *Design and management of service processes* (Addison-Wesley, Reading, 1996)
J. Rasmussen, A. Pejtersen, L. Goodstein, *Cognitive systems engineering* (Wiley, New York, 1994)
E. Rechtin, *Systems architecting, creating and building complex systems* (Prentice-Hall, Englewood Cliffs, 1991)
G. Rizzolatti, L. Fogassi, V. Gallese, Neurophysiological mechanisms underlying the understanding and the imitation of action. Nat. Rev. Neurosci. **2**, 661–670 (2001)
T. Sakao, Y. Shimomura, Service engineering: A novel engineering discipline for producers to increase value combing service and product. J. Clean Prod. **15**, 590–604 (2007)
E. Salas, M.A. Rosen, C.S. Burke, G.F. Goodwin, S.M. Fiore, in *The making of a dream team: When expert teams do best*, ed. by K.A. Ericsson et al. Handbook of expertise and expert performance (Cambridge University Press, Cambridge, 2006)

M. Shaw, D. Garlan, *Software architecture: Perspectives on an emerging discipline* (Prentice-Hall, Englewood Cliffs, 1996)

L.G. Shostack, How to design a service. Eur. J. Mark. **16**(1), 49–63 (1982)

D.J. Siegel, *The developing mind* (The Guildford Press, New York, 1999)

D.J. Siegel, *The mindful brain* (W W Norton & Company, New York, 2007)

N. Stanton, P. Salmon, G. Walker, C. Baber, D. Jenkins, *Human factors methods: A practical guide for engineering and design* (Ashgate Publishing, Aldershot, 2005)

US Department of Defense, DoD architectures framework version 2.0. www.defenselink.mil/cio-nii/docs/DoDAF%20V2%20-%20Volume%201.pdf. Accessed 6 Nov 2009 (2009)

G. Vickers, *The art of judgement* (Chapman and Hall, London, 1965)

S. Williams, Business process modeling improves administrative control. Automat Dec, pp 44–50 (1967)

Chapter 22
Towards Integrative Modelling of Service Systems

Peter J. Wild

Abstract The chapter is concerned with presenting an approach to the high-level and integrative modelling of service systems. The challenge of such an approach is to provide a high-level view, without overloading the representation with too many theoretical concepts, or too much detail of the final implementations of a service system. The concern is with feasible or desired configurations and the potential trade-offs implied by specific configurations; rather than an optimised specific implementation.

22.1 Introduction

Recent years have seen growing interest in services that are complex, enacted over long time-periods, and interrelate and interact with complex products (e.g., Mont 2002; Goedkoop et al. 1999). In the UK, projects such as S4T, IPAS and KIM have examined products and services in combination. Monikers such as (Industrial) Product-Service Systems (Goedkoop et al. 1999; Aurich et al. 2007) and Functional Products (Alonso-Rasgado et al. 2004) attempt to get to grips with the combined nature of product and services. Within industry monikers such as sustainment, supportability, Power-by-the-Hour and related terms such as Smarter Planet[1] indicate practitioner and corporate interest in this area.

[1] Power by the hour is copyright Rolls-Royce and Smarter Planet is copyright IBM.

P. J. Wild (✉)
Psychology and Communication Technology Laboratory,
Department of Psychology, Northumbria University,
Northumberland Building, Newcastle upon Tyne, NE1 8ST, UK
e-mail: peter.j.wild@gmail.com

I. Ng et al. (eds.), *Complex Engineering Service Systems*, Decision Engineering,
DOI: 10.1007/978-0-85729-189-9_22, © Springer-Verlag London Limited 2011

The contemporary services landscape can be seen to be rich, diverse, and fragmented (Chesbrough and Spohrer 2006; IfM and IBM 2008), owing in part to the multiple relevant academic and practitioner disciplines; rapid growth in the research into the area; and an array of existing techniques that are applicable to services (Wild et al. 2009c; IfM and IBM 2008).

Previously we have argued (Wild et al. 2009a) that the abstractions these approaches embody are not enough to demonstrate how products, services and people work together in different ways to provide value, or how we can distinguish different kinds of service systems. As part of this argument, we demonstrated that a range of common approaches did not consider all aspects that could be identified in a service system (Wild et al. 2009a).

We argue that what is missing in these approaches is an attempt to identify recurring key components across different services systems and consideration of how these components interact and relate to each other in a systemic matter. The concern of this chapter to is to present a developing approach to the integrative modelling of service systems (Wild et al. 2009a, c; Wild 2010). The approach aims to avoid viewing service delivery through a 'single' lens, for example as a technology management problem; as an information management problem or as a human resources organisational design problem. Rather, modelling service systems are viewed as a systemic activity that needs to glue or bring together different components.

The challenge of such an approach is to provide a high-level view, without overloading the representation with too many theoretical concepts, or too much detail of the final implementations of a service system. The concern is with feasible or desired configurations and the potential trade-offs implied by specific configurations; rather than an optimised specific implementation.

This chapter presents a conceptual framework for the Integrative modelling of Service Systems. The framework known as the Activity Based Framework for Services (ABFS) was first presented as an approach to relating a number of disparate and cross disciplinary definitions of service, such as Service Blueprinting, the Service-Dominant Logic and Product-Service Systems (Wild et al. 2009a). Since then it has been used to represent high-level models of service systems (Wild 2010), and it is this capability that is of concern to this chapter.

The chapter continues emphasise on thought the following sections:

- Section 22.2 provides a high-level overview of what it meant by Integrative Modelling
- Section 22.3 introduces the Activity Based Framework for Services modelling framework
- Section 22.4 covers additional elements of the ABFS approach, including: Different Types of Services and Domains; the relationship between a core and service system; The Emergent and Co-Created Nature of Value
- Section 22.5 considers various sources of complexity
- Section 22.6 reviews the chapters and relates the ABFS to the Common Integrative Framework (see Chap. 23).

22.2 Integrative Modelling

Different disciplines, technical functions and organisations can use the same terms in different ways. As a term critical to this chapter, we present a brief overview of modelling, and what we mean by Integrative Modelling.

After Collins we maintain that a model is

> a structure in one domain used to represent an object in some other domain, for the purpose of understanding or controlling it (1994).

As such, an integrative service model is a representation of a service system that aims to provide a high-level representation of the key entities within a service system, including products, service activities information, organisational roles and structures, service goals and the values by which people make effective judgements.

Curtis et al. (1992) note that models embody some kind of *abstraction* or *idealisation*. Frigg and Hartmann (2006) distinguished between Aristotelian and Galilean idealisation:

- Aristotelian idealisation strips away properties from the represented entity that are believed to be irrelevant to the problem at hand.
- Galilean idealisation involves deliberate distortions; for example, physicists build models consisting of point masses moving on frictionless planes; economists assume that agents are omniscient; biologists study isolated populations.

Both forms of idealisation can be present within modelling approaches, and are often difficult to discern. The lack of exceptions in many task and process modelling approaches could reflect a concern for high-level abstractions, or it could reflect a deliberate distortion to build normative and ideal models—for example, training purposes.

Thus, motivation for modelling has a role to play; Minsky (1965/1995) takes a pragmatic view in arguing a model of something is useful when it helps their users in resolving questions they ask themselves about the thing modelled. If the aim is to compare models across different domains then Aristotelian idealisation strips away elements that are not being compared. If the aim is to model the basics of a task for training then Galilean idealisation makes sense; rather than overloading the trainee with exceptions, the Galilean method allows simplifications for training purposes.

We can also distinguish between *first-* and *second-class* modelling concepts (Wild et al. 2009a).

- *First class* concepts are those concepts considered 'native' to the modelling approach; for example, tasks/processes and parameters in processes modelling.
- *Second class* concepts are elements which are not supported but are somehow represented through annotation and the adoption of naming conventions; for example, prefixing DOC or DB on process modelling parameters to indicate that the information comes from a document or database.

Modelling approaches of the same 'type' can embody different *perspectives* (e.g., Melão and Pidd 2000). These perspectives can be formalised into distinct theoretical positions or can exist more informally within different communities of practice. Examples of perspectives include:

- the distinction between hard and soft systems modelling (Mingers 2006; Checkland and Poulter 2006; Jackson and Keys 1984);
- different positions on process modelling such as: deterministic machines, complex dynamic systems; interacting feedback loops and socially constructed entities (Melão and Pidd 2000).

As noted, modelling approaches are defined as embodying abstractions; however, they differ in the level of detail they represent (Frigg and Hartman 2006). Compared to many modelling approaches, simulation models require a lot of detail about the domain being modelled but still make idealisations and abstractions about it. Queues and behaviours are considered a core concept in many simulations, but the size, shape and colour of the buildings are generally ignored.

In relation to abstraction, Daniels (2002) reminds us of the classic software and systems engineering distinction between *Conceptual*, *Specification* and *Implementation* models and provides simple definitions.

- Implementation model: how something is implemented (e.g., specific staff in specific roles)
- Specification model: a more abstract model that explains what should be implemented (e.g., specific roles);
- Conceptual model: describes a situation of interest in the world (e.g., broad roles needed).

The framework presented in this chapter is intended to be conceptual, which contrasts with the work of Shimomura and associates (Shimomura et al. 2006; Tomiyama 2005). Their approach, while still embodying abstractions, can be viewed as a specification model of a specific service system. Implementation models are particularly used in Software and are less applicable for representation of the complexity of an enacted service system and may simply replicate representations used in best practice in functions, such as Information Systems and Human Resources.

Terms such as Service Science and Service Systems represent an aspiration towards an interdisciplinary agenda for Services research and practice, recognising long-standing interest in Services across a range of academic disciplines and practitioner functions. Integrative/interdisciplinary research is an old term (e.g., Piaget 1970) that has had a resurgence of interest in the last few years (Winder 2005a, b; Szostak 2007; Tress et al. 2005).

We maintain that *Integrative Models* are representations providing abstractions that act to show the links between the different concepts in different modelling approaches. Integrative Models go for breadth rather than depth and will tend to be descriptive and systemic rather than predictive or prescriptive. So, while Information (data given meaning in context) and Actants (people and organisations) are first class entities within the ABFS, neither is explored with the depth of related work (see Chaps. 3, 10, 12, 20). Rather, as a *conceptual* representation of service systems the focus is on: the high-level systemic relationships between the different parts of a service system; the identification of different systems and subsystems; and recurring trade-offs between different possible service system configurations [see also Daniels (2002); Woodfield (1997)].

It remains an open question as to whether integrative models could operate as specification models. The ABFS was conceived as a conceptual model and ABFS models should be seen as a general set of concepts for understanding a research area, not tightly organised enough to be a full specification or implementation model. In its current stage of evolution, the ABFS embodies abstractions that would need additional translations for use as a specification of a Service System. Several outputs from S4T and other projects could serve as detailed specification models for a service system; for example, Enterprise Imaging (Chap. 3) and the 12 Box Model (Chap. 19).

22.3 The Activity Based Framework for Services

Three aspects re-occur across service definitions:

1. that services are *activities* (Hill 1977; Lovelock 1983; Vargo and Lusch 2004a);
2. that services activities can be *transferred* between people/economic units (Hill 1977; Lovelock 1983; Lovelock and Gummesson 2004; Vargo and Lusch 2004a);
3. that services exist in and interact with a *context or system* that includes people, tools, products, goals, values etc. (Lovelock 1983; Mont 2002; McAloone and Andreasen 2002; Wild et al. 2009a, c).

The ABFS builds on these three observations. Working from the view that services are consistently defined as *activities*—rather than objects or artefacts—the concepts of the ABFS are drawn from activity modelling approaches, such as task analysis, domain and process modelling and soft systems methodology. This synthesis produced a framework that can relate together the disparate streams of service research (Wild et al. 2009a) and help classify the design foci (i.e., what is being designed) of service design approaches (Wild et al. 2009c). The core concepts of the ABFS are represented schematically in Fig. 22.1, and discussed in turn.

22.3.1 *Domain*

The 'world' whose possibilities and constraints are organised in relation to specific goals. A domain is conceptualised as being composed of concrete and abstract

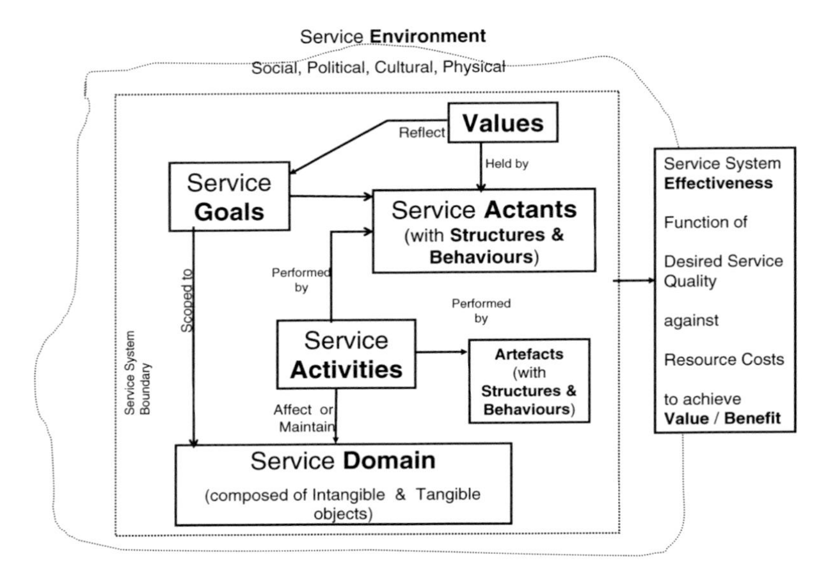

Fig. 22.1 Schematic of the activity based framework for services

objects (Dowell and Long 1998; Wild 2010). Objects are the elemental focus of services, such as heads and hairstyles or engines and aeroplanes. Objects can be abstract/intangible (i.e., informational, social) or concrete and can emerge at different levels of analysis (e.g., physical, cognitive, affective and socio-cultural). Representing domains in this manner allows for differences between different service domains and contexts to be understood and represented.

22.3.2 Goals

The specification of desired or needed changes to domain objects. They are carried out by *actants*, and sophisticated artefacts through automation, but can only be 'held' by *actants*. Classically, they are represented hierarchically, each goal mapping to a finer description of changes to a *domain* (Dowell and Long 1998; Diaper 2004). As service objects can be either concrete or abstract, goals also vary in their concreteness or abstractness. Other work suggests that goals can be heterarchical, being embedded in a complex of higher and lower level goals (Diaper 2004; Wild et al. 2004). Some activities can achieve or contradict one or more of these higher level goals simultaneously. Goals are also assumed to be public or private, as well as being cooperative (i.e., complementary) or collaborative (i.e., shared).

22.3.3 Activities

The sequence and type of actions (physical/non physical) carried out in order to achieve *goals*. Activities are concerned with changing (education, surgery, technology, upgrades) or maintaining (preventative healthcare, hardware maintenance) the states of service *domain* object attributes, when they are carried out they should achieve all, or part of a *goal*.

22.3.4 Actants

Those entities capable of carrying out activities; the term covers people, and groups of people. This element of the approach covers different teams, groups, organisation and communities as well as the overlap between them and the roles that individual and collective actants play. Key roles include service provider and service recipient, but other stakeholders are also relevant.

22.3.5 Artefacts and Technologies

The tools used to carry out activities in a *domain*. The creativity and innovation of people has enabled a huge range of technologies that enable new activities, magnify and replace human abilities and skills. People can also be in a co-evolutionary process of technological possibilities with new artefacts suggesting new activities, which suggest new artefacts (Carroll et al. 1991).

22.3.6 Values

The criteria with which judgements are made about other entities (Checkland and Poulter 2006). There is little evidence that there is a set of universally applicable values and norms for the many different stakeholders involved in activities. While the purchasing and outsourcing of services imply some kind of benefit exchange for example, saving time for cleaning services; or complementary competences for more complex services, this does not imply an automatic alignment of values. As a key component in effectiveness judgements, values affect how a system is judged by different stakeholders.

22.3.7 Environment

The world outside the service system—other than its *domain*—that has physical and socio-cultural impacts on the system, as well as being affected by it (beneficially and negatively).

22.3.8 *Structures and Behaviours*

Structures provide capabilities (knowledge, skills, information) in reference to a *domain*, while 'behaviours' are the activation of these structures to perform tasks. Resource can relate to be effort (conative), emotion (affective), physical (1986) and socio-cultural (Tiger 2000; Hall 1959; Elster 2007).

22.3.9 *Service Effectiveness*

Represented as a function of the Service Quality *goals* against the Service System setup and execution resource costs. This includes those considered as concerning the domain's reason for existence, and increasingly 'good' management also considers the wider implications for socio-cultural states, the physical environment and the affective reactions of service recipients, even in domains considered technical in nature (see Chap. 21).

22.4 Additional Elements of the ABFS Approach to Service System Modelling

22.4.1 *Different Types of Service and Domains*

The ABFS expands the domain concept beyond its traditional association with informational and physical changes (Rasmussen et al. 1994; Dowell and Long 1998) to take in affective and socio-cultural changes, in addition to the inclusion of structures and behaviours that can be assessed as conative, affective and socio-cognitive resource costs (Wild 2010). Because of these additional concepts, we are in a position to start representing significant differences between service system configurations. The patterns of how each element of the ABFS is instantiated start to highlight differences between different service systems.

- Theme parks use heavily engineered artefacts to induce visceral and emotional reactions, as well as physical changes to the client (e.g., gravitational shifts, exposure to water). The client's interest and ability to withstand such phenomena leads to variable experiences of the services (e.g., exhilaration vs. nausea).
- Film, music and entertainment use people and artefacts to produce and/or embody (i.e., record and play back) an abstract product (Hill 1999), such as a song, book, play or film to produce visceral, emotional social and intellectual reactions. Physical changes beyond the perceptual/physiological are not enacted.
- Recording, broadcasting and publishing services take abstract products (Hill 1999), and create either a physical media for the them (e.g., books, recordings) or transmit it via an analogue or digital signal.

- Maintenance services undertake a range of physical and informational activities to ensure a product runs, and can be supplied with consumables and replacements when needed.
- Education and training using artefacts such as books, paper, computers, simulations and other domain-relevant artefacts to change the service client's cognitive and physical knowledge and skills.
- Counselling services aim to change cognitive, effective, social and physical behaviours. The behaviours promoted are associated with a value set considered more functional than the one currently enacted by the client.

While artefacts and activities both provide an element of service (Vargo and Lusch 2004b), artefacts and activities provide it in different ways in different contexts (Stauss 2005).

22.4.2 The Emergent and Co-Created Nature of Value

Within Services Research a position has emerged that views benefit as being determined by the customer (Vargo and Lusch 2004a), and in turn entails the customer co-creating value with service suppliers. Vargo and Lusch presented a 10th Foundation Principle for the SDL that states "*value is always uniquely and phenomenological determined by the beneficiary* (Vargo and Lusch 2008)." This appeal to the often subjective and intersubjective nature of value is pre-dated by work in Economics such as that of the 'Austrian' school (see Menger 1871/1976, in Heskett 2009). In the ABFS, the values held by actants are a component in making evaluations of Service effectiveness (i.e., benefit). These in turn shape the public and private goals held by actants in the core and service system.

In the ABFS, value or benefit is modelled as an emergent property dependent on two or more parties carrying out activities, rather than an explicit entity within the Service System. In terms of the ABFS, this value or benefit is reliant on both parties having the *structures and behaviours* necessary to co-create value. Values are specifically included in the ABFS as a concept to indicate the criteria that people bring to bear on judgements of value. Often, values are traded off against each other; for example, a small premium for renewable energy versus lowest cost. Even this process is not always rational; impulse buys of clothing can circumvent values held; for example, to buy fair-trade or buying local produce. Approaches such as Spiral Dynamics (Beck and Cowan 1996; Cowan and Todorovic 2000) provide a deeper consideration of values that could act as a *specification* model of the different value sets held by actants involved in a service.[2] The ABFS includes values as a concept in order to model at a high level how values affect service effectiveness and how different value sets can be held by different partners, as well as the relationship between values and goals.

[2] Other value schemas have been identified (Hall 1959; Lages and Fernandes 2005).

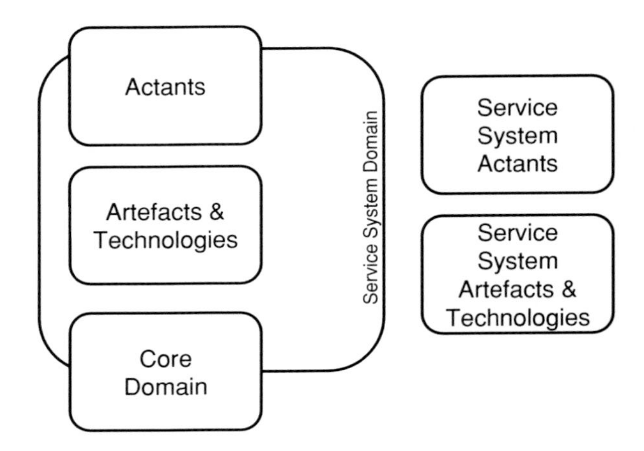

Fig. 22.2 Simplified relationship between core and service system

22.4.3 Core and Service Systems

The ABFS can be considered to outline a general high-level 'architecture' for Human Activity Systems (Checkland and Poulter 2006). Following Vargo and Lusch (2004a), we acknowledge that both products and services serve to create value for people, but in different ways and in different contexts. However, agreeing with Stauss (2005), we do not elide products and services and argue that products and services—both tangible and intangible exist within a system, and provide different and sometimes overlapping functions, depending on how the overall function is allocated to actants and artefacts (Sheridan 1988). Following on from the ABFS acting as a generic representation of an activity system, we suggest that service systems are in a fundamental relationship with a core system. A simplified view of this is illustrated in Fig. 22.2. Depending on the scope of the service design and/or contract, the domain of the service system can embrace the actants, artefacts and the informational objects of a *Core* Human Activity System. Actants would be the focus of training or education services, while maintenance services would focus on artefacts and planning support would draw upon the informational objects in the domain (e.g., fleet plans).

In this example, the goal of the core service system could be force projection (Kerr et al. 2006; Friedman 2009); the domain would be sea land and air territory, targets and the information represented about them. Artefacts could include a range of options, such as the carrier strike fleet, comprising an aircraft carrier and its air fleet, destroyer submarine support and supply vessels, as well as long distance reconnaissance artefacts (e.g., satellite, AWACS). Core system activities embrace tasks such as target identification, tracking and elimination, command and control or evacuation of personnel, and are enacted through and enabled through the artefacts of the carrier group.

Table 22.1 provides an illustrative example of an ABFS model in relation to the ATTAC and ROCET availability programmes. This is meant to be illustrative of

Table 22.1 Elements of the ABFS in relation to ATTAC and ROCET

ABFS	ROCET and ATTAC
Domain	CORE SYSTEM: Airspace, targets SERVICE SYSTEM: Physical objects such as aircraft and their major subsystems (e.g., engine, airframe, weaponry, avionics and ejector seat) COMMON TO BOTH Abstract objects such as usage patterns (e.g., Peacetime, Storage, Theatre, Fleets within Fleets). RAF staff (training, emotion and articulation work)
Goals	CORE SYSTEM: The overall goals of the RAF broken down to map to specific fastjet instantiation (e.g., reconnaissance, bombing, interception and training) SERVICE SYSTEM: The overall goals of the ATTAC/ROCET service system, both formally defined and contracted (i.e., KPIs) monitored (PIs), and less formally through ongoing discussion, collaboration and evolution
Activities	CORE SYSTEM: Maintaining the security of UK and UK-controlled/ protected airspace; reconnaissance; projecting munitions at targets SERVICE SYSTEM: Concrete and abstract activities. The former being heavily concerned with physical maintenance of the platform and the movement and storage of components and modules. The later concern flight and fleet planning, lifecycle costing, reporting and information/ knowledge management
Actants	CORE SYSTEM: Relevant RAF pilot, co-pilot. Ground crew, 1st line SERVICE SYSTEM: RAF (IPT, flight and ground crew); BAE Systems (Project (ATTAC) and Capabilities (e.g., Engineering for Support); and Rolls-Royce Project (ROCET) and Capabilities (e.g., Various service lines)
Artefacts	CORE SYSTEM: Aircraft and their major sub-systems (e.g., engine, airframe, weaponry, avionics and ejector seat), runways, weapons, fuel SERVICE SYSTEM: Ground Support Equipment, Hangers, Bays, HUMS, Specific LCC costing software. General computers and software, general artefacts, forms engine logs etc
Service effectiveness	CORE SYSTEM: % target id, % target eliminated, % successful take off and landing, availability of assets SERVICE SYSTEM: The costs of maintaining the structures and behaviours for the Tornado platform. Can be costed in financial terms, but broader issues are present, such as changes in levels of trust and openness to service innovation or taking on more non-traditional service areas.
Values	CORE and SERVICE SYSTEM: ATTAC and ROCET are driven by contracted and high-level goals (e.g., Defence Industrial Strategy). The latter are concerned with the values held by each organisation involved in the contracts, but also with the value placed upon retaining UK military and general engineering capability. Values associated with value-for-money; availability in generally; dependence/independence; interdependence; duty-of-care;
Environment	CORE and SERVICE SYSTEM: Socio-cultural: Covers the culture within the MOD, RAF, RAF-IPT and industrial partners. Attitudes to risk, the rate and reasons for changes to culture. The political attitudes to armed services; costs of armed services; risk of conflict.

the broad relationship between the framework and a domain and is not a definitive account or report of case studies of ATTAC or ROCET [see Chaps. 2–5, and (Wild 2008)].

22.4.4 Changes to Service Systems

The economist Hill was keen to stress the role of exchange in defining services, and to distinguish between activities that can and cannot be solely performed by oneself. He noted that

> if an individual grows his own vegetables or repairs his own car, he is engaged in the production of goods and services. On the other hand, if he runs a mile to keep fit, he is not engaged because he can neither buy nor sell the fitness he acquires, nor pay someone else to keep fit for him (1977)

This indicates that there are transferable activities, and that this exchange is subject to constraints, such as physical laws and socio-cultural practices. The former include Hill's exercise example along with other biological based-processes.

These activities could be modelled using the concept of Enabling activities (Whitefield et al. 1993). Enabling activities are activities that place or maintain artefacts and people, in a particular state for normal or enhanced use (e.g., exercise, food growing and preparation). For complex services, physical constraints still operate (e.g., the size and location of air or naval bases) but they are less inherently resistant to transfer than biological functions. Decisions to outsource services should centre on the nature of the outsourcing organisation's core mission and the resources needed to enact that mission. The general process of service outsourcing assumes that the external partner has structures and behaviours that enable it to provide complementary or enhanced service quality and/or lower cost. This can be due to economies of scale, reduced facilities replication costs, additional capabilities such as greater numbers of personnel and assets to deal with surges, higher levels of product knowledge or better capability to model asset usage (McIvor 2005; Mol 2006). It is also assumed that this is done in a manner that reduces factors such as trust security, or the ability of the core organisation to maintain its core mission in the face of changes.

The end of the Cold War and recent economic events have placed restrictions on military and non-military economic resources. Several key projects have been enacted to find ways of both reducing normal operating budgets and to increase asset availability. Terry et al. (2007) discuss the contracts for the support of the UK's fleet of Panavia Tornados, which is now undertaken through collaboration with industrially-based partners (BAE Systems, Rolls-Royce and additional supply chain organisations). The core activity of system's domain concerns transformations such as maintaining the security of the UK and UK-controlled airspace, engaging in reconnaissance, and, where necessary, projecting munitions at targets.

Thus, the domain embraces airspace, threats and allies, both as actual physical objects and higher-level information about them. In executing tasks in this domain, additional abstract objects are created to represent pilots, fleets and operations. One artefact to enact this mission is the Panavia Tornado fast jet. As highly complex engineered artefacts, Tornados require substantial maintenance support and upgrade, alongside routine activities such as refuelling. In recent years, large amounts of these activities have been taken on by industrial partners. RAF maintenance staff retain responsibility for day-to-day physical changes, but with the support from industrial staff in roles such as programme, supply chain and fleet management. There is a move to shared planning activities so that the same information is not generated two or three times for different parties (e.g., RAF, BAE Systems and Rolls-Royce) and so that maintenance plans do not clash for different service systems. This information is vital to the maintenance planning and execution, as are the RAF base staff and facilities used within the service system.

Using Whitefield's (1993) terminology, the RAF's previous arrangements utilised a series of enabling entities (activities, actants, artefacts) within the 'original' RAF core system (see Fig. 22.3). In contrast, the post ROCET/ATTAC has greater resemblance to Fig. 22.4, with the service sub-system becoming a service system in its own right.

Effectiveness measures such as cost and availability have changed in the new arrangement, with the former going down and the latter increasing. In contrast, some resources such as trained maintenance staff have decreased.

What is not represented in Figs. 22.3 and 22.4 are the values held by actants within different systems. Before ROCET and ATTAC, we assume that the boundaries between the core and support system were softer; the 'manager' in

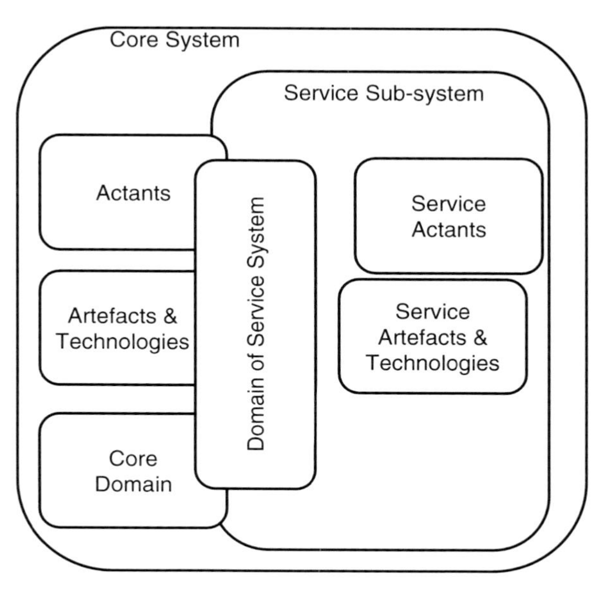

Fig. 22.3 Pre ROCET and ATTAC, maintenance is a sub-system of the core system

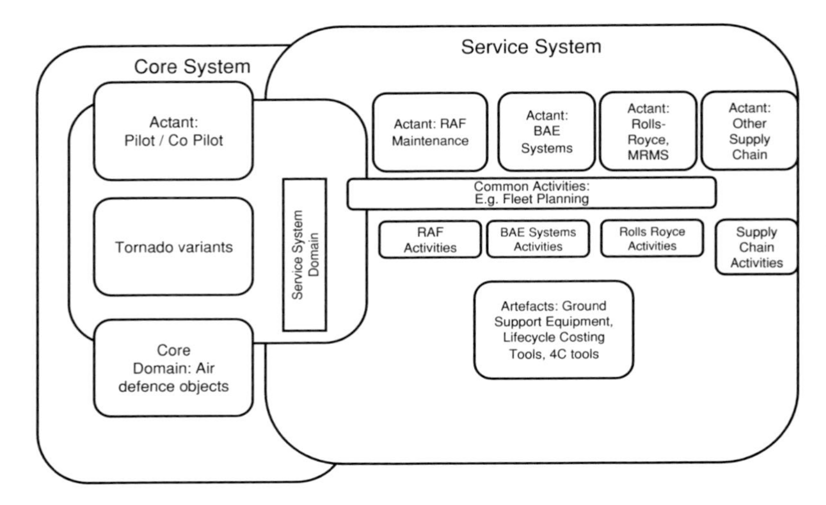

Fig. 22.4 Post ROCET and ATTAC, maintenance is a separate service system

charge of the core system was responsible for the service system. Both groups were recruited, trained and employed by the same organisation and could be assumed to work with a value set and organisational culture more closely aligned. Post-ROCET and ATTAC, we can assume that boundaries are more firmly set, formally through the contract and more informally through the different tools used, varying organisational identities, the buildings worked from and the lines of management they report to.

However, it is not always as clear-cut as 'them and us.' As staff are transferred from the MoD to industry and there is greater interaction between the different service recipients and providers, there is also the potential for the development of greater understanding between partners through articulation and emotional work (Hochschild 1979; Schmidt and Bannon 1992). While such concerns may never become formalised in contracts, they become an explicit part of ongoing relationship management activities, both explicit and implicit. In turn, the relationship is not one way; in ROCET, the programme manager was keen to impress the need to embed with the RAF, culturally and physically to better understand their service requirements and learn about their culture and practices (Wild 2008).

Because service systems are socio-technical and include people, such changes cannot be modelled deterministically. Transferred staff, however well they are treated in the transfer, can still end up lacking motivation due to a conflict in the their values of the organisation they end up in. In contrast, they could also learn to embrace the view that there is nothing inherent with pursuing reasonable profit and revenue and that it can be an effective way for maintaining military capability in financially restricted times. After all, it was these same profit making companies that provided the artefacts that support the core mission.

22.5 Other Modelling Issues

Within this section we consider how issues such as complexity, adaptivity and the emergent nature of value relate to the concepts of the ABFS.

22.5.1 Complexity due to differing and conflicting values

Values affect how actants view the effectiveness of service system configurations that are acceptable and unacceptable (e.g., labour and timesaving vs. social cohesion). For example, placing high value on one's carbon footprint can lead to transport choices such as cycling and walking, which may be in a trade-off situation with time use goals and values. Alternatively, certain values and their trade-offs may lead to alternative behaviours such as carbon offsetting, using renewable fuel and lift/car share schemes.

Values are a key aspect of scoping the other elements of a service system. The form of the service system can reflect the values held by its actants. While a high-level goal of restaurants is to provide food and generate profit, the values held by the owners, staff and patrons could drive radically different manifestations of eating location, menu and experience. Comparing a high-class restaurant with a roadside catering outlet without reference to values would be meaningless, yet their basic transformations and domain objects remain remarkably similar: the preparation and serving of foodstuffs.

Some individuals and groups see their values as true and objective; those who do not share your values are classed as having none (Beck and Cowan 1996; Goodwin and Darley 2008). With respect to services, one key trend in recent years has been the emergence of availability and capability contracts (Terry et al. 2007; Tukker 2004). These go beyond outsourcing to contractual arrangements where two or more partners work together to deliver services. In many contexts, this brings commercial and non-commercial organisations together with a potential for clashes of values, the most obvious being when public sector services interact with commercially oriented organisations. Furthermore, these arrangements rely on the service recipient providing facilities back to its supplier, with both parties acting as supplier and recipient of services.

In other contexts, service design is tackling the design of public services, another situation that can bring together different actants and values, from the efficiency driven targets beloved by bureaucrats and politicians to those concerned with retaining or promoting broad and difficult and abstract goals, such as community cohesion and community participation (Seddon 2008; Parker and Heapy 2006). Values can entail complexity because values are often difficult to articulate and can result in implicit or tacit behaviours being performed that actually go against stated or believed values.

22.5.2 Complexity due to multiple and overlapping actants

One reason for complexity in service systems is the need for multiple organisations (in ABFS terms, collective actants) to work together. Given the size and scope of many service contracts, this can also entail organisations which are competing in other markets to work together. While process alignment and complementary competencies are seen as key parts of service co-creation (Chap. 6), this tension between organisations adds to the complexity of assessing service system success. While public goals can be developed and discussed, most actants will retain private goals that will affect how they carry out service activities and ultimately affect how they will judge the success of a service system.

22.5.3 Complexity due to multiple, overlapping and private goals

Service providers' motivations in providing complex services are varied. Sometimes they complement national policies (MOD 1998) while at other times they are concerned with keeping competitors at bay. The contract bidding goals are often driven by cost-reduction concerns rather than longer-term considerations, such as maintaining front line capability in the core organisation through readily available staff. The ABFS offers the potential for representing this complexity by being able to distinguish between goals of the core domain and service domain; higher-level goals concerning the required resource costs for setting up and maintaining as service system; and the possibility that goals remain private, but still affect behaviour. Through the acknowledgement of public and private, shared and non shared goals (Wild 2010), we have a mechanism for articulating, representing and discussing some of the complexities multiple and overlapping goals.

22.5.4 Complexity due to multiple roles for a service system

The relationships between service systems can be complex. One source of complexity is where the core system provides 'services' or assets usage back to the support system. In contracts such as ROCET and ATTAC, this is referred to as GFX or 'government furnished assets' (see Chap. 13), and within these contracts the industrial partner makes use of government assets in order to provide the service, with these assets embracing office space, maintenance workspace, tools and personnel.

In some cases, such as military ones, service contracts reflect the need to maintain the ability to service equipment in-theatre, as well as reflecting regulations on non-military personnel working in-theatre. In other contexts, a services customer may wish to retain some aspect of support in-house (McIvor 2005; Mol

2006). A core system providing services back to the service system is not restricted to the provision of physical assets. Activities such as providing asset usage information and plans can be viewed as services back to the service provider, as well as a key facet of co-creation.

The core system can also be considered a service system itself. Defence services are set up to provide services to its native populace and government or the populace and government of a defended territory. There can be a repeated and complex pattern of relationships between different core and service systems. Human activity systems are in numerous co-creational relationships.

22.5.5 *Complexities of Service Aims*

The ABFS represents the distinction between when a change is the 'goal of the system' and when it is a 'resource cost.' Whether or not wider forms of resource cost are explicitly taken into account in the Service Design process, all service systems have such resource costs. They are ecological, socio-cultural, emotional, as well as our usual measures of 'cost' such as time, throughput and financial costs. The ABFS allow us to distinguish between situations where, for example,

- Affective issues are an evaluative criterion alongside others (e.g., work applications)
- Where affective states such as 'fun' are the goal of the activity (games and theme parks)
- Where affective states such as 'fun' are balanced against other factors such as 'knowledge gained' (e.g., modern interactive museums)

Thus, the same factor can act in different ways in different service contexts. In the S4T context, while the maintenance or change of affective and socio-cultural behaviours is unlikely to be a core part of the contracted process, both remain important (see Chaps. 2, 5, 6, 9). It is naive to assume that services, which in general require greater interaction than a product development process (Parker and Heapy 2006, Chapter 20; Bitner et al. 1997), do not require emotional (Hochschild 1979) and articulation work (Schmidt and Bannon 1992) to maintain the ongoing relationship.

22.6 Conclusions

The Activity Based Framework for Services' development can be characterised in three movements.

1. To situate and relate different disciplinary definitions of services (Wild et al. 2009a).

2. To understand the difference between the design foci of different service design approaches (Wild et al. 2009c)
3. To produce models of service systems (Wild 2010).

This last issue has been the concern in this chapter and we have presented a high-level approach to the modelling of service systems. In modelling mode, the ABFS produces high-level models of service systems and can be seen to 'sit' on top of a number of modelling approaches, both generic and developed specifically within the S4T programme (e.g., Chaps. 3, 10, 12, 20). We have suggested a foundational relationship between a Core system and one or more Service systems, and presented examples of how concepts within the ABFS represent issues of transformation and complexity.

Given that this volume has already presented a framework that makes claim to provide an Integrative perspective for Complex Services, the Common Integrative Framework (see Chap. 23), we consider how the ABFS relates to the CIF.

As the nascent discipline of Service Science/Systems matures, we can hope that more and more approaches will emerge that are a synthesis of approaches from the many fields (IfM and IBM 2008) that can contribute knowledge to services. A less ambitious vision is to show systematically the links and relationships between the concepts and strands of research (Kagan 2009). Such synthesis can be furthered by work that explores the connections between knowledge communities and abstractions common to different disciplinary knowledge bases (Wild et al. 2009a, c).

As a field of study emerges, the level or strength of integration can vary. Many frameworks and approaches to Services can be found (Wild et al. 2009a; Sampson and Froehle 2006; Vargo and Lusch 2004a; Lovelock and Gummesson 2004; Hill 1999). These are in addition to 'locally' developed approaches within organisations and companies providing services and service design capabilities (Vanguard Consulting 2005; Seddon 2008; Parker and Heapy 2006; Guardian 2010; Cabinet Office 2010). Many of these approaches make the claim that they are generic to *all* services (Sampson and Froehle 2006; Vargo and Lusch 2004a; Lovelock and Gummesson 2004), in effect claiming to be integrative.

Elsewhere (Wild et al. 2009a) we have argued that frameworks are a general set of concepts for understanding a research area; that they are not tightly organised enough to be a predictive theory and that they sketch out the general concepts of a field of enquiry and the possible relationships between them. In turn we have noted that frameworks could occur at one or more of three different levels (Wild et al. 2009b), see also (Rao 2007).

- The Strategic level is normative and is concerned with action and sequencing commitments and issues such as what should a project recommend to partners as well as making sense of the broad contributions to Services Science a project can make.
- The Tactical level is descriptive. It is concerned with abstractions applicable to all Service Systems; that is, how do we model recurring elements of a service system?

- The Operational is the methodological level. It is concerned with the actions that realise strategy, and maps how research project outputs relate to specific methods.

Overall, we consider the CIF to sit predominantly at the Strategic level while the ABFS sits at the Tactical level. The CIF is a high-level, normative and strategically oriented framework that discusses high-level relationships between Information, Material and People, in the co-creation of value, and the broad transformations an Engineering Product organisation would have to undergo to provide services to complex products. In contrast, the ABFS, while relatively high level, is descriptive in nature.

This relationship between a core and one or more service systems also relates to the relationship between the ABFS and the Common Integrative Framework (CIF), which is illustrated in the next figure. Here, we demonstrate the mapping between the three transformations of the CIF and the concepts of the ABFS (Fig. 22.5).

In contrast to the CIF, the ABFS goes beyond stating that service systems transform material, information and people, to provide more detail about what is being transformed. For example, is the service system transforming people's physical structure and behaviours (sports/exercise coaching); minds (education, counselling, and entertainment); a combination (e.g., theme park); their physical assets (e.g., decorating, car or other equipment servicing)? In turn, what artefacts are we using, what activities are undertaken by the service system and by the core system? The ABFS, by offering a set of concepts common to all service systems, can start to tease out the complexities of the service system design and the distinction between different 'kinds' of service activities within the same contract or service system.

However, Service Science is still evolving and the challenges of developing designing and engineering services around complex engineered products entail many different tools and perspectives. We have argued elsewhere that frameworks can exist at different levels for different purposes, and that they can relate and

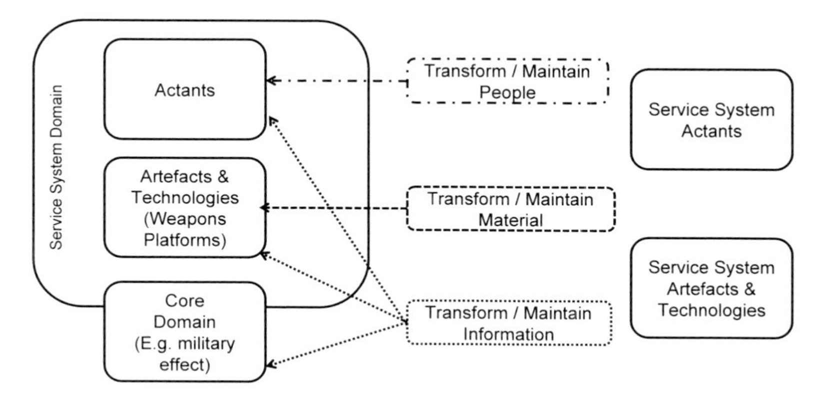

Fig. 22.5 Outline of the relationships between core concepts of the ABFS and the CIF

complement each other (Wild et al. 2009b). We expect both the CIF and the ABFS to continue to evolve, to interact, but to retain a focus at different levels of concern.

22.7 Chapter Summary Questions

We conclude with a number of questions about modelling approaches in general and in relation to Service Systems.

- What decisions about the types and scope of the abstractions are made in modelling approaches?
- What explicit and implicit links between modelling approaches can be identified and can they be pushed upwards into lightweight integrative models?
- Are modelling approaches aiming to model concepts, specifications, implementations or some combination of all three?

References

T. Alonso-Rasgado, G. Thompson, B.-O. Elfstrom, The design of functional (total care) products. J. Eng. Des. **15**(6), 515–540 (2004)

J.C. Aurich, E. Schweitzer, C. Fuchs, Life-cycle oriented planning of industrial product-service systems. in *ICMR 2007 De Montfort/Inderscience, Leicester*, 11–13 September, pp 270–274 (2007)

D. Beck, C. Cowan, *Spiral dynamics: Mastering values leadership and change* (Blackwell Business, London, 1996)

M.J. Bitner, W.T. Faranda, A.R. Hubbert, V.A. Zeithaml, Customer contributions and roles in service delivery. Int. J. Serv. Ind. Manag. **8**, 193–205 (1997)

Cabinet Office, *Successful service design: Turning innovation into practice* (Cabinet Office, London, 2010)

J.M. Carroll, W.A. Kellogg, M.B. Rosson, The task-artifact cycle, in *Designing interaction*, ed. by J.M. Carroll (Cambridge University Press, Cambridge, 1991), pp. 74–102

P.B. Checkland, J. Poulter, *Learning for action* (Wiley, Chichester, 2006)

H. Chesbrough, J. Spohrer, A research manifesto for services science. CACM **49**, 35–40 (2006)

C.C. Cowan, N. Todorovic, Spiral dynamics: The layers of human values in strategy. Strategy Leadersh. **28**, 4–12 (2000)

B. Curtis, K. Marc, J. Over, Process modeling. Commun. ACM **35**, 75–90 (1992)

J. Daniels, Modeling with a sense of purpose. IEEE Softw, pp 8–10 (2002)

D. Diaper, Understanding task analysis for human–computer interaction, in *The handbook of task analysis for HCI*, ed. by D. Diaper, N.A. Stanton (LEA, Mahwah, 2004), pp. 5–47

J. Dowell, J.B. Long, Conception of the cognitive engineering design problem. Ergonomics **41**, 126–139 (1998)

J. Elster, *Explaining social behavior* (Cambridge University Press, Cambridge, 2007)

N. Friedman, The carrier and the Royal Navy. The Third Annual Corbett Lecture Kings College London, 9 Sep (2009)

R. Frigg, S. Hartmann, in *Stanford Encyclopaedia of Philosophy,* ed. by E.N. Zalta. Models in science (Stanford University, Stanford, 2006)

M. Goedkoop, C. van Halen, H. te Riele, P. Rommens, Product service systems, ecological and economic basics. PRé Consultants, Amersfoort, p 132 (1999)

G.P. Goodwin, J.M. Darley, The psychology of meta-ethics: Exploring objectivism. Cognition **106**, 1339–1366 (2008)

Guardian, Service design, 15th March The Guardian, Guardian News & Media Group, London (2010)

E.T. Hall, *The silent language* (Doubleday, New York, 1959)

J. Heskett, Creating economic value by design. Int. J. Design **3**, 71–84 (2009)

P. Hill, On goods and services. Rev. Income Wealth **23**, 315–338 (1977)

P. Hill, Tangibles, intangibles and services. Can. J. Econ. **32**, 426 (1999)

A.R. Hochschild, Emotion work, feeling rules and social structure. Am. J. Soc. **85**, 551–557 (1979)/

IfM, IBM, *Succeeding through service innovation* (Institute of Manufacturing, University of Cambridge, Cambridge, 2008)

M. Jackson, P. Keys, Towards a system of systems methodology. J. Oper. Res. Soc. **33**, 473–486 (1984)

J. Kagan, *The three cultures* (Cambridge University Press, Cambridge, 2009)

C. Kerr, R. Phaal, D. Probert, in *ICCRTS 2006*. A framework for strategic military capabilities in defense transformation, Cambridge, England (2006)

C.H. Lovelock, Classifying services to gain strategic marketing insights. J. Mark. **7**, 9–20 (1983)

C.H. Lovelock, E. Gummesson, Whither services marketing? J. Serv. Res. **7**, 20–41 (2004)

L. Lages, J. Fernandes, The SERPVAL scale: A multi-item instrument for measuring service personal values. J. Bus. Res. **58**, 1562–1572 (2005)

T.C. McAloone, M.M. Andreasen, in *Design for X, TU Erlangen* ed. by H. Meerkamm. Defining product service systems, pp 51–60 (2002)

R. McIvor, *The outsourcing process: Strategies for evaluation and management* (Cambridge University Press, Cambridge, 2005)

N. Melão, M. Pidd, A conceptual framework for understanding business processes and business process modelling. Inf. Syst. J. **10**, 105–129 (2000)

J. Mingers, *Realising systems thinking* (Springer, New York, 2006)

Ministry of Defence (MoD), *Strategic defence review* (Ministry of Defence, London, 1998)

M.J. Mol, *Outsourcing: Design, process and performance* (Cambridge University Press, Cambridge, 2006)

O.K. Mont, Clarifying the concept of product-service system. J. Clean. Prod. **10**, 237–245 (2002)

J. Piaget, *Main trends in inter-disciplinary research* (Allen & Unwin, London, 1970)

V. Rao, Strategy, tactics, operations and doctrine: A decision-language tutorial. Available at: http://www.ribbonfarm.com/2007/09/24/strategy-tactics/print/ (2007)

J. Rasmussen, A. Pejtersen, L. Goodstein, *Cognitive Systems Engineering* (Wiley, New York, 1994)

S. Parker, J. Heapy, The Journey to the Interface. Report, DEMOS, London, 2006

S.E. Sampson, C.M. Froehle, Foundations and implications of a proposed unified services theory. Prod. Oper. Manag. **15**, 329–343 (2006)

K. Schmidt, L. Bannon, Taking CSCW seriously: Supporting articulation work. Comput. Support. Coop. Work **1**(1–2), 7–40 (1992)

J. Seddon, *Systems thinking in the public sector* (Triarchy Press, Axminster, 2008)

T.B. Sheridan, Task allocation and supervisory control, in *Handbook of human–computer interaction*, ed. by M. Helander (North-Holland, New York, 1988), pp. 159–173

Y. Shimomura, T. Sakao, E. Sundin, M. Lindahl, in *Proceedings of the 9th International Design Conference DESIGN 2006*, ed. by D. Marjanovic. Service engineering: A novel engineering discipline for high added value creation. Faculty of Mechanical Engineering and Naval Architecture, University of Zagreb, Croatia (2006)

B. Stauss, A Pyrrhic victory: The implications of an unlimited broadening of the concept of services. Manag. Serv. Qual. **15**, 219–229 (2005)

R. Szostak, How and why to teach interdisciplinary research practice. J. Res. Practice. **3**(2), (2007)

T. Tomiyama, A design methodology of services, in *ICED05 Melbourne*, Aug 15–18 (2005)

A. Terry, D. Jenkins, T. Khow, K. Summersgill, P. Bishop, M. Andrews, *Transforming logistics support for fast jets* (National Audit Office, London, 2007)

L. Tiger, *The pursuit of pleasure* (Transaction Press, New Brunswick, 2000)

G. Tress, B. Tress, G. Fry, Clarifying integrative research concepts in landscape ecology. Landsc. Ecol. **20**, 479–493 (2005)

A. Tukker, Eight types of product-service system. Bus Strategy Environ. **13**, 246–260 (2004)

Vanguard Consulting, *A systematic approach to service improvement: Evaluating systems thinking in housing* (Office of the Deputy Prime Minister, London, 2005)

S.L. Vargo, R.F. Lusch, Evolving to a new dominant logic for marketing. J. Mark. **68**, 1–17 (2004a)

S.L. Vargo, R.F. Lusch, The four service marketing myths. J. Serv. Res. **6**, 324–335 (2004b)

S. Vargo, R. Lusch, Why "service"? J. Acad. Mark. Sci. **36**, 25–38 (2008)

A. Whitefield, A. Esgate, I. Denley, P. Byerley, On distinguishing work tasks and enabling tasks. Interact. Comput. **5**, 333–347 (1993)

P.J. Wild, P. Johnson, H. Johnson, in *TAMODIA'04, Vol. 17–24* ed. by P. Palanque, P. Salvik, M. Winckler. Towards a composite model for multitasking (ACM Press, Prague, 2004), Nov 15–16

P.J. Wild, in *workshop on HCI and the analysis, design, and evaluation of services. HCI 2008*, Liverpool. Implicit user centred design in aerospace service contract creation, Sep 1–5 (2008)

P.J. Wild, P.J. Clarkson, D. McFarlane, in *Industrial Product Service Systems*, ed. by R. Roy, E. Shehab. A framework for cross disciplinary efforts in services research (Cranfield University Press, Cranfield, 2009a), pp 145–152

P.J. Wild, I. Ng, D.C. McFarlane, in *Frontiers in Service,* Honolulu, Hawaii, Oct 29–Nov 1. "Taming" the ecology of transdisciplinary services research: Combining strategic, tactical, and operational views in integrative services research (2009b)

P. J. Wild, G. Pezzotta, S. Cavalieri, D.C. McFarlane, Towards a classification of service design foci, activities, phases, perspectives and participants, in *MITIP'09 Bergamo* (2009c)

P.J. Wild, Longing for service: Bringing the UCL conception towards services research. Interact. Comput. **22**, 28–42 (2010)

N. Winder, Breaking the phoenix cycle: An integrative approach to innovation and cultural ecodynamics. Available at: www.tigress.ac/reports/final/Phoenix.pdf (2005a)

N. Winder, Integrative research as appreciative system. Syst. Res. Behav. Sci. **22**, 299–309 (2005)

S.N. Woodfield, in *Proceedings of the ER'97 Workshop on Behavioral Models and Design Transformations: Issues and Opportunities in Conceptual Modeling*. The impedance mismatch between conceptual models and implementation environments. Los Angeles, CA (1997)

Chapter 23
Complex Engineering Service Systems: A Grand Challenge

Irene Ng, Glenn Parry, Roger Maull and Duncan McFarlane

Abstract This chapter examines the contributions made within this book and seeks to add to the development of the framework for complex engineering service systems. Particular focus is placed upon emergent value, which is co-created as customer variety permeates processes.

In this chapter we reflect on the contributions made by the authors and further develop the framework for complex engineering service systems to include value co-creation with state-dependent outcomes where customer variety pushes into the multi-organisation processes.

Managers and service researchers describe the "moment of truth" as the defining period when the interaction between the firm and buyer is of crucial importance to determine customer satisfaction (Bitner et al. 1990; Churchill and Surprenant 1982; Anderson and Sullivan 1993). The service encounter embodies value-in-use, value which is jointly co-created between the customer and the firm for mutual benefits (Payne et al. 2008; Prahalad and Ramaswamy 2003). The concept of value co-creation subsumes previous service research in operations and strategy that has emphasised the role of the customer within a service system such as the customer contact model (Chase and Apte 2007; Chase and Tansik 1983), customer interactions (Johnson et al. 2005) and value co-production with the

I. Ng (✉) · R. Maull
University of Exeter Business School, Streatham Court, Exeter, Devon EX4 4PU, UK
e-mail: Irene.Ng@exeter.ac.uk

G. Parry
Bristol Business School, University of the West of England, Frenchay Campus, Coldharbour Lane, Bristol, BS15 1QY, UK
e-mail: glenn.parry@uwe.ac.uk

D. McFarlane
Institute for Manufacturing, Alan Reece Building, University of Cambridge, Charles Babbage Road 17, Cambridge, CB3 0FS, UK
e-mail: dcm@eng.cam.ac.uk

I. Ng et al. (eds.), *Complex Engineering Service Systems*, Decision Engineering,
DOI: 10.1007/978-0-85729-189-9_23,

customer (Ramirez 1999). In exploring these concepts, much of the literature has taken the dyadic firm/customer relationship as a unit of analysis and not articulated the broader enterprise nature of complex engineering service delivery. The work presented has extended this thinking into consideration of broader, multi-organisational enterprise service systems.

Within this book we have sought to explore the concept and meaning of service within the context of complex engineering service provision. In the Introduction chapter we presented a framework to illustrate the core transformations delivered by complex engineering service systems. We proposed that to meet the full value-in-use of the firm's offering, three integrated simultaneous transformations are required; people, material/equipment and information. Through this book we have begun to explore the transformations in more depth and the work demonstrates the integrated nature of the transformations, showing clearly that for complex engineering service each transformation cannot be considered in isolation from the other. Mastery of the three simultaneous transformations within a multi-organisational context and the interactions between them will provide a firm with sustainable returns and competitive advantage.

23.1 Key Theoretical implications for a Value-Centric approach in Complex Engineering Service Systems

23.1.1 Complex Engineering Service Systems to Deliver Outcomes Changes Boundaries

At the beginning of the book we proposed that the purpose of complex engineering service systems is for the integration of equipment, people and information transformation to achieve excellent outcomes, as opposed to merely high-performance outputs from complex engineered equipment. Several chapters within the book have dealt with the notion of contracting for such outcomes, rather than for resources (such as time) and materials. The exploration of outcome-based contracting (OBC) for the provision of capability via the Tornado ATTAC contract between BAE Systems and the MoD has been the focus of a great deal of the work and provides a detailed and complex example of a contracting mechanism where the firm is tasked to deliver outcomes rather than merely assets or activities. As a further example, Rolls Royce offer "TotalCare®" contracting for their civil aerospace engines, where the continuous maintenance and servicing of the engine is not paid according to the spares, repairs or activities rendered to the customer, but by how many hours the customer gets power from the engine. Hockley et al. (Chaps. 13 and 14) give a comprehensive description of OBC in the defence environment and compared the issues across a range of contracts, while Caldwell and Settle (Chap. 8) explored the role of incentives in such contracts. Mills et al. (Chap. 3) discuss the need for enterprise-level management to align behaviours

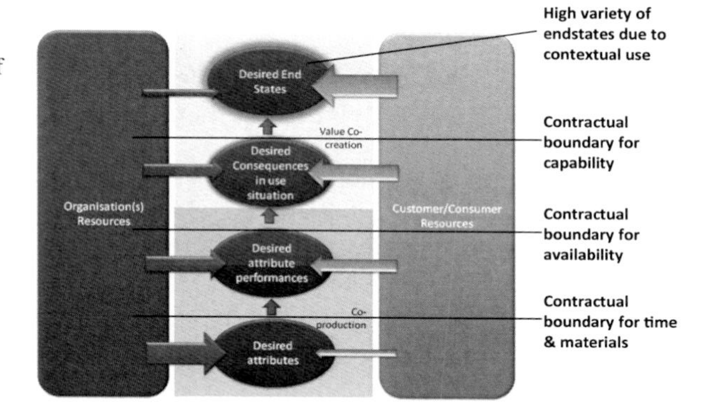

Fig. 23.1 Hierarchy of outcomes in complex engineering service (width of *arrows* indicative contribution of resources)

between the customer and the firm to achieve such outcomes. From a service perspective, McKay and Kundu in Chap. 12, describe how service information blueprints allow service definitions to be created which can identify information requirements for both current and future state service systems.

A higher level of outcomes (availability and capability) requires greater dependency on the customer and its resources (Ng et al. 2009a). In essence, a firm that contracts on outcomes has to have the capability to manage the customer, an area over which the firm may or may not have much control. Improvements in the flow of information may help improve this situation. In their work on the role of information on service strategies, Cuthbert et al. (in Chaps. 10 and 11) show how services may be improved as a result of providing better equipment condition information feedback to the service provider. To realise the full value of this information feedback requires customer input and transformations which support it. Consequently, the firm has to be motivated to think about its capability as that which *includes* aspects of the customers' materials and equipment, information and people. The alignment of the two organisations in co-creating value is paramount.

Theoretically, this is consistent with Lapierre et al.'s (2008) study, which proposes that value is a multi-layered hierarchical concept where the customer realises the firm's value-proposition to achieve higher level 'end-states'. Insofar as availability or capability is an end-state, the use of customer resources to achieve higher level end-states is undoubtedly necessary, which of course increases risk to the firm as there is less control of the achievement of such outcomes. Figure 23.1 depicts the hierarchy of outcomes as the firm takes over greater responsibility from the customer and offers higher level outcomes.

Where delivering availability is a desired attribute, increasing availability levels may become an attribute of performance. Availability level contributes to customer desire in terms of their use, which in turn contributes to the customer end-states. Contracting for capability may therefore be the desired consequence for value-in-use situations. To deliver a service closer to higher levels (end-states), the combination of resources by suppliers, providers and customer takes centre stage. Labelled as the co-creation of value (Prahalad and Ramaswamy 2004), Ng et al. (Chap. 6 describe a way to develop that capability in an organisation. Similarly,

Sect. 23.1 of the book suggests an enterprise perspective, stating that "complex engineering service systems inevitably involve complex organisational solutions which go beyond the boundary of a single organisation" (Purchase et al. Chap. 2). Cuthbert et al. in Chap. 10 provide a framework which may be used to determine information requirements for the design, delivery and evaluation of different service offerings. The framework provides a high-level assessment of the information required for the delivery of service value-propositions and develops understanding of the gap between proposition and the available information.

End-state was also discussed in Chap. 21 by Wood and Tasker, who proposed that it is a mental concept highly influenced by the context in which it was achieved. The notion of end-state here includes the emotional state of mind, the perception and pre-conceptions of the customer. In Chap. 5, Mills et al. look at how gaining an understanding of the client's aspirations and fears is a value driver for co-creation, even when their full requirement may lie outside of contractual bounds. Such a concept challenges the core transformation of people in the delivery of complex engineering systems to include customer transformation processes and practices within its domain. Towards that end, Guo and Ng (Chap. 9) emphasise interpersonal relationships between them, with cooperation moved from reciprocal to communal, leading to a common identity and with both parties working towards their collective goals.

More concretely, the move into end-states has an impact on material/equipment transformation. From this perspective, the firm can no longer be content with merely providing a repair solution, but has to consider the *effect* or *consequence* of the asset through its usable life. As Kelly and Ratchev (Chap. 19) suggest, the firm has to have "an intrinsic role in through life management" of the assets. Since parties are aligned in wanting better outcomes (i.e., lower maintenance cost, reduced occurrence of failures and increase availability), Wang and Carr (Chaps. 16 and 17) suggest the use of condition-monitoring processes and condition-based models to achieve this. Paradoxically, the move into end-states could also compel the firm to redesign and reengineer an asset and its corresponding activities to better achieve outcomes as Butterfield and McEwan (Chap. 18) have shown. In their chapter, they propose digital simulation as an animated method for the replacement of an aerospace structural panel providing a simple illustration of a method that addresses not only implementation of design changes to the asset, but also all activities surrounding it so that the entire subsystem can be optimised for delivering the work and its corresponding costs. The chapter demonstrates how the line between design, engineering and manufacturing of an asset and the service provision becomes blurred when the firm begins to contract for outcomes. Indeed, as Caldwell and Settle suggest (citing Normann and Ramirez 1993) in Chap. 8, firms do not really 'add value' but 'reinvent value'. This also impacts upon costs, as shown by Erkoyuncu et al. in Chap. 7.

In summary, the book chapters have highlighted that contracting for outcomes for complex engineering service systems blurs (a) the boundaries between the firm and the customer, (b) the boundary between the processes of all three core value transformation (material/equipment, information and people), (c) the responsibility for the core value transformations towards described end-states and (d) the lines of accountability within a system that contribute towards costs for both customer and provider firm. The higher the end-states, the greater the blurring and the more complex is the system.

23.1.2 Complex Engineering Service System has to Focus on Links, Interactions and Alignments which Requires a Systems Approach

The industrial era of dividing tasks between functions (and academically, between disciplines) has worked when there was an obvious value chain towards a concrete outcome such as the production of cars or televisions. Within such environments, a modular 'click and play' approach was appropriate as the interactions between the functions (and disciplines) were minimal. One could therefore 'lean' the system to make it more efficient to produce six-sigma outcomes and the like. Within complex engineering service systems, processes, people, engineered assets and automated technologies function oftentimes with the customer *within* the system (e.g., airports, healthcare). Functional departments and academic disciplines have to account not merely for the tasks they need to perform (for functions) and the knowledge within its domain (for disciplines), but for the *interactions* between them (Ng et al. 2009b). Thus, those working in complex engineering service systems must move away from linear, cause and effect thinking and consider a systems approach.

Systems thinking have a long academic tradition dating back to the open systems concepts of von Bertalanffy (1968) and the control systems work of Wiener (1948). Much academic literature has taken this approach as an alternative to the reductionist view which has dominated much of management as well as engineering research. Reductionism breaks a problem down into its component parts and seeking to optimise each part. At the core of such a reductionist approach are three fundamental assumptions (Ng et al. 2010b):

1. The connections between the parts must be very weak;
2. The relationship between the parts must be linear so that the parts can be summed together to make the whole;
3. Optimising each part will optimise the whole.

In the study of complex engineering service systems, particularly outcome-based systems that we have discussed previously, these assumptions do not hold. Complex engineering service systems involve tightly coupled parts; changing one component (e.g., one of the core value transformations) affects many others, leading to unintended consequences. The interactions between them are often highly complex and non linear, as Parry et al. (Chap. 4) have elaborated. Forrester (1968) points to the importance of time delays, amplification and structure on the dynamic behaviour of the system. Lipsey and Lancaster (1956) in their theory of the second best show that achieving the optimal condition for a system could result in sub-optimality of component parts. Buckley (1980) in his masterly essay on the problems of causality in organisations summarises it thus;

> Of particular importance are those kinds of mutual relations that make up circular causal chains; the effect of an event or a variable returns indirectly to influence the original event itself by way of one or more intermediate events or variables.

Thus, complex engineering service systems research has to reject the linear perspective on causality for the richer insights that can be gained from the systems view. Yet, such a rejection does not imply the rejection of the analytical approach. Both analytical and synthetic research perspectives could be employed in the scientific method. However, it is important to understand that certain properties of the system (such as customer experience), could be an emergent property that is not present in any of the component parts but in the dynamic relationship of the parts. In such cases, the design of the system should consider interventionistic approaches rather than believing that the system could be pre-determined.

Given this orientation, a firm tasked to manage complex systems needs to not only understand the nature of work and its responsibility at a functional (or disciplinary) level, but also the nature of the interactions and links between them. Such linkages and interactions are crucial for the success of systems that are able to consistently deliver service excellence. From an interdisciplinary perspective, a complex engineering service system is not about finding a weak 'glue' between various functions or disciplines who have an unwavering adherence to core principles. Rather, it is how the same 'glue' becomes critical to deliver value and superior customer experience and how the disciplines themselves adapt and move forward to deliver new collective knowledge and excellent value outcomes.

Systems thinking and approaches are not new in engineering. Particularly in defence systems, system engineering has become an accepted domain for the interdisciplinary field of engineering that focuses on how complex engineering projects (such as the Apollo programme by NASA) should be designed and managed, integrating processes and including technical and human-centered activities. However, system engineering and indeed engineering itself struggles with social and behavioural aspects of a system as it has always been associated with "assembling pieces that work in specific ways" (Ottino 2004) and "a process of precise composition to achieve a predictable purpose and function" (Fromm 2006, p. 2). To achieve "a predictable purpose and function", deemed to be core to engineering activity, becomes seriously challenged when the component parts of the system are *people* whose activities may not be easily controlled by predictable processes, and whom exhibit autonomous human behaviours and yet are component parts of the value creating system. There is clearly an opportunity for further study into the transformation of people as a core part of the system transformation. This text notably does not contain a section specifically addressing the 'people transformation'

A research agenda for the value transformation of people is currently under development.

As Wood and Tasker (Chap. 21) put it:

> The system engineering thinking style is deeply embedded in the development of complex systems, especially defence systems where it has become a universal language; systems of regulation and governance are built on it. Notwithstanding the development of soft systems and similar approaches, system thinking is perceived to be inextricably rooted in the product paradigm ensuring a cultural rigidity inhibiting development of the method to accommodate design in the social space necessary for the attainment of service excellence.

While inspired by systems thinking, systems engineering also struggles with how to deal with the concept of emergent properties and its related concept of interventions. Emergent properties are often described as properties exhibited by a whole system that do not exist within its component parts, i.e., properties that cannot be identified through functional or component-based decomposition. Emergent properties epitomise the idea that a whole system is greater than the sum of its parts. Emergence is often a result of a complex system and it has been a thorn on the side of system engineering.

In his article about the significant issues that face the US National Academy of Engineering, Wulf (2000) declares that "the key point is that we are increasingly building engineered systems that, because of their inherent complexity, have the potential for behaviours that are impossible to predict in advance". Within the last decade, academic literature in engineering and information systems has attempted to make sense of this phenomenon and in Chap. 4, Parry et al. discuss complexity management and the factors that contribute to it.

When "predictable purpose and function" of complex engineering service systems is oriented towards emergent outcomes (i.e., a customer experience), *as well as* deterministic outputs (turnaround time for repair), engineering as a field of study is further challenged. The challenge of contracting, designing, engineering and delivering end-states that include both deterministic outputs and emergent outcomes sets the agenda for future research in complex engineering service systems.

One suggested approach when delivering excellent outcomes that are unpredictable due to behavioural uncertainty and emergent properties is the use of systemic interventions. Midgley (2000) defines systemic interventions as "purposeful action by an agent to create change in relation to reflection on boundaries" (p. 129). In achieving better outcomes through technology insertion for example, Wang and Carr (Chap. 17) propose four different types of maintenance interventions through a theoretical model that could produce substantial savings. Future research into interventions could also result in an 'intelligent design' of complex engineering service systems that could allow for the achievement of deterministic outputs (attribute performances) as well as emergent outcomes (end-states). While engineered assets are called to deliver higher availability (lowered asset failure) and lower through life cost across contexts, they are also called to answer the challenge of higher *capability* in the same environment. This is clearly an important factor (Kerr et al. Chap. 20) and higher availability, lower through life cost and better capability is often achieved through technology insertion.

From the resource perspective, the combination of resources to achieve end-states therefore necessitates a combination of human activities and complex engineered assets. It is important to know how each component resource impacts on value or whether it is not merely the components but the *interactions* between them that is the value driver. Hence in determining the value offer, the firm has to understand what resources are contributed by both customers and the firm to achieve the outcomes and outputs realised within the customer experience. In Chap. 3, Mills et al. propose an enterprise imaging methodology that provides a visualisation of the resources used to deliver value and also gives insight to the interactions between providers and customer. Only by understanding the value co-creation resources contributed by both parties can an assessment be made as to whether they are appropriate substitutes for achieving the best outcomes for customers in the target market, at the lowest costs. The substitutability of resources contributed by firm, customer and technology/assets must therefore be evaluated not merely from the cost perspectives, but with the possibility that it could also lead to better outcomes, resulting in the firm being able to either increase price, demand for the service and delivery of excellence within the system (Ng 2007).

To deliver excellent service and be economically viable, firms need to develop the customer as a core competence within the system, a point echoed by Prahalad and Ramaswamy (2000). The customer's failure to co-create value results in the firm not being able to achieve the outcomes they have been contracted to deliver. Hence, the customer's capability to co-create value is now the firm's responsibility under complex engineering service system for outcomes. In Chap. 6, Ng et al. propose that the capability of an organisation to exhibit seven value co-creation attributes suggests they may deliver better service outcomes. Firms with these attributes achieve effective behavioural transformation of the customer which enables them to deliver a greater variety of service end-states. This is a requirement arising when the customers' experience manifests in a variety of ways as a result of the different contexts they are in when using a service.

23.1.3 Value-In-Use is Contextual and Variety-Laden Across All Interactions in the System

At the beginning of the book we proposed that a value-centric approach in complex engineering service systems must put value-in-use at the centre of what the firm needs to deliver, in partnership with the customer. The previous section described how a complex engineering service system for outcomes has the customer within the boundaries of the system. In this section, we elaborate on further challenges of customer presence within the system.

To achieve value-in-use, the firm has to ask how value is created and understand the role of the customer within that space. The chapters presented in the book show that value-in-use (i.e., contextual value) is not a static concept. The notion of 'use' is dependent on the 'state of the world' and customer's use in different states (i.e., contexts) has a tremendous impact on the firm delivering to outcomes (c.f. Karni 1983; Fishburn 1974; Ng 2008; Shugan and Xie 2000; Xie and Shugan 2001). In other words, as many of the chapters in this book demonstrate, even when the firm is required to do the same activities each time to co-create value within the system, the state of the world changes and together with it the resources and capabilities of both parties. A simple analogy can be used to illustrate this. A helicopter produced by a firm would consume various resources of the firm to manufacture it. The customer would value the helicopter through the employment of its own resources to 'use' it, thus achieving the co-creation of value. Yet, if the weather was not good on a given day (i.e., a change in the 'state of the world') the contextual value-in-use of the helicopter would be diminished, resulting in the customer experiencing the service (or consuming the asset) differently with each changing context. This implies that if the firm is serious about delivering higher level end-states, it has to be capable of responding to the variability introduced by the customer and the environment.

It is important to distinguish between the concepts of variability and variety. Frei (2006) proposes that 'throwing the customer into the works' introduces five types of variability:

- request variability (different requirements for each customer)
- arrival variability (peaks and troughs in service demand)
- capability variability (customers have differing skill levels)
- effort variability (some services require customer input/participation and customers will have differing willingness to make effort)
- subjective preference variability (different and contradictory views of what constitutes good service) (Godsiff 2010)

There is a difference between known and unknown variability. Some variability introduced by the customer is predictable e.g., arrival variability for many call centres is a known calculable using known functions, request variability may be limited by a tariff sheet or catalogue. Other variability is unknown and a source of shock or disturbance to the firm. This is often the case when the 'state of the world' changes.

When considering the firm, we can distinguish between resources and competencies. The theoretical perspective of the resource-based view of the firm considers resources as what we term properties that carry out the transformation. They can be physical, human, technological or organisational. Competencies are the capacity of a group of resources when well-managed to carry out an activity. This has echoes of the Service Dominant Logic perspective; "that resources are not, *they become*" (Lusch and Vargo 2006, p. 65), meaning that they are only resources if used, otherwise they add no value and may be seen as waste in the system (Ohno 1988). The process through which such resources *become, in context,* is the capability or competence of the producer system. A third consideration has been added by Teece et al. (1997), who consider the dynamic capabilities view which looks at the ability to renew competencies. This emphasises the managerial role in adapting, integrating and reconfiguring resources and capabilities. Resources and therefore capabilities have limitations in what they can do. Precisely defining these capabilities is notoriously difficult, particularly in service environments where human knowledge is frequently the transforming resource.

The coming together of customer variability and producer system capability is described in Ashby's Law of Requisite Variety. This in its simplest form asserts that 'only variety can absorb variety'. Ashby (1956) is stating that in order for an entity to be viable it must have an internal state which responds to the external stimulus in such a way as to produce a 'survivable outcome'. It is likely that the producer system has been designed to respond to only a limited set of customer variabilities. Correspondingly, the customer's use context that introduces variability may encounter the firm's limited set of capabilities. In a world where customer requirements are continuously changing even beyond the original specification, the producer system needs to have the resources and competencies and in particular the dynamic capabilities to be able to absorb the variety and produce an acceptable outcome. In his articles on the viable systems model, Beer (1981) recognises that in practice most organisations will attenuate the variety they offer to the marketplace. In the case of complex engineering service systems for *outcomes,* that attenuation may be unacceptable to the customer. The message for the producer is that it must not just match the variety demanded in the original specification but also be capable of matching the variety as the user requirements change due to the use of the product in the varied contextual states throughout the product life. This implies considerable inbuilt redundancy, required to absorb variety. Yet, it may also provide opportunities for innovation and new business models.

This book has identified how variety impacts on the determination of outcomes, particularly in the defence industry. As Hockley et al. put it in Chap. 13,

> When considering the phrase "under stated conditions," it is primarily about environmental conditions but there can be many facets, such as:

- In which geographic environment it will be used (e.g., hot, dusty, cold and icy),
- Different conditions will be seen by the same item when fitted on different platforms (e.g., a helicopter or vehicle),
- Storage conditions (e.g., humidity controlled or open air),
- Maintenance may be completed in diverse environments,
- Handling and transportation will expose items to varied conditions (e.g., vibration and orientations).

In Chap. 19, Kelly and Ratchev discuss process resilience and focus upon three adaptive considerations that the impact on variety:

- Adaptation to changes in platform capability requirements during the life cycle.
- Adaptation to platform changes due to technology insertion.
- Integration of past and future maintenance information.

Their maintenance dashboard incorporates variety through accurate asset maintenance information and analysis of capability trade-offs to ensure optimised performance of assets. Similarly, Wang and Carr in Chap. 16 suggest the use of condition-monitoring (CM) techniques and condition-based maintenance (CBM) models to deal with usage challenges particularly where historical data on use is not available. In Chap. 15, McNaught and Zagorecki also propose practitioner guidelines for CBM and prognostic modelling, discussing the challenges posed on information transformation in complex engineering service systems.

Contextual variety therefore challenges the firm to consistently deliver functional outputs across contextual states. In Chap. 20, Kerr et al. suggest that such changes come in the form of:

- The threat of obsolescence
- The requirement for additional or new capability
- The challenge of affordability

Certainly the cost implication is non-trivial. In Chap. 7, Erkoyuncu et al.'s treatment of uncertainty in evaluating costs is a direct consequence of changes in boundaries and also the incorporation of customer contextual usage (and varieties) within the delivery costs.

We may learn from manufacturing as the 'late customisation' or 'postponement', in which a process is standardised as much as possible and variety capabilities are introduced very close to the point of customer delivery. Even at the late stage, a key challenge is to achieve high levels of product variety with a limited set of standard process capabilities (Yang et al. 2004). In automotive manufacturing operations, delivery of customer variety via the combination of modular units late in the production process allows extensive customer variety to be delivered while using standard processes and a controlled set of reusable constituent elements (Parry and Graves 2008). Yet such late customisation assumes customer variability enters through a limited set of 'ports' (cf. Godsiff 2010; Weinberg 2001), representing limited interfaces that connect them. In a complex system for outcomes, the boundaries between production and consumption are blurred and overlapping. Weinberg describes the interface between the customer and the firm no longer as a 'port' but as a 'membrane' where "an obvious example is the cell wall that may be penetrated *at almost every point on its surface, but not by everything and* not at all times." This implies that customer variety permeates through the entire system. Most disturbingly, variety permeation into the system often by-passes the firm's designed processes, disrupting them and creating complexity. This is paradoxically a consequence of the firm's original and implicit assumption of low customer usage variability when designing the system. Consequently, rigidities and the creation of closed systems are realised as a result of underestimating the number of possible states for the system in requirements engineering.

As Ashby's law suggests, only variety can destroy variety and thus from a systems perspective, a complex engineering service system can only be viable (cf. Beer 1981) if the firm *attenuates* customer variety to match the rigidities of its system, or the firm *absorbs* the customer's variety by designing a system that can do so while retaining economic viability.

Systems with high human resource can absorb more variety since human beings are capable of a greater number of behavioural modifications and may be able to match the customer's contextual variety (e.g., being empathetic, smiling). Paradoxically, such interactions by autonomous human behaviours that absorb variety or persuade the customer to attenuate their variety (with a smile) actually result in an improved customer experience (Ng et al. 2010a), thus increasing system viability and stability. Yet, the use of human resources within a system to absorb variety is often a challenge to replicate and scale. This challenge can be viewed from three perspectives. On one hand, one can argue that human factors supporting a product or equipment can be viewed as a resource to absorb variety to achieve service and support excellence but may not be easily scalable or replicable. On the other hand, the very product that human resources are supporting could be designed in such a way that the human resource skill set required to support it *is* more scalable or replicable even for the absorption of variety. Finally, the product could be designed in such a way that the human resource to achieve absorption of variety could be the customers themselves, such as how the iPhone is a platform for customers' usage of apps to co-create value for themselves contextually across infinite states of use (which we term as hyper variety). In all three scenarios, the central issue is around design and resource configuration of a complex engineered service system towards outcomes.

23.2 Complex Engineering Service Systems: Concepts and Research

Developing the complex engineering service systems framework, we have to factor in customer variety that could either radically alter what was designed (if the variety is absorbed) or create dissatisfaction (if the variety is attenuated). As an example, the business traveller may favour an airline and choose the very same airline for their family holiday, expecting the same level of service excellence. That airline may have set processes for the business traveller who has few bags, seeks speed and wants no distractions. How does the same process cope with large numbers of bags and potentially small children who require distractions? Do they absorb the variety or attenuate it? Customer variety is context dependent such that the same customer may approach a service from a number of contexts, placing different demands upon the system and expecting the same level of excellence in outcomes. Figure 23.2 integrates these concepts, illustrating the nature of a complex engineering service system.

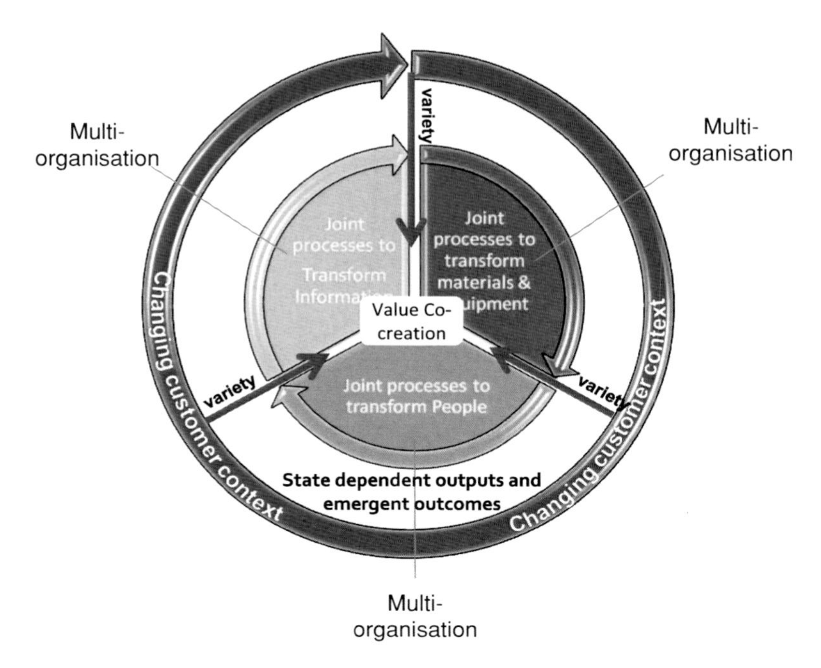

Fig. 23.2 Complex service system of core transformations and value co-creation with state-dependent outcomes

Value co-creation with the customer is at the heart of the system. Transformation of materials and equipment, information and people is undertaken by multi-organisational structures that may include the customer and their resources. The transformations occur to deliver service outputs, but these are state dependent. Changing customer context introduces variety which pushes-back into the firm, impacting on and across processes of transformation, leading to emergent outcomes. Chap. 6 (Ng et al.) suggests seven attributes the system should exhibit to ensure that variety could be absorbed with minimal disruption.

By delivering to an outcome, the firm is committed to the same outcomes even when the context of usage changes. Thus, both the customer and the firm may not know the contingency nature of the context that could change how the service is delivered or co-produced at that point in time. For example, it may not be possible to predict when the customer will increase their demand for service on a particular day. Thus, delivering to a context dependent value-in-use suggests that the firm has to be prepared and capable of absorbing customer variety and still deliver satisfaction when committed to delivering outcomes.

There is a tendency to attempt to de-risk complex systems to minimise emergence by establishing centralised command and control through reducing autonomous human activity. This may be useful if the objective of the system is to achieve deterministic outputs, but may be erroneous if the system has also to attain emergent outcomes such as customer experience (or perception of service excellence), where the very interactions that are autonomous and unpredictable may be key to contributing to that emergence. Worse, when applied to outcome-based systems as described in this book, the customer within the system that introduces a high degree of variety and exhibits autonomous behaviour is the very same customer that is to be transformed by the firm as part of an expected emergent outcome. The concept of command and control management becomes illusory if systems are to adapt to absorb variety and satisfy customer demand. Revenue will be lost in the long term if rigidities are established to provide control that attenuates demand to such a degree that outcomes fail to meet customer expectation. The way managers plan or react to push-back of variety will have subsequent effects upon the costs and the delivery of core transformation processes and over the long-term sustainability of the enterprise.

This book has only begun to address the challenge of complex engineering service. Research may broaden to provide detailed analysis of a range of complex engineering service organisations. For example, it would be instructive to understand the challenges faced by lower tier or third party complex engineering service providers dealing with equipment, materials and potentially information from different suppliers, such as those charged with oil refinery management. In addition, there remains a great deal of scope to address one of the core transformations, the transformation of people. The long-term goal is the further development of theory and practice on how sustainable revenues may be gained by enterprises providing complex engineering services that incorporate the intelligent design of the system for achieving contextual and variety-laden outcomes, both of the kind that is deterministic and that which is emergent, and that includes the transformation of the customer co-creating that value. This, then is the grand challenge of complex engineering service systems.

References

E. Anderson, M. Sullivan, The antecedents and consequences of customer satisfaction for firms. Mark. Sci. **12**(2), 125–143 (1993)

W.R. Ashby, An introduction to cybernetics. Chapman and Hall, London. Available on the Internet (1999) http://pcp.vub.ac.be/books/IntroCber.pdf (1956)

S. Beer, *Brain of the firm*, 2nd edn. (John Wiley, London and New York, 1981). (much extended, reprinted 1986)

M.J. Bitner, B.H. Booms, M.S. Tetreault, The service encounter: Diagnosing favorable and unfavorable incidents. J. Mark. **54**, 71–84 (1990)

W. Buckley, Systems, in *Organisations as systems*, ed. by M. Lockett, R. Spear (Open University Press, Milton Keynes, 1980)

R.B. Chase, U.M. Apte, A history of research in service operations: What's the big idea? J. Oper. Manag. **25**(2), 375–386 (2007)

R.B. Chase, D.A. Tansik, The customer contact model for organisation design. Manag. Sci. **29**(9), 1037–1050 (1983)

G.A. Churchill, C. Surprenant, An investigation into the determinants of customer satisfaction. J. Mark. Res. **19**, 491–504 (1982)

P.C. Fishburn, On the foundations of decision making under uncertainty, in *Essays on economic behavior under uncertainty*, ed. by M. Balch, D. McFadden, S. Wu (North Holland Publishing Co, Amsterdam, 1974)
J.W. Forrester, *Principles of systems* (Productivity Press, Portland, 1968)
F.X. Frei, Breaking the trade off between efficiency and service. Harv. Bus. Rev. **84**(11), 92–101 (2006)
J. Fromm, On engineering and emergence. nlin.AO/0601002 (2006)
J.W. Forrester, *Principles of systems* (Productivity Press, Portland, 1968)
P. Godsiff, Service systems and requisite variety. Serv. Sci. **2**(1/2), 92–101 (2010)
B.C. Johnson, J.M. Manyika, L.A. Yee, The next revolution in interactions. McKinsey Q. **4**, 20–33 (2005)
E. Karni, Risk aversion for state-dependent utility functions: Measurement and applications. Int. Econ. Rev. **24**(3), 637–647 (1983)
J. Lapierre, A. Tran-Khanh, J. Skelling, Antecedents of customers' desired value change in a business-to-business context: Theoretical model and empirical assessment. Serv. Mark. Q. **29**(3), 114–148 (2008)
R.G. Lipsey, K. Lancaster, The general theory of second best. Rev. Econ. Stud. **25**(63), 11–32 (1956)
R.F. Lusch, S.L. Vargo, *The service dominant logic of marketing: Dialog, debate and directions* (ME Sharpe, Armonk, 2006)
G. Midgley, *Systemic intervention: Philosophy, methodology, and practice* (Kluwer-Academic/Plenum, New York, 2000)
I.C.L. Ng, Advanced demand and a critical analysis of revenue management. Serv. Ind. J. **27**(5), 525–548 (2007)
I.C.L. Ng, *The pricing and revenue management of services: A strategic approach* (Routledge, London, 2008)
I.C.L. Ng, R. Maull, N. Yip, Outcome-based contracts as a driver for systems thinking and service-dominant logic in service science: Evidence from the defence industry. Eur. Manag. J. **27**(6), 377–387 (2009)
I.C.L. Ng, J. Williams, A. Neely, Outcome-based contracting: Changing the boundaries of B2B customer relationships. Advanced Institute of Management (AIM) Research Executive Briefing Series. Available at: http://www.aimresearch.org/index.php?page=alias-3 (2009a)
I.C.L. Ng, S.S. Nudurupati, P. Tasker, Value co-creation in the delivery of outcome-based contracts for business-to-business service. AIM working paper series, WP No 77—May 2010 http://www.aimresearch.org/index.php?page=wp-no-77 (2010a)
I.C.L. Ng, R.S. Maull, L. Smith, in *2010 volume in Service Science: Research and Innovations (SSRI) in the Service Economy Book Series*, ed. by H. Demirkan, H. Spohrer, V. Krishna. Embedding the new discipline of service science (Springer, New York, ISSN: 1865-4924, forthcoming) (2010b)
R. Normann, R. Ramirez, From value chain to value constellation: Designing interactive strategy. Harv. Bus. Rev., July–August: (1993), pp. 65–77
T. Ohno, *Toyota production system: Beyond large-scale production* (Productivity Press, London, 1988)
J.M. Ottino, Engineering complex systems. Nature **427**, 399 (2004)
G. Parry, A. Graves, *Build to order: The road to the 5-day car.* (Springer, London, 2008)
A. Payne, K. Storbacka, P. Frow, Managing the co-creation of value. Acad. Mark. Sci. **36**(1), 83–96 (2008)
C.K. Prahalad, V. Ramaswamy, The new frontier of experience innovation. MIT Sloan Manag. Rev. **44**(4), 12–18 (2003)
C.K. Prahalad, V. Ramaswamy, *The future of competition: Co-creating unique value with customers* (Harvard Business School Press, Boston, 2004)
C.K. Prahalad, V. Ramaswamy, Co-opting customer competence. Harv. Bus. Rev. **78**, 79–87 (2000). (Jan/Feb)

R. Ramirez, Value co-production: Intellectual origins and implications for practice and research. Strateg. Manag. J. **20**(1), 49–65 (1999)

S. Shugan, J. Xie, Advance pricing of services and other implications of separating purchase and consumption. J. Serv. Res. **2**(3), 227–239 (2000)

D.J. Teece, G. Pisano, A. Shuen, Dynamic capabilities and strategic management. Strateg. Manag. J. **18**(7), 509–533 (1997)

L. von Bertalanffy, *General system theory* (Penguin, Harmondsworth, 1968)

W.A. Wulf, Great achievements and grand challenges. The Bridge, US. Nat. Acad. Eng. (30)3&4, Fall/Winter (2000)

G.M. Weinberg, An introduction to general systems thinking—silver anniversary edition. (Dorset House, New York, 2001)

N. Wiener, *Cybernetics* (Wiley, New York, 1948)

J. Xie, S.M. Shugan, Electronic tickets, smart cards, and online prepayments: When and how to advance sell. Mark. Sci. **20**(3), 219–243 (2001)

B. Yang, N.D. Burns, C.J. Backhouse, Postponement: A review and an integrated framework. Int. J. Oper. Prod. Manag. **24**(5), 468–487 (2004)

About the Authors

Joe Butterfield is a lecturer in Digital Manufacturing in the school of Mechanical and Aerospace Engineering at Queen's University Belfast. He is currently working on the application of digital manufacturing methods to the development of a whole life approach to the manufacture, operation and disposal of sustainable transport systems. Previous work has included the utilisation of simulation methods for aerospace assembly process design for a period of six years. Joe has successfully completed collaborative research programs in both the Virtual Engineering Centre and the Centre of Excellence for Integrated Aircraft Technologies at Queen's, where he worked on integrated design technologies for the optimisation of aerospace engine components.

Dr Matthew Carr is currently a Research Associate within the Health Methodology Research Group that forms part of the University of Manchester's Faculty of Medicine. His previous research role was with the Centre for Operational Research and Applied Statistics at the University of Salford, where he was involved in the development of techniques for complex system maintenance and prognostic modelling in the context of condition-based replacement decision analysis. Matthew holds a BSc in Business Operation and Control, an MSc in Operational Research and Applied Statistics, and a PhD in Mathematics from the University of Salford. His PhD thesis was entitled 'State and Parameter Estimation Techniques for Stochastic Systems'. His research interests include survival analysis, prognostic modelling and Bayesian analysis.

Dr Nigel Caldwell is a Research Fellow at the University of Bath School of Management. His research examines contracting for complex bundles of products and services, the risks inherent in such complex performance and governance and incentives for such contracts. Nigel is an Associate Editor of the Journal of Purchasing and Supply Management and the author, with Mickey Howard, of the book "Procuring Complex Performance: Studies of Innovation in Product-Service Management".

Rachel Cuthbert graduated in 2000 with an MA and MEng in Engineering from Cambridge University, with a specialisation in fluid mechanics and

I. Ng et al. (eds.), *Complex Engineering Service Systems*, Decision Engineering,
DOI: 10.1007/978-0-85729-189-9, © Springer-Verlag London Limited 2011

thermodynamics. Following this, she successfully completed the Advanced Course in Design, Manufacture and Management (ACDMM, now ISMM) in 2001 at the Institute for Manufacturing, Cambridge University. She has gained significant industrial experience from several years working in the chemical process and inkjet industries from research and development, manufacturing and supply chain roles. Rachel is a Chartered Engineer and a Member of the Institution of Mechanical Engineers (2004). She is a Research Associate in the Distributed Information and Automation Laboratory and has just completed an MPhil entitled 'The Information Requirements for Service Development'.

Dr Partha Priya Datta is an Assistant Professor in Operations Management at the Indian Institute of Management in Calcutta, India. He completed his PhD at Cranfield School of Management, where his research involved resilient supply chain design using agent based modelling, and received the James Cooper memorial cup for best PhD from Chartered Institute of Logistics and Transport, UK. He also holds an MBA from Lancaster University and a BTech (Hons) in Mechanical Engineering from the Indian Institute of Technology Kharagpur. While at Cranfield, Partha coordinated the risk and cost assessment work within the Support Service Solutions: Strategy and Transition (S4T) programme. His current areas of research are supply chain risk, resilience, robustness, uncertainty management, supply chain strategy and planning.

John Ahmet Erkoyuncu is a PhD researcher at the Decision Engineering Centre at Cranfield University, having begun his PhD studies after completing an MSc in Econometrics and Applied Business Management from Imperial College, London. His research concentrates on integrating uncertainty into cost estimation at the bidding phase of availability type contracts within the defence and aerospace industries. The current focus of the research involves establishing links between uncertainties and cost drivers with a view to develop mathematical representations of cost drivers. The research is part of the Product-Service System (PSS)-Cost project, which ended in September 2010. The project aims to better understand whole life cycle cost and affordability management in availability type contracts in the defence and aerospace industries.

Dr. Lei Guo obtained her PhD in Management in 2010 from the University of Exeter Business School. Prior to her doctoral studies, she spent nine years as a business practitioner in the higher education and manufacturing industries. She had held various marketing positions with organisations such as the National University of Singapore Business School, Singapore Chinese Chamber of Commerce and Industry and Har Paw Corporation, Singapore. Lei's research interest is in service and relationship marketing, especially customer relationships at individual levels. Drawing on theories of marketing and social psychology, she explores the impact of interpersonal relationships between salespeople/service providers and their customers on their behaviours in marketing transactions and service provision.

Chris Hockley joined Cranfield University in 2003 after 36 years in the Royal Air Force (RAF) where he specialised in aircraft maintenance, including commanding an Engineering Wing providing support for two aircraft squadrons.

He served in the Ministry of Defence in several appointments before joining the Department responsible for improving the reliability and maintainability (R&M) of defence equipment. He completed a Defence Fellowship to study R&M and has commanded the RAF's R&M policy department. Chris has presented on R&M at international conferences and contributed to much of the existing Defence R&M policy. His research interests are in Health and Usage Monitoring Systems (HUMS) Prognostics Health Management (PHM) Condition Based Maintenance (CBM) and delivering availability and support contracts; he organises and chairs annual Symposia covering these subjects.

Dr Emma Kelly is a Senior Research Fellow in the Precision Manufacturing Centre at the University of Nottingham. She obtained her PhD from the Department of Aeronautical Engineering at Loughborough University where her research was focussed on the development and application of system fault diagnostics. Emma gained industrial reliability and safety experience working on both nuclear defence and nuclear power generation projects before returning to work in research at the University of Nottingham. Her current interests lie in the life cycle aspect of services engineering, in particular the role of maintenance considering both preventative and diagnostic techniques.

Dr Clive Kerr is a Research Associate at the Centre for Technology Management at the University of Cambridge. His current research interests are visual strategy, roadmapping, technology intelligence, technology insertion and through-life capability management. Prior to joining Cambridge, he was a Research Officer in Engineering Design at the Enterprise Integration Department of Cranfield University. Clive has a First Class Honours degree in Electrical and Mechanical Engineering, a Diploma degree in Economics, a Postgraduate Certificate in the Social Sciences and a Doctorate in Engineering.

Saikat Kundu is a Research Fellow in the Institute of Engineering Systems and Design in the School of Mechanical Engineering at the University of Leeds. His research is focused on design, representation and engineering of product-service systems, extended enterprise supply networks and knowledge and information management.

Laura Lacey joined Cranfield University in 2006 after 21 years in the Royal Air Force, where she gained experience in the engineering, maintenance and support of defence equipment. She provided technical expertise and acted as engineering assessor for Ministry of Defence maintenance contracts and actively managed critical aspects of these contracts including liaising with contractors both in the UK and abroad. During her last three years of service she lectured on propulsion technology for an Aerosystems Engineering MSc programme. Now working for the Cranfield Defence and Security in the Centre for Systems Engineering, Laura contributes to a variety of courses, including MSc for Systems Engineering for Defence Capability, for which she is the Academic Leader, and has undertaken lecturing commitments in Australia and the USA.

Professor Roger Maull is a Professor of Management in the University of Exeter Business School. He has a BA in Economics, and an MSc in Management Information Systems. He gained his PhD in 1986 in the use of business process modelling (BPM) in manufacturing. His major research interests are in applying systems thinking within the management of business operations. In the last 5 years, Roger has moved away from traditional manufacturing-based Operations Management to working more closely with the services and public sectors, and has written in numerous publications about managing processes in sectors such as computing, banking, telecoms and logistics. He regularly acts as a BPM advisor for a wide ranging group of companies and public bodies including Vodafone, LloydsTSB, Britannia Building Society, Compaq, IBM, hospital trusts, Society of British Aerospace Companies (SBAC) and the Met Police.

William McEwan is a Research Fellow at the school of Mechanical and Aerospace Engineering, Queens University Belfast. He obtained his PhD in experimental aerodynamics from Queen's in 2007. Subsequent research activities have included an investigation of friction stir welding on aircraft structures, predictive methods for aerospace cost modelling and digital simulation of maintenance methods for complex engineering systems. He is currently working on the application of process simulation methods for the life cycle management of sustainable transport systems.

Professor Alison McKay is Professor of Design Systems in the School of Mechanical Engineering at the University of Leeds. Her research is positioned in the context of stage gate processes that typify current industry practice. Recent work has included the application of product data engineering principles for the representation and evaluation of service, process and extended enterprise structures, and studies exploring the use of shape grammars to support computer-based design synthesis.

Dr Ken McNaught is a lecturer in Operational Research and Statistics in the Department of Engineering Systems and Management at Cranfield University's School of Defence and Security at the Defence Academy of the UK. He has an MSc in Operational Research from Strathclyde University and a PhD in Operational Research from Cranfield. His PhD and earlier research work focused on stochastic combat modelling. Current research interests include decision analysis, simulation modelling and reliability and maintenance modelling. Ken is particularly interested in probabilistic graphical models such as Bayesian networks and influence diagram decision networks. Application areas of interest include reliability and maintenance, security and intelligence and emergency preparedness. He is currently involved in two Engineering and Physical Sciences Research Council (EPSRC) projects which involve the use and development of probabilistic graphical models.

John Mills spent 20 years with Shell International and Philips Electronics prior to joining the University of Cambridge in 1992. He has wide industrial experience in the simulation of worldwide oil distribution alternatives, location decisions for chemical plants, reservoir engineering, factory management, as well as business management in the consumer electronics, white goods and mobile communications sectors. John's research has concentrated on the development of practical processes for the formation and implementation of manufacturing strategies, the design of coherent performance measurement systems and a focus on competence analysis, development and exploitation. These approaches are currently being applied to study manufacturing companies transitioning to service provision, such as the defence support contracts being struck between the UK Ministry of Defence and defence OEMs (original equipment manufacturers) for the support of military platforms.

Frank Murphy is Head of Business and Solutions Modelling in Military Air Systems. In his current role, he aims to help managers in deciding the right form of modelling from performance and value perspectives. He graduated from City of London Polytechnic in 1976 and since then has had various roles within the Defence and Aerospace industries.

Professor Andy Neely is widely recognised as one of the world's leading authorities on organisational performance, measurement and management. He holds joint posts at Cambridge University and Cranfield School of Management, and is also Deputy Director of the Advanced Institute of Management Research. Previously he has held appointments at London Business School, Cambridge University, Nottingham University where he completed his PhD and British Aerospace. He was elected as a Fellow of the Sunningdale Institute in 2005, a Fellow of the British Academy of Management in 2007, an Academician of the Academy of Social Sciences in 2008 and a Fellow of the European Operations Management Association in 2009. Currently Andy is researching issues of performance and innovation in services, with a particular focus on Enterprise Performance Management and the servitization of manufacturing.

Dr Sai Nudurupati is a senior lecturer in Manchester Metropolitan University Business School. His research interests include operations management with special focus on performance measurement, and service science specifically on value co-production and co-creation. He holds a PhD from University of Strathclyde in the implementation of performance measurement and studying its impact on business and management, and won Outstanding Doctoral Award for his PhD in 2005 in the category of Operations Management. Prior to joining MMU, Sai was a Research Fellow on the EPSRC-sponsored S4T project at the University of Exeter Business School, where he explored and tested the factors influencing the value co-creation and co-production in the service delivery. He had 6 years of industrial experience prior to joining academia.

Dr Robert Phaal is a Senior Research Associate at the Centre for Technology Management at the University of Cambridge. He conducts research in the area of strategic technology management—particular interests include technology evaluation, the emergence of technology-based industry, the use of visual techniques for strategy and the development of practical management tools. Rob has a mechanical engineering background, with a PhD in computational mechanics and industrial experience in technical consulting, contract research and software development.

David Probert is a Reader in Technology Management and the Director of the Centre for Technology Management at the Engineering Department of the University of Cambridge. David's research interests include technology and innovation strategy, technology management processes, make or buy, technology acquisition and software sourcing. He had an industrial career with Marks and Spencer and Philips for 18 years before returning to Cambridge. His experience covers a wide range of industrial engineering and management disciplines in the UK and overseas.

Dr Valerie Purchase is a lecturer in Organisational Communication at the University of Ulster. She lectured for 7 years in the School of Communication, where she obtained a PhD in Communication, before specialising as a consultant in the management of organisational change. Her experience and interest in managing change led to her joining the University of Bath where she managed a number of national and international research programmes targeting transformation in the aerospace and manufacturing sectors. Valerie's research over 10 years examined factors influencing successful organisational transformation including managerial, cultural and communication issues within individual and across collaborating organisations. Her current interests lie in managing and leading change across 'multi-organisational service enterprises', and in particular, the role of communication in both framing and facilitating change.

Professor Svetan Ratchev is a Professor in Manufacturing Engineering at the University of Nottingham, where he is also Head of the Manufacturing Division of the Faculty of Engineering and Director of the Nottingham Innovative Manufacturing Research Centre and the Precision Manufacturing Centre. He researches and consults in key areas of precision manufacture including assembly automation, precision machining and distributed design and manufacture. Svetan is a Fellow of the Institute of Mechanical Engineers, member of several IFAC technical committees and the founding chair of the International Precision Assembly Seminar IPAS. As coordinator of several EU FP6 and FP7 projects he has been involved in defining the EU research agenda in Micro- and Nano- Manufacturing, and is also a co-founder of the European Sub-Technology Platform in Micro- and Nano- Manufacturing.

Professor Rajkumar Roy led Competitive Design research at Cranfield for over 10 years before becoming Head of Manufacturing Department in 2009. He is known for his qualitative cost modelling, requirements management and design optimisation research, and is currently the Principal Investigator of EPSRC and Innovative Manufacturing Research Centre-funded PSS projects in the areas of whole life cost and service knowledge capture. He is also leading the new initiative on Competitive Creative Design at Cranfield in collaboration with University of the Arts London, and has developed an innovative PSS Futures Lab at Cranfield. A Chartered Engineer and the President and Fellow of the Association of Cost Engineers, Rajkumar has a PhD in design optimisation using soft computing techniques. He is also the Editor in Chief of the Applied Soft Computing Journal.

Vince Settle has been with BAE Systems and its predecessor companies for over 40 years, working in Commercial Management. He has worked with many Governments and Air Forces during this time, to tailor the company's products and services to their requirements. Notable among these are the £40 Billion Al-Yamamah defence deal with Saudi Arabia and the setting up of a trination support service for the Tornado fleets of the UK, Germany and Italy, the first such arrangement in the world. In the last 10 years Vince has specialised in Contracting for Availability and developed innovative, commercial and pricing approaches for these long-term contracts.

Jeremy Smith joined Cranfield University in 2008, following a 25 year career as a logistician and Ammunition Technical Officer in the British Army. During his service he commanded a variety of EOD, supply and transport units, including on operations in Northern Ireland, the Balkans and the Middle East, and he completed a variety of weapons and equipment management staff appointments. Jeremy underwent command, staff and technical training at the Royal Military College of Science, Shrivenham, and at the Army Command and Staff College in Camberley. He now works for Cranfield Defence and Security in the Centre for Defence Acquisition where he lectures in Through Life Support, Integrated Logistic Support and Logistics and Supply Chain Management. His principal research interest is in Contracting for Availability.

Professor Wenbin Wang is Professor in Operational Research and Applied Statistics at Salford Business School, University of Salford. Prior to this, he lectured at Harbin Institute of Technology where he still holds an overseas professorship and is part-time Associate Dean of the School of Management. Wenbin holds a PhD in Operational Research and Applied Statistics from the University of Salford, an MSc in Operations Management from Xian and a BSc in Mechanical Engineering from Harbin University of Science and Technology. His main research areas are in maintenance modelling, particularly with applications to inspection and condition-based maintenance, supply chain modelling, health care and international manufacturing strategy. He is a Fellow of the Royal Statistics Society, and an executive member of the International Foundation for Research in Maintenance.

Phil Wardle joined BAE Systems in 2001 after working mainly in the defence industry since leaving the University of Manchester in 1977. He has worked in the UK and USA in engineering and management roles on military and civil projects from three man-months to 100 man-years. He has developed Cost Engineering as a business improvement theme in BAE Systems' businesses, and has been promoting it as a candidate for capability development initiatives involving customers, suppliers, academia, and tools vendors. Phil is a Fellow of the British Computer Society and the Institution of Engineering and Technology, and a Visiting Fellow at the School of Applied Sciences, Cranfield University.

Jason Williams is a Service Transformation Analyst working on a collaborative project between the University of Exeter and Harmonic Ltd to develop a frame-work for service transformation. His interests lie in dealing with the implications of value co-creation and the challenges it presents in organisational design. Jason holds a BA in Business and Management from the University of Exeter and a Diploma in Programme and Project Management. He is currently undertaking an MPhil in Management supervised by Professor Irene Ng.

Adam Zagorecki joined Cranfield University in 2005 as a post doctoral Research Fellow. His interest in decision analysis began with an MSc in Computer Science, and continued during his PhD studies at the University of Pittsburgh, USA. Adam's research involved uncertainty in artificial intelligence, specialising in applications of Bayesian networks to hardware diagnosis and prognosis. His industrial experience includes Intel Research, HRL Laboratories (formerly Hughes Laboratories) and consulting for Rockwell Scientific. He now works for Cranfield Defence and Security in the Department of Engineering Systems and Management, where among other interests he conducts research on modelling and simulation with applications to prognostics and other problems within the Decision Analysis and Risk Modelling Laboratory.

About the Editors

Professor Irene Ng is the Professor of Marketing Science at the University of Exeter Business School, the ESRC/Advanced Institute of Management Research (AIM) Services Fellow and Honorary Senior Visiting Fellow at the University of Cambridge. Irene was a business practitioner for more than 10 years before switching to an academic career. During her time in industry she occupied a number of senior positions, rising to become CEO of SA Tours group of companies. She also founded Empress Cruise Lines, a company with an annual turnover of USD250m. Irene's research interest lies in value; understanding, delivering, designing, pricing, contracting and innovating based on value, as well as in value-based complex service systems. She has published in numerous international journals in the domain of engineering, management, marketing, information systems, economics, education and sociology, and is also the author of the book The Pricing and Revenue Management of Services: A Strategic Approach. Irene holds a PhD in Marketing and a BSc in Physics and Applied Physics both from the National University of Singapore.

Dr Glenn Parry is Principal Lecturer in Strategy and Operations Management at Bristol Business School, University of the West of England. Glenn's work is characterised by developing creative solutions to business challenges, and he has managed and contributed towards research consortia within the automotive, aerospace and construction industries. Prior to joining the S4T project, Glenn was a leader of a £16m European automotive project that looked at developing the transition from 'mass manufacture' to 'build-to-order;' he also edited the book Build To Order: The Road to the 5 day Car which resulted from the project. His interests include service, enterprise transformation, complexity, costing, core competence, and lean. He has published in numerous international journals, and is editor of the text book Service: Design and Delivery. Glenn holds a PhD in Materials Science from Cambridge University, a BSc in Chemistry and Business and an MPhil in Materials Science from Swansea University, and a Diploma in Psychotherapy from Warwick University. He is a Senior Visiting Fellow at the

I. Ng et al. (eds.), *Complex Engineering Service Systems*, Decision Engineering,
DOI: 10.1007/978-0-85729-189-9, © Springer-Verlag London Limited 2011

University of Bath and has previously worked for British Steel, LEK Consulting and Warwick University.

Peter J Wild is a Senior Research Associate in the Psychology and Communication Technology Lab (PaCTLab), Department of Psychology, University of Northumbria. His original training was in Human-Computer Interaction (HCI) / Interaction Design (BSc Psychology and Computing, MSc in HCI). Over the years he has worked on research projects embracing Information Technology artefact evolution, multitasking and collaboration, Information Management in Engineering Design, Service Design in Engineering and Integrative Research in Services. A common theme among these seemingly diverse works has been a user-centred perspective; task/activity modelling; and cross-disciplinary research. Peter has held research positions, at Brunel, Bath and Cambridge, and has co-published with psychologists, engineers, designers, architects and computer scientists.

Professor Duncan McFarlane is Professor of Service and Support Engineering at the Cambridge University Engineering Department, and head of the Distributed Information and Automation Laboratory within the Institute for Manufacturing. Duncan completed a B Eng degree at Melbourne University in 1984, a PhD in the design of robust control systems at Cambridge in 1988, and worked industrially with BHP Australia in engineering and research positions between 1980 and 1994. His research work is focused in the areas of distributed industrial automation, reconfigurable systems, RFID integration and valuing industrial information. Most recently he has been examining the role of automation and information solutions in supporting service environments. In 2004, Duncan became head of the Cambridge Auto ID Lab and co-founded the Aero ID Programme, examining the role of RFID in the aerospace industry. He was appointed to the Professorship of Service and Support Engineering on 1 October 2006.

Professor Paul Tasker is Royal Academy of Engineering Visiting Professor in Integrated System Design at Cranfield University and the University of Kent, as well as being an Industrial Fellow with the University of Cambridge Institute for Manufacturing. Now working as an independent consultant, he has considerable engineering management experience in the defence sector, working in both government and industry. Most recently he has been Engineering Director for BAE Systems Naval Support business, before developing BAE Systems research partnership with the University of Cambridge and initiating the BAE Systems/ EPSRC S4T programme on which this book is largely based. Paul is a Chartered Engineer, Fellow of the Royal Institution of Naval Architects and a member of the Royal Corp of Naval Constructors.

About the Editorial Review Board

The following Editorial Review Board members, comprising esteemed academics and practitioners in the field of engineering and service systems, provided peer review for the book to ensure an outstanding level of scholarship, innovation and practices.

I. Ng et al. (eds.), *Complex Engineering Service Systems*, Decision Engineering,
DOI: 10.1007/978-0-85729-189-9, © Springer-Verlag London Limited 2011

Glossary of Terms

2D	Two Dimensional
3D	Three Dimensional
AACE	Association for the Advancement of Cost Engineering
AAS (off)	Asset Availability Service (off Balance Sheet)
AAS (on)	Asset Availability Service (on Balance Sheet)
ABFS	Activity Based Framework for Services
ABM	Agent Based Modelling
ACQ	Acquisition cost
AFV	Armoured Fighting Vehicle
AHP	Analytic Hierarchy Process
AIM	Advanced Institute for Management
ALDT	Administrative and Logistic Delay Time
AOF	Acquisition Operating Framework
ATS	Automated Transport and Sorting system
ATTAC	Availability Transformation: Tornado Aircraft Contract
AVC	Attributes of value co-creation
AVST	Armoured Vehicle Support Transformation
AW	Agusta Westland
BATUS	British Army Training Unit in Suffield
BN	Bayesian Network
BOM	Bill of Materials
BTIA	Business Transformation Incentivisation Agreement
C130J	A variant of the Lockheed C-130 Hercules
C130K	A variant of the Lockheed C-130 Hercules
CAD	Computer Aided Design
CADMID	Concept Assessment Demonstration Manufacture In-Service Disposal
CADMIT	Concept, Assessment, Demonstration, Migration, In-Service, Termination
CAE	Computer Aided Engineering

I. Ng et al. (eds.), *Complex Engineering Service Systems*, Decision Engineering,

DOI: 10.1007/978-0-85729-189-9,

CALCE	Center for Advanced Lifecycle Engineering (University of Maryland)
CATIA	A 3D CAD modelling environment
CBM	Condition Based Maintenance
CBS	Cost Breakdown Structure
CE	Concurrent Engineering
CES	Complex Engineering Service
CfA	Contracting for Availability
CfC	Contracting for Capability
CIF	Core Integrative Framework
CLS	Contractor Logistics Support
CM	Condition Maintenance
CMU	Combined Maintenance and Upgrade
CS (off)	Capability Service (off Balance Sheet)
CS (on)	Capability Service (on Balance Sheet)
CSE	Cognitive Systems Engineering
CVR	Combat Vehicle Reconnaissance
CVR(T)	Combat Vehicle Reconnaissance (Tracked)
DA	Design Authority
DAO	Data Access Objects
DARA	Defence Aviation and Repair Agency
DBN	Dynamic Bayesian Networks
DELMIA	An integrated, 3D CAD based environment used for engineering process design and optimisation
DE and S	Defence Equipment and Support
DES	Discrete Event Simulation
DIS	Defence Industrial Strategy
DLO	Defence Logistics Organisation
DLoD	Defence Lines of Development
DoD	Department of Defence (United States)
DoDAF	DoD Architectural Framework
DPM	Digital Process for Manufacture
DSDA	Defence Storage and Distribution Agency
EPSRC	Engineering and Physical Sciences Research Council
ESA	Equipment Support Agreement
ETLC	Enabling Through-Life Capability
FMECA	Failure Modes, Effects and Criticality Analysis
GDL/G-D Logic	Goods-Dominant Logic
GFX	Government Furnished Assets
GSE	Ground Support Equipment
HMM	Hidden Markov Models
HPAC	Harrier Platform Availability Contract
HUMS	Health Usage and Monitoring Systems
IBM	International Business Machines
ILS	Integrated Logistic Support

IMOS	Integrated Merlin Operational Support
IOS	Integrated Operational Support
IPAS	Integrated Products and Services
IPSS	Industrial Product-Service System(s)
IPT	Integrated Project Team
IRI	Incentivised Reliability Improvement
ISCL	Integrated Service CAD and Lifecycle simulator
ITT	Invitation to Tender
IUCR	Incentivised Upkeep Cost Reduction
JAMES	Joint Asset Management and Engineering System
JSF	Joint Strike Fighter
JUMP	Harrier Joint Upgrade and Maintenance Programme
KBS	Knowledge-Based System
KIM	Knowledge and Information Management
KPIs	Key Performance Indicators
LCC	Life Cycle Costing
MACMT	Mean active corrective maintenance time
MART	Mean Active Repair Time
MLFP	Mid Life Fatigue Programme
MRO	Maintenance, Repair and Overhaul
MSG	Maintenance Steering Group
MTBF	Mean Time Between Failure
MTTR	Mean Time to Repair
NAO	National Audit Office
NATO	North Atlantic Treaty Organisation
NUSAP	Numerical, Unit, Spread, Assessment and Pedigree
OAS	Operation and Support
OEM	Original Equipment Manufacturer
OPV	Offshore Patrol Vessel
PBL	Performance Based Logistics
PCP	Procuring Complex Performance
PDF	Probability Density Function
PFI	Private Finance Initiative
PGM	Prognostic Model
PHM	Prognostics and Health Management
PIs	Performance Indicators
PMC	Project Management Complexity
PoF	Physics of Failure
PPR	Product, Process and Resource
PSS	Product-Service System(s)
QC	Queen's Counsel
QCD	Quality, Cost, Delivery
QFD	Quality Function Deployment
QUB	Queens University Belfast
R&M	Reliability and Maintainability

RAAF	Royal Australian Air Force
RAF	Royal Air Force (UK)
RCM	Reliability Centred Maintenance
RDO	Remote Data Objects
RN	Royal Navy (UK)
ROCET	RB199 Operational Contract for Engine Transformation
RR	Rolls-Royce
S4T	Support Service Solutions: Strategy and Transition
SAE	Society of Automotive Engineers
SAR	Search and Rescue
SCAD	Service Computer Aided Design
SCH	Schedule
SDL/S-D Logic	Service-Dominant Logic
SEU	Spares Exclusive Upkeep
SIU	Spares Inclusive Upkeep
SKIOS	Sea King Integrated Operational Support
SLA	Service Level Agreement
SOA	Service-Oriented Architecture
SOM	Support Options Matrix
SPA	Strategic Partnering Arrangement
TAS	Typhoon Availability Service
TCM	Total Corrective Maintenance Time
TEC	Technical difficulty
TIPSS	Technical Product-Service System(s)
TLC	Through-Life Capability
TLM	Through-Life Management
UAV	Unmanned Aerial Vehicle
UK MoD	United Kingdom Ministry of Defence
VEA	Virtual Enterprise Architecture
VT	Vosper Thorneycroft
WBS	Work Breakdown Structure
WFM	Whole Fleet Management

MIX
Papier aus verantwortungsvollen Quellen
Paper from responsible sources
FSC® C105338

Printed by Books on Demand, Germany